Viral and Antiviral Nanomaterials

Emerging Materials and Technologies

Series Editor: Boris I. Kharissov

Semiconducting Black Phosphorus: From 2D Nanomaterial to Emerging 3D Architecture
Han Zhang, Nasir Mahmood Abbasi, and Bing Wang

Biomass for Bioenergy and Biomaterials
Nidhi Adlakha, Rakesh Bhatnagar, and Syed Shams Yazdani

Energy Storage and Conversion Devices: Supercapacitors, Batteries, and Hydroelectric Cell
Anurag Gaur, A.L. Sharma, and Anil Arya

Nanomaterials for Water Treatment and Remediation
Srabanti Ghosh, Aziz Habibi-Yangjeh, Swati Sharma, and Ashok Kumar Nadda

2D Materials for Surface Plasmon Resonance-Based Sensors
Sanjeev Kumar Raghuwanshi, Santosh Kumar, and Yadvendra Singh

Functional Nanomaterials for Regenerative Tissue Medicines
Mariappan Rajan

Uncertainty Quantification of Stochastic Defects in Materials
Liu Chu

Recycling of Plastics, Metals, and Their Composites
R.A. Ilyas, S.M. Sapuan, and Emin Bayraktar

Viral and Antiviral Nanomaterials: Synthesis, Properties, Characterization, and Application
Devarajan Thangadurai, Saher Islam, and Charles Oluwaseun Adetunji

Drug Delivery Using Nanomaterials
Yasser Shahzad, Syed A.A. Rizvi, Abid Mehmood Yousaf, and Talib Hussain

Nanomaterials for Environmental Applications
Mohamed Abou El-Fetouh Barakat and Rajeev Kumar

For more information about this series, please visit: https://www.routledge.com/Emerging-Materials-and-Technologies/book-series/CRCEMT

Viral and Antiviral Nanomaterials

Synthesis, Properties, Characterization, and Application

Edited by

Devarajan Thangadurai, Saher Islam, and Charles Oluwaseun Adetunji

CRC Press
Taylor & Francis Group
Boca Raton London New York

CRC Press is an imprint of the
Taylor & Francis Group, an **informa** business

First edition published 2022
by CRC Press
6000 Broken Sound Parkway NW, Suite 300, Boca Raton, FL 33487-2742

and by CRC Press
4 Park Square, Milton Park, Abingdon, Oxon, OX14 4RN

CRC Press is an imprint of Taylor & Francis Group, LLC

Library of Congress Cataloging-in-Publication Data
Names: Thangadurai, D. (Devarajan), 1976- editor. | Islam, Saher, editor. |
Oluwaseun Adetunji, Charles, editor.
Title: Viral and antiviral nanomaterials : synthesis, properties, characterization, and application / edited by Devarajan Thangadurai, Saher Islam, Charles Oluwaseun Adetunji.
Description: First edition. | Boca Raton : CRC Press, 2022. | Series: Emerging materials and technology | Includes index.
Identifiers: LCCN 2021042077 (print) | LCCN 2021042078 (ebook) |
ISBN 9780367682576 (hardback) | ISBN 9780367682583 (paperback) |
ISBN 9781003136644 (ebook)
Subjects: LCSH: Anti-infective agents. |
Nanostructured materials.
Classification: LCC RM267 .V57 2022 (print) |
LCC RM267 (ebook) | DDC 615.7/92--dc23/eng/20211027
LC record available at https://lccn.loc.gov/2021042077
LC ebook record available at https://lccn.loc.gov/2021042078

ISBN: 978-0-367-68257-6 (hbk)
ISBN: 978-0-367-68258-3 (pbk)
ISBN: 978-1-003-13664-4 (ebk)

DOI: 10.1201/9781003136644

Typeset in Times
by MPS Limited, Dehradun

Dedication

In fond memory of a great teacher, an affable mentor, and a social reformer,

Mr. Cullan Devarajan, father of Dr. Devarajan Thangadurai

Contents

Preface

Microbial infections are the primary cause of the high mortality rate around the globe, particularly viruses making worldwide impact on socioeconomic development and healthcare. The unavailability of functional antiviral therapeutics for a variety of life-threatening viral diseases is due to limited solubility of antiviral drugs in aqueous media, inadequate access to required anatomic compartments, and short lifetime. Besides this, rapid multidrug-resistance development to existing therapies and side effects as a result of prolonged drug usage are causing serious health concerns. Accordingly, novel strategies and continuous efforts are crucial to improve the physico-chemical characteristics of available antiviral drugs for the treatment of viral diseases.

Interaction of nanomaterials with microbes is rapidly revolutionizing the medical era by proposing several advantages in diagnostic and therapeutic applications. Nanostructures offer distinctive physical characteristics, including smaller sizes that affect circulation and bioavailability time, larger surface area that enhances solubility, adjustable surface charge that enables encapsulation, and higher drug payloads. All these characteristics, which are distinct from larger-size materials of the same composition, make nanomaterials an ideal candidate for drug delivery to explore and to improve therapeutic outcomes.

Different nanostructures, such as carbon nanotubes, quantum dots, silica, graphene oxide, and metal/metal oxide nanoparticles, are presently manipulating as biosensors to detect and identify the plant, animal, and human viral diseases. Nanomaterials have great potential due to their vast applications in diagnostics, including analytical reagents preparation and advanced imaging tools for detecting infectious agents and diseases. The field of viral nanotechnology is recent and quickly growing due to its immediate necessity to restrain viral infections and address the increased demand for applications in prevention, diagnostics, and treatment.

The book is divided into three sections and presents a comprehensive overview on the synthesis, properties, characterization, and application of nanostructures for therapy of viral infections and cancer treatment. Section I provides recent findings and developments in the rapidly emerging discipline of virus nanotechnology. It also provides the insight of synthesis and characterization techniques of various nanomaterials for viral and antiviral applications. Section II is exclusive for viral nanomaterials and their role in biomedical applications. Section III has special emphasis on using antiviral nanostructures as powerful tools for developing drugs and vaccines to improve the public health status. The book generally explores how antiviral nanomaterials can be utilized in developing targeted drug-delivery practices and how these nano-sized structures enable pharmaceutical industries to reformulate available drugs on market, thereby encompassing the products' lifetime and improving performance by enhancing patient adherence, safety, effectiveness, ultimately minimizing health-care cost.

This book encompasses a larger number of scientific areas, including virology, microbiology, immunology, cell biology, molecular biology, biotechnology, chemistry, and statistical modelling. Reflecting its interdisciplinary subject matter nature, this book embraces contributions by leading practitioners, experts, and researchers from diverse fields, making it an indispensable and comprehensive source for research and study. This book serves as a practical guidance and reference material source to postgraduate students, scientists, and academic researchers working in nanomedicine, nanotechnology, materials science, pharmaceutical technology, and biomedical science.

Devarajan Thangadurai
Saher Islam
Charles Oluwaseun Adetunji

Editors

Devarajan Thangadurai is an associate professor at Karnatak University, Dharwad, in India and did his postdoctoral research at the University of Madeira, Portugal, University of Delhi, India, and ICAR National Research Centre for Banana, India. He is the recipient of Best Young Scientist Award with Gold Medal from Acharya Nagarjuna University, India, and the VGST-SMYSR Young Scientist Award of the government of Karnataka, India. He has interest and expertise in the fields of biochemistry, microbiology, biotechnology, and nanotechnology for sustainable future. He has authored/edited more than 25 books, including *Genetic Resources and Biotechnology* (3 Vols.), *Genes, Genomes and Genomics* (2 Vols.), *Genomics and Proteomics, Biotechnology of Microorganisms, Mycorrhizal Biotechnology, Nanotechnology for Food, Agriculture and Environment, Functional Bionanomaterials,* and *Algal Nanotechnology* with international publishers in USA, Canada, Switzerland, and India. He has authored/coauthored 230 publications, inclusive of journal articles, book chapters, books, and invited presentations. He has extensively traveled to many universities and institutes in Africa, Asia, Europe, and the Middle East for academic works, scientific meetings, and international collaborations.

Saher Islam is a visiting lecturer in the Department of Biotechnology, Lahore College for Women University, Pakistan. She earned a BS, an MPhil, and a PhD in molecular biology, biotechnology, and bioinformatics. She has been a research scholar at Cornell University, New York, and a visiting scholar at West Virginia State University, West Virginia, in the United States. She has keen research interests in microbial genetics, molecular biology, and bioinformatics and has ample hands-on experience in molecular-marker analysis, whole-genome sequencing, and RNA sequencing. She is the recipient of Boehringer Ingelheim Fonds Travel Grant from European Molecular Biology Laboratory, Germany. She is an author/coauthor of 60 publications, including journal articles, book chapters, books, and conference presentations.

Charles Oluwaseun Adetunji is presently the Ag. Director of Intellectual Properties and Technology Transfer, chairman of the Committee on Research Grants, and an associate professor of Microbiology and Biotechnology at EUI. He has won several scientific awards and grants from renowned academic bodies like the Council of Scientific and Industrial Research (CSIR) India and the Department of Biotechnology (DBT) India, The World Academy of Science (TWAS) Italy, Netherlands Fellowship Programme (NPF) Netherlands, The Agency for International Development Cooperation Israel, and the Royal Academy of Engineering UK, among many others. He has filed several scientific patents on nanobiosurfactants, nanobiopesticides, and many more. He has published over 150 scientific journals and conference proceedings in international and local-refereed journals. His research interest includes microbiology, biotechnology, post-harvest management, and nanotechnology. He is an editorial board member of many international journals and serves as a reviewer to many double-blind peer review journals of Elsevier, Springer, Taylor & Francis, Wiley, PLoS One, Nature, American Chemistry Society, Bentham Science Publishers, etc. He is a member of many scientific and professional bodies, including American Society for Microbiology, Nigerian Young Academy, Biotechnology Society of Nigeria, and Nigerian Society for Microbiology. He has gained international recognition and also acted as a keynote speaker delivering invited talk/position paper at various universities, research institutes, and several centers of excellence that span several continents of the globe. He has over the last 15 years built strong working collaborations with reputable research groups in numerous and leading universities across the globe. He is the convener for Recent Advances in Biotechnology, which is an annual international conference where renowned microbiologists and biotechnologists come together to share their latest discoveries. He is also the Founder & CEO of BECTIK Biotechnology and Nanotechnology Company.

Contributors

Semra Akgönüllü
Department of Chemistry
Hacettepe University
Ankara, Turkey

Mohammad Al Farroukh
Department of Virology
Federal State Budgetary Scientific Institution "Institute of Experimental Medicine"
and
Faculty of Dental Medicine and Medical Technologies
St. Petersburg State University
St. Petersburg, Russia

Maleeha Azam
Department of Biosciences
Faculty of Sciences
COMSATS University
Islamabad, Pakistan

Malik Badshah
Department of Microbiology
Faculty of Biological Sciences
Quaid-i-Azam University
Islamabad, Pakistan

Monireh Bakhshpour
Department of Chemistry
Hacettepe University
Ankara, Turkey

Karan Chaudhary
Department of Chemistry
University of Delhi
Delhi, India

Bruna Galdorfini Chiari-Andréo
Department of Biological and Health Sciences
University of Araraquara – UNIARA
São Paulo, Brazil

Zaira Zaman Chowdhury
Nanotechnology and Catalysis Research Center (NANOCAT)
Institute of Advanced Studies (IAS)
University of Malaya
Kuala Lumpur, Malaysia

Kammila Martins Nicolau Costa
Pharmaceutical Sciences Postgraduate Program
Center for Biological and Health Sciences
State University of Paraíba (UEPB)
Campina Grande, Brazil

Chaomeng Dai
Department of Hydraulic Engineering
College of Civil Engineering
Tongji University
Shanghai, China

Adil Denizli
Department of Chemistry
Hacettepe University
Ankara, Turkey

Dushyant R. Dudhagara
Department of Life Sciences
Bhakta Kavi Narsinh Mehta University
Junagadh, India

Abu Nasser Faisal
Nanotechnology and Catalysis Research Center (NANOCAT)
Institute of Advanced Studies (IAS)
University of Malaya
Kuala Lumpur, Malaysia

Isadora Frigieri
Department of Biological and Health Sciences
University of Araraquara – UNIARA
São Paulo, Brazil

Megha S. Gadhvi
Department of Life Sciences
Bhakta Kavi Narsinh Mehta University
Junagadh, India

Tajamal Hussain
Institute of Chemistry
University of Punjab
Lahore, Pakistan

Ekaterina Ilyechova
Laboratory of Trace Elements Metabolism
ITMO University
and
Department of Molecular Genetics
Federal State Budgetary Scientific Institution "Institute of Experimental Medicine"
St. Petersburg, Russia

Arshad Islam
Department of Pathology
Government Lady Reading Hospital
Medical Teaching Institution
Peshawar, Pakistan

Saher Islam
Department of Biotechnology
Lahore College for Women University
Lahore, Pakistan

Mayuri Iyer
Epigeneres Biotech Private Ltd.
Mumbai, India

Fabeena Jahan
School of Chemical Sciences
Kannur University, Swami Anandatheertha Campus
Payyanur, India

Monika Kaurav
Institute of Pharmaceutical Sciences
Guru Ghasidas Vishwavidyalaya (A Central University)
Bilaspur, India

Samiullah Khan
Department of Microbiology
Faculty of Biological Sciences
Quaid-i-Azam University
Islamabad, Pakistan

Irina Kiseleva
Department of Virology
Federal State Budgetary Scientific Institution "Institute of Experimental Medicine"
and
Faculty of Dental Medicine and Medical Technologies
Saint Petersburg State University
and
Educational Department
Smorodintsev Research Institute of Influenza
St. Petersburg, Russia

Anusha Konatala
University of Windsor
Windsor, Ontario, Canada

Kah Hon Leong
Department of Environmental Engineering
Faculty of Engineering and Green Technology
Universiti Tunku Abdul Rahman
Kampar, Malaysia

Sindhu Thalappan Manikkoth
School of Chemical Sciences
Kannur University, Swami Anandatheertha Campus
Payyanur, India

Dhanraj T. Masram
Department of Chemistry
University of Delhi
Delhi, India

Sunita Minz
Indira Gandhi National Tribal University Amarkantak (Central University)
Anuppur, India

Ghulam Mustafa
Department of Chemistry
University of Okara
Okara, Pakistan

Zubera Naseem
Department of Chemistry
University of Agriculture
Faisalabad, Pakistan

Jit Jang Ng
Department of Environmental Engineering
Faculty of Engineering and Green Technology
Universiti Tunku Abdul Rahman
Kampar, Malaysia

João Augusto Oshiro-Júnior
Pharmaceutical Sciences Postgraduate Program
Center for Biological and Health Sciences
State University of Paraíba (UEPB)
Campina Grande, Brazil

Dhanalakshmi Padi
Department of Biotechnology
Andhra University
Visakhapatnam, India

Kavita Pal
Advanced Centre for Treatment
Research and Education in Cancer
Tata Memorial Centre
Mumbai, India

Deepthi Panoth
School of Chemical Sciences
Kannur University, Swami Anandatheertha Campus
Payyanur, India

Fain Parackel
University of Windsor
Windsor, Ontario, Canada

Anjali Paravannoor
School of Chemical Sciences
Kannur University, Swami Anandatheertha Campus
Payyanur, India

Sai Brahma Penugonda
Department of Operations
North Carolina State University
Raleigh, North Carolina, USA

Sudhakar Pola
Department of Biotechnology
Andhra University
Visakhapatnam, India

Madhulika Pradhan
Rungta College of Pharmaceutical Sciences and Research
Bhilai, India

Badma Priya
Molecular Biophysics Unit
Indian Institute of Science
Bangalore, India

Ludmila Puchkova
Laboratory of Trace Elements Metabolism
ITMO University and Department of Biophysics
Peter the Great St. Petersburg Polytechnic University
and
Department of Molecular Genetics
Federal State Budgetary Scientific Institution "Institute of Experimental Medicine"
St. Petersburg, Russia

Kantrol Kumar Sahu
Institute of Pharmaceutical Sciences
Guru Ghasidas Vishwavidyalaya (A Central University)
Bilaspur, India

Pavani Sanapala
Department of Biotechnology
Andhra University
Visakhapatnam, India

Jeyabalan Sangeetha
Department of Environmental Science
Central University of Kerala
Kasaragod, India

Pichiah Saravanan
Environmental Nanotechnology Laboratory
Department of Environmental Science and Engineering
Indian Institute of Technology (ISM)
Dhanbad, India

Yeşeren Saylan
Department of Chemistry
Hacettepe University
Ankara, Turkey

Janani Semalaiyappan
ICAR-Indian Institute of Millet Research
Hyderabad, India

Haripriya Shanmugam
Department of Nano Science and Technology
Tamil Nadu Agricultural University
Coimbatore, India

Shahjalal Mohammad Shibly
Nanotechnology and Catalysis Research Center (NANOCAT)
Institute of Advanced Studies (IAS)
University of Malaya
Kuala Lumpur, Malaysia

Maria Shoukat
Department of Microbiology
Faculty of Biological Sciences
Quaid-i-Azam University
and
National Institute of Health
Islamabad, Pakistan

Lan Ching Sim
Department of Chemical Engineering
Lee Kong Chian Faculty of Engineering and Science
Universiti Tunku Abdul Rahman
Kajang, Malaysia

Shien Yoong Siow
Department of Chemical Engineering
Lee Kong Chian Faculty of Engineering and Science
Universiti Tunku Abdul Rahman
Cheras, Malaysia

Manickam Senguttuvan Swetha
Department of Nano Science and Technology
Tamil Nadu Agricultural University
Coimbatore, India

Noor Tahir
Department of Chemistry
University of Agriculture
Faisalabad, Pakistan

Suman Tahir
Department of Chemistry
University of Agriculture
Faisalabad, Pakistan

Bo Tan
State Key Laboratory of Hydraulics and Mountain River Engineering
College of Water Resource and Hydropower
Sichuan University
Sichuan, China

Devarajan Thangadurai
Department of Botany
Karnatak University
Dharwad, India

Kunnambeth Madam Thulasi
School of Chemical Sciences
Kannur University, Swami Anandatheertha Campus
Payyanur, India

Anjana K. Vala
Department of Life Sciences
Maharaja Krishnakumarsinhji Bhavnagar University
Bhavnagar, India

Baiju Kizhakkekilikoodayil Vijayan
School of Chemical Sciences
Kannur University, Swami Anandatheertha Campus
Payyanur, India

Wei Qing Wee
Department of Chemical Engineering
Lee Kong Chian Faculty of Engineering and Science
Universiti Tunku Abdul Rahman
Cheras, Malaysia

Krishna Yadav
University Institute of Pharmacy
Pt. Ravishankar Shukla University
Raipur, India

Muhammad Zahid
Department of Chemistry
University of Agriculture
Faisalabad, Pakistan

Section I

Nanosized Materials in Virology

1 Viruses as Nanomaterials

Dushyant R. Dudhagara, Megha S. Gadhvi, and Anjana K. Vala

CONTENTS

DOI: 10.1201/9781003136644-2

1.1 INTRODUCTION

In the 21st Century, nanotechnology has become one of the most rapidly developing fields of science and technology (Mangematin and Walsh 2012). Nanotechnology is a multidisciplinary field that combines applied science and engineering to focus on ways to control and measure objects at the molecular level (Zhang 2003). Materials with at least one dimension within the range of 1–100 nm are commonly used in the field of nanotechnology. The fabrication of biomimetic materials has received a lot of attention in recent years. Viruses are the best example of highly organized nanoscale biological materials and structures that have motivated researchers to develop new methods for producing novel nanomaterials. Bio-nanotechnology encompasses a multidisciplinary approach and can be very broad. To analyze and use biological molecules at the nanoscale, scientists have been combining concepts from biology, chemistry, physics, materials science, and engineering into a unified pattern. Bio-templated nanomaterials, biomimetics, bio-nanopatterning, and nano-toxicology are only a few of the topics covered by this mixture of biology and nanotechnology (Soto and Ratna 2010).

Bionanoscience is a rapidly developing area of nanoscience that uses biological building blocks such as DNA, peptides, and proteins (Pinheiro et al. 2011). For example, any 2D or 3D DNA-based nanostructure that can be used as a drug-delivery vehicle can be designed using a particular DNA sequence (Zheng et al. 2009; Wei et al. 2012). By designing specific peptide sequences, it becomes easy to create self-assembled nanospheres and nanofibers, which are useful for drug carrier development and tissue-regenerative nanofibrous scaffolds (Koutsopoulos et al. 2009). Modified modular-based protein machinery, such as effective protein-based motors and modular-based tissue-engineering scaffolds used to repair damaged soft and hard tissue, can be achieved by designing particular proteins.

Virus-based nanomaterial designs have recently received a lot of attention (Pinheiro et al. 2011). Viruses are biochemical structures that contain both genetic and protein content. The physical, chemical, and biological functions of viral particles are all described by the genetic sequences found within them. The natural function of a virus is to infect host cells and to interrupt their metabolism to produce a high number of viral particles and to spread the viral population. Viral particles have been used as a versatile method in recent biotechnology and nanotechnology developments to design novel functional nanomaterials such as semiconductor materials, energy-producing materials, tissue-engineering materials, and drug-delivery materials (Ghosh et al. 2012; Lee et al. 2012).

Viruses contains several copies of one or a few capsid protein subunits that are arranged in icosahedral or helical symmetry. Such repeating components can be used as a medium for attaching or combining different functional groups to multiple sites, allowing for a wide range of nanomaterial fabrication options.

Virus capsids can be designed as containers for targeted delivery of drugs and other therapeutic agents, or changed to create nano-sized models for material deposition (i.e. metals) (Steinmetz et al. 2009a; Kobayashi et al. 2012; Aljabali et al. 2013; Zeng et al. 2013).

This chapter highlights the advancements made in the production of various nanomaterials/nanostructures using animal, bacterial, and plant viruses through different techniques, such as controlled self-assembly, layer-by-layer assembly (LbL), biomineralization, biotemplating, encapsulation, infusion, and bioconjugation strategies, as well as their potential applications in the field of nanotechnological and nanomedicine. Furthermore, this chapter also highlights detection and characterization techniques used for nanomaterial configuration.

1.2 PLANT VIRUSES AS NANOMATERIALS

Viruses have historically been studied as animal, human, and plant pathogens. Now, due to advances in genetic and chemical engineering, they can also serve as an effective tool for the development of nanomaterials. Recent studies include a great body of knowledge about the biological, chemical, and physical properties of viruses, which detail specific genetic and chemical modifications to their capsids (Steinmetz et al. 2009a). Although viruses, in general, have a lot of potential for nanotechnology application, plant viruses have certain advantages; for instance, they are not pathogenic to animals. Another advantage is that they have noninfectious capsids (empty virus-like particles; eVLP) that can be easily and rapidly replicated in high yields using plants as hosts (Saunders et al. 2009; Fischer et al. 2004). This approach of producing viruses and virus-like structures is generally considered as being secure and cost effective compared to other host-expression systems, such as yeast, insect cells, and bacteria (Rybicki 2010; Penney et al. 2011). The VNPs of icosahedral plant viruses, such as VLPs of Cowpea mosaic virus (CPMV), Tobacco mosaic virus (TMV), Brome mosaic virus (BMV), and Hibiscus chlorotic ringspot virus (HCRSV), are commonly used in the development of various nanomaterials and nanostructures.

1.2.1 Cowpea Mosaic Virus (CPMV)

As mentioned above, plant viruses can be used as building blocks for synthesis of nanomaterials. The natural properties of CPMV make it an attractive nanoscale building block for medical and material science applications. Generally, CPMV has a 28 nm diameter and an icosahedral capsid that contains 5.9 and 3.5 kb of positive-sense genetic material, as RNA1 and RNA2, respectively. These RNAs have a single open reading frame and are expressed through the synthesis and successive processing of precursor polyproteins. A cofactor of proteinase, 24 K proteinase, helicase, and RNA-dependent RNA polymerase are mostly found in RNA-1. RNA-2 codes for cell-to-cell and long-distance movement proteins are found in plants, as well as large (L) and small (S) coat proteins. The two domains of the L-coat protein connect with the S-coat protein, which forms asymmetric units. These 60 asymmetrical units are in control of forming the CPMV capsid

(Sainsbury et al. 2010). The surface of CPMV particles is exposed and produces addressable amines (lysine), carboxylates (aspartic acid and glutamic acid), and hydroxyl (tyrosine) groups of amino acids, which can be used for various applications in selectively connected elements like redox-active molecules, fluorescent dyes, metallic and semi-conducting nanoparticles, carbohydrates, DNA, proteins, and antibodies (Steinmetz et al. 2009b).

Additionally, these functional groups have promoted the electroless deposition (ELD) of metal ions, such as cobalt, nickel, iron, and platinum, on the external surface of CPMV (Aljabali et al. 2010a). ELD is the process that uses chemicals for the reduction of metals on the external surface; it is not conductive for electrical current deposition. The ELD method can be used to study interaction of the CPMV surface amine groups with a palladium "activator," which can form clusters after treatment with a reducing agent (i.e. dimethyl amine borane); these clusters act as nucleation points for the absorbance and fusion of the metals' ions. This process allows the formation of metal-coated CPMV nanoparticles with a size of 35 nm in diameter, although deposition steps can be increased for the synthesis of larger metallized spheres (Aljabali et al. 2010a). In another study, Aljabali et al. (2011a) used 1-ethyl-3-(3-dimethlyaminopropyl) carbodiimide hydrochloride (EDC) chemical-linkage techniques to connect externally exposed amine groups with peptides, which control specific mineralization techniques. In this study, the peptides supported the coating of CPMV with metals like cobalt, platinum, iron, and zinc sulphide after being subjected to a reducing agent; this process successfully synthesized 32 nm diameter metal-coated spherical particles. This specific type of CPMV metal-coated nanoparticle has significance in catalysis and ultra-high density magnetic data-storage devices (FePt and CoPt; Hyeon 2003); semiconducting zinc sulphide nanoparticle having importance in optoelectronic devices, such as solar cells (Gao et al. 2004). Another approach for CPMV metallization is the amplification of the negative charge on the surface of CPMV by striking at least 240 externally subjected amine groups with succinic anhydride (Aljabali et al. 2011b). The higher negative charge promotes the binding of cobalt or iron ions, which can be reduced to cobalt metal or iron oxide under mild ELD conditions; therefore, 32 nm in diameter metal ion-coated CPMV particles are formed.

Aljabali et al. (2011b) further illustrated that the iron oxide coated CPMV can later be operationalized with a thiolated oligosaccharide, a probable association with iron or persisting hydroxyl groups, hence extending the range of modifications that may be carried out. Such particles have magnetic properties, which can permit their use as magnetic resonance imaging (MRI) contrast agents or as molecules that can attach to cancer cells and kill them by producing heat over the application of an oscillating magnetic field (hyperthermia therapy) (Chatterjee et al. 2011). In addition to MRI, CPMV can also be complemented with Gd^{3+} ions and clinically accepted Gd tetraazacyclododecane-tetraacetic acid (DOTA) paramagnetic complexes. Previous studies have reported that Gd^{3+} ions can equitably attach to CPMV nucleoprotein, or, alternately, Gd (DOTA) complexes can bind through Cu-mediated azide–alkyne cycloaddition to linkers that have been EDC associated to the CPMV external surface subjected amine groups. It has been reported that gadolinium bound to CPMV had a three-fold improvement in relaxivity compared to free Gd (DOTA), therefore showing that the changed CPMV has capacity to work as an improved contrast agent. The utility can be purified by selecting the CPMV-Gd complexes to specific tissue types of interest through the association of recognition motifs, which will help to increase the local concentration of Gd (Prasuhn et al. 2007).

Genetic modification on CPMV can increase the exterior surface affinity with metals and metalloids. For example, Steinmetz et al. (2009b) explained that CPMV can be genetically modified to surface display a dodecapeptide that allows silica attachment and mineralization. The peptide particles stimulated the deposition of a 2 nm layer of silica on the modified CPMV surface after treatment with the tetraethylorthosilicate and aminopropyltriethoxysilane as precursor agents. In another case, a chimeric CPMV was modified to contain a surface-subjected metal-attaching peptide, which promoted metallization through iron and platinum after treatment with salts and successive subjection to a reducing agent (Shah et al. 2009). In this study, the chimeric CPMV was covered with iron and platinum to a thickness of 1 nm. Genetic modification of CPMV was performed to insert capsid surface-exposed cysteine residues (that are not generally present on the external surface of wild type CPMV) at defined sites (Blum et al. 2004; 2005); these can connect preformed gold nanoparticles with their thiol groups. The attached gold nanoparticles were connected with each other via di-thiolated linkers containing metal particles. Eventually, the surface lysine residues of the CPMV were chemically attached to biotin by EDC carbodiimide conjugation. Because of the metal coating, the modified CPMV particles were observed to be highly conductive; furthermore, this conductance caused the interaction and binding of avidin to biotin, hence making an effective biosensor (Blum et al. 2011).

Additionally, for the surface modification of CPMV, there has also been an interest in stuffing the internal structure with metals and other functional chemicals. Previously, icosahedral plant viruses like Cowpea chlorotic mottle virus (CCMV) were observed to be capable of encapsulating nanoparticles like those made up of vanadate, tungstate, titanium, and Prussian blue; this process occurs when the nucleic acid-free empty particles use in vitro assembly (Liepold et al. 2007; Klem et al. 2008). It is tough to get RNA-free capsids of CPMV that can be filled because they present in small numbers in infected plants. Utilizing cutting edge plant-expression technology, empty CPMV particles can be achieved in high quantity by expressing the coat protein precursor with the 24 K proteinase in plants; in this study, the proteinase split the coat protein precursors and permitted their self-assembly in empty capsids (Saunders et al. 2009; Montague et al. 2011). The pure empty CPMV particles (eCPMV) can be filled with any cargo, such as drugs, fluorescent dyes, or metals. Aljabali et al. (2010b) auspiciously filled the internal cavity of eCPMV with cobalt or iron ions earlier to reduction with sodium borohydride, and that action led to the development of a cobalt metal or iron oxide core. This perspective has not changed the structure of the particles or affected their ability to be functionalized on their exterior side. The capacity to encapsulate materials within eCPMV and modify the exterior surface increases the range of how that CPMV can be customized for a certain nanotechnological intention.

Another objective is to exploit these diversified purposes in the manufacturing process of new devices. CPMV should be properly patterned into 2D and 3D structures at the nanoscales. Earlier work by Fang et al. (2002) illustrated that CPMV could form parallel lines and cross lines like patterns when dried on acid-treated mica. These designs were observed to be 9 capsid layers with 250 nm deep, 600 nm wide, and of 7 mm in length. CPMV can also be organized in highly ordered single-layer membranes that can be formed at the liquid interphase of oil droplets in aqueous solutions and stabilized by cross-linking (Russell et al. 2005). In a second perspective, chemoselective protein-surface linkers were designed onto a gold plate using scanning probe nano lithography into 30–50 nm width; subsequently, genetically modified CPMV that has special cysteine residues at designated locations was covalently connected onto these linker patterns (Cheung et al. 2003). Genetically modified CPMV, including surface-subjected cysteine residues, has been used to form monolayers on gold surfaces by a gold–thiol interaction. As another approach, monolayers can be created using biotinylated-linker-functionalized CPMV that can bind with gold surfaces accessorized with streptavidin. The significance of a layer of streptavidin on the higher surface of the biotinylated CPMV mono layer allows the subsequent joining of a second biotinylated CPMV monolayer (Steinmetz et al. 2006).

This method can be extended to create multilayered arrays of CPMV. The length of the linker that is joining the biotin molecule and the CPMV, and the number of biotin molecules per CPMV, can be modified to change the density and porosity of the CPMV array. Shorter lengths of linkers and higher amounts of biotin functionalization can lead to dense CPMV arrays, while less biotin and long linkers leads to channels in the array; therefore, it is important to consider flexibility in the types of applications in determining how they can be used. For instance, condensed packing of internally metal-coated or electroactive CPMV in the array will favor electron-transfer events in the study of electronic applications. In contrast, the CPMV array with channels may permit liquid movement into the network, which would have functions in catalysis and biosensors (Evans 2010). In one another study, Falkner et al. (2005) formed macroscale 3D structures from CPMV crystals. These crystals are body-centered cubic crystals with cavities; the authors demonstrated that these are solvent obtainable and can be used for the controlled growth of palladium and platinum. CPMV not only has great probabilities for the synthesis of new unique monodisperse metal-coated nanoparticles with many functions, but it can be used as a highly flexible building block for the formation of functional 2D and 3D macro- and nanoscale devices.

1.2.2 Tobacco Mosaic Virus (TMV)

TMV virions are rod-shaped virions that encapsulate a positive-sense ssRNA genome having 6395 nucleotides in 2130 similar coat protein subunits of 17.5 kDa each, making a hard helical tube-like structure with a length of 300 nm, a diameter of 18 nm, and a width of 4nm in the central channel (Clare and Orlova 2010). This structure shows some discrete and repeated functional groups on the interior and exterior surfaces (Knez et al. 2004); it is appropriate for different changes, such as metal coating or specific biochemical conjugation reactions. For instance, the external surface has the COOH terminal of the coat proteins with serine and threonine as an amino acid residue (OH groups), while the central channel has OH groups with primary amines. Like CPMV, the TMV surface chemistry permits binding of a various metal ions, which can be decreased into a metal coating through techniques like electroless deposition (ELD). To illustrate, one of the earlier reports on the metal coating of TMV included utilizing the external surface glutamate and aspartate particles for nucleating the co-precipitation of CdS or PbS from H_2S gas and metal salt precursors, or for the accumulation of iron oxide developed via oxidative hydrolysis (Shenton et al. 1999). While the authors illustrated the metal coating of the exterior surface of TMV, they have also suggested that the surface chemistry of the central channel might be useful for metal precipitation. Utilizing ELD decreasing baths made of dimethylamine borane or ascorbate, Knez et al. (2004) found that metal could be precipitate in the central channel or alternatively on the exterior surface through differing the pH, various metal ion, and changing the phosphate concentration of the reactions. These techniques have permitted the formation of TMV-based nanowires that contain different metals and alloys of copper, iron, nickel, cobalt, and palladium (Balci et al. 2012). Balci et al. (2012) explained that the small TMV alloy nanowires (>200 nm length) could be significant in high-density data storage, imaging, sensing, catalysis, and drug delivery. For instance, palladium-coated TMV can be used in chemical catalysis for the reduction of dichromate, a system that can be upgraded by adding the palladium-TMV into easily obtainable hydrogel microparticles (Yang et al. 2013).

In addition to TMV, TiO_2 can be used as sensors for the examination of liquefied petroleum gases (Rong et al. 2009), and if ZnO is covered on the viral surface, it can be used as semiconducting transistor (Atanasova et al. 2011). Furthermore, study by Gorzny et al. (2010) illustrated that platinized TMV rods synthesized using methanol ELD had high stability and were 65% more effective as anodes in direct methanol fuel cells (DMFCs) than conventional nanoparticles of a similar size. Hence, DMFCs provide 10-fold the high-energy density storage that traditional lithium-ion batteries provide; the authors also explained that this would be upgraded by utilizing TMV-Pt particles as the anode.

Moreover, controlled precipitation of metals on the surface of unmodified TMV, utilizing the ELD perspective, as mentioned earlier, few research groups have tried to attach already synthesized nanoparticles to the surface of TMV. Khan et al. (2013) utilized citrate-coated negatively charged gold and iron oxide nanoparticles in an acidic solution and applied this approach to wild type TMV (WT TMV). In this study, binding of the nanoparticles occurred when the natural negative charge of TMV was removed using the acidic protonation method; this method allowed electrostatic attraction within the nanoparticles and the surface of TMV. Other studies observed that TMV may be connected with iron nanoparticles in Ferro fluids to make a complex of straight TMV/nanoparticle scaffolds. This organization is regarded to report the significantly increased magneto viscosity upon the application of a magnetic field when differentiating the ferrofluids requiring the TMV (Wu et al. 2010). The authors explained that the upgraded magneto viscosity of the TMV "doped" ferrofluids can have significance in the engineering and transport sectors.

Another report stated that filtered TMV coat proteins, which can self-build into ring-like disk structures under specific salt and pH conditions, can be used as a scaffold for the binding of previously synthesized gold nanoparticles (Zahr and Blum 2012). Zahr and Blum (2012) reported that the gold nanoparticles were previously treated with bis (p-sulfonato phenyl) phenyl phosphine to increase the electrostatic association of the nanoparticles in the presence of arginine residues on the top margin of the TMV coat protein disk. This process resulted in the production of a ring of gold nanoparticles on the external boundary of the TMV disk. The authors also illustrated that the internal channel of the TMV-nanoparticle cluster could also attach to a gold nanoparticle in a pH-adjustable manner, where the pH-dependent protonation of the negatively charged arginine and carboxyl groups in this site permits the attachment. The TMV-nanoparticle rings contained very fascinating local surface plasmon-resonance spectra that may be easily tuned through the insertion of a gold nanoparticle in the central core. The authors explained that such optically harmonized structures could have broad significance in the field of nonlinear optics and in sensor sand laser technology (Zahr and Blum 2012).

TMV can be genetically engineered to incorporate metal-binding motifs in a surface subjected to the C-terminal region of the coat protein to enhance its applications. When differentiated to WT TMV, a titanium-attaching motif was connected to the C-terminal point of the coat protein; that action increased the amount, but reduced the size, of the connected TMV-bound gold nanoparticles synthesized from gold salts and $NaBH_4$ baths (Kobayashi et al. 2012). This research also disclosed that these TMV-gold nanoparticle hybrid materials had interesting spectral characteristics in the UV and visible light region, indicating that they could be used in optics. A hexa-histidine tag can be introduced in the coat protein in a different sample, and this will enhance the growth density of gold precipitation differentiated to WT TMV (Nan et al. 2012). TMV CP with hexa-histidine tag can also be used to attach previously synthesized gold nanoparticles, which can work as nucleation points for further metal precipitation, if required (Wnek et al. 2013). As differentiated to precisely developed gold nanowires, these techniques form gold TMV nanowires having a very low electrical resistivity, recommending that gold TMV wires would be useful in the electronics industry. TMV was changed to have a new CP cysteine residue (TMV1cys), which permitted concisely packed and vertically directed binding to gold surfaces (Royston et al. 2008) as a basic component for the formation of 3D battery components. After previous activation by a palladium catalyst, the vertically arrayed TMV1cys were exposed to dimethyl amino-borane mediated ELD of cobalt and nickel, resulting in the production of a 40 nm thick metallization. The structure was created to be extremely stable under a variety of vacuum, aqueous, and electrical conditions, and it had two-fold increase in electrode capacity in a nickel-zinc battery, as differentiated to controls that did not have a virus (Royston et al. 2008). The authors suggested that this method is simple to perform, claiming that only 450 mg of virus is needed to coat a square metre of electrode surface.

The cysteine-changed TMV has been used in a new process for the formation of battery elements (Gerasopoulos et al. 2010). The modified TMV was prepared on a gold-coated substrate and activated by palladium above a 20 nm thick nickel layer; it was precipitated on the virus surface using dimethylamino borane. An Atomic Layer Deposition (ALD) technique has been utilized to coat the nickel layer with 20 nm of TiO_2. ALD is a method for controlling thin film precipitation on surfaces; this method includes injecting two or more chemical vapours or gaseous precursors into a gaseous medium having the target substrate in short pulses. The TMV-based anode was found to have a two-fold higher discharge capacity and can outperform the control planar films at very high currents and cycle numbers; uncommonly, the TMV-based structures remained perfect during such harsh electrical testing. Gerasopoulos et al. (2010) and Royston et al. (2008) hypothesized that increased electrical characteristics can occur because of factors like high surface areas being transmitted through TMV templates.

TMV assemblage is managed by the RNA Origin of Assembly (OA) motif's high compatibility for 20S disc coat protein accumulates (Butler and Lomonossoff 1980), which could be utilized to affect the arraying and structural features of the collected TMV rods in a different way. Refined WT or modified TMV CP subunits (obtained from TMV separated from plants and synthesized from yeast or bacterial expression systems) could be mixed with synthetic RNA having multiple OAs, resulting in the construction of branched TMV structures. These structures include star, tree, and boomerang patterns that might be free or connected RNA or metal-coated nanoparticle centrals (Eber et al. 2013). These are a novel division of metallized TMV hybrids with a wide surface-area protein canopy that can be applicable further.

1.2.3 Brome Mosaic Virus (BMV)

Brome mosaic virus (BMV) is a typical member of the genus *Bromovirus* in the family *Bromoviridae* and has a plus-sense ssRNA separately encapsidated in nonenveloped isometric particles of ~28 nm in diameter. It infects both monocots and dicots, causing yellow or white spots and streaks on leaves, as well as subsequent yellowing that results in slightly stunted growth and shrunk grain. BMVs are T = 3 (180 subunits) icosahedral particles that shed their hexameric capsomeres and condense into T = 1 (60 subunits) icosahedral particles with pentameric capsomeres in some situations (Larson et al. 2005).

1.2.3.1 BMV in Encapsulation Strategies for Imaging/Therapeutics

VNPs can encapsulate foreign molecules or nanoparticles. This capability has potential applications in drug delivery and catalysis. Chen et al. (2005) fabricated, in vitro, the nucleus of negatively charged gold nanoparticles surrounded by reassembled BMV proteins. Synthetic DNA nucleotide sequences that copy the high-compatibility capsid binding site of native RNA in BMV were used to observe the formation of a closed shell-encasing nanoparticles. The shell/core interactions can also be tuned by using coatings on the nanoparticle core (Loo et al. 2007). In general, the VLP cage interior provides an attractive compartment for encapsulating non-native cargo (Schwarz and Douglas 2015). Dragnea et al. (2003) showed that the generation of bioinspired architectures for nongenomic materials packaging can be attributed to plant-infecting icosahedral viruses (Chen et al. 2006). Brome mosaic virus (BMV) is an icosahedral RNA plant virus that forms virus-like particles (VLPs) by self-assembling BMV coat protein around functionalized gold nanoparticles. The electrostatic behavior of nucleic acid components of the native virus is mimicked by encapsulating gold nanoparticles (foreign material) inside virus cages. These VLPs give researchers access to materials that can be used to make new optical and functional probes for biosensors and biomedical imaging (Chen et al. 2006). The electrostatic interaction between metal particles and the positively charged inner surface of the BMV capsid has been driven by the encapsulation of metals within viral capsid cages. Gold nanoparticles (AuNPs) incorporated in BMV capsids have been used as spectroscopic markers. Furthermore, Rayleigh resonance (RR) spectroscopy on single viruses encapsulated with AuNPs also opens up new possibilities for in vivo viral capsid transformation monitoring (Dragnea et al. 2003). Similarly, when BMV viral coat proteins self-assembled around iron oxide nanoparticles (IONPs) functionalized with PEGylated phospholipids, highly symmetrical structures were formed, even when the IONPs were larger than the native BMV capsid's inner cavity scale.

Furthermore, these VLPs have superparamagnetic properties and may be used in MRI (Huang et al. 2007). Viral capsids also act as constrained reaction environments for the synthesis of nanomaterials. In the synthesis of nanomaterials, viral capsids often serve as restricted reaction conditions. Organic self-assembly in combination with inorganic photo-responsive nanoparticles of II–VI semiconductors offers a method for fabricating quantum-confined semiconductor arrays. The self-assembly of VLPs of BMV virus around quantum dots (QDs) (CdSe/ZnS nanocrystals) functionalized with HS-poly (ethylene glycol) (PEG)–COOH and DNA have been demonstrated. However, dihydrolipoic acid (DHLA)-coated QDs and phospholipid micelle-coated QDs did not induce the BMV capsid assembly. VLPs with HSPEG–COOH coated QDs were found to be photostable to long-wavelength visible light and to have the greatest ability to induce BMV capsid assembly (Huang et al. 2007). Furthermore, nanoparticle-biomaterial conjugates have many other important biomedical applications, such as for molecular imaging, drug delivery, and biosensors. In particular, the luminescence property of encapsulated QDs in VLPs can be used to fabricate luminescent bioprobes (Dixit et al. 2006). Similarly, the encapsulation of ~19 nm cubic iron-oxide nanoparticles (IONPs) coated with carboxyl-terminated PEGylated phospholipids (HOOC-PEG-PL) by BMV capsid shells yielded hybrid encapsulated structures of mean diameter ~37 nm. In *Nicotiana benthamiana* leaves, these quasi-spherical virus-based nanoparticles (VNPs) exhibited cell-to-cell and long-distance transit activity, and also an excellent T2-weighted MRI contrast agent response, and were found to be useful for monitoring physiological processes important for plant growth and development (Huang et al. 2011).

The majority of supramolecular structures involving protein subunits self-assemble to form cage structures that can encapsulate foreign molecules, which could be useful for therapeutic delivery, catalysis, and cell-targeting imaging. Interestingly, virus capsids that self-assemble into cages may undergo reversible structural modifications that open or close gated pores, permitting switchable entry to their interior. Kusters et al. (2015) presented a model for the encapsulation of spherical gold nanoparticles by BMV CP if the nanoparticle surface receives a large enough negative charge to resolve repulsive interactions between positively charged RNA-binding domains of the proteins. This finding indicates that, in addition to nanoparticle size, the capsid protein's size has a major impact on encapsulation performance. For tissue-specific imaging and/or drug delivery, structurally interesting nanoparticle formulations have also been developed. Gupta et al. (2013) used a nano-construct made up of an FDA-approved near-infrared (NIR) chromophore called indocyanine green (ICG) encapsulated by a BMV protein shell as a contrast agent for NIR photoacoustic imaging (PAI). As optical viral ghosts (OVGs), this virus-mimicking nanoconstruct could provide a flexible forum for the development of a multifunctional nanoconstruct with potential for site-specific and combined deep tissue imaging and phototherapy. Furthermore, over the wavelength range of 760–820 nm, these OVGs released far stronger photoacoustic signals than blood. Interestingly, Indocyanine green (ICG), an organic NIR chromophore, can be encapsulated by genome-depleted BMV. These nontoxic and non-cytotoxic optical viral ghosts (OVGs) have been used as nanoprobes for intracellular optical imaging of human bronchial epithelial (HBE) cells with no toxicity (Jung et al. 2011).

1.2.4 Hibiscus Chlorotic Ringspot Virus (HCRSV)

Hibiscus chlorotic ringspot virus (HCRSV) is a nonenveloped plant virus with a positive-sense, monopartite ssRNA genome. It belongs to the Carmovirus class (family *Tombusviridae*). The Malvaceae flowering plant family is infected by HCRSV (Liang et al. 2002). The virion has a 30 nm icosahedral capsid with T = 3 quasi-symmetries, 180 coat protein subunits of approximately 38 kDa, and 3.9 kb of genomic RNA encapsidated (Cheng et al. 2009). HCRSV coat protein reassembles to form empty VLPs, which can be used as nanosized systems for cancer chemotherapy drug delivery. Ren et al. (2006) demonstrated that HCRSV-derived VLPs can be used as a transport vehicle for anionic polyacid (PC) drug molecules, such as polystyrenesulfonic acid (PSA) and polyacrylic acid (PAA) with an upper-limit molecular mass threshold of ~13 kDa. Polyacid-loaded VLPs were found to be comparable to native genomic HCRSV in terms of size, morphology, and surface-charge density. Similarly, Ren et al. (2007) HCRSV protein cages of 30 nm nanoscale, conjugated with folic acid (a targeting molecule) and encapsulating the anticancer drug doxorubicin, were prepared and shown to have cytotoxicity in human ovarian cancer cells (Figure 1.1).

1.3 BACTERIAL VIRUSES (BACTERIOPHAGES) AS NANOMATERIALS

1.3.1 M13 Bacteriophage

The M13 phage has a long-rod, filamentous shape that is approximately 880 nm long and 6.6 nm wide and contains circular, single-stranded DNA that is covered

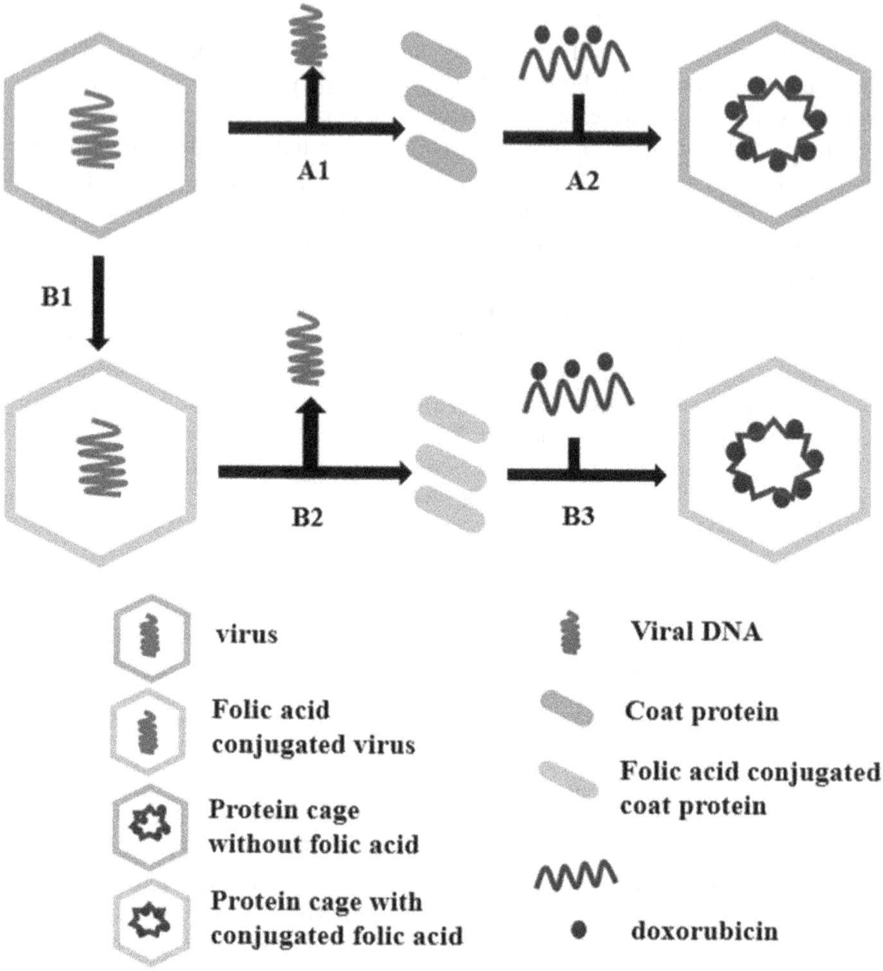

FIGURE 1.1 The preparation of doxorubicin-loaded Hibiscus chlorotic ringspot virus (HCRSV) (Ren et al. 2007). A1 and B2 are steps to remove the viral RNA from the plant virus and purification of coat proteins. Steps A2 and B3 indicate the encapsulation of polyacid and doxorubicin during the reassembly of protein cage. Step B1 shows the conjugation of folic acid onto the viral protein coat.

by approximately 2800 copies of the major coat protein p8 and capped by five copies of each of the four minor coat proteins (p9, p7, p6, and p3) at the ends, giving the phage particle distinguishable ends that can be changed independently. Functional peptides can be displayed on the p3 and p9 minor coat proteins, as well as the p8 main coat protein, by genetic modification (Huang et al. 2005).

M13 bacteriophages can be assembled layer by layer to form a film to develop "advanced nanoporous materials." The layer-by-layer assembly was achieved using covalent interaction, which was conducted through 1-ethyl-3-(3-dimethylaminopropyl) carbodiimide (EDC) crosslinking. The bacteriophage film had a surface porosity of 59% and a roughness of 40%. The substrates were dipped in a $TiCl_4$ solution after layer-by-layer assembly to deposit Ti ions, which were then hydrolyzed to form TiO_2 nanoparticles using the sol-gel process. The as-prepared TiO_2 film's surface morphology revealed a mesh-like assembly of nanowires 10 nm in diameter. Pore sizes on the surface and within these meshes ranged from 4 to 15 nm. The reaction temperature and time have been shown to be crucial in interconnecting the wires and preventing overgrowth. Since TiO_2 can be sensitized with Au for optoelectronic applications, the capacity of the bacteriophage film to embed Au nanoparticles was investigated using dip-and-drop coating. Because of capillary forces, drop coating increased nanoparticle penetration into the network. M13 bacteriophages were used as models for the synthesis of nano-barium titanate (BTO) nanostructures, in contrast to the virus-based, layer-by-layer assembly. Genetically engineered M13 bacteriophages with three glutamates on the N-terminus of the major capsid p8 were incubated with barium glycolates, followed by titanium glycolates. The interaction of viruses with barium glycolates to form a barium containing complex 1 (virus + barium glycolate) was aided by electrostatic interaction. Through electrostatic and hydrogen bonding interactions, complex 1 was eventually coated with titanium glycolates to form titanium- and barium-containing complex 2 (virus + barium glycolate + titanium glycolate). After that, Complex 2 was calcined to form a perovskite crystal structure (BTO). Despite the fact that the viruses were calcined, their fibrous morphology (lengths of 50–100 nm) was preserved. Following that, BTO was spin-coated on a polydimethylsiloxane layer, and indium tin oxide coated with polyethylene terephthalate was applied. The indium tin oxide coated polyethylene terephthalate was used as the top electrode, which formed a sandwich structure for piezoelectric evaluations. The resulting flexible nanogenerator, with a surface area of 6.25 cm^2, generated 300 nA of short-circuit current and 6 V of open-circuit voltage. Because of the dispersed configuration of the BTO nanoparticles, which formed from viral templates, the nanogenerator was also more efficient than the BTO nanoparticles assembled with the C nanotubes. The viral template assembly remained stable for up to 21,000 stress cycles (Luo et al. 2015).

1.3.2 T4 Bacteriophage

Bacteriophage T4 is one of the most complex viruses. The mature virion, which consists of a protein shell encasing a 172 kbp double-stranded genomic DNA, a 'tail', and fibres connected to the distal end of the tail, is made up of more than 40 different proteins. The host cell recognition sensors are carried by the fibres and tail, which are necessary for phage attachment to the cell surface. The tail also acts as a conduit for phage DNA transmission from the head into the cytoplasm of the host cell. The tail is connected to the head's special 'portal' vertex, which is used to package phage DNA during head assembly. The unique vertex is dominated by a dodecameric portal protein, which is involved in DNA packaging, similar to other phages and herpes viruses (Leiman et al. 2003).

1.3.2.1 Synthesis of Fe Based Nanocomposite by T4 Bacteriophage Virus

The T4 bacteriophage virus was used and collected for the synthesis of nanoparticle composite. To obtain "abundant T4 virions, a concentrated crude lysate of bacteriophage T4 was multiplied on E. coli B cells and cultured overnight in a lysogeny broth (LB) medium. The suspension of enrichment T4 virus was centrifuged and purified. The purified T4 virions were distributed in deionized water before being stored at 4°C for further use. The T4-supported Fe@Au composite was made by mixing a freshly formulated aqueous solution of $FeCl_3$ (10 mM, 200 L) with a stock T4 suspension (300 L) kept at a pH of 8.5–9.0 for 3 hours. The solution was thoroughly mixed and incubated for 20 h at room temperature (RT). The system was diluted tenfold with deionized water and centrifuged for 2.5 h at 11,000 g. The pellet was dispersed in 500 μL of deionized water after the supernatant was removed. The system was reduced by adding a drop wise freshly prepared aqueous solution of $NaBH_4$ (10 mM, 45 μL) and then incubated at RT for 5 h. The reaction solution and reductant were purged with ultrapure nitrogen to prevent oxidation, and the system was sealed in a microcentrifuge tube under nitrogen. In the reaction method, an aqueous solution of $AuCl_3$ (10 mM, 100 L) was applied. The mixtures were vortically agitated for 5 minutes before being incubated at room temperature for 10 hours. Finally, dropwise $NaBH_4$ solution (10 mM, 75 L) was added to the reaction mixtures, which were then incubated at room temperature for 5 h. The system was centrifuged at the high speed, and then deionized water was used to re-disperse the deposition. Different amounts of Au solution (10 mM, 50 μL, 70 μL, 80 μL, and 90 μL) were applied to regulate gold shell growth with different thicknesses, and the amount of reductant was consequently adjusted (10 mM, 37.5 μL, 52.5 μL, 60 μL, and 67.5 μL) (Xu et al. 2012). The Fe/T4 complex's detailed structure as shown by TEM. Figure 1.2 depicts a simplified schematic sketch map that depicts the nanocomplexes' basic preparation method. With a T4 core, a thin iron inner layer, and a thick gold outer shell (noted as Fe@Au/T4), this complex is a T4-supported Fe-core/Au-shell nanostructure (Xu et al. 2012) (Figure 1.2).

1.4 ANIMAL VIRUSES AS NANOMATERIALS

1.4.1 Encapsulation of Gold Nanoparticles by Simian Virus 40 Capsids

Viral capsid-nanoparticle hybrid structures are a novel type of nanoarchitecture that can be utilized for different applications. Viral capsids have captivated great interest in the field of nanotechnology because of their nano in size, regular structural organization, loading capacity, controlled self-assembly, and ability to be easily modified (Sapsford et al. 2006; Soto et al. 2006). A new class of bio-nanomaterials called viral capsid-nanoparticle hybrid structures, which merge the bioactivities of virus capsids through the applications of nanoparticles, includes many potential applications such as therapeutic and diagnostic vectors, imaging agents, and advanced nanomaterial-producing reactors (Aniagyei et al. 2008; Kwak et al. 2010). In humans, the viral capsid-nanoparticle hybrid mechanism can be used for therapeutic and diagnostic purposes. SV40 has a nonenveloped capsid with a diameter of around 45 nm, which contains the histone associated 5.2 kb

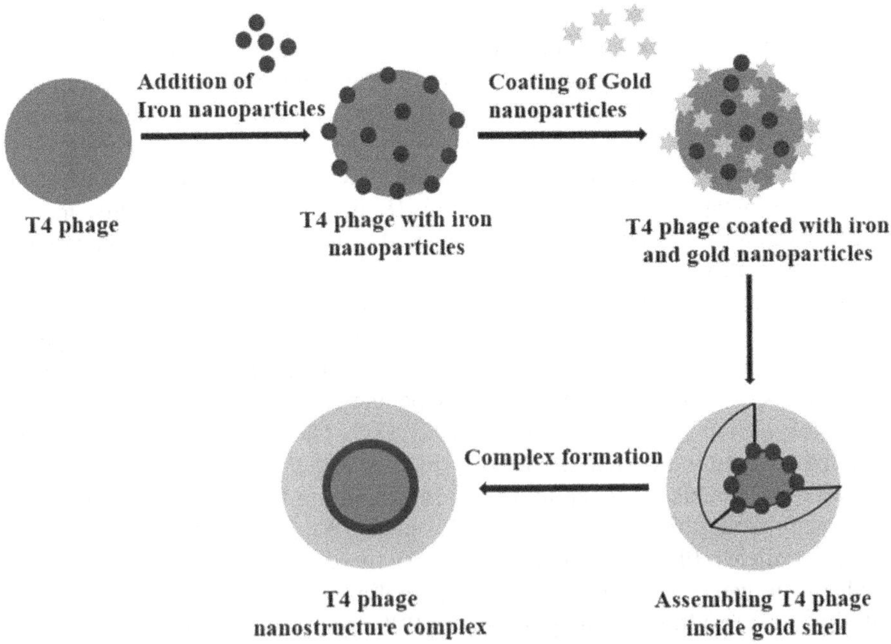

FIGURE 1.2 Schematic diagram of the preparation T4 phage Nano complex (Xu et al. 2012).

genome DNA (minichromosome) that encodes six proteins including three structural proteins, the major capsid protein VP1, and the minor capsid proteins VP2 and VP3. The outer shell of the icosahedral structure is made up of 72 copies of the VP1 pentamer, while VP2 or VP3 attaches to the inner surface cavity of each pentamer to link it to the minichromosome (Stehle et al. 1996). In vitro self-assembly of VP1 pentamers can be used to make SV40 virus-like particles (VLPs). Because VP1 can assemble into a variety of structures, including wild-type VLPs and intermediate, small, and tubular particles, SV40 VLPs could be useful as vectors for artificial cargos. Furthermore, SV40 capsids can infect a wide range of human cells via endocytosis, including hematopoietic stem cells (Lund et al. 2009). In addition to the method of encapsulating nanoparticles via electrostatic interactions that has been documented for other viral capsids, it was recently discovered that SV40 capsids could encapsulate QDs with different surface charges (Li et al. 2010). This means that SV40 capsids can encapsulate nanoparticles more effectively than other viral capsids, making them ideal for creating multifunctional hybrid nanostructures. However, the encapsulation mechanism for SV40 capsids is still unknown, and whether SV40 can encapsulate other inorganic nanoparticles is also unknown. In this case, the encapsulation of gold nanoparticles (AuNPs) with different sizes and surface coatings by SV40 capsids has been explained in detail to examine the encapsulation mechanism of SV40 capsids. The present study showed that AuNPs could be successfully encapsulated in SV40 capsids. AuNPs of various sizes and coated with various ligands were encapsulated into SV40 capsids to investigate the interesting issues concerning the effects of particle size and surface decoration of nanoparticles on encapsulation into viral capsids. Liu et al. (2007) explained the encapsulation of mPEG-AuNPs with different sizes by SV40 VP1 pentamers. To replace the citric acid ligand and reduce the negative surface charge of AuNPs, a neutral ligand of thiolated mPEG was decorated on the surface. To investigate the effect of ligand size on the encapsulation of AuNPs by VP1, different mPEG ligands, namely mPEG 750 and mPEG2000, were conjugated onto AuNPs. When compared to mPEG750, mPEG2000 was supposed to improve AuNP stability in solution. Meanwhile, mPEG2000 will reduce the interaction between AuNP and the VP1 pentamer by increasing the distance between them (Liu et al. 2007). VP1 pentamers effectively encapsulated mPEG750-AuNPs of 15, 20, and 30 nm, but not 5 nm and 10 nm mPEG750-AuNPs. VP1 pentamers were unable to encapsulate AuNPs of various sizes coated with mPEG2000. The stability of AuNPs is determined by the total potential energy, which includes van der Waals attractions and electrical double layer repulsive forces, according to the Derjaguin–Landau–Verwey–Overbeek theory. Because the van der Waals attraction energy is reduced, smaller mPEG-AuNPs are generally more stable. By increasing steric repulsion, longer surface ligands and higher ligand coverage would favour the stability. As a result, as the length of the mPEG increases and the diameter of the nanoparticle decreases, the stability of the mPEG-AuNPs should increase. As a result, the mPEG2000-coated AuNPs may have been too stable to absorb VP1 pentamers and trigger self-assembly. Similarly, due to the weak van der Waals force caused by the small particle sizes, 5 nm and 10 nm mPEG750-AuNPs were unable to assemble with VP1 pentamers. However, the fact that mPEG750-AuNPs with diameters of 15, 20, and 30 nm were successfully encapsulated may be attributable to the fact that nonspecific protein absorption by AuNPs increases with increasing diameter, at least between 10 and 60 nm (Liu et al. 2007). As a result, the minimum size of AuNPs coated with mPEG750 that formed SV40-Au VLPs in this study was 15 nm. SV40-Au VLPs containing 30 nm mPEG750-AuNPs, whose diameter (44.7 ± 3 nm) is similar to that of wild-type SV40, had the highest encapsulation efficiency of 89.3% (Liu et al. 2007).

The effects of electrostatic interactions and particle size on the encapsulation of DNA-AuNPs with negatively charged surfaces were also investigated. Providing biological functions to organic or inorganic nanomaterials, such as DNA functionalized nanoparticles, has enormous potential for molecular diagnostics and drug delivery. The conformation of the DNA strands determines the thickness of the DNA layer on AuNPs. DNA strands with fewer than 30 bases are completely extended and aligned perpendicular to the surface. Longer DNA strands (over 30 bases) have a fully extended inner section and a randomly coiled outer section (Parak et al. 2003). In this study, VP1 pentamers did not encapsulate 5 nm AuNPs conjugated to 10T or 50T DNA. When 10 nm 10T-AuNPs were used as the cargo, SV40-Au VLPs were formed with a 32% encapsulation efficiency. 5 nm 50T-AuNPs are actually larger than 10 nm 10T-AuNPs in terms of total size of functionalized particles. The inability of the VP1 pentamers to encapsulate the former suggests that the size of the Au core is more important than the overall particle size in the construction of SV40-Au VLPs. Assuming that 10 nm mPEG750-AuNPs were not encapsulated by VP1 pentamers, the efficient encapsulation of 10 nm 10T-AuNPs demonstrates that the electrostatic effect derived from surface decorations will aid the assembly process when the AuNPs are small. The encapsidation efficiency improved as the AuNPs' diameter was increased to 15, 20, and 30 nm, similar to what was seen with mPEG750-AuNPs. The results indicate that particle size and surface decoration play important roles in VP1 pentamer encapsulation of AuNP. The van der Waals force should be one of the major encapsulation drives for mPEG750-AuNPs. The electrostatic interaction between the DNA ligand and VP1 will facilitate the encapsulation of DNA-AuNPs.

3It is interesting to note that VLPs, including cowpea mosaic virus, CCMV, canine parvovirus, and MS2, all have a diameter between 25 and 30 nm, which may only be sufficient to encapsulate inorganic nanoparticles less than 20 nm in diameter (Manchester and Singh 2006). Encapsulation of inorganic nanoparticles of various sizes can be required for specific functions in practical applications. In this context, it's interesting to note that VP1 pentamers can encapsulate AuNPs with diameters of up to 30 nm. Encapsidation of both mPEG750-AuNPs and 10T-AuNPs was observed when the AuNPs' diameter was increased to 50 nm. In the assembly buffer used in this analysis, the majority of the empty SV40 VLPs made from VP1 pentamers had diameters of around 20 nm. When different sizes of AuNPs were applied, however, the diameters of the SV40-Au VLPs compared with those of the AuNPs. The capacity of VP1 pentamers to encapsulate a broad range of AuNPs of various sizes could be due to the SV40 VLPs' versatile native configuration. VP1 is unusual in that it can form a variety of polymorphic aggregates, including wild-type VLPs, intermediate, short, and tubular particles, which may be linked to the bonding specificity switching required to construct the virus capsid. Similar to the packaging of DNA by VP1 pentamers, AuNPs may act as templates during the self-assembly of VP1 pentamers (Mukherjee et al. 2007). The assembly process begins when VP1 pentamers are absorbed onto the surface of AuNPs and reach a certain local concentration. The formation of SV40 VLPs is then carried out by adding individual VP1 pentamers to the artificial cargos one at a time. The SV40-Au VLPs containing 15 nm mPEG750-AuNPs were able to infect Vero cells, while the bare 15 nm mPEG750-AuNPs were not. This may be due to the fact that the main determinants for SV40 cell attachment and entry are found in VP1 pentamers. The TEM images revealed that SV40-Au VLPs have a similar cellular entry behavior to SVLP-QDs, which enter living cells through caveolae-mediated endocytosis and accumulate within the ER (Li et al. 2009). As a result, the SV40 capsid can carry AuNPs into host cells, and SV40 VLPs may be used as a new vector for delivering artificial cargos to cells.

1.4.2 Encapsulation of Fe_3O_4 Nanoparticles into Genetically Modified Hepatitis B Core Virus-Like Particles

Viruses are quite effective at identifying cells and delivering cargo into them. As a result, simulating a virus can be a very effective way to transmit drugs or imaging probes to cancerous cells. VLP proteins, which are a subclass of proteins that form symmetric protein shells around an empty interior space, may self-assemble into organized macromolecular structures that are identical or close to homologous virion structures (Grgacic and Anderson 2006). For multifunctional imaging, therapeutic, or diagnostic agent delivery, nanoparticles entrapped by in vitro interaction have recently been achieved by using various VLPs (Capehart et al. 2013). It has been documented that certain viruses and metal nanoparticles form conjugates with antimicrobial and antiviral properties against viruses (Vauthier and Bouchemal 2009). On the inner side of VLPs, intermolecular associations dependent on electrostatic attraction have been used to connect various negatively charged nanoparticles (Chen et al. 2006). Another approach used the origin of assembly (the RNA sequence of a virion) as a catalyst to package nanoparticles based on noncovalent interactions (Loo et al. 2006). However, certain processes are either complicated or have a poor encapsulation yield. The inner surface of VLPs can be modified to provide new affinity properties for specific capture of foreign molecules, potentially increasing loading performance. The affinity of histidine tags, which can bind to metal-ion complexes including nickel-nitrilotriacetic acid (nickel-NTA) chelate, is used to create a simple and high-yielding process for encapsulating magnetic nanoparticles into VLPs (Kim et al. 2010).

Shen et al. (2014) suggested that the particular affinity of histidine tags to the nickel chelate, a novel and simple method for efficiently encapsulating nanoparticles into genetically engineered hepatitis B core VLPs has been created. The author theorizes that a mutant HBc144 VLP with six consecutive histidine residues (His-tag) incorporated at the N-terminal ends and shown on the inner surface of the VLPs was able to efficiently encapsidate the nanoparticles by exploiting histidine tags' affinity for the nickel-NTA chelate. Monodispersed, uniformly sized magnetic cores containing VLPs can be obtained with high efficiency using this process. It's also worth noting that the core size has a significant impact on the formation of these magnetic VLPs. Shen et al. (2014) also demonstrates that in comparison to simple Fe_3O_4 nanoparticles, the resulting Fe_3O_4 -NTA-Ni^{2+} core-containing HBc-144-His VLPs had a higher cellular uptake, indicating that these core-containing VLPs could be used as highly effective cellular T2 contrast agents for MRI applications.

Shen et al. (2014) suggested a unique, straightforward method to efficiently encapsulate nanoparticles into genetically engineered hepatitis B core VLPs through the specific affinity of histidine tags to the nickel-NTA chelate. The resulting HBc-144-His VLPs encapsulated Fe_3O_4 nanoparticles were highly monodisperse and uniform. The size effect is believed to be the result of a correlation between the T-number and the self-assembly efficiency, and the efficiency of the VLPs encapsulation was found to be highly dependent on the core size. As compared to pure Fe_3O_4 nanoparticles, the obtained core-containing VLPs had higher cellular uptake and higher R^2 values, and thus could be used as high-efficiency cellular T2 contrast agents for MR imaging applications (Shen et al. 2014).

1.5 CHARACTERIZATION OF NANOMATERIALS

The size and shape are the two main parameters for characterization of nanomaterials. Size, size distribution, and organic ligands present on the surface of the particles may influence other properties and possible applications of the NPs. Furthermore, the crystal structure of the NPs and their chemical composition are thoroughly investigated as a first step after nanoparticle synthesis due to their lack of standard protocol for the analysis and identification. However, there are important challenges in the analysis of nanomaterials because of the interdisciplinary nature of the work, the absence of suitable reference materials for the calibration and standardization of analytical tools, the difficulties to the sample preparation for analysis, and the interpretation of the data. Therefore, many challenges exist in categorization of nanomaterials.

1.5.1 Phase Characterization of Nanomaterials

In chemistry phase or matter of substance is a distinct state characterized by homogeneous composition and properties and the possession of a clearly defined boundary; in physics, phase is a particular stage in a periodic process. Phase characterization after biological synthesis is important to determine the composition, purity, and morphology of nanomaterials.

1.5.1.1 X-Ray Diffraction of Synthesized Nanomaterials

X-ray diffraction (XRD) is a popular analytical technique that has been used for the analysis of both molecular and crystal structures (Waseda et al. 2011), qualitative identification of various compounds, quantitative resolution of chemical species, measuring the degree of crystallinity, isomorphous substitutions, particle sizes, etc. When X-ray light reflects on any crystal, it leads to the formation of many diffraction patterns, and the patterns reflect the physico-chemical characteristics of the crystal structures. In a powder form, diffracted beams typically come from the sample and reflect its structural physico-chemical features. Thus, XRD can analyze the structural features of a wide range of materials, such as inorganic catalysts, superconductors, biomolecules, glasses, polymers, and so on (Waseda et al. 2011; Singh et al. 2013). Analysis of these materials largely depends on the formation of diffraction patterns. Each material has a unique diffraction beam that can define and identify patterns by comparing the diffracted beams with the reference database in the Joint Committee on Powder Diffraction Standards (JCPDS) library. The diffracted patterns also explain whether the sample materials are pure or contain impurities. Therefore, XRD has long been used to define and identify both bulk and nanomaterials, forensic specimens, industrial materials, and geochemical materials (Wu et al. 2011; Yazdian et al. 2013).

XRD is a primary technique for the identification of the crystalline nature at the atomic scale (Sapsford et al. 2011; Waseda et al. 2011). X-ray powder diffraction is a nondestructive technique with great potential for the characterization of both organic and inorganic crystalline materials. This method has been used to measure phase identification, conduct quantitative analysis, and determine structural imperfections in samples from various disciplines, such as geological, polymer, environmental, pharmaceutical, and forensic sciences. Recently, the applications have extended to the characterization of various nanomaterials and their properties (Das et al. 2014). The working principle of X-ray diffraction follows Bragg's law.

Typically, XRD techniques are worked based on the wide-angle elastic scattering of X-rays (Lin et al. 2014). Although XRD has several advantages, it has limited disadvantages, including difficulty in growing the crystals and the ability to get results pertaining only to a single conformation/binding state (Sapsford et al. 2011). Another drawback of XRD is the low intensity of diffracted X-rays compared to electron diffractions (Chapman et al. 2011).

1.5.2 Morphological Characterization of Nanomaterials

Morphology of nanoparticles refers to their size distribution dimensions and shape characteristics in materials science and engineering; it can be analyzed by various of microscopical techniques.

1.5.2.1 Transmission Electron Microscopy (TEM)

TEM is an important technique for the characterization of nanomaterials, used to obtain quantitative measures of particle and/or grain size, size distribution, and morphology (Joshi et al. 2008). TEM exploits the interaction between a uniform current-density electron beam (i.e. a range within the 60 to 150 keV) and a thin sample. When the electron beam reaches the sample, some of the electrons are transmitted, while the rest are elastically or inelastically scattered (Reimer and Kohl 2008). The magnification of TEM is mainly determined by the ratio of the distance between the objective lens and the specimen and the distance between objective lens and its image plane. TEM is the most common technique to analyze nanoparticle size and shape since it provides not only direct images of the sample but also the most accurate estimation of the nanoparticle homogeneity. However, some limitations have to be considered when using this technique, such as the difficulty in quantifying a large number of particles or misleading images due to orientation effects.

Nanoparticle physical properties not only depend on their size and morphology but also other factors, such as interparticle distance. For instance, when two metal NPs are brought in close proximity, their plasmons couple, red-shifting their plasmon band and changing their colour. Therefore, TEM has been used to characterize the nanoparticle aggregation for different biomedical applications such as (1) sensing and diagnostics, where the aggregation depends on the presence of a biomarker or analyte; (2) therapy, where the aggregation causes an increase of the nanoparticle therapeutic effect; and (3) imaging, where the aggregation improves the response signal (Thanh and Rosenzweig 2002; Nam et al. 2009; Perrault and Chan 2010).

1.5.2.2 Scanning Electron Microscopy (SEM)

Recently, the field of nanoscience and nanotechnology has provided a driving force in the development of various high-resolution microscopy techniques to learn more about nanomaterials using a beam of highly energetic electrons to probe objects on a very fine scale (Yao and Kimura 2007). Among various electron microscopy techniques, SEM is a surface-imaging method, fully capable of resolving different particle sizes, size distributions, nanomaterial shapes, and the surface morphology of the synthesized particles at the micro and nanoscales

(Hall et al. 2007; Lin et al. 2014). SEM is a widely used method for the high-resolution imaging of surfaces that can be employed to also characterize nanoscale materials. Using SEM, we can probe the morphology of particles and derive a histogram from the images by either measuring and counting the particles manually, or by using specific software (Fissan et al. 2014). The combination of SEM with energy-dispersive X-ray spectroscopy (EDX) can be used to examine silver powder morphology and also conduct chemical composition analysis. The limitation of SEM is that it is not able to resolve the internal structure, but it can provide valuable information regarding the purity and the degree of particle aggregation. The modern high-resolution SEM can identify the morphology of nanoparticles below the level of 10 nm.

1.5.2.3 Atomic Force Microscopy (AFM)

Instead of using SEM and TEM, atomic force microscopy (AFM) is a scanning probe technique that can reveal a range of information about the NM, the biomolecule, and the NM-biomolecule interaction on a single particle basis. Although once considered a fairly specialized technique, recent advances in both instrumentation and software have not only expanded the technique's capabilities but have also made it more user-friendly. Unlike TEM and SEM, which are best performed with conducting or semiconducting samples under vacuum conditions, AFM can be applied to nonconductive, wet, and soft samples, allowing for many different types of materials to be analyzed in physiological environments (Cohen and Bitler 2008). Samples can be analyzed using various scanning modes, including static (noncontact) and dynamic/tapping (contact and intermittent sample contact), which provide information about various sample parameters including morphological information about the NP (size and shape, etc), information about the biomolecule (adhesion, elasticity, Young's modulus, and stretching parameters), and interactions between the individual particles and biomolecules (protein folding-unfolding, biomolecular attachment to an NM surface, NM–NM agglomeration, etc.)(Sitterberg et al. 2010). Moreover, AFM has the advantage that it does not require any surface modification or coating prior to imaging. Thus, the topological analysis of small NPs (≤6 nm), such as ion-doped Y_2O_3, has been performed by AFM without any special treatment. As compare to SEM and TEM, AFM provided better results in terms of NP sizes.

1.5.3 Spectroscopical Characterization of Nanomaterials

Spectroscopy is the most widely used analytical technique for surface chemical analysis, also employed for the characterization of nanoscale materials. It is a powerful technique for characterization of structure, composition, size and other basic features of the NPs. In nanotechnology, spectroscopy is useful for the determination of nanomaterials phase, morphology, and optical property characteristics.

1.5.3.1 UV-Visible Spectroscopy

UV-visible spectroscopy is a very useful and reliable technique for the primary characterization of synthesized nanoparticles; it is also used to monitor the synthesis and stability of nanomaterials. It measures the intensity of light reflected from a sample and compares it to the intensity of light reflected from a reference material. In addition, UV-vis spectroscopy is easy, fast, simple, and selective technique for different types of nanomaterials; there is no need to calibrate for particle characterization of colloidal suspensions (Tomaszewska et al. 2013). In nanoparticles, the conduction band and valence band lie very close to each other in which electrons move freely. These free electrons give rise to a surface plasmon resonance (SPR) absorption band, occurring due to the collective oscillation of electrons of nanoparticles in resonance with the light wave (Das et al. 2014). The absorption of nanoparticles depends on the particle size, dielectric medium, and chemical surroundings. Observation of the peak assigned to a surface plasmon is well documented for various metal nanoparticles with sizes ranging from 2 to 100 nm (Noginov et al. 2006).

1.5.3.2 Fourier Transformed Infrared (FTIR) Spectroscopy

FTIR is able to provide accuracy, reproducibility, and also a favourable signal-to-noise ratio. By using FTIR spectroscopy, it becomes possible to detect small absorbance changes, which helps to perform difference spectroscopy, which can distinguish the small absorption bands of functionally active residues from the large background absorption of the entire protein (Kim and Barry 2001; Zscherp and Barth 2001). FTIR spectroscopy is frequently used to find out whether biomolecules are involved in the synthesis of nanoparticles, which is more pronounced in academic and industrial research (Gurunathan et al. 2014). Furthermore, FTIR has also been extended to the study of nano-scaled materials, such as confirmation of functional molecules covalently grafted onto silver, carbon nanotubes, graphene and gold nanoparticles, or interactions occurring between enzyme and substrate during the catalytic process (Baudot et al. 2010). Additionally, it is a noninvasive technique. Finally, the advantages of FTIR spectrometers over dispersive ones are rapid data collection, strong signal, large signal-to-noise ratio, and less sample heat-up (Kumar and Barth, 2010). Recently, further advancement has been made in an FTIR method called attenuated total reflection (ATR)-FTIR spectroscopy (Hind et al. 2001). Using ATR-FTIR, the chemical properties on the polymer surface may be determined, and sample preparation is easy compared to conventional FTIR (Liu et al. 2007). Therefore, FTIR is a valuable, noninvasive, suitable, cost-effective, and simple technique to identify the nanoparticles.

1.5.3.3 X-Ray Photoelectron Spectroscopy (XPS)

XPS is the most commonly used analytical technique for surface chemical analysis, which is often used to classify nanoscale materials. Its underlying physical principle is the photoelectric effect. XPS is an effective quantitative method for determining the physical properties of elements, such as their electronic structure, elemental composition, and oxidation states of the material. It can also examine ligand exchange interactions, surface functionalization, and core/shell structures of NPs, and it works in ultra-high vacuum. The environment-dependent crystal structure tuning of metal chalcogenide NPs of different sizes has been investigated using XPS (Sarma et al. 2013). It can also distinguish between core/shell and homogeneous alloy structures, as well as determine the bonding mode of ligands including trioctylphosphine oxide (TOPO) on metal chalcogenide NPs surfaces. XPS provides the depth information, similar to the size of NPs, i.e. up to 10 nm depth from the surface, and it does not cause any harm to the samples (Shard 2012). In XPS analysis, on the other hand, sample preparation (i.e. dry and solid sample without contamination needed for analysis) and data interpretation are extremely difficult. As a result, the technique is underutilised in nanomaterial characterization.

1.5.3.4 Nuclear Magnetic Resonance (NMR) Spectroscopy

Another important analytical technique for determining the quantitative and structural properties of nanoscale materials is NMR spectroscopy. It is based on the NMR phenomenon, which occurs when nonzero spin nuclei are put in a strong magnetic field, resulting in a slight energy difference between the spin-up and spin-down states. Electromagnetic radiation in the radio wave spectrum can be used to investigate transitions between these states. The interaction or coordination between the ligand and the surface of diamagnetic or antiferromagnetic NPs is usually studied using NMR. It is not appropriate for characterizing ferri- or ferromagnetic materials, however, because the high-saturation magnetization of such materials induces changes in a local magnetic field, resulting in signal frequency shifts and dramatic reductions in relaxation times. As a consequence, the signal peaks broaden significantly, rendering the measurements practically useless and indecipherable (Mourdikoudis et al. 2018). NMR spectroscopy can help toward the routine, straightforward, molecular-scale investigation of NP formation and morphology in situ, in both solution and solid phase. It is particularly useful for studying both the formation and final structure of metal oxide NPs (Table 1.1).

1.6 APPLICATIONS OF VIRUSES AS NANOMATERIALS

1.6.1 Biomedical Applications

Virus-based nanoparticles (VNPs) have received great attention in recent years; plant viruses have been used in materials science research and have a potential for application in nanomedicine (Czapar and Steinmetz 2017). As viruses possess the property of self-assembling around the cargo (the genome) and delivering this cargo to specific cells and tissues, they are ideal candidates for specific delivery sites for therapeutic or contrast agents. In fact, several VNP technologies for clinical trials of gene delivery and oncolytic virus therapy are being developed and applied. Due to their high degree of symmetry, multivalency, monodispersity, and genetic or chemical programmability, VNPs have become promising materials. Most VNP structures have been resolved to allow a high degree of spatial control of the tailored atomic resolution. VNPs can be modified by imaging contrast agents, therapeutic moieties, or targeting ligands (e.g. peptides or antibodies) in a chemical selective bioconjugation reaction (Lee et al. 2012).

TABLE 1.1
Various Analytical Techniques Used for the Characterization of Nanomaterials

Methods	Analytical Techniques	Advantages	References
Phase characterization	X-ray Diffraction	Primary technique for the identification of crystalline structure and characterization of various nanomaterials and viruses.	Sapsford et al. (2011)
Morphological characterization	Scanning Electron Microscopy (SEM)	SEM is a surface-imaging method used to identify size, shape, and other morphological characteristics of viruses as well as viral nanoparticles.	Fissan et al. (2014); Lin et al. (2014)
	Transmission Electron Microscopy (TEM)	Useful to visualize internal assembly of viral nanoparticles.	Lin et al. (2014)
	Atomic Force Microscopy (AFM)	Useful to determine biomolecular interactions between molecules. Useful for 3D mapping of the sample surface. It can be applied to nonconducting wet and soft samples on various substrates.	Cohen and Bitler (2008); Sitterberg et al. (2010)
Spectroscopical characterization	UV-Visible Spectroscopy	Cost-effective, simple, and fast technique. Intrinsic optical properties of biomolecules can be used to determine concentration.	Tomaszewska et al. (2013)
	Fourier Transformed Infrared (FTIR) Spectroscopy	Provides conformational information about the biomolecule and Nanomaterials bioconjugation. Possible to detect small absorbance changes.	Baudot et al. (2010)

1.6.2 Magnetic Resonance Imaging

MRI is a technique that visualizes the structure and metabolites within the capsid and therefore has the potential to detect physiological processes in vivo. Among them, the contrast agent and macromolecular platform can further enhance imaging sensitivity. Paramagnetic gadolinium (Gd^{3+}) is frequently applied as a MRI contrast agent; however, it has to be used in a chelated form to reduce the toxicity, usually with DOTA or diethylenetriamine pentaacetic acid (DTPA) (Qazi et al. 2012). After conjugation with such nanoparticles, a decrease in the molecular tumbling rate of gadolinium ions results in an increase in the longitudinal relaxation rate. These reagents, determined as nontoxic in vivo, were used for the MRI imaging of live mice (Singh et al. 2007). The increase in local concentration can be caused by the multivalent display, and both of these events can increase the sensitivity. Since Gd^{3+} or other lanthanides can increase the relaxation rate, differences in MRI signals can be detected regardless of where they are present, and the contrast level by many factors can be finally obtained. A multifunctional MRI contrast and photodynamic therapy agents (chelated Gd^{3+} and Zn^{2+} phthalocyanine dye) have been incorporated inside capsids of the CCMV protein; this represents the first step toward the consecution of fully self-assembled protein cage nanoparticles for multimodal imaging and therapy (Millán et al. 2014).

The covalent attachment of chelated gadolinium ions to the TMV leads to the enhanced ionic relaxivity of the Gd^+ ions based on reduced tumbling rates. Multivalent display leads to a relaxivity per nanoparticle that is four times higher than other VNP contrast agents. Steinmetz et al. (2009a, b) developed a high contrast-to-noise ratio, molecularly targeted MRI contrast agent using TMV nanoparticles (Bruckman et al. 2015). To achieve dual optical and MRI, the TMV was modified to carry near-infrared dyes and chelated Gd^+ ions. The probe targeted atherosclerotic plaques. On the basis of the multivalency and multifunctionality, the TMV-based MR probe showed an enhanced detection limit, 400 times lower than the typical clinical dose (Bruckman et al. 2014).

1.6.3 Drug Delivery by Virus Based Nanoparticles

The formation of VNPs that target particular cell types has permitted for the insertion of toxic loads by conjugation, infusion, or encapsulation so that the target cells are killed, therefore permitting the selective exclusion of cancer cells, or another diseased cells, without off-target effects. As discussed earlier, conjugation involves the selective covalent insertion of load molecules to specific amino acid residues of the coat protein. Infusion is obtained by incubating the intact VNP in a solution having the cargo (Yildiz et al. 2013), and encapsulation needs the carrier to be constructed around the load (Brown et al. 2002). Many different types of therapeutic cargo can be delivered, containing genes and short interfering RNAs (Galaway and Stockley 2013), traditional small-molecule drugs, photoactive molecules that support photodynamic therapy, and heterologous viral genomes for gene therapy, like an alphavirus genome encapsulated in a VNP based on CCMV (Azizgolshani et al. 2013).

Toxic cargos have been loaded into the VNP cavity, alternatively coated onto the exterior surface, to save them from enzymatic and chemical deterioration in vivo and avoid interconnections with nontarget cells. The capacity of VNPs and the ability of loading are generally upgraded by discarding the native viral genome and can be obtained by expressing the coat proteins from a plasmid (for the formation of bacteriophage VNPs) or from a transgene (for the synthesis of plant VNPs) so that the viral nucleic acid is never present; the resulting empty particle is explained as a virus-like particle (VLP). Instead, the viral genome can be discarded by selective chemical or enzymatic deterioration.

The covalent binding of toxic cargo molecules to internally subjected side chains ensures that there is no early release, but noncovalent methods generally permit for higher loading capacity because there is space in the VNP for more cargo if the complete cavity is utilized rather than only the interior surface. For instance, up to 300 doxorubicin molecules can be conjugated to the capsid surface of cowpea mosaic virus (CPMV) (Aljabali et al. 2013), but up to ~950 doxorubicin molecules could be loaded into the interior cavity of hibiscus chlorotic ringspotvirus (HCRSV) by encapsulation (Ren et al. 2007). Both viruses contain icosahedral capsids 30 nm in diameter. Polymerization may achieve the best by providing a branching network of functionalized groups for load binding, which extends from the exterior surface of the VNP or penetrates the interior (Hovlid et al. 2014). However, most cases have been focused on the structural feature of VNPs and their in vitro toxicity; the preclinical testing of a VNP-based drug-delivery vehicle has explained in vivo effectiveness and decreased cardiotoxicity of a doxorubicin-loaded VNP, particularly, cucumber mosaic virus (CMV) modified by folic acid to target ovarian cancer (Zeng et al. 2013).

Furthermore, standard chemotherapy, VNPs have been packed through photosensitizers for photodynamic therapy applications. For instance, a VLP based on bacteriophage Qβ can be loaded with a metalloporphyrin derivative for photodynamic therapy and a glycan ligand targeting cells bearing the CD22 receptor. Additionally, a multifunctional MRI contrast and photodynamic therapy agent (chelated Gd^{3+} and Zn^{2+} phthalocyanine) can be successfully encapsulated in CCMV as a first illustration of theragnostic VNPs. Hybrid VNP-based materials carrying metal nanoparticles for photothermal therapy have been explored (Millán et al. 2014).

1.6.4 Virus as Catalysts

Enzymes can also be loaded into the virus capsid to create nanoreactors. In nature, enzymes are found in the limited space of cells. Encapsulation of the enzyme within the viral capsid provides the possibility of studying enzymatic behavior in a cell-mimicking environment. In artificial nanoreactors, the envelopment of enzymes as a biomimetic system are expected to help us understand the interaction of enzymes in active and confined spaces. A number of groups have used phages to improve the efficiency of enzyme or chemical catalyst-driven reactions. In these cases, higher-order structures are not assembled, but the modified capsid alone provides sufficient structure to improve the catalysis, although the reason for the improvement is not always clear. Consequently, while these materials may not strictly fit into our definition of a nanostructured material, they are still worth noting. All three of the phage-derived catalysts have been used to produce hydrogen gas from either water or ethanol. Nam et al. (2009) used the light-absorbing Zn porphyrin phage, as described earlier, complexed with iridium oxide nanocrystals, to create a water oxidation catalyst. Water oxidation to hydrogen and oxygen gas by the iridium oxide was improved by the proximity of the photoexcited Zn porphyrin. A similar approach was explained by Neltner et al. (2010), who used M13 modified to bind nickel oxide, rhodium oxide, and ceria (cerium oxide) ions. After heating to fuse the bound particles and destroy the organic material (the phage), they found that the catalytic material improved rates of hydrogen production from breakdown of ethanol. The reaction using the phage-derived catalyst was equally efficient at lower temperatures compared to the unassembled metal oxide ions at higher temperatures. Overall, the phage-derived catalyst had a better functional temperature profile.

A different approach was recently demonstrated by Jordan et al. (2016) who used enzyme catalysts and the P22 capsid to create a self-assembling, high-efficiency material. This work, from Trevor Douglas' group at Indiana University, builds on their experience using capsids of viruses of animals, plants, and bacteria as assembly frames for various inorganic materials, such as metal crystals (Douglas and Young 1999). In the hydrogen-producing catalyst work, they fused two subunits of a bacterial hydrogenase enzyme to two copies of the P22 scaffold protein that normally directs the assembly of the major coat protein to form a procapsid before genome packaging. They allowed the scaffold protein-hydrogenases to assemble, presumably guided by the hydrogenase dimerization, and then activated expression of the capsid protein. The assembled procapsids formed around the internal scaffold, creating capsid-sized particles internally coated with the active hydrogenase dimer. These complexes released hydrogen gas at about a 100 times greater rate than the free hydrogenase enzyme or the scaffold protein-hydrogenase without the capsid. While the reason for this increase is not clear, they hypothesize that there is a synergistic effect of the close proximity of many hydrogenase proteins in a small volume. Furthermore, they also showed that the complex was more heat stable than the free enzyme, as well as resistant to inactivation by trypsin or exposure to air. This protective effect of the capsid presumably mirrors the capsid's normal function of protecting the internally located phage genome while in the extracellular environment.

1.6.5 Light-Harvesting System Based on the TMV

Viral capsids provide a means to precisely position chromophores with spatial control at the sub-nanometre level. The tobacco mosaic virus coat protein can serve as a convenient template for the synthesis of chromophores into light-harvesting arrays (Delor et al. 2018). An interesting aspect of this virus is that when the ionic strength and pH of a solution changes, the monomers can self-assemble into different nanoscale structures and form a double-layer disk composition comprised of 17 monomers per layer or an extended helical rod comprised of 16.33 monomers per turn. The disk and rod structure can provide a rigid frame for establishing and maintaining an accurate distance relationship between several microns of monomer (for rods), which is a very challenging technique for protein without capsid synthesis methods. TMV coat proteins were covalently labelled with either a donor or an acceptor from a Förster resonance energy transfer (FRET) pair and then assembled into disk or rod structures at a controlled ratio of donors to acceptors. During the assembly process, it is possible to adjust the TMV capsid by simply adjusting the ratio of monomers tagged for each component (Ma et al. 2008). For instance, the TMV virus was covalently modified at both the N-terminus and at an introduced cysteine residue with two different chromophores that undergo FRET to produce a "light harvesting rod" (Dedeo et al. 2010).

Although TMV is the first structurally characterized virus, research on the preparation of new materials for covalent modification has not been reported since 2005. In the first report on the subject, Francis described the use of diazotization chemistry to link tyrosine residues on the outer surface of a native capsid with carbodiimide chemistry to link an amine to an internal channel surface. According to this method, many different chromophores, ion chelators, and polymer chains have been added to the viral capsid surface (Schlick et al. 2005). Through the use of cysteine modification chemistry, porphyrins have been attached to individual capsid monomers to yield conjugates that self-assemble into efficient light-harvesting systems containing thousands of interacting groups (Endo et al. 2007).

1.6.6 Vaccines for Cancer

Cancer vaccines are based on our knowledge of the molecular identities of tumour-associated antigens (Yaddanapudi et al. 2013), associated with the capacity to provoke effective immune responses against these self-antigens in spite of active immunosuppression by the tumour. Anti-tumour vaccination has certain benefits over chemotherapy, containing a simplified outpatient process, lesser side effects, the prevention of drug resistance, the ability to guide the immune system to remove residual drug-resistant cells, and the initiation of long-term immunological memory to protect against metastases and relapse (Ryan et al. 2010).

Since tumour antigens are obtained from the host, it is important to break immunological tolerance to develop immune responses against nonimmunogenic or weakly immunogenic targets. Some VNP-based cancer vaccine perspectives, which include the patterned display of tumour-associated carbohydrate or peptide antigens, have been analyzed. CPMV particles were utilized as scaffolds to chemically display the Tn antigen, a glycoprotein that is overexpressed on numerous cancer cells involving breast, colon, and prostate cancer cells. Association to the virus-based scaffold and multivalent display resulted in formation of high titre of Tn-specific antibodies. Likewise, Tn antigen bound to TMV can induce antigen-specific IgG and IgM responses (Yin et al. 2012). The organization of cancer epitopes on virus-based scaffolds allows the arrangement of these self-epitopes in a non-native molecular environment; that is a hopeful procedure to control self-tolerance, as shown for human epidermal growth factor receptor 2 (HER2)-positive breast cancer patients (Shukla et al. 2014).

1.6.7 Vaccines for Neurological Diseases and Addiction

VLPs can be utilized as scaffolds to display the amyloid beta (Aβ) protein involved in the development of Alzheimer's disease. Papillomavirus and Qβ VLPs present Aβ antigens extracted anti-Aβ antibodies in the absence of adjuvant, and with controlled T cell responses. The antibody subclasses varied according to the use of

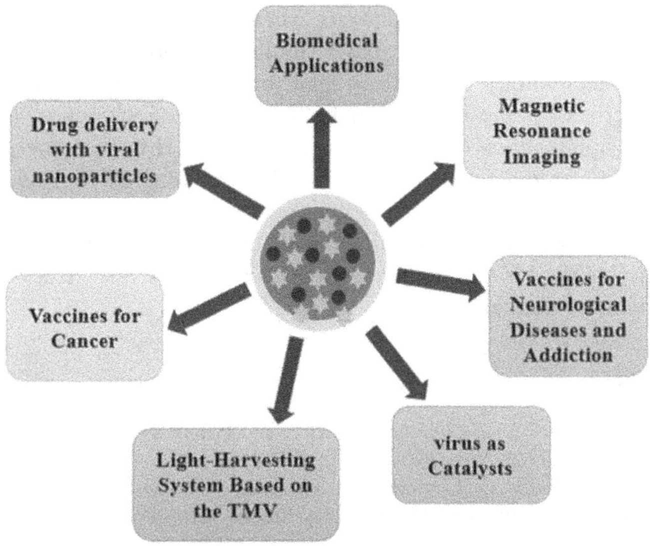

FIGURE 1.3 Applications of viruses as nanomaterials in various fields.

the whole antigen or peptide antigens (Chackerian et al. 2006). An HBV core antigen (HBcAg) containing N-terminal epitopes of Aβ has been produced as a vaccine and illustrated capacity in a transgenic mouse model of Alzheimer's disease (Jin et al. 2014).

A potential vaccine against nicotine addiction has been newly synthesized, utilizing the 30 nm icosahedral capsid of bacteriophage Qβ chemically modified to present nicotine in a multivalent manner. The multivalent and particulate nature of the Qβ-based vaccine boosts the formation of antinicotine neutralizing antibodies, thus decreasing blood nicotine levels and controlling transport across the blood-brain barrier. The nicotine vaccine is currently under examination in clinical trials with the goal to decrease smoking addiction (Cornuz et al. 2008) (Figure 1.3).

1.7 CONCLUSION

Developed viral nanoparticles and viral-like particles from animal, bacterial, and plant viruses are generally safe and provide highly variable platforms, which can be upgraded in terms of specificity for nanotechnological applications. Developments of novel bionanomaterials using new viruses are being promoted to exploit the abilities of new viral multivalent structures for various biotechnological and biomedical applications. Due to diverse capsid sizes, stabilities, and bioconjugation potentials of VNPs and VLPs, there exists an interesting range of bionanomaterials derived from various viruses. Different analytical techniques like X-ray diffraction, SEM, TEM, AFM, UV-Visible, FTIR, XPS, and NMR spectroscopy are very helpful for the characterization and detection of developed bionanomaterials. This chapter will help to expand our understanding of the effects of viral nanoparticles and virus-like particles on the applications in targeted-drug delivery, cell imaging, light-harvesting systems, and as catalysts, vaccines for cancer, neurological disorders, addiction, etc., which may have significant implications in the growing field of nanobiotechnology and help to lead the development of functional-based virus nanomaterials for the biomedical field. The use of viruses as nanoparticles for biomedical applications is still in the early stages. However, the properties of structurally characterized viruses are extremely well-adapted for nanotechnology. The combination of a defined, multivalent structure that can be modified for specific interactions and targeting in vivo, along with the natural container-like shape of virus particles, makes viruses highly appropriate tools for further developments in the field of nanobioscience.

REFERENCES

Aljabali, A.A.A., Barclay, J.E., Lomonossoff, G.P. and Evans, D.J. 2010 a. Virus templated metallic nanoparticles. *Nanoscale* 2(12):2596–2600.

Aljabali, A.A.A., Sainsbury, F., Lomonossoff, G.P. and Evans D.J. 2010 b. Cowpea mosaic virus unmodified empty viruslike particles loaded with metal and metal oxide. *Small* 6(7):818–821.

Aljabali, A.A.A., Shah, S.N., Gowing, R.E., Lomonossoff, G.P. and Evans D.J. 2011 a. Chemically-coupled-peptide-promoted virus nanoparticle templated mineralization. *Integrative Biology* 3(2):119–125.

Aljabali, A.A.A., Barclay, J.E., Cespedes, O., et al. 2011 b. Charge modified cowpea mosaic virus particles for templated mineralization. *Advanced Functional Materials* 21(21): 4137–4142.

Aljabali, A.A.A. Shukla, S., Lomonossof, G.P., Steinmetz, N.F. and Evans, D.J. 2013. CPMV-DOX delivers. *Molecular Pharmaceutics* 10(1):3–10.

Azizgolshani, O., Garmann R.F., Cadena-Nava, R., Knobler, C.M. and Gelbart, W.M. 2013. Reconstituted plant viral capsids can release genes to mammalian cells. *Virology* 441(1):12–17.

Atanasova, P., Rothenstein, D., Schneider, J.J., et al. 2011. Virus-templated synthesis of ZnO nanostructures and formation of field-effect transistors. *Advanced Materials* 23(42):4918–4922.

Aniagyei, S.E., Dufort, C., Kao, C.C. and Dragnea, B. 2008. Self-assembly approaches to nanomaterial encapsulation in viral protein cages. *Journal of Materials Chemistry* 18(32):3763–3774.

Blum, A., Soto, C.M., Wilson, C.D., et al. 2004. Cowpea mosaic virus as a scaffold for 3-D patterning of gold nanoparticles. *Nano Letters* 4(5):867–870.

Blum, A.S., Soto, C.M., Sapsford, K.E., Wilson, C.D., Moore, M.H. and Ratna, B.R. 2011. Molecular electronics based nanosensors on a viral scaffold. *Biosensors and Bioelectronics* 26(6):2852–2857.

Blum, A.S., Soto, C.M., Wilson, C.D., et al. 2005. An engineered virus as a scaffold for three-dimensional self-assembly on the nanoscale. *Small* 1(7):702–706.

Baudot, C., Tan, C.M. and Kong, J.C. 2010. FTIR spectroscopy as a tool for nano-material characterization. *Infrared Physics and Technology* 53(6):434–438.

Bruckman, M.A., Jiang, K., Simpson, E.J., et al. 2014. Dual-modal magnetic resonance and fluorescence imaging of atherosclerotic plaques in vivo using VCAM-1 targeted tobacco mosaic virus. *Nano Letters* 14(3):1551–1558.

Bruckman, M.A., Randolph, L.N., Gulati, N.M., Stewart, P.L. and Steinmetz, N.F. 2015. Silica-coated Gd (DOTA)-loaded protein nanoparticles enable magnetic resonance imaging of macrophages. *Journal of Materials Chemistry B* 3(38):7503–7510.

Balci, S., Hahn, K., Kopold, P., et al. 2012. Electroless synthesis of 3nm wide alloy nanowires inside Tobacco mosaic virus. *Nanotechnology* 23(4):045603.

Brown, W.L., Mastico, R.A., Wu, M., et al. 2002. RNA bacteriophage capsid-mediated drug delivery and epitope presentation. *Intervirology* 45(4–6):371–380.

Capehart, S.L. Coyle, M.P., Glasgow, J.E. and Francis, M.B. 2013. Controlled integration of gold nanoparticles and organic fluorophores using synthetically modified MS2 viral capsids. *Journal of the American Chemical Society* 135(8):3011–3016.

Chackerian, B., Rangel, M., Hunter, Z. and Peabody, D.S. 2006. Virus and virus-like particle-based immunogens for Alzheimer's disease induce antibody responses against amyloid-β without concomitant T cell responses. *Vaccine* 24(37–39):6321–6331.

Chapman, H.N., Chapman, H.N., Fromme, P., et al. 2011. Femtosecond X-ray protein nanocrystallography. *Nature* 470(7332):73–78.

Chatterjee, D.K., Diagaradjane, P. and Krishnan, S. 2011. Nanoparticle-mediated hyperthermia in cancer therapy. *Therapeutic Delivery* 2(8):1001–1014.

Chen, C., Daniel, M.C., Quinkert, Z.T., et al. 2006. Nanoparticle-templated assembly of viral protein cages. *Nano Letters* 6(4):611–615.

Chen, C., Kwak, E.S., Stein, B., Kao, C.C. and Dragnea, B. 2005. Packaging of gold particles in viral capsids. *Journal of Nanoscience and Nanotechnology* 5(12):2029–2033.

Cheng, A., Speir, J.A., Yuan, Y.A., Johnson, J.E. and Wong, S.M. 2009. Preliminary X-ray data analysis of crystalline hibiscus chlorotic ringspot virus. *Acta Crystallographica Section F: Structural Biology and Crystallization Communications* 65(6):589–593.

Cheung, C.L., Camarero, J.A., Woods, B.W., Lin, T.W., Johnson, J.E. and DeYoreo, J.J. 2003. Fabrication of assembled virus nanostructures on templates of chemoselective linkers formed by scanning probe nanolithography. *Journal of the American Chemical Society* 125(23):6848–6849.

Clare, D.K. and Orlova, E.V. 2010. 4.6Å Cryo-EM reconstruction of tobacco mosaic virus from images recorded at 300keV on a 4k × 4k CCD camera. *Journal of Structural Biology* 171(3):303–308.

Cohen, S.R. and Bitler, A. 2008. Use of AFM in bio-related systems. *Current Opinion in Colloid and Interface Science* 13(5):316–325.

Cornuz, J., Zwahlen, S., Jungi, W.F., et al. 2008. A vaccine against nicotine for smoking cessation: A randomized controlled trial. *PLoS One* 3(6):e2547.

Czapar, A.E. and Steinmetz, N.F. 2017. Plant viruses and bacteriophages for delivery in medicine and biotechnology. *Current Opinion in Chemical Biology* 38:108–116.

Dragnea, B., Chen, C., Kwak, E.S., Stein, B. and Kao, C.C. 2003. Gold nanoparticles as spectroscopic enhancers for in vitro studies on single viruses. *Journal of the American Chemical Society* 125(21):6374–6375.

Delor, M., Dai, J., Roberts, T.D., et al. 2018. Exploiting chromophore-protein interactions through linker engineering to tune photoinduced dynamics in a biomimetic light-harvesting platform. *Journal of the American Chemical Society* 140(20):6278–6287.

Dedeo, M.T., Duderstadt, K.E., Berger, J.M. and Francis, M.B. 2010. Nanoscale protein assemblies from a circular permutant of the tobacco mosaic virus. *Nano Letters* 10(1):181–186.

Das, R., Ali, M.E. and Hamid, S.B.A. 2014. Current applications of x-ray powder diffraction - A review. *Reviews on Advanced Materials Science* 38(2):95–109.

Dixit, S.K., Goicochea, N.L., Daniel, M.C. et al. 2006. Quantum dot encapsulation in viral capsids. *Nano Letters* 6(9):1993–1999.

Douglas, T. and Young, M. 1999. Virus particles as templates for materials synthesis. *Advanced Materials* 11(8):679–681.

Eber, F.J., Eiben, S., Jaske, H. and Wege, C. 2013. Bottom-Up-assembled nanostar colloids of gold cores and tubes derived from tobacco mosaic virus. *Angewandte Chemie - International Edition* 52(28):7203–7207.

Endo, M., Fujitsuka, M. and Majima, T. 2007. Porphyrin light-harvesting arrays constructed in the recombinant tobacco mosaic virus scaffold. *Chemistry - A European Journal* 13(31):8660–8666.

Evans, D.J. 2010. Bionanoscience at the plant virus-inorganic chemistry interface. *Inorganica Chimica Acta* 363(6):1070–1076.

Falkner, J.C., Turner, M.E., Bosworth, J.K., et al. 2005. Virus crystals as nanocomposite scaffolds. *Journal of the American Chemical Society* 127(15):5274–5275.

Fang, J., Soto, C.M., Lin, T., Johnson, J.E. and Ratna B. 2002. Complex pattern formation by cowpea mosaic virus nanoparticles. *Langmuir* 18(2):308–310.

Fischer, R. Stoger, E., Schillberg, S., Christou, P. and Twyman, R.M. 2004. Plant-based production of biopharmaceuticals. *Current Opinion in Plant Biology* 7(2):152–158.

Fissan, H., Ristig, S., Kaminski, H., Asbach, C. and Epple, M. 2014. Comparison of different characterization methods for nanoparticle dispersions before and after aerosolization. *Analytical Methods* 6(18):7324–7334.

Galaway, F.A. and Stockley, P.G. 2013. MS2 viruslike particles: A robust, semisynthetic targeted drug delivery platform. *Molecular Pharmaceutics* 10(1):59–68.

Gao, X.D., Li, X.M. and Yu, W.D. 2004. Morphology and optical properties of amorphous ZnS films deposited by ultrasonic-assisted successive ionic layer adsorption and reaction method. *Thin Solid Films* 468(1–2):43–47.

Gerasopoulos, K., Mccarthy, M., Banerjee, P., Fan, X., Culver, J.N. and Ghodssi, R. 2010. Biofabrication methods for the patterned assembly and synthesis of viral nanotemplates. *Nanotechnology* 21(5):055304.

Ghosh, D., Lee Y., Thomas, S., et al. 2012. M13-templated magnetic nanoparticles for targeted in vivo imaging of prostate cancer. *Nature Nanotechnology* 7(10):677–682.

Gorzny, M.L., Walton, A.S. and Evans, S.D. 2010. Synthesis of high-surface-area platinum nanotubes using a viral template. *Advanced Functional Materials* 20(8): 1295–1300.

Grgacic, E.V.L. and Anderson, D.A. 2006. Virus-like particles: Passport to immune recognition. *Methods* 40(1):60–65.

Gupta, S., Chatni, M.R., Rao, A.L.N., Vullev, V.I., Wang, L.V. and Anvari, B. 2013. Virus-mimicking nano-constructs as a contrast agent for near infrared photoacoustic imaging. *Nanoscale* 5(5):1772–1776.

Gurunathan, S., Han, J., Kwon, D.N. and Kim, J.H. 2014. Enhanced antibacterial and antibiofilm activities of silver nanoparticles against Gram-negative and Gram-positive bacteria. *Nanoscale Research Letters* 9(1):373.

Hall, J.B., Dobrovolskaia, M.A., Patri, A.K. and McNeil, S.E. 2007. Characterization of nanoparticles for therapeutics: Physicochemical characterization. *Future Medicine* 2(6):789–803.

Hind, A.R., Bhargava, S.K. and McKinnon, A. 2001. At the solid/liquid interface: FTIR/ATR - The tool of choice. *Advances in Colloid and Interface Science* 93(1–3): 91–114.

Hovlid, M.L., Lau, J.L., Breitenkamp, K., et al. 2014. Encapsidated atom-transfer radical polymerization in Qβ virus-like nanoparticles. *ACS Nano* 8(8):8003–8014.

Huang, X., Bronstein, L.M., Retrum J., et al. 2007. Self-assembled virus-like particles with magnetic cores. *Nano Letters* 7(8):2407–2416.

Huang, X., Stein, B.D., Cheng, H., et al. 2011. Magnetic virus-like nanoparticles in *N. benthamiana* plants: A new paradigm for environmental and agronomic biotechnological research. *ACS Nano* 5(5):4037–4045.

Huang, Y., Chiang, C.Y., Lee, S.K., et al. 2005. Programmable assembly of nanoarchitectures using genetically engineered viruses. *Nano Letters* 5(7):1429–1434.

Hyeon, T. 2003. Chemical synthesis of magnetic nanoparticles. *Chemical Communications* 3(8):927–934.

Jin, H., Wang, W., Zhao, S., et al. 2014. Aβ-HBc virus-like particles immunization without additional adjuvant ameliorates the learning and memory and reduces Aβ deposit in PDAPP mice. *Vaccine* 32(35):4450–4456.

Jordan, P.C., Patterson, D.P., Saboda, K.N., et al. 2016. Self-assembling biomolecular catalysts for hydrogen production. *Nature Chemistry* 8(2):179–185.

Joshi, M., Bhattacharyya, A. and Ali, S.W. 2008. Characterization techniques for nanotechnology applications in textiles. *Indian Journal of Fibre and Textile Research* 33(3):304–317.

Jung, B., Rao, A.L.N. and Anvari, B. 2011. Optical nano-constructs composed of genome-depleted brome mosaic virus doped with a near infrared chromophore for potential biomedical applications. *ACS Nano* 5(2):1243–1252.

Khan, A.A., Fox, E.K., Gorzny, M.L., et al. 2013. PH control of the electrostatic binding of gold and iron oxide nanoparticles to tobacco mosaic virus. *Langmuir* 29(7):2094–2098.

Kim, J., Piao, Y., Lee N., et al. 2010. Magnetic nanocomposite spheres decorated with NiO nanoparticles for a magnetically recyclable protein separation system. *Advanced Materials* 22(1):57–60.

Kim, S. and Barry, B.A. 2001. Reaction-induced FT-IR spectroscopic studies of biological energy conversion in oxygenic photosynthesis and transport. *Journal of Physical Chemistry B* 105(19):4072–4083.

Klem, M.T., Young, M. and Douglas, T. 2008. Biomimetic synthesis of β-TiO_2 inside a viral capsid. *Journal of Materials Chemistry* 18(32):3821–3823.

Knez, M., Sumser, M., Bittner, A.M., et al. 2004. Spatially selective nucleation of metal clusters on the tobacco mosaic virus. *Advanced Functional Materials* 14(2): 116–124.

Kobayashi, M., Tomita, S., Sawada, K., et al. 2012. Chiral meta-molecules consisting of gold nanoparticles and genetically engineered tobacco mosaic virus. *Optics Express* 20(22):24856–24863.

Koutsopoulos, S., Unsworth, L.D., Nagai, Y. and Zhang, S. 2009. Controlled release of functional proteins through designer self-assembling peptide nanofiber hydrogel scaffold. *Proceedings of the National Academy of Sciences USA* 106(12):4623–4628.

Kusters, R., Lin, H.-K., Zandi, R., Tsvetkova, I., Dragnea, B. and van der Schoot, P. 2015. Role of charge regulation and size polydispersity in nanoparticle encapsulation by viral coat proteins. *The Journal of Physical Chemistry B* 119(5):1869–1880.

Kwak, M., Minten, I.J., Anaya, D.-M., et al. 2010. Virus-like particles templated by DNA micelles: A general method for loading virus nanocarriers. *Journal of the American Chemical Society* 132(23):7834–7835.

Larson, S.B., Lucas, R.W. and McPherson, A. 2005. Crystallographic structure of the T=1 particle of brome mosaic virus. *Journal of Molecular Biology* 346(3):815–831.

Lee, B.Y., Zhang, J., Zueger, C., et al. 2012. Virus-based piezoelectric energy generation. *Nature Nanotechnology* 7(6):351–356.

Leiman, P.G., Kanamaru, S., Mesyanzhinov, V.V., Arisaka, F. and Rossmann, M.G. 2003. Structure and morphogenesis of bacteriophage T4. *Cellular and Molecular Life Sciences* 60(11):2356–2370.

Li, F., Li, K., Cui, Z.Q., et al. 2010. Viral coat proteins as flexible nano-building blocks for nanoparticle encapsulation. *Small* 6(20):2301–2308.

Li, F., Zhang, Z.P., Peng, J., et al. 2009. Imaging viral behavior in mammalian cells with self-assembled capsid-quantum-dot hybrid particles. *Small* 5(6):718–726.

Liang, X.Z., Lee, B.T.K. and Wong, S.M. 2002. Covariation in the capsid protein of hibiscus chlorotic ringspot virus induced by serial passaging in a host that restricts movement leads to avirulence in its systemic host. *Journal of Virology* 76(23):12320–12324.

Liepold, L., Anderson, S., Willits, D., et al. 2007. Viral capsids as MRI contrast agents. *Magnetic Resonance in Medicine* 58(5):871–879.

Lin, P.C., Lin, S., Wang, P.C. and Sridhar, R. 2014. Techniques for physicochemical characterization of nanomaterials. *Biotechnology Advances* 32(4):711–726.

Liu, Y., Shipton, M.K., Ryan, J., Kaufman, E.D., Franzen, S. and Feldheim, D.L. 2007. Synthesis, stability, and cellular internalization of gold nanoparticles containing mixed peptide–poly (ethylene glycol) monolayers. *Analytical Chemistry* 79(6):2221–2229.

Loo, L., Guenther, R.H., Basnayake, V.R., Lommel, S.A. and Franzen, S. 2006. Controlled encapsidation of gold nanoparticles by a viral protein shell. *Journal of the American Chemical Society* 128(14):4502–4503.

Loo, L., Guenther, R.H., Lommel, S.A. and Franzen, S. 2007. Encapsidation of nanoparticles by red clover necrotic mosaic virus. *Journal of the American Chemical Society* 129(36):11111–11117.

Lund, P.E., Hunt, R.C., Gottesman, M.M. and Kimchi-Sarfaty, C. 2009. Pseudovirions as vehicles for the delivery of siRNA. *Pharmaceutical Research* 27(3):400–420.

Luo, C.H., Shanmugam, V. and Yeh, C.S. 2015. Nanoparticle biosynthesis using unicellular and subcellular supports. *NPG Asia Materials* 7(8):e209–e211.

Ma, Y.Z., Miller, R.A., Fleming, G.R. and Francis, M.B. 2008. Energy transfer dynamics in light-harvesting assemblies templated by the tobacco mosaic virus coat protein. *Journal of Physical Chemistry B* 112(22):6887–6892.

Manchester, M. and Singh, P. 2006. Virus-based nanoparticles (VNPs): Platform technologies for diagnostic imaging. *Advanced Drug Delivery Reviews* 58(14):1505–1522.

Mangematin, V. and Walsh, S. 2012. The future of nanotechnologies. *Technovation* 32(3–4): 157–160.

Millán, J.G., Brasch, M., Anaya-Plaza, E., et al. 2014. Self-assembly triggered by self-assembly: Optically active, paramagnetic micelles encapsulated in protein cage nanoparticles. *Journal of Inorganic Biochemistry* 136:140–146.

Montague, N.P., Thuenemann, E.C., Saxena, P., Saunders, K., Lenzi, P. and Lomonossoff, G.P. 2011. Recent advances of Cowpea mosaic virus-based particle technology. *Human Vaccines* 7(3):383–390.

Mourdikoudis, S., Pallares, R.M. and Thanh, N.T.K. 2018. Characterization techniques for nanoparticles: comparison and complementarity upon studying nanoparticle properties. *Nanoscale* 10(27):12871–12934.

Mukherjee, S., Abd-El-Latif, M., Bronstein, M., Ben-nun-Shaul, O., Kler, S. and Oppenheim, A. 2007. High Cooperativity of the SV40 Major Capsid Protein VP1 in Virus Assembly. *PLoS One* 2(8):765.

Nam, J., Won, N., Jin, H., Chung, H. and Kim, S. 2009. pH-induced aggregation of gold nanoparticles for photothermal cancer therapy. *Journal of the American Chemical Society*131(38):13639–13645.

Nan, L., Wang, C., Zhang, W., et al. 2012. Au nanocrystals grown on a better-defined one-dimensional tobacco mosaic virus coated protein template genetically modified by a hexahistidine tag. *Nanotechnology* 23(33):1–6.

Neltner, B., Peddie, B., Xu, A., et al. 2010. Production of hydrogen using nanocrystalline protein-templated catalysts on M13 Phage. *ACS Nano* 4(6):3227–3235.

Noginov, M.A., Zhu, G., Bahoura, M., et al. 2006. The effect of gain and absorption on surface plasmons in metal nanoparticles. *Applied Physics B* 86(3):455–460.

Parak, W.J., Pellegrino, T., Micheel, C.M., Gerion, D., Williams, S.C. and Alivisatos, A.P. 2003. Conformation of oligonucleotides attached to gold nanocrystals probed by gel electrophoresis. *Nano Letters* 3(1):33–36.

Penney, C.A., Thomas, D.R., Deen, S.S. and Walmsley, A.M. 2011. Plant-made vaccines in support of the Millennium Development Goals. *Plant Cell Reports* 30(5):789–798.

Perrault, S.D. and Chan, W.C.W. 2010. In vivo assembly of nanoparticle components to improve targeted cancer imaging. *Proceedings of the National Academy of Sciences USA* 107(25):11194–11199.

Pinheiro, A.V., Han, D., Shih, W.M. and Yan, H. 2011. Challenges and opportunities for structural DNA nanotechnology. *Nature Nanotechnology* 6(12):763–772.

Prasuhn, D.E., Yeh, R.M., Obenaus, A., Manchester, M. and Finn, M.G. 2007. Viral MRI contrast agents: Coordination of Gd by native virions and attachment of Gd complexes by azide–alkyne cycloaddition. *Chem. Commun.* (12):1269–1271.

Qazi, S., Liepold, L.O., Abedin, M.J., et al. 2012. P22 Viral capsids as nanocomposite high-relaxivity MRI contrast agents. *Molecular Pharmaceutics* 10(1):11–17.

Reimer, L. and Kohl, H. 2008. *Transmission electron microscopy physics of image formation*. New York: Springer-Verlag, 590.

Ren, Y., Sek, M.W. and Lim, L.Y. 2007. Folic acid-conjugated protein cages of a plant virus: A novel delivery platform for doxorubicin. *Bioconjugate Chemistry* 18(3):836–843.

Ren, Y., Wong, S.M. and Lim, L.Y. 2006. In vitro-reassembled plant virus-like particles for loading of polyacids. *Journal of General Virology* 87(9):2749–2754.

Rong, J., Oberbeck, F., Wang, X., et al. 2009. Tobacco mosaic virus templated synthesis of one dimensional inorganic–polymer hybrid fibres. *Journal of Materials Chemistry* 19(18):2841–2845.

Royston, E., Ghosh, A., Kofinas, P., Harris, M.T. and Culver, J.N. 2008. Self-assembly of virus-structured high surface area nanomaterials and their application as battery electrodes. *Langmuir* 24(3):906–912.

Russell, J.T., Lin, Y., Böker, A., et al. 2005. Self-assembly and cross-linking of bionanoparticles at liquid-liquid interfaces. *Angewandte Chemie International Edition* 44(16): 2420–2426.

Ryan, S.O., Turner, M.S., Geriepy, J. and Finn, O.J. 2010. Tumor antigen epitopes interpreted by the immune system as self or abnormal-self differentially affect cancer vaccine responses. *Cancer Research* 70(14):5788–5796.

Rybicki, E.P. 2010. Plant-made vaccines for humans and animals. *Plant Biotechnology Journal* 8(5):620–637.

Sainsbury, F., Cañizares, M.C. and Lomonossoff, G.P. 2010. Cowpea mosaic virus: The plant virus-based biotechnology workhorse. *Annual Review of Phytopathology* 48: 437–455.

Sapsford, K.E., Soto, C.M., Blum, A.S., et al. 2006. A cowpea mosaic virus nanoscaffold for multiplexed antibody conjugation: Application as an immunoassay tracer. *Biosensors and Bioelectronics* 21(8):1668–1673.

Sapsford, K.E., Tyner, K.M., Dair, B.J., Deschamps, J.R. and Medintz, I.L. 2011. Analyzing nanomaterial bioconjugates: A review of current and emerging purification and characterization techniques. *Analytical Chemistry* 83(12):4453–4488.

Sarma, D.D., Santra, P.K., Mukherjee, S. and Nag, A. 2013. X-ray photoelectron spectroscopy: a unique tool to determine the internal heterostructure of nanoparticles. *Chemistry of Materials* 25(8):1222–1232.

Saunders, K., Sainsbury, F. and Lomonossoff, G.P. 2009. Efficient generation of cowpea mosaic virus empty virus-like particles by the proteolytic processing of precursors in insect cells and plants. *Virology* 393(2):329–337.

Schlick, T.L., Ding, Z., Kovacs, E.W. and Francis, M.B. 2005. Dual-surface modification of the tobacco mosaic virus. *Journal of the American Chemical Society* 127(11): 3718–3723.

Schwarz, B. and Douglas, T. 2015. Development of virus-like particles for diagnostic and prophylactic biomedical applications. *Wiley Interdisciplinary Reviews: Nanomedicine and Nanobiotechnology* 7(5):722–735.

Shah, S.N., Steinmetz, N.F., Aljabali, A.A.A., Lomonossoff, G.P. and Evans, D.J. 2009. Environmentally benign synthesis of virus-templated, monodisperse, iron-platinum nanoparticles. *Dalton Transactions* 40:8479–8480.

Shard, A.G. 2012. A straightforward method for interpreting XPS data from core–shell nanoparticles. *The Journal of Physical Chemistry C* 116(31):16806–16813.

Shen, L., Zhou, J., Wang, Y., et al. 2014. Efficient encapsulation of Fe_3O_4 nanoparticles into genetically engineered hepatitis B Core virus-like particles through a specific interaction for potential bioapplications. *Small* 11(9–10):1190–1196.

Shenton, W., Douglas, T., Young, M., Stubbs, G. and Mann, S. 1999. Inorganic-organic nanotube composites from template mineralization of tobacco mosaic virus. *Advanced Materials* 11(3):253–256.

Shukla, S., Wen, A.M., Commandeur, U. and Steinmetz, N.F. 2014. Presentation of HER2 epitopes using a filamentous plant virus-based vaccination platform. *Journal of Materials Chemistry* B2(37):6249–6258.

Singh, D.K., Pandey, D.K., Yadav, R.R. and Singh, D. 2013. A study of ZnO nanoparticles and ZnO-EG nanofluid. *Journal of Experimental Nanoscience* 8(5):731–741.

Singh, P., Prasuhn, D., Yeh, R.M., et al. 2007. Bio-distribution, toxicity and pathology of cowpea mosaic virus nanoparticles in vivo. *Journal of Controlled Release* 120(1–2): 41–50.

Sitterberg, J., Özcetin, A., Ehrhardt, C. and Bakowsky, U. 2010. Utilising atomic force microscopy for the characterisation of nanoscale drug delivery systems. *European Journal of Pharmaceutics and Biopharmaceutics* 74(1):2–13.

Soto, C.M., Blum, A.S., Vora, G.J., et al. 2006. Fluorescent signal amplification of carbocyanine dyes using engineered viral nanoparticles. *Journal of the American Chemical Society* 128(15):5184–5189.

Soto, C.M. and Ratna, B.R. 2010. Virus hybrids as nanomaterials for biotechnology. *Current Opinion in Biotechnology* 21(4):426–438.

Stehle, T., Gamblin, S.J., Yan, Y. and Harrison, S.C. 1996. The structure of simian virus 40 refined at 3.1 Å resolution. *Structure* 4(2):165–182.

Steinmetz, N.F., Calder, G., Lomonossoff, G.P. and Evans, D.J. 2006. Plant viral capsids as nanobuilding blocks: construction of arrays on solid supports. *Langmuir* 22(24): 10032–10037.

Steinmetz, N.F., Lin, T., Lomonossoff, G.P. and Johnson, J.E. 2009a. Structure-based engineering of an icosahedral virus for nanomedicine and nanotechnology. *Current Topics in Microbiology and Immunology* 327:23–58.

Steinmetz, N.F., Shah, S.N., Barclay, J.E., Rallapalli, G., Lomonossoff, G.P. and Evans, D.J. 2009b. Virus-templated silica nanoparticles. *Small* 5(7):813–816.

Thanh, N.T.K. and Rosenzweig, Z. 2002. Development of an aggregation-based immunoassay for anti-protein using gold nanoparticles. *Analytical Chemistry* 74(7):1624–1628.

Tomaszewska, E., Soliwoda, K., Kadziola, K., et al. 2013. Detection limits of DLS and UV-Vis spectroscopy in characterization of polydisperse nanoparticles colloids. *Journal of Nanomaterials* 2013:1–10.

Vauthier, C. and Bouchemal, K. 2009. Methods for the preparation and manufacture of polymeric nanoparticles. *Pharmaceutical Research* 26(5):1025–1058.

Waseda, Y., Matsubara, E. and Shinoda, K. 2011. *X-ray Diffraction Crystallography: Introduction, Examples and Solved Problems*. Berlin, Germany: Springer Verlag, 310.

Wei, B., Dai, M. and Yin, P. 2012. Complex shapes self-assembled from single-stranded DNA tiles. *Nature* 485(7400):623–626.

Wnek, M., Gorzny, M.L., Ward, M.B., et al. 2013. Fabrication and characterization of gold nano-wires templated on virus-like arrays of tobacco mosaic virus coat proteins. *Nanotechnology* 24(2):025605.

Wu, H., He, L., Gao, M., Gao, S., Liao, X. and Shi, B. 2011. One-step in situ assembly of size-controlled silver nanoparticles on polyphenol-grafted collagen fiber with enhanced antibacterial properties. *New Journal of Chemistry* 35(12):2902–2909.

Wu, Z., Mueller, A., Degenhard, S., et al. 2010. Enhancing the magnetoviscosity of ferrofluids by the addition of biological nanotubes. *ACS Nano* 4(8):4531–4538.

Xu, Z., Sun, H., Gao, F., Hou, L. and Li, N. 2012. Synthesis and magnetic property of T4 virus-supported gold-coated iron ternary nanocomposite. *Journal of Nanoparticle Research* 14(12):1267. https://doi.org/10.1007/s11051-012-1267-9.

Yaddanapudi, K., Mitchell, R.A. and Eaton, J.W. 2013. Cancer vaccines: Looking to the future. *OncoImmunology* 2(3):e23403.

Yang, C., Choi, C.H., Lee, C.S. and Yi, H. 2013. A facile synthesis–fabrication strategy for integration of catalytically active viral-palladium nanostructures into polymeric hydrogel microparticles via replica molding. *ACS Nano* 7(6):5032–5044.

Yazdian, N., Karimzadeh, F. and Enayati, M.H. 2013. In-situ fabrication of Al_3V/Al_2O_3 nanocomposite through mechanochemical synthesis and evaluation of its mechanism. *Advanced Powder Technology* 24(1):106–112.

Yildiz, I., Lee, K.L., Chen, K., Shukla, S. and Steinmetz, N.F. 2013. Infusion of imaging and therapeutic molecules into the plant virus-based carrier cowpea mosaic virus: Cargo-loading and delivery. *Journal of Controlled Release* 172(2):568–578.

Yin, Z., Nguyen, H.G., Chowdhury, S., et al. 2012. Tobacco mosaic virus as a new carrier for tumor associated carbohydrate antigens. *Bioconjugate Chemistry* 23(8):1694–1703.

Zahr, O.K. and Blum, A.S. 2012. Solution phase gold nanorings on a viral protein template. *Nano Letters* 12(2):629–633.

Zeng, Q., Wen, H., Wen, Q., et al. 2013. Cucumber mosaic virus as drug delivery vehicle for doxorubicin. *Biomaterials* 34(19):4632–4642.

Zhang, S. 2003. Fabrication of novel biomaterials through molecular self-assembly. *Nature Biotechnology* 21(10):1171–1178.

Zheng, J., Birktoft, J. J., Chen, Y., et al. 2009. From molecular to macroscopic via the rational design of a self-assembled 3D DNA crystal. *Nature* 461(7260):74–77.

Zscherp, C. and Barth, A. 2001. Reaction-induced infrared difference spectroscopy for the study of protein reaction mechanisms. *Biochemistry* 40(7):1875–1883.

2 Synthesis and Characterization of Nanoparticles as Potential Viral and Antiviral Agents

Deepthi Panoth, Sindhu Thalappan Manikkoth, Fabeena Jahan, Kunnambeth Madam Thulasi, Anjali Paravannoor, and Baiju Kizhakkekilikoodayil Vijayan

CONTENTS

DOI: 10.1201/9781003136644-3

2.1 INTRODUCTION

New viruses are spreading worldwide at a faster pace as a result of high globalization, infecting people and even causing death to millions of human beings throughout the world. The continued emergence of novel viruses has posed a major threat to public health, and these sudden uncontrollable outbreaks have the potential to hamper the health-care system and derail the global economy. In 2020, the severe acute respiratory syndrome Coronavirus-2 (SARs Cov-2) pandemic spread all over the world, taking the lives of millions of people and dramatically changing our day-to-day life. The modern history has witnessed the spread of various viral pandemics, including influenza A Virus subtype H2N2 (H2N2 flu, 1956), H3N3 flu (1968), human immunodeficiency virus (HIV, 2005-2012), severe acute respiratory syndrome (SARS, 2009), whereas Middle East respiratory syndrome (MERS) and the Ebola viruses are still in the pre-pandemic phase (Zhou et al. 2021). Currently, one of the major challenges of the medical, biotechnological, and pharmaceutical sectors is the reemergence of resistance in different viruses and newly-emerging viruses, causing a global threat to human health. To reduce or avoid the risk of new pandemics, there is an urgency to develop new viral treatments and antiviral tools to efficiently stop the spread of the viruses, boost immunization, and enhance the survival of infected people (Reina et al. 2020).

Nanotechnology, an emerging technology, has brought many revolutionary changes to the fields of medicine, biotechnology, and pharmaceuticals. Advancement in the field of nanotechnology provides a unique platform for the development of effective and potential viral and antiviral agents for various applications via modifying the nanomaterials; salient features include easy surface modification, high surface-to-volume ratio, physiochemical stability, etc (Alimardani et al. 2021). Currently, a wide variety of nanoparticles are being utilized as potential viral and antiviral agents in the form of viral nanoparticles, as well as to address the issues related to newly emerging viruses, resistance, untreatable viral infections, and stain-specific targets (Maduray and Parboosing, 2020). Viral nanoparticles are virus-based nanoparticle formulations, which are naturally occurring plant or animal viruses or bacteriophages that are tailored for therapeutic and/or imaging applications. To fabricate metallic nanoparticles and nanotube structures via mineralization technique, viruses were used as scaffolds or templates (Steinmetz, 2010). For noninvasive imaging techniques like PET (positron emission tomography) and MRI (magnetic resonance imaging), viral nanoparticles are used as imaging reagents. Antiviral nanoparticles are potential antiviral agents that act by inhibiting viral replication either inside or outside the host by hindering the entry of viral particles (Delshadi et al. 2021).

Viral diseases are the greatest threat to human health and economy, and thus, it's essential to address the issues emerging viruses cause. This chapter highlights the recent developments of the synthesis of viral and antiviral nanoparticles, including inorganic nanoparticles like metal nanoparticles and metal-oxide nanoparticles and organic nanoparticles like polymeric, liposomes, dendrimers, etc. The characterization and antiviral mechanism of nanoparticles will be discussed as well.

2.2 NANOPARTICLES USED AS VIRAL AND ANTIVIRAL AGENTS

Nanoparticles are categorized as viral and antiviral nanoparticles based on their actions and applications. They are effectively used for drug delivery to the targeted site, detect viral diseases at an early stage using sensitive imaging, and can cross cellular barriers to deliver vaccines to the diseased organs. Additionally, they interact with the molecules present in blood or tissue organ and help in the inactivation of viruses or the inhibition of viral binding with the host cell receptor. Viral nanoparticles are used for intravital vascular imaging because they can be covalently conjugated to a wide variety of imaging reagents, such as fluorescent dyes. The fluorescent CPMV (cowpea mosaic virus) sensors were utilized for the visualization of vasculature and blood flow in chick embryos and living mice to a depth of nearly 500 μm for 72 hours (Yildiz et al. 2011). In several ways, the structure of the viral nanoparticles can be 'programmed' such that the internal cavity can be filled with imaging reagents, drug molecules, other nanoparticles, and quantum dots, while targeting ligands are used to decorate external surfaces allowing cell-specific delivery. For clinical purposes, several platforms have been approved; these include nanoparticle-based contrast agents for MRI and PET, albumin-bound paclitaxel (abraxane), drug delivery using poly (lactic-co-glycolic acid) (PLGA) formulations, and vaccines based on viruses.

Nanoparticles have been extensively developed and studied for antiviral therapy over the last decade. One of the mechanisms commonly adopted by most antiviral agents is immediate and direct interference with the viral replication at the first phase itself (Figure 2.1). Nanomaterials possess similar or lesser dimensions when compared with the target viruses, which enables intimate interaction. Antiviral nanoparticles may be either inorganic, such as metal nanoparticles (gold, silver, copper, etc.) or metal-oxide nanoparticles, or organic like liposomes, polymeric, ferritin, micelles (Maduray and Parboosing 2020). Among them, metal nanoparticles are widely utilized as antiviral agents owing to their ability to attack multi-targets on viruses with minimum effect on the development of the reemergence of resistance. Silver nanoparticles are a broad-spectrum antiviral observed to inhibit in vitro replication of several viruses, including monkeypox, HIV, hepatitis, and norovirus with less toxicity (Vecitis 2021). The cell proliferation of H1N1 influenza is suppressed by zinc oxide size ranging from 16 to 20 nm, thus exhibiting antiviral property. The viral transcripts of the H1N1 influenza A virus changed within 24 hours using iron oxide nanoparticles (Sharmin et al. 2021).

2.3 ANTIVIRAL NANOPARTICLES

Antiviral nanoparticles are antiviral agents used to control or treat viral infections. The antiviral agents primarily target the different stages of a virus life cycle, and thus, the antiviral drugs perform by inhibiting the viral replication cycle at various stages. Here in this section, we will discuss the synthesis and

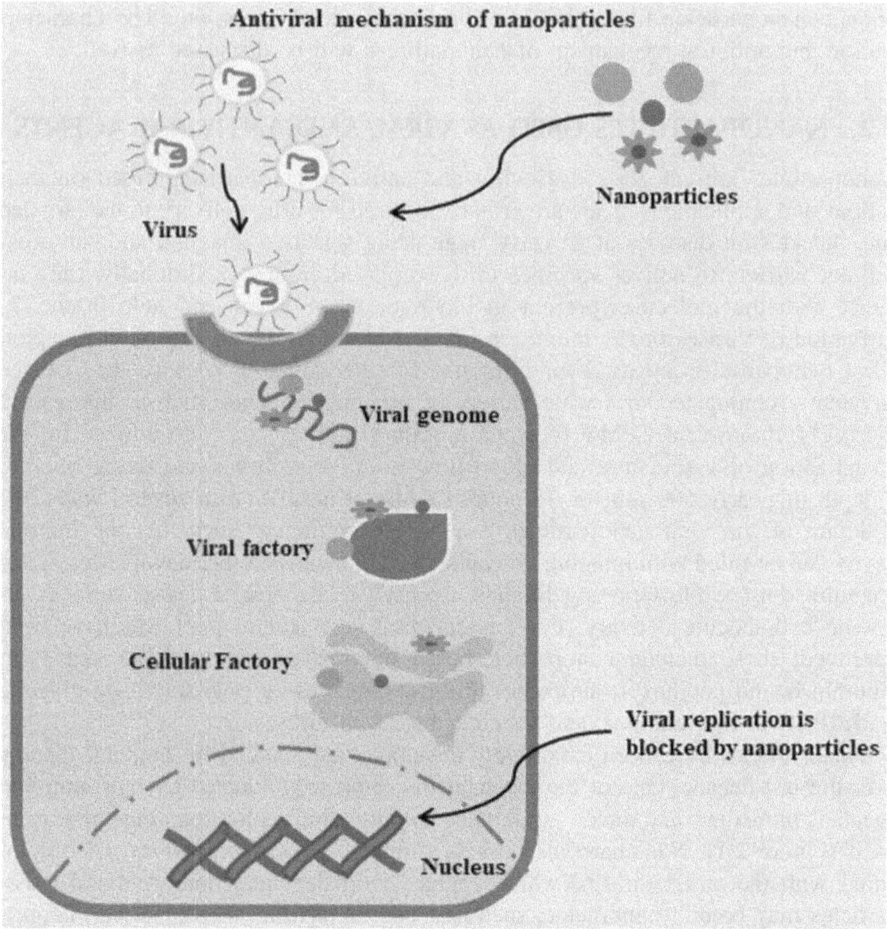

FIGURE 2.1 Schematic illustration of the antiviral mechanism of nanoparticles. Nanoparticles exhibit their antiviral activity directly after entering into the host cell by binding themselves to the viral and host cellular factors, thereby inhibiting the viral replication.

characterization of different types of antiviral nanoparticles that are utilized as a potential antiviral agent.

2.3.1 Metal Nanoparticles

In recent decades, researchers have gained interest in studying mainly metal nanoparticles, including gold nanoparticles, silver nanoparticles, platinum nanoparticles, copper nanoparticles, etc. and also metal-oxide nanoparticles, which resulted in the advancement of the biomedical field. Metal nanoparticles were found to exhibit great bioactivity comprising antibacterial, antiviral, and antibiotic effect (Khandel et al. 2018). This chapter pays most attention to the synthesis and characterization of these metal nanoparticles that can act as potential antiviral agents.

2.3.1.1 Synthesis of Metal Nanoparticles

Various physical, chemical, and biological methods that utilize the three-component reaction of precursors, capping/stabilizing agents, and reducing agents are employed for the synthesis of metal nanoparticles (Khandel et al. 2018; Odularu 2018). These three methods are discussed in detail in this section.

The main physical methods for the synthesis of metal nanoparticles comprise techniques such as ultraviolet/microwave irradiation, sonochemistry, thermal/photochemical decomposition, radical induction, laser ablation, and so on (Dhand et al. 2015; Khandel et al. 2018). These physical methods mainly utilize the top-down approach of nanoparticle synthesis, which involves the breaking down of bulk materials into nanosized structures by using mechanical techniques of size reduction. During the physical process of synthesis, metal atoms evaporate and condensate on various support as smaller aggregates of metallic nanoparticles. These physical methods yield monodisperse nanoparticles of the desired size and shape with no solvent contamination. The need for highly sophisticated instruments and expensive chemicals, high operating cost, abundant waste generation, etc. make the physical methods of nanoparticle synthesis economically unfavourable (Dhand et al. 2015). There are limited reports on the physical methods of metal nanoparticle synthesis. Mazyar et al. reported a novel sonochemical method for the synthesis of silver nanorods conjugated with sodium 2-mercaptoethane sulphonate (Ag-MES), which exhibited antiviral activities against HIV and HSV viruses (Etemadzade et al. 2016). Pfaff et al. (2019) demonstrated the synthesis of tungsten carbide nanoparticles through the plasma atomization method, which showed virucidal activity against enveloped and nonenveloped model viruses. Very recently, Balagna et al. (2020) elucidated the antiviral effect of a silver nanocluster/silica composite coating deposited by a radio frequency co-sputtering process against Coronavirus SARS-CoV-2 (Figure 2.2).

Sol-gel method, hydrothermal synthesis, microemulsion method, chemical vapour synthesis, polyol synthesis, etc. are some chemical methods of nanoparticle synthesis (Dhand et al. 2015). These chemical methods utilize the bottom-up

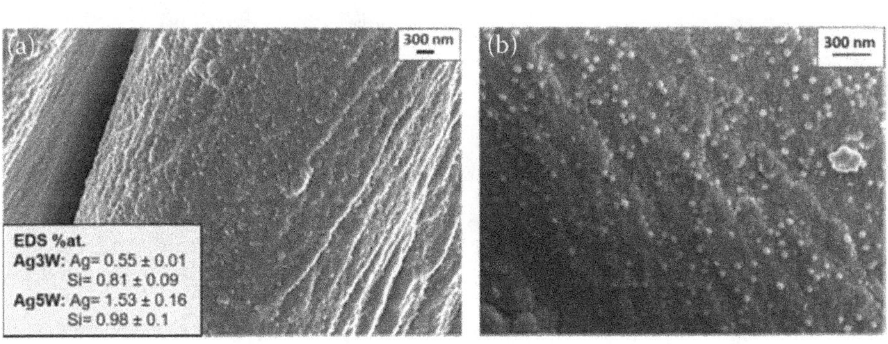

FIGURE 2.2 Field Emission Scanning Electron Microscope (FESEM) images of Ag_3W coated facial FFP3 mask surface, which shows the morphology of silica matrix embedded with silver nanoclusters at (a) lower magnification and (b) higher magnification (Reproduced with permission from Balagna et al. 2020).

approach of nanoparticle synthesis and involve the assembly of atoms or molecules to nanoscale structures. The mechanism of nanoparticle generation involves reducing metal ions by reducing agents or decomposition of metal precursors in the presence of a stabilizer with extra energy. The chemicals such as hydrazine, sodium borohydride, hydrogen, etc. are generally used as reducing agents (Khandel et al. 2018). Organic solvents, natural/synthetic polymers, surfactants, etc. are used as stabilizing agents to avoid agglomeration of metal nanoparticles. Even though nanoparticles of well-defined dimension, size, shape, and composition have resulted through chemical methods (Yu et al. 2008), major disadvantages such as environmental pollution, use of toxic and nonbiodegradable chemicals, lack of specificity in biomedical applications, etc. limit the use of chemical methods for the synthesis of metal nanoparticles. Some reports on chemical methods of metal nanoparticle synthesis are discussed here in short. Chen et al. (2013) reported the synthesis of silver nanoparticles by a chemical reduction method, which shows inhibitory effects against type 3 adenovirus (Figure 2.3). Lysenko et al. synthesized

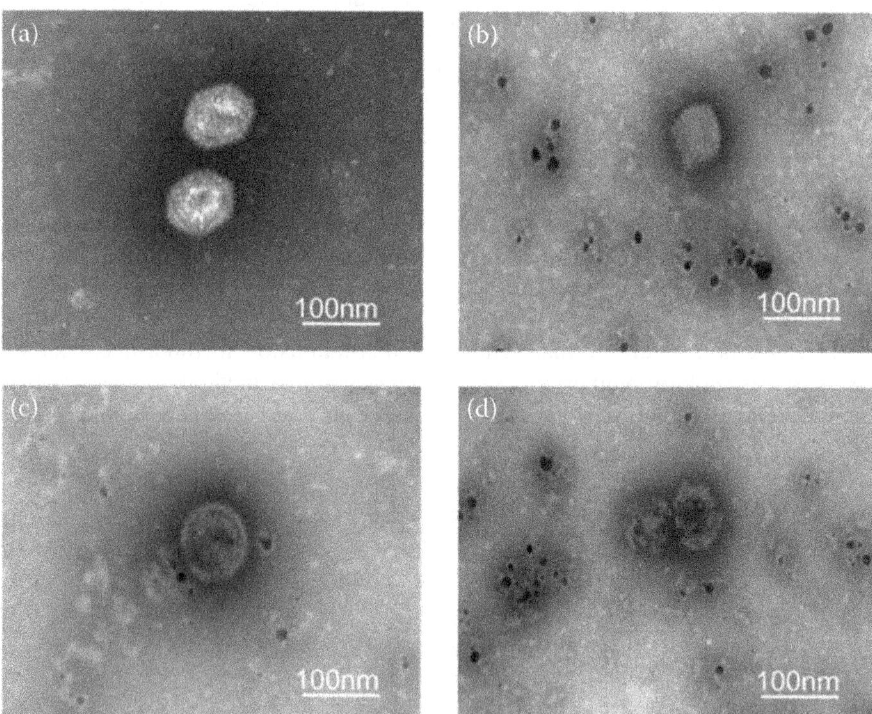

FIGURE 2.3 TEM images reveals the damage created by silver nanoparticles to Adenovirus type 3 (Ad3) particles at different time-period, (a) Samples without silver nanoparticles (control sample); Samples after incubation with 50 μg/mL of silver nanoparticle for (b) 30 minutes, (c) 90 minutes and (d) 120 minutes. The morphology exhibits noticable damage to the Ad3 viral particles compared to the control sample (Reproduced with permission from Chen et al. 2013).

gold nanoparticles covered with silicon dioxide shells through the sol-gel method and used them as antiviral agents against adenoviruses.

Biological methods for the synthesis of metal nanoparticles employ unicellular and multicellular biological organisms such as virus, fungus, bacteria, actinomycetes, yeast, algae, plants, etc. and are essentially based on the bottom-up approach (Ingale 2013; Pantidos 2014). Several controlling factors such as temperature, pH, synthesis time, reactant concentration, etc. influence the, size, nucleation, morphology, and stability of the biogenic nanoparticles (Figure 2.4) (Shah et al. 2015). Either intracellular or extracellular routes, which depend on the transportation and trapping of metal ions on the microbial surface followed by enzymatic reduction, were used for metal nanoparticle synthesis (Fang et al. 2019). Generally, for the synthesis of nanoparticles, plant-based viruses such as cowpea mosaic virus (CPMV), cowpea chlorotic mottle virus (CCMV), tobacco mosaic virus (TMV), brome mosaic virus (BMV), etc. are used as biological templates. Their symmetric structure, monodispersity, ease of functionalization, small size, ability to self-assemble, noninfectious nature to humans and animals, etc. are the advantages of using plant viruses for nanoparticle synthesis (Zhang et al. 2018). The existing metabolites, such as terpenoids, alkaloids, sugars, proteins, phenolic acids, etc., in plant extracts, enable the bioreduction of metal ions forming stable metal nanoparticles and thus are considered safe and eco-friendly for metal nanoparticle synthesis (Shah et al. 2015). Biogenic metal nanoparticles can be easily produced in large quantities and are biocompatible,

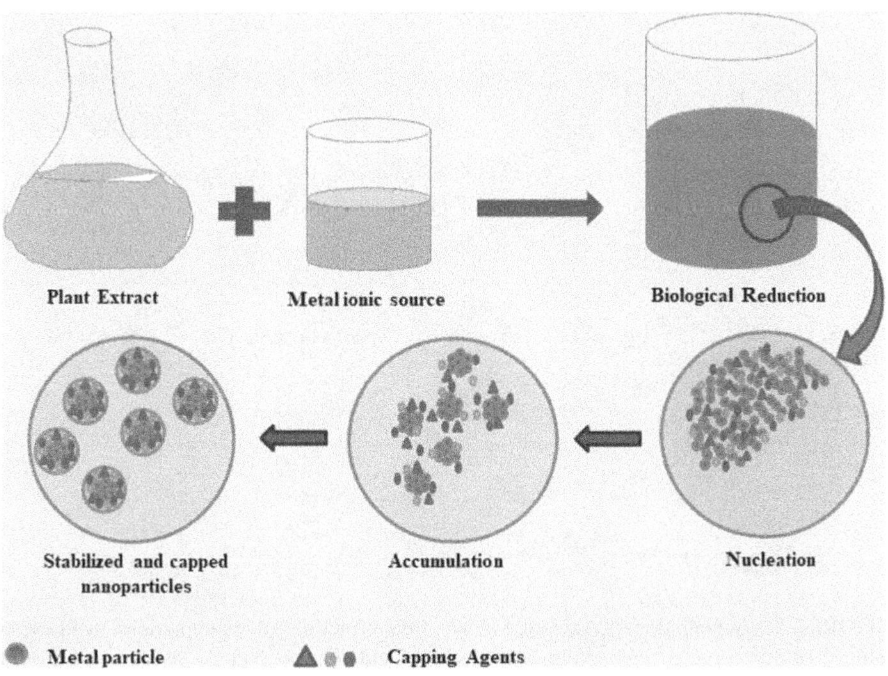

FIGURE 2.4 Synthesis of metal nanoparticles using plant extract via biological method.

environmentally friendly, and possess well-defined size and morphology. Biological methods are preferred over other methods to avoid environmental toxicity, carcinogenicity, and cytotoxicity (Biazar et al. 2011).

There are several reports on biological methods of metal nanoparticle synthesis and some are discussed here in short. Salem et al. (2012) reported the biosynthesis of silver nanoparticles derived from plant extract, which exhibited antiviral activity against coxsackievirus b3. Dhanasezhian et al. (2019) reported the biosynthesis of gold and silver nanoparticles using seaweed *Sargassum wightii*, which exhibited antiviral activity against herpes simplex virus. Meléndez-Villanueva et al. (2019) synthesized gold nanoparticles with garlic extract and evaluated the antiviral activity against the measles virus (MeV) (Figure 2.5). Over the past few years, there have been numerous reports on the synthesis of metal nanoparticles that exhibited antiviral activity, and Maduray et al. discussed in detail (Maduray and Parboosing 2020).

2.3.1.2 Characterization of Metal Nanoparticles

Characterization of synthesized metal nanoparticles is very essential to generate the requisite with unique physical/chemical/biological properties and to determine their bio-distribution, behaviour, and efficacy. Analytical techniques such as field emission scanning electron microscopy (FESEM), localized surface plasmon

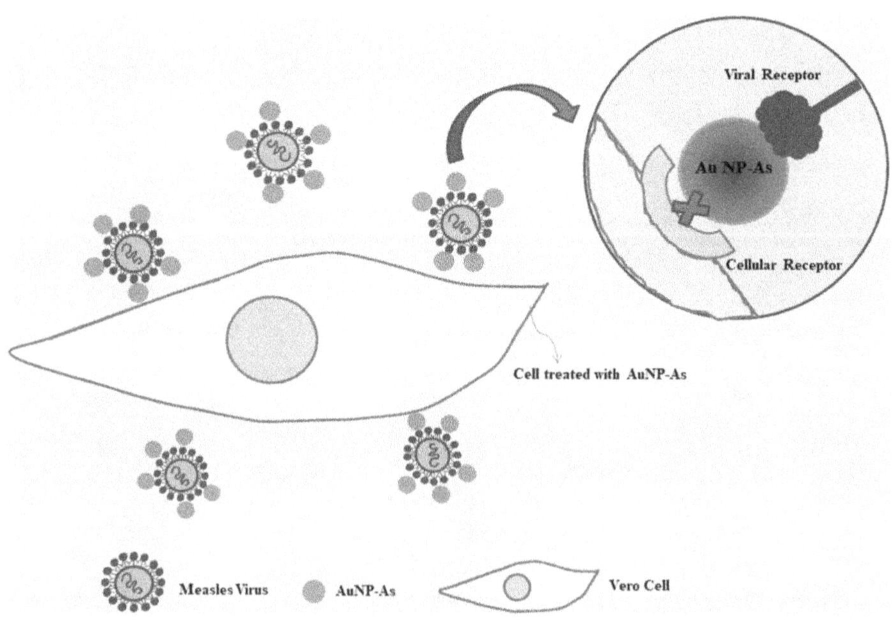

FIGURE 2.5 Schematic representation of virucidal effect of gold nanoparticles synthesized using garlic extracts (*Allium sativa*) as reducing agent. The AuNPs-As blocks the union of viral particle with the cell receptors by binding the viral particle and, thus, inhibits the viral replication.

resonance (LSPR), atomic force microscopy (AFM), Fourier transform infrared (FTIR) spectroscopy, X-ray photoelectron spectroscopy (XPS), electrophoresis laser doppler, and energy-dispersive X-ray (EDX) spectroscopy were used to determine the size and size distribution, to measure the zeta potential and to estimate the surface morphology of nanoparticles. X-ray diffraction (XRD) is used to determine the crystal structures and using UV-Vis Spectroscopy the stability of synthesized NPs is observed. Narrow particle size distributions can be measured using dynamic light scattering (DLS). Transmission electron microscopy (TEM) and scanning electron microscopy (SEM) are the major techniques used for the characterization of surface morphology (Zhang et al. 2016). Fouad listed the techniques utilized for the characterization of metal nanoparticles for effective antiviral treatment in detail (Ibrahim Fouad 2021).

2.3.1.3 Mechanism Underlying the Antiviral Action of Metal Nanoparticles

The approved mechanisms of antiviral action for metal nanoparticles can happen either inside or outside the host cells. The important mechanism underlying the antiviral action of nanoparticles is based on the direct interaction of nanoparticles with the viral surface glycoproteins (Galdiero et al. 2011). The nanoparticles adsorb onto the viral surfaces and induce subsequent local transformations, such as agglutination of glycoproteins, blocking the viral penetration into the host cells.

The inhibition of generic expression by metal nanoparticles is another proposed mechanism behind the antiviral action of metal nanoparticles. An infected cell is abundant in virally encoded factors required for viral replication in the intracellular compartment. Metal nanoparticles interact with these replication factors, preventing their multiplication. Murugan et al. (2015) demonstrated biosynthesized silver nanoparticles, which blocked the expression and production of the envelope protein of Dengue Virus in vitro.

Silver and gold nanoparticles exhibited antiviral activity against different viruses. Gold nanoparticles blocks glycoprotein gp120 attachment with CD4 to hinder viral entry. The antiviral mechanism of silver NPs involves inhibition of CD4-dependant virion binding and/or inhibition of viral replication. Copper nanoparticles inhibit the viral entry into host cells by blocking the virus attachment. Copper NPs are also capable of destroying the viral genome and thus preventing viral replication. The preferential binding of iron NPs to the virus inhibits the virion binding to the host cells (Aderibigbe 2017).

2.3.2 Organic Nanoparticles

Organic nanoparticles are the most widely employed drug-transport agents and antiviral therapeutic agents against different variability of pathological diseases among human beings; this is because of their high biocompatibility, biodegradability, and easy surface modification. Here we discuss some of the most common organic nanoparticles, like polymeric nanoparticles, dendrimers, liposomes, etc. Organic nanoparticles exhibit a wide spectrum of antiviral action by obstructing the entry of virus and further infection (Singh et al. 2017).

Polymeric nanoparticles are long chain colloids of sizes ranging from 10 to 1000 nm, which facilitate the capillary penetration and thus enhance uptake efficacy by cells so that drug concentration at the target could be higher. They possess high antiviral activity because of their branched chain and flexible structure of the molecule. They are utilized not only as active antiviral drugs but also as co-factors for the curing of viral diseases. The special benefits of polymers are the constant discharge of antiviral agents and the boosted metabolic stability of the incorporated drug. These properties validate the immense potential of polymer for their effective delivery of antiviral agents. Poly(D,L-lactide-co-glycolide) (PLG) and poly(lactide-co-glycolides) (PLGA)-based nanoparticles are the most widely used polymeric nanoparticles due to their greater biodegradability and biocompatibility background. To lessen the nonspecific interactions with serum proteins and to give way for the uptake by phagocytosis, a surface modification with hydrophilic polymers is necessary; this modification makes the drug half-life longlasting by making changes in the biodistribution of the drug (Gurunathan et al. 2020). In 2012, Baram-Pinto developed surface-modified bovine serum albumin nanospheres using a simple single-step sonochemical process, which can be effectively used as potential antiviral agents that could inhibit the HSV-1 virus. The surface of the protein is covalently conjugated with mercaptoethane sulphonate that functions similarly to cellular heparan sulphate (Baram-Pinto et al. 2012). Dey et al. synthesized polyglycerol-based sulfated nanogels of different forms having varied flexibility using inverse nanoprecipitation method where bio-orthogonal cross-linking chemistry that permits effective gelation is employed. Characterization was carried out by DLS, AFM, nanosight-tracking analysis (NTA), proton NMR spectroscopy, and cryogenic TEM. This flexible nanogel is found to have great antiviral potential and thus inhibits the binding of various virus species; also, the antiviral activity is found to be higher for nanogels having greater flexibility (Dey et al. 2018). Chun et al. (2018) developed an amphiphilic copolymer consisting of methoxy-poly (ethylene glycol)-block poly(phenylalanine), encapsulated with mir-323a in the inner core and favipiravir at the outer layer as antiviral agents. This polymer acts as a drug-distributing agent in the treatment of the Influenza A virus.

Dendrimers are highly branched, globular, symmetrical, and macromolecular structures having regions of a central core, branching units, and functional groups (function as targeting/interacting sites). Fullerenes, a class of dendrimers having a well-defined and established composition, shape, and functionalization chemistry have been widely used as anti-HIV agents for different viruses using different derivatives of fullerene with promising results. They provide these results because they can produce an enormous amount of singlet oxygen that will destroy the viruses and hence sanitize the surfaces; but, in this case, a careful study of cytotoxicity is also necessary (Innocenzi and Stagi 2020). Cyclosporine A-encapsulated PLGA nanoparticles conjugated with liver-targeting peptide by an oil-in-water emulsification solvent evaporation accompanied by lyophilization were developed. Morphology of the resultant nanoparticle is studied through SEM analysis, and the size distribution and ζ-potential were characterized by DLS. Results found that the nanoparticles prepared could successfully arrest viral replication in vitro and in vivo in a Hepatitis C virus-infected mouse (Kim et al. 2015).

2.3.3 Antiviral Potential of Nanoparticles against Various Viruses

Nanotechnology has become a central attraction of recent research advancements in biological applications for human healthcare. The exploration of various types of nanoscale materials for their preventive, diagnostic, and therapeutic potential activity against various viruses, since the interaction of nanostructures with viruses is quick compared to the bulk materials of the same, makes the nanoparticle-based therapy a promising factor for biomedical applications in the current scenario. The nano-sized materials possess exceptional and intriguing physico-chemical properties; these are mainly due to their smaller size, high surface area-to-volume ratio, and a surface charge that can be changed to be encapsulated. In addition, large drug loads may enhance their bioavailability, solubility, and circulation time compared to bulk particles and thus provide benefits of drug delivery, making them an ideal tool for viral treatment. Through the surface modifications of these nanoparticles, diverse biological and antimicrobial properties can be explored to improve their therapeutic effects (Cojocaru et al. 2020).

Viral diseases are becoming a serious global health concern due to their high mortality rate, which greatly impacts socio-economic life. The drug resistance and anomalous replication of viruses made the use of nanotechnology important in antiviral therapies. Currently, various types of nanomaterials, like nanospheres, nanoparticles, nanosuspensions, nanogels nano emulsions, etc., are used for drug delivery because they possess antimicrobial activities against different viruses as antiviral agents and their dimensions are similar to the biomolecules (Ghaffari et al. 2019; Sharma et al. 2019). The antiviral potential of nanoparticles have been reported against several viruses like vaccinia virus, chikungunya virus, influenza virus, herpes simplex virus, monkeypox virus, HIV, hepatitis B virus, etc. as the nanoparticles inhibit the attack of the virus to host cell and thus prevent their entry to the cell. Nanostructures thus can be used to function as either a delivery aid for specific vaccine to acquire immunization of the host, or as nanocarriers to provide diverse therapeutics to the target site, enhancing circulation time by safeguarding the therapeutics from deprivation. Here we present a brief summary of the application of nanosized materials for the treatment of viral infections like the Chikungunya virus, Coronaviruses (CoVs), HIV, H1N1 influenza virus (Heinrich et al. 2020).

CoVs are one of the severe pathogens that have the RNA (ribonucleic acid) sheathe, and they widely affect all mammals including humans and birds. So far, this century has faced a pandemic of three CoVs: SARS-CoV in 2002, the middle-east respiratory syndrome Coronavirus (MERS-CoV) in 2012, and the newly identified SARS-CoV-2/COVID-19 in December 2019, first reported in the city of Wuhan, China. CoVs originated were associated with chronic respiratory diseases with symptoms of fever, pneumonia, bronchiolitis. and cough. The novel SARS CoV-2 is causing deadly infection worldwide, and a severe pandemic is still ongoing. Various therapeutics and vaccines are currently being developed, of which some have already been administered to people, with hopes for a fast solution and recovery from this drastic situation (Muhammad et al. 2020). A unique feature of CoVs is that the angiotensin-converting enzyme 2 (ACE2) receptor site of the host cell plays a major role in binding the CoVs to host cells where the viral replication occurs. Thus, decreasing the ACE2 levels will help to fight against infection, as well as help the antibody's development against ACE2. Hence, it is recommended to prevent the infection, rather than fight it after infection. An important feature of nanomaterials is that they could impede entry of the virus to the host cell by inhibiting the binding at the surface receptor. Nanotechnology could be used for the production of masks, clothes, nose filters, gloves, and chewing gums by utilizing ACE2-coated nanoflowers and QDs to prevent the spread of these viruses into humans. They can also serve as biodetection probes. A nanoparticle-intermediated intranasal delivery system having the benefits of long-period persistence, exact targeting of cells by the interaction of receptor and ligand, boosting of the immune system, and absence of enzyme degradation will be an effective and benign method to deliver vaccines, antibodies, drugs, etc. (Gurunathan et al. 2020).

CHIKV is an enveloped arbovirus and is responsible for the fever called chikungunya fever; it is transmitted in humans by female *Aedes albopictus* and *Aedes aegypti* mosquitoes. it is a major public health concern because it causes many epidemics in Asian countries. CHIKV has symptoms of high fever, headaches, rashes, nausea, fatigue, and severe joint pain, which can cause weakness. Even though it is a self-limiting disease, neurological, heart, and gastrointestinal complications are also seen in severe cases. Until now, no vaccines or antiviral drugs existed, so the treatment is mainly focused on supportive care to relieve symptoms and pain (Kumar et al. 2018). Kumar et al. (2018) synthesized ZnO nanoparticles (ZnONP) by the precipitation method; they are characterized by TEM, zeta dynamic light scattering, and FTIR. On determining the cytotoxicity of ZnONP in MA104 cells, and on treating with CHIKV cells and further conducting RT-PCR testing, it is found that there are 10-fold reductions in the virus when treated with ZnONP. Thus, ZnONP is exhibiting antiviral activity toward CHIKV. In 2018, Sharma et al. and co-workers biologically synthesized silver nanoparticles (AgNPs) from various medicinal plants and studied their antiviral properties against CHIKV; they found that it could be used for the treatment of CHIKV for which vaccines had not been developed yet (Sharma et al. 2019).

Influenza viruses affect the human respiratory tract. This seasonal epidemic is responsible for pandemics around the world. Currently, only two classes of drugs are available against different influenza virus: matrix-2 protein ion channel blockers and neuraminidase inhibitors. Also, drug resistance is increasing, so it becomes urgent to explore anti-influenza agents that can serve as an alternative to the conventional drugs available; alternative agents possess a different action mechanism. In vitro experimentations have proved that the nanoparticles possess inhibitory activity against the influenza virus only after the entry of virus to the host cells (Ghaffari et al. 2019). A study by Papp et al. (2010) demonstrated the size-dependant, antiviral potential of sialic acid (SA)-modified gold nanoparticles (AuNPs), which block the entry of the influenza virus H1N1 into the cell membrane and observed a higher antiviral potential for larger sized NPs. Another study by Ghaffari et al. (2019) observed the antiviral activity of polyethylene glycol (PEG)-coated ZnO-NPs (PEG-ZnO NPs), against the H1N1 influenza virus that stimulates the antiviral and antioxidant potential, resulting in viral inhibition along with minimal toxicity. The results show that PEG-ZnO NPs have a superior anti-influenza action and lesser cytotoxicity than bare ZnONPs on MDCK-SIAT1 cells (Brandelli et al. 2017). In 2017, Lin et al. evaluated the antiviral potential of zanamivir-modified selenium NPs (SeNPs-ZNV) against H1N1, which is found to be greater than that of bare SeNPs or zanamivir (Lin et al. 2017). It was observed by Levina et al. that DNA tagged-titanium NPs inhibited the viral replication of H1N1 and H5N1 influenza viruses (Levina et al. 2014).

HIV is a chronic disease considered one of the major burdens all over the world. This lifelong illness requires long-term care and support for those who die, which demands high treatment costs for the HIV-affected households and greatly affects the economy. Elechiguerra et al. in 2005 synthesized AgNPs, by polyol method, using glycerine as the reducing agent and solvent; characterization is carried out by TEM and UV-Visible spectral analysis. They studied the antiviral potential of AgNPs and their interaction with HIV-1, and showed that the binding of AgNPs to gp-120 glycoprotein knobs inhibits the virus from entering the host cells; otherwise, they could inhibit the replication of the virus. This was demonstrated as in vitro (Elechiguerra et al. 2005). Another study regarding in vitro antiviral potential of AgNPs was conducted by Lara et al. in 2010 and suggested that these NPs exert anti-HIV-1 action in the first phase by inhibiting the viral replication and later on inhibit their binding to the host cell (Lara et al. 2010).

Current antiviral drugs are available for only a few viral diseases, including HIV, influenza, HSV, and hepatitis; thus, novel strategies for the antiviral therapeutics are needed in the future to address secondary effects of conventional medical practice and drug resistance of certain viruses. Future studies should also investigate the actual antiviral potential mechanism of the nanoparticles against various viruses, with the aid of modern techniques like TEM, and also be able to determine viral gene change occurring in presence of nanoparticles.

2.4 VIRAL NANOPARTICLES

The development and applications of synthetic biology and nanotechnology have influenced medical science significantly, especially in drug delivery, diagnosis, and imaging. A distinct class of materials such as dendrimers, quantum dots (QDs), liposomes, polymer vesicles, and metallic nanoparticles has opened new avenues for medical research, owing to their unique properties that are suitable for diverse biomedical applications. However, each system is vulnerable to some merits and demerits in respect of their biocompatibility, toxicity, pharmacokinetics, immunogenicity, and specificity for the target tissue. For instance, dendrimers are inexpensive and easy to synthesize; their cytotoxicity is a major obstacle for in vivo applications (Svenson and Tomalia 2005). QDs are recognized as a promising imaging tool in terms of their broad bandwidth absorption, narrow bandwidth emission, and long-lasting fluorescence; they also display in vivo toxicity (Portney and Ozkan 2006). Liposomes provide an approved platform for clinical usage in terms of organ avoidance and moderate drug release, and hence, they are used to deliver the anticancer drug doxorubicin; however, its high production cost and difficulties in sterilization restrict its wide use (Murphy et al. 2008). In this context, viral nanoparticles (VNPs) are promising candidates regarding their natural occurrence, low cost, biocompatibility, and biodegradability.

Viruses are naturally evolved creatures that carry the cargo of genetic material to specific host cells with greater efficiency. Thus, viruses are an epitome for the targeted delivery of molecules such as imaging reagents and drugs. Also, prodigious replication of viruses allows inexpensive and industrial-scale production of virus-derived materials. Therefore, advancements and applications of virus-based materials such as viral nanoparticles (VNPs) and virus-like particles (VLPs) are gaining keen interest and impact of chemists, biologists, and biomedical engineers. VNPs are nanoparticle formulations of viruses and are employed as an essential entity for several materials having diverse features. Several types of viruses are used for the synthesis of VNPs, including bacteriophages and animal and plant viruses; they can be infectious or noninfectious. VLPs are identified as a subset of VNPs in heterologous systems and, the exception of genomic nucleic acid makes them noninfectious. VNPs are genetically encoded and self-assembled into monodisperse and discrete structural designs with a high degree of atomic resolution. This kind of top-level structural engineering and quality control is exclusive for VNPs and cannot be accomplished with synthetic materials. The high symmetry, polyvalency, and robustness render them superior to the synthetic counterparts. The highly advanced and versatile materials exhibit programmable scaffolds where the outer surface can be functionalized with targeting ligands for cyto-specific delivery and the inner cavity can be impregnated with imaging reagents, quantum dots, drug molecules, etc.

2.4.1 Synthesis of VNPs

Although VNPs are obtained in large amounts in their innate hosts, VLPs are synthesized in heterologous expression systems. Plant virus-based VNPs such as cowpea chlorotic mottle virus (CCMV), tobacco mosaic virus (TMV), cowpea mosaic virus (CPMV), brome mosaic virus (BMV), and potato virus X (PVX) can be easily produced in their host in high titre. VNPs based on bacteriophages, such as HK97, Qβ, M13, and MS2, can be produced in similar quantities using cultures of Escherichia coli bacteria. The same cultures can also be used as heterologous expression systems for the construction of chimeric VNPs and VLPs. Direct synthesis of VLPs from VNPs can be achieved with successive pH-induced swelling and alkaline hydrolysis of the released nucleic acids. On the other hand, the whole VNP structure can be dismantled into discrete coat-protein units and reunited into VLPs by the removal of nucleic acids. Eukaryotic virus-based VLPs are produced by using either adenovirus or baculovirus vectors in mammalian and insect cells, respectively (Garnier et al. 1994; Schneemann and Young 2003).

2.4.2 Modification of VNPs and VLPs

A series of methods can be used to customize and functionalize VNPs for anticipated applications, including bioconjugation, encapsulation, infusion, biomineralization, and genetic engineering. The flexibility offers different possibilities, such as reversible binding of active molecules, protection inside the proteinaceous scaffolds, and targeting to specific sites. A few of the methods are discussed below.

2.4.2.1 Bioconjugation

The use of bioconjugate chemistry is the most powerful approach for the tailoring of VNPs. Targeting ligands, drugs, or contrast agents can be linked to capsid shell or interior cavity by applying bioconjugation protocols. Particular amino acids with reactive side chains, such as reactive amines on lysine residues, carboxylate functional groups on glutamic and aspartic acid residues, phenol groups on tyrosine residues, and sulfhydryl groups on cysteine residues, can be functionalized with oligonucleotides, antibodies, carbohydrates, proteins, peptides, fluorescent reagents, drugs with an N-hydroxysuccinimidyl ester (NHS) group, isothiocyanate, maleimide, and carbodiimide. These groups can be either directly conjugated to specific molecules or tuned to unveil functional groups for more improved and advanced conjugation approaches. Incorporation of synthetic functional groups can also be realized with bioconjugation permitting the use of oxime-hydrozone ligation, tyrosinediazonium coupling, and click chemistry (Pokorski and Steinmetz 2011).

2.4.2.2 Encapsulation

The genetic material in the interior of the virus capsid is wrapped through an uncomplicated supramolecular self-assembly method, and the same theory can be utilized to pack functional materials inside the virus. Disassembly/reassembly and swelling protocols of the self-assembled structures enable the encapsulation of drugs and therapeutic agents in VNPs. Under physiological conditions, virus coat proteins self-assembled around the nucleic acids. The disassembling and reassembling enable the formation of a variety of structures around the cargo molecules. Two principles are mainly used to stimulate the cargo encapsulation; surface charge or electrostatic interactions and exclusive binding interactions take place during self-assembly. For instance, MS2, a bacteriophage, has translational repression (TR) operator protein that binds with an RNA stem loop. The TR protein is made to carry smaller molecules. However, modification of the TR operator facilitates diffusion inside the VNPs and stable binding with the capsid. Large drug molecules, such as ricin with a chain and 5-fluorouridine unit, have been effectively integrated into MS2 particles. The cargo delivery and successful target cell demolition are confirmed with in vitro studies that validate the potential of the design strategies (Brown et al. 2002).

2.4.2.3 Biomineralization

The assembling of virus-coat proteins around the mineral nucleus is referred as Biomineralization. The engineering strategy endows VNPs as a versatile contrasting agent in medicine. For example, the CCM virus with electrostatically modified scaffold expedites the oxidation and nucleation of Fe(II) payloads to iron-oxide nanocrystals in a spatially constrained reaction environment and are employed for hyperthermia treatment and magnetic resonance imaging (MRI) (Douglas et al. 2002). A large variety of mineral cargos have been perfectly incorporated into VNPs via this approach.

2.4.2.4 Genetic Engineering

Modifications in the basic structure of virus-coat proteins can be accomplished with genetic engineering, either by removing, substituting, or inserting amino acid residues. The introduction of such reproducible and precise modifications helps to manufacture identical particles in higher amounts, displaying targeting ligands or ligation handles for additional modifications via bioconjugation for desired applications. The terminal extensions, surface loop formation, and side-chain functionalization by genetic engineering may alter the physico-chemical properties of VNPs, enabling their use in distinct applications. For example, the introduction of epitome sequence allows the use of VNPs for vaccine purposes. Chimeric VNPs containing extra cysteine residues are fabricated to allow covalent modification via thiol selective chemistry, and the incorporation of His6 tags, biotin binding peptides, and the B domain of Staphylococcal protein A (SPAB) simplifies the labelling and purification procedures (Park et al. 2009). The incorporation of peculiar amino acids as unique handles for ensuing chemical reactions is also credible with analogous recombinant-expression strategies. Genetic engineering offers around 100% efficient modification as compared to chemical methods.

2.4.3 Applications of VNPs and VLPs

VNPs are employed for a broad range of applications, including immunotherapies, cancer therapies, antimicrobial therapies, gene therapies, cardiovascular therapies, imaging, vaccines, and theragnostics. Although there is a great improvement in the in vitro research of VNPs as drug-delivery agents, constant and adequate study is needed to render their use in in vivo studies. Some of the potential applications of VNPs are discussed below.

2.4.3.1 Vaccines

Vaccines hold a great promise for the treatment of chronic diseases such as cancer and for the prevention of infectious diseases. In the current COVID-19 outbreak, the swift development of a safe and effective vaccination platform is immediately required. Viruses act as a convenient tool for the development of novel vaccines, owing to their repetitive proteinaceous structure, pathogen-related molecular patterns, and particulate nature. These significant characteristics evoke immune responses, thus making them advantageous for the development of vaccines. VNPs are engineered for the vaccine purpose by three basic strategies: (i) VNPs are altered to be noninfectious by chemical treatment, (ii) noninfectious, genome-free VLPs are developed in heterologous expression systems, and finally, (iii) Chimeric VNPs are fabricated via genetic engineering. Even though immunogenicity is a major requirement for vaccine development, it decreases the efficiency of VNPs for drug delivery and imaging necessities. Thus, shielding of VNPs from the immune system must be achieved for their effective use as vaccines or immunomodulators. The most familiar strategy is the binding of polyethylene glycol chains to the outer surface, known as PEGlyation. The approach increases stability and plasma circulation time and reduces immunogenicity and bispecific interactions (Steinmetz and

Manchester 2009). Chimeric VLPs with complex antigen structures have been established based on Flock House virus (FHV), a virus that infects insects. A portion of anthrax toxin receptor (ANTXR2), which acts as a scaffold to display the Bacillus anthracis protective antigen, has been included with the multivalent display system of FHV. This virus-antigen complex trigger shielded immune responses in the absence of adjuvant, even after a single dose (Manayani et al. 2007). Many efforts have been taken for the development of VNP-based cancer vaccines that encompass a patterned display of tumour-associated peptide or carbohydrate antigens. For instance, a glycoprotein Tn antigen that is over expressed on several cancer cells (like colon, prostate. and breast cells) is chemically displayed by CPMV scaffolds. Bioconjugation of the scaffold and multivalent display results in the formation of Tn specific antibodies in high titre (Miermont et al. 2008).

2.4.3.2 Imaging

VNPs are covalently conjugated to several imaging reagents, such as fluorescent dyes used for intravital vascular imaging. Fluorescent CPMV sensors facilitate the visualization of vasculature and blood flow in living mammalian cells (Lewis et al. 2006). Hybrid VNPs with fluorescent cores were designed by encapsulation techniques for cellular imaging. For instance, quantum dot encapsulated simian virus 40 (SV40) was synthesized and confocal microscopy was used to track them. Recently, VNPs such as CCMV, CPMV and Qβ have been established as contrast agents for MRI. The noninvasive scanning technique involving VNPs with functional entities can carry hundreds of chelated Gd complexes, which can covalently bind to interior and exterior surfaces (Cai et al. 2009).

2.4.3.3 Targeted Delivery

Targeted delivery of VNPs with toxic payloads through infusion, encapsulation, and conjugation enables the killing of selective cancer cells or other diseased cells. Several therapeutic cargos can be delivered this way, including RNAs or genes. Photoactive molecules that support photodynamic therapy, conventional small drug molecules, viral genes for gene therapy (like alphavirus genome encapsulated in a VNP-based CCMV) can easily be delivered through this targeted strategy (Azizgolshani et al. 2013).

2.5 CONCLUSION

Nanoparticles' unique properties, such as the high surface area-to-volume ratio, compatible size, surface chemistry, etc., offer great potential as an antiviral agent for disinfection, therapeutics, and filtration. It's highly challenging to cultivate, characterize, and inactivate the viral nanoparticles (especially pathogenic), and thus, the lack of research has hindered the advancement in this area. Currently, the range of viruses studied and evaluated has been very narrow; therefore, it's important to study the various viruses to develop high potential nanomaterial with a broad-spectrum antiviral property. A better fundamental understanding of virus nanoparticle interactions and reactions can make it possible to develop a smart design of antiviral nanoparticles.

REFERENCES

Aderibigbe, B. 2017. Metal-based nanoparticles for the treatment of infectious diseases. *Molecules* 22(8):1370. 10.3390/molecules22081370

Ai, J., Biazar, E., Jafarpour, M., et al. 2011. Nanotoxicology and nanoparticle safety in biomedical designs. *International Journal of Nanomedicine* 6:1117–1127. 10.2147/IJN.S16603

Alimardani, V., Abolmaali, S.S. and Tamaddon, A.M. 2021. Recent advances on nanotechnology-based strategies for prevention, diagnosis, and treatment of coronavirus infections. *Journal of Nanomaterials* 2021:1–20. 10.1155/2021/9495126

Azizgolshani, O., Garmann, R.F., Cadena-Nava, R., Knobler, C.M. and Gelbart, W.M. 2013. Reconstituted plant viral capsids can release genes to mammalian cells. *Virology* 441(1):12–17.10.1016/j.virol.2013.03.001

Balagna, C., Perero, S., Percivalle, E., Nepita, E.V. and Ferraris, M. 2020. Virucidal effect against coronavirus SARS-CoV-2 of a silver nanocluster/silica composite sputtered coating. *Open Ceramics* 1:100006. 10.1016/j.oceram.2020.100006

Baram-Pinto, D., Shukla, S., Richman, M., Gedanken, A., Rahimipour, S. and Sarid, R. 2012. Surface-modified protein nanospheres as potential antiviral agents. *Chemical Communications* 48(67):8359. 10.1039/c2cc33448h

Brandelli, A., Ritter, A.C. and Veras, F.F. 2017. Antimicrobial activities of metal nanoparticles. In *Metal Nanoparticles in Pharma*, eds. M. Rai and R. Shegokar, 337–363. Cham, Switzerland: Springer International Publishing. 10.1007/978-3-319-63790-7_15

Brown, W.L., Mastico, R.A., Wu, M., et al. 2002. RNA bacteriophage capsid-mediated drug delivery and epitope presentation. *Intervirology* 45(4–6):371–380. 10.1159/000067930

Cai, J., Shapiro, E.M. and Hamilton, A.D. 2009. Self-assembling DNA Quadruplex conjugated to MRI contrast agents. *Bioconjugate Chemistry* 20(2):205–208. 10.1021/bc8004182

Chen, N., Zheng, Y., Yin, J., Li, X. and Zheng, C. 2013. Inhibitory effects of silver nanoparticles against adenovirus type 3 in vitro. *Journal of Virological Methods* 193(2):470–477. 10.1016/j.jviromet.2013.07.020

Chun, H., Yeom, M., Kim, H.O., et al. 2018. Efficient antiviral co-delivery using polymersomes by controlling the surface density of cell-targeting groups for influenza A virus treatment. *Polymer Chemistry* 9(16):2116–2123. 10.1039/C8PY00116B

Cojocaru, F.D., Botezat, D., Gardikiotis, I., et al. 2020. Nanomaterials designed for antiviral drug delivery transport across biological barriers. *Pharmaceutics* 12(2):171. 10.3390/pharmaceutics12020171

Delshadi, R., Bahrami, A., McClements, D.J., Moore, M.D. and Williams, L. 2021. Development of nanoparticle-delivery systems for antiviral agents: A review. *Journal of Controlled Release* 331:30–44. 10.1016/j.jconrel.2021.01.017

Dey, P., Bergmann, T., Cuellar-Camacho, J.L., et al. 2018. Multivalent flexible nanogels exhibit broad-spectrum antiviral activity by blocking virus entry. *ACS Nano* 12(7): 6429–6442. 10.1021/acsnano.8b01616

Dhanasezhian, A., Srivani, S., Govindaraju, K., Parija, P., Sasikala, S. and Kumar, M.R.R. 2019. Anti-Herpes Simplex Virus (HSV-1 and HSV-2) activity of biogenic gold and silver nanoparticles using seaweed *Sargassum wightii*. *Indian Journal of Geo-Marine Sciences* 48(8):1252–1257. http://nopr.niscair.res.in/handle/123456789/49704

Dhand, C., Dwivedi, N., Loh, X.J., et al. 2015. Methods and strategies for the synthesis of diverse nanoparticles and their applications: A comprehensive overview. *RSC Advances* 5(127):105003–105037. 10.1039/C5RA19388E

Douglas, T., Strable, E., Willits, D., Aitouchen, A., Libera, M. and Young, M. 2002. Protein engineering of a viral cage for constrained nanomaterials synthesis. *Advanced Materials* 6:4.

Elechiguerra, J., Burt, J.L., Morones, J.R., et al. 2005. Interaction of silver nanoparticles with HIV-1. *Journal of Nanobiotechnology* 3(1):6. 10.1186/1477-3155-3-6

Etemadzade, M., Ghamarypour, A., Zabihollahi, R., et al. 2016. Synthesis and evaluation of antiviral activities of novel sonochemical silver nanorods against HIV and HSV viruses. *Asian Pacific Journal of Tropical Disease* 6(11):854–858. 10.1016/S2222-1808(16)61145-3

Fang, X., Wang, Y., Wang, Z., Jiang, Z. and Dong, M. 2019. Microorganism assisted synthesized nanoparticles for catalytic applications. *Energies* 12(1):190. 10.3390/en12010190

Galdiero, S., Falanga, A., Vitiello, M., Cantisani, M., Marra, V. and Galdiero, M. 2011. Silver nanoparticles as potential antiviral agents. *Molecules* 16(10):8894–8918. 10.3390/molecules16108894

Garnier, A., Côté, J., Nadeau, I., Kamen, A. and Massie, B. 1994. Scale-up of the adenovirus expression system for the production of recombinant protein in human 293S cells. *Cytotechnology* 15(1–3):145–155. 10.1007/BF00762389

Ghaffari, H., Tavakoli, A., Moradi, A., et al. 2019. Inhibition of H1N1 influenza virus infection by zinc oxide nanoparticles: Another emerging application of nanomedicine. *Journal of Biomedical Science* 26(1):70. 10.1186/s12929-019-0563-4

Gurunathan, S., Qasim, M., Choi, Y., et al. 2020. Antiviral potential of nanoparticles - Can nanoparticles fight against coronaviruses? *Nanomaterials* 10(9):1645. 10.3390/nano10091645

Heinrich, M.A., Martina, B. and Prakash, J. 2020. Nanomedicine strategies to target coronavirus. *Nano Today* 35:100961. 10.1016/j.nantod.2020.100961

Ibrahim Fouad, G. 2021. A proposed insight into the anti-viral potential of metallic nanoparticles against novel coronavirus disease-19 (COVID-19). *Bulletin of the National Research Centre* 45(1):36. 10.1186/s42269-021-00487-0

Ingale, A. G. and Chaudhari, A.N. 2013. Biogenic synthesis of nanoparticles and potential applications: An eco- friendly approach. *Journal of Nanomedicine and Nanotechnology* 4(2). 10.4172/2157-7439.1000165

Innocenzi, P. Stagi, L. 2020. Carbon-based antiviral nanomaterials: Graphene, C-dots, and fullerenes. A perspective. *Chemical Science* 11(26):6606–6622. 10.1039/D0SC02658A

Khandel, P., Yadaw, R.K., Soni, D.K., Kanwar, L. and Shahi, S.K. 2018. Biogenesis of metal nanoparticles and their pharmacological applications: Present status and application prospects. *Journal of Nanostructure in Chemistry* 8(3):217–254. 10.1007/s40097-018-0267-4

Kim, S.S., Jyothi, K.R., Beloor, J., et al. 2015. Liver-targeted cyclosporine A-encapsulated poly (lactic-co-glycolic) acid nanoparticles inhibit hepatitis C virus replication. *International Journal of Nanomedicine* 903. 10.2147/IJN.S74723

Kumar, R., Sahoo, G., Pandey, K., et al. 2018. Virostatic potential of zinc oxide (ZnO) nanoparticles on capsid protein of cytoplasmic side of chikungunya virus. *International Journal of Infectious Diseases* 73:368. 10.1016/j.ijid.2018.04.4247

Lara, H.H., Ayala-Nuñez, N.V., Ixtepan-Turrent, L. and Rodriguez-Padilla, C. 2010. Mode of antiviral action of silver nanoparticles against HIV-1. *Journal of Nanobiotechnology* 8(1):1. 10.1186/1477-3155-8-1

Levina, A.S., Repkova, M.N., Ismagilov, Z.R., Shikina, N.V., Mazurkova, N.A. and Zarytova, V.F. 2014. Efficient inhibition of human influenza a virus by oligonucleotides electrostatically fixed on polylysine-containing TiO_2 nanoparticles. *Russian Journal of Bioorganic Chemistry* 40(2):179–184. 10.1134/S1068162014020095

Lewis, J.D., Destito, G., Zijlstra, A., et al. 2006. Viral nanoparticles as tools for intravital vascular imaging. *Nature Medicine* 12(3):354–360. 10.1038/nm1368

Lin, Z., Li, Y., Guo, M., et al. 2017. Inhibition of H1N1 influenza virus by selenium nanoparticles loaded with zanamivir through p38 and JNK signaling pathways. *RSC Advances* 7(56):35290–35296. 10.1039/C7RA06477B

Maduray, K. and Parboosing, R. 2020. Metal nanoparticles: a promising treatment for viral and arboviral infections. *Biological Trace Element Research.* 10.1007/s12011-020-02414-2

Manayani, D.J., Thomas, D., Dryden, K.A., et al. 2007. A viral nanoparticle with dual function as an anthrax antitoxin and vaccine. *PLoS Pathogens* 3(10):e142. 10.1371/journal.ppat.0030142

Meléndez-Villanueva, M.A., Morán-Santibañez, K., Martínez-Sanmiguel, J.J., et al. 2019. Virucidal activity of gold nanoparticles synthesized by green chemistry using garlic extract. *Viruses* 11(12):1111. 10.3390/v11121111

Miermont, A., Barnhill, H., Strable, E., et al. 2008. Cowpea mosaic virus capsid: A promising carrier for the development of carbohydrate based antitumor vaccines. *Chemistry - A European Journal* 14(16):4939–4947. 10.1002/chem.200800203

Muhammad, W., Zhai, Z. and Gao, C. 2020. Antiviral activity of nanomaterials against coronaviruses. *Macromolecular Bioscience* 20(10):2000196. 10.1002/mabi.202000196

Murphy, E.A., Majeti, B.K., Barnes, L.A., et al. 2008. Nanoparticle-mediated drug delivery to tumor vasculature suppresses metastasis. *Proceedings of the National Academy of Sciences of the USA* 105(27):9343–9348. 10.1073/pnas.0803728105

Murugan, K., Dinesh, D., Paulpandi, M., et al. 2015. Nanoparticles in the fight against mosquito-borne diseases: Bioactivity of *Bruguiera cylindrica*-synthesized nanoparticles against dengue virus DEN-2 (in vitro) and its mosquito vector *Aedes aegypti* (Diptera: Culicidae). *Parasitology Research* 114(12):4349–4361. 10.1007/s00436-015-4676-8

Odularu, A.T. 2018. Metal nanoparticles: thermal decomposition, biomedicinal applications to cancer treatment, and future perspectives. *Bioinorganic Chemistry and Applications* 2018:1–6. 10.1155/2018/9354708

Pantidos, N. 2014. Biological synthesis of metallic nanoparticles by bacteria, fungi and plants. *Journal of Nanomedicine and Nanotechnology* 5(5). 10.4172/2157-7439.1000233

Papp, I., Sieben, C., Ludwig, K., et al. 2010. Inhibition of influenza virus infection by multivalent sialic-acid-functionalized gold nanoparticles. *Small* 6(24):2900–2906. 10.1002/smll.201001349

Park, J.S., Cho, M.K., Lee, E.J., et al. 2009. A highly sensitive and selective diagnostic assay based on virus nanoparticles. *Nature Nanotechnology* 4(4):259–264. 10.1038/nnano.2009.38

Pfaff, F., Glück, B., Hoyer, T., Rohländer, D., Sauerbrei, A. and Zell, R. 2019. Tungsten carbide nanoparticles show a broad spectrum virucidal activity against enveloped and nonenveloped model viruses using a guideline-standardized in vitro test. *Letters in Applied Microbiology* 69(4):302–309. 10.1111/lam.13208

Pokorski, J.K. and Steinmetz, N.F. 2011. The art of engineering viral nanoparticles. *Molecular Pharmaceutics* 8(1):29–43. 10.1021/mp100225y

Portney, N.G. and Ozkan, M. 2006. Nano-oncology: Drug delivery, imaging, and sensing. *Analytical and Bioanalytical Chemistry* 384(3):620–630. 10.1007/s00216-005-0247-7

Reina, G., Peng, S., Jacquemin, L., Andrade, A.F. and Bianco, A. 2020. Hard nanomaterials in time of viral pandemics. *ACS Nano* 14(8):9364–9388. 10.1021/acsnano.0c04117

Salem, A.N.B., Zyed, R., Lassoued, M.A., Nidhal, S., Sfar, S. and Mahjoub, A. 2012. Plant-derived nanoparticles enhance antiviral activity against coxsakievirus b3 by acting on virus particles and vero cells. *Digest Journal of Nanomaterials and Biostructures* 7(2):737–744.

Schneemann, A. and Young, M.J. 2003. Viral assembly using heterologous expression systems and cell extracts. In *Advances in Protein Chemistry*, Vol. 64, 1–36. Amsterdam, Netherlands: Elsevier . 10.1016/S0065-3233(03)01001-5

Shah, M., Fawcett, D., Sharma, S., Tripathy, S. and Poinern, G. 2015. Green synthesis of metallic nanoparticles via biological entities. *Materials* 8(11):7278–7308. 10.3390/ma8115377

Sharma, V., Kaushik, S., Pandit, P., Dhull, D., Yadav, J.P. and Kaushik, S. 2019. Green synthesis of silver nanoparticles from medicinal plants and evaluation of their antiviral potential against chikungunya virus. *Applied Microbiology and Biotechnology* 103(2):881–891. 10.1007/s00253-018-9488-1

Sharmin, S., Rahaman, Md.M., Sarkar, C., Atolani, O., Islam, M.T. and Adeyemi, O.S. 2021. Nanoparticles as antimicrobial and antiviral agents: A literature-based perspective study. *Heliyon* 7(3):e06456. 10.1016/j.heliyon.2021.e06456

Singh, L., Kruger, H.G., Maguire, G.E.M., Govender, T. and Parboosing, R. 2017. The role of nanotechnology in the treatment of viral infections. *Therapeutic Advances in Infectious Disease 4*(4):105–131. 10.1177/2049936117713593

Steinmetz, N.F. 2010. Viral nanoparticles as platforms for next-generation therapeutics and imaging devices. *Nanomedicine: Nanotechnology, Biology and Medicine* 6(5):634–641. 10.1016/j.nano.2010.04.005

Steinmetz, N.F. and Manchester, M. 2009. PEGylated viral nanoparticles for biomedicine: The impact of PEG chain length on VNP cell interactions in vitro and ex vivo. *Biomacromolecules* 10(4):784–792. 10.1021/bm8012742

Svenson, S. and Tomalia, D. 2005. Dendrimers in biomedical applications - Reflections on the field. *Advanced Drug Delivery Reviews* 57(15):2106–2129. 10.1016/j.addr.2005.09.018

Vecitis, C.D. 2021. Antiviral-nanoparticle interactions and reactions. *Environmental Science: Nano* 8(1):11–19. 10.1039/D0EN00980F

Yildiz, I., Shukla, S. and Steinmetz, N.F. 2011. Applications of viral nanoparticles in medicine. *Current Opinion in Biotechnology* 22(6):901–908. 10.1016/j.copbio.2011.04.020

Yu, C.-H., Tam, K. and Tsang, E.S.C. 2008. Chemical methods for preparation of nanoparticles in solution. In *Handbook of Metal Physics*, Vol. 5, 113–141. Amsterdam, Netherlands: Elsevier . 10.1016/S1570-002X(08)00205-X

Zhang, X.F., Liu, Z.G., Shen, W. and Gurunathan, S. 2016. Silver nanoparticles: Synthesis, characterization, properties, applications, and therapeutic approaches. *International Journal of Molecular Sciences* 17(9):1534. 10.3390/ijms17091534

Zhang, Y., Dong, Y., Zhou, J., Li, X. and Wang, F. 2018. Application of plant viruses as a biotemplate for nanomaterial fabrication. *Molecules* 23(9):2311. 10.3390/molecules23092311

Zhou, J., Krishnan, N., Jiang, Y., Fang, R.H. and Zhang, L. 2021. Nanotechnology for virus treatment. *Nano Today* 36:101031. 10.1016/j.nantod.2020.101031

3 Biogenic Synthesis of Nanoparticulate Materials for Antiviral Applications

Kah Hon Leong, Jit Jang Ng, Lan Ching Sim, Pichiah Saravanan, Chaomeng Dai, and Bo Tan

CONTENTS

DOI: 10.1201/9781003136644-4

3.1 INTRODUCTION

Various severe viruses have emerged, causing multiple infectious diseases and triggering a global health concern. The current epidemic of atypical pneumonia instigated by the 2019 Coronavirus spread to the entire world, resulting in more than 3 million deaths; an immediate solution is needed to curb this phenomenon. Besides the Coronavirus, many other viruses exist, too, such as human immunodeficiency virus (HIV), human papillomavirus (HPV), norovirus, hepatitis viruses, and many others (Delshadi et al. 2021). Viruses can be transmitted through body fluids and aerosol particles from an infected person to a healthy person. Aerosols show a dynamic role in dispersing viruses in the air, making them more critical and dangerous for humankind. Therefore, many attempts have been made to develop antivirus drugs to curb and tackle the many different viruses present in the atmosphere. However, studies have found that many of these drugs have inevitable side effects on human tissues and cells, particularly for younger people in the community. Thus, the evolution of nanotechnology has created an alternative pathway, and it shows the prevailing ability in improving the effectiveness against virus contagion and imitation (Chaudhuri et al. 2018).

Nanotechnology is nothing new to virus research. Previously, it was applied in the virus-detection technology that produced many promising discoveries, such as the fabrication of various biosensors and bioelectronics (Chen et al. 2012). Also, various nanomaterials have shown excellent efficacy; for instance, it's been effective to develop virions and viruses like particles as templates for improving biocompatibility (Luo et al. 2018). The unique advantages of the nano-sized particles enable nanomaterials to achieve a positive outcome in combating viruses. Nanoparticles often refer to the invisible particulate material with a range between 1–100 nm in size that plays a significant role in modern science application (Song et al. 2021). This material is principally used as a potential viral and antiviral agent. This is mainly due to the advanced nanotechnology approach of using nano-sized materials in attacking the virus or bacteria. Nanoparticles proved to overwhelm the cell entry and viral imitation through their surface-binding sites effect that enables the relations with the target molecules concurrently incapacitating the viruses (Li et al. 2021). Hence, this chapter discusses various types of nanoparticle synthesis methods and their characterization for antiviral application.

3.2 SYNTHESIS OF NANOPARTICLES

There are two distinct types of synthesis of nanoparticles: top-down and bottom-up approach, as shown in Figure 3.1. The top-down technique is used when the bulk material breaks and creates the nanoparticles through various physical forces. These physical forces include crushing and grinding, ball milling, arc-discharge

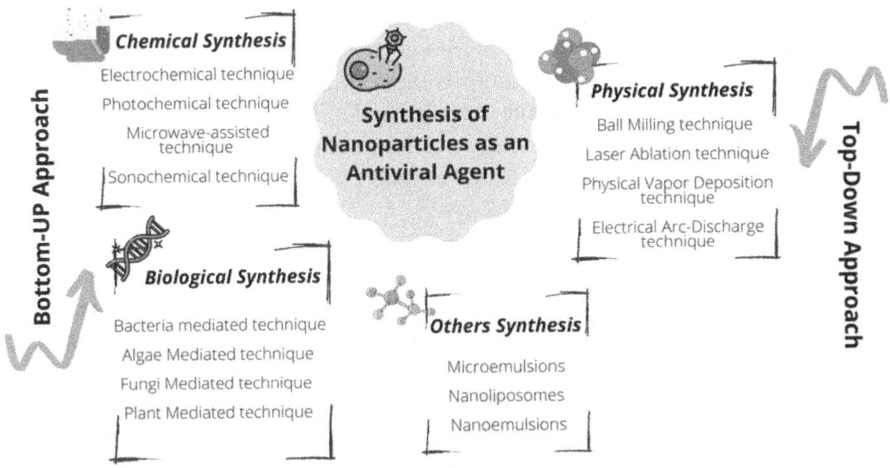

FIGURE 3.1 Different nanoparticles synthesis techniques for antiviral applications.

technique, laser ablation technique, and vapour condensation (Ju-Nam and Lead 2008). Through the top-down technique, the size of the nanoparticles can be controlled between 10 nm to 100 nm through the physical method. Meanwhile, the bottom-up technique is used to form multifaceted clusters to attain nanoparticles from the molecular components. This process is carried out through the nucleation and growth process. This approach is commonly used for chemical and biological synthesis through reduction of the precursor salt. The chemical synthesis typically utilizes alternative energies, such as sonochemical, microwave-assisted, electrochemical, and photochemical technique (Zheng et al. 2016; Chung et al. 2018; Kuntyi et al. 2019). The chemical synthesis route will apply many chemical additives, limiting the application as an antiviral agent. Therefore, a biological synthesis route that uses macromolecular substances such as bacteria, algae and fungi will possess minimum toxicity and be environmentally friendly.

3.2.1 Physical Method

Vapour-based and mechanical processes are the standard technique in the physical synthesis of nanoparticles. Various energies are required during the formation of nanoparticles through the physical route. There are light energy, mechanical energy, thermal energy, and electrical energy (Amendola and Meneghetti 2009; Khayati and Janghorban 2012). The physical synthesis route enables a massive production of nanoparticles; however, it leads to a high possibility of agglomeration that might affect the application as an antiviral agent. Therefore, a stabilizer such as polyvinyl pyrrolidone (PVP) shall be added to reduce the agglomeration during the synthesis process.

3.2.1.1 Ball Milling Technique

The ball milling method uses the rotation technique where milling balls and metal materials are added to a container with inert gas and subject to a high rotation speed.

The parameters such as the milling time, atmospheric medium, and rotating speed play a vital part in determining the morphology structure of the nanoparticles. In addition, the milling time directly controls the formation of the nanoparticles' size, while the phase formation is influenced by the ball milling powder temperature. It is reported that higher powder temperatures tend to result in an intermetallic nature, while lower powder temperatures result in amorphous and nanocrystalline phases (Xu et al. 2020).

3.2.1.2 Laser Ablation Technique

The laser ablation technique uses a pulsed laser that applies instantaneous heat in targeting a bulk metal material engrossed in organic solvent or water. This material then will undergo nucleation and particles growth through the plasma-cooling procedure. This process results in the formation of metal nanoparticles. The nanoparticles can absorb photons during the laser ablation process through plasmon excitation, multiphoton absorption, and interband transitions. During this process, the parameters that will influence the formation of nanoparticles are pulse timing, laser fluence, and laser wavelength. Moreover, the type of aqueous medium might disturb the formed nanoparticles' features, especially in terms of particle size. The inclusion of organic stabilizers like PVP and cetyltrimethylammonium bromide can improve the nanoparticles dispersibility (Diaz-Nunez et al. 2017).

3.2.1.3 Physical Vapour Deposition Technique

The physical vapour deposition technique is alienated into two different techniques of sputtering and arc evaporation. The arc evaporation technique utilizes a cathodic arc source in an inert atmosphere; the nanoparticles are then deposited onto a coating material. During this formation, the highly ionized nanoparticles vapour will produce plasma (Xu et al. 2020). Then, it uses high energy to enable the coating of the nanoparticles to take place. Hence, this technique, which requires complicated equipment and external energy, will form only a small control nanoparticle.

3.2.1.4 Electrical Arc-Discharge Technique

The electrical arc-discharge technique involves a direct current power source attached to two silver rods engrossed in dielectric liquids. Nanoparticles are formed when the silver electrode is etched, and the silver surface will be vaporized due to the high temperature near the electrode. Then, the silver vapour condensation forms nanoparticles suspended in the dielectric liquid (Tien et al. 2008).

3.2.2 Chemical Method

The chemical method has been found to be the most common technique in the synthesis of nanoparticles for antiviral agents. The chemical process involves the reduction of the precursor with the presence of energy. This technique requires the electrons to be transferred and reduced with the presence of a reducing agent. Various techniques use a chemical approach, including electrochemical, photochemical, microwave-assisted, and sonochemical. Using the chemical method in nanoparticles synthesis has been shown to result in a high yield with a controllable route. However, this process will cause environmental pollution due to the high usage of chemicals.

3.2.2.1 Electrochemical Technique

The electrochemical technique involves forming electric potential in the electrolyte to decrease metal nanoparticles (A^+) to (A^0). The formation of nanoparticles will occur instantaneously under the external electric field through the nucleation and growth process (Khaydarov et al. 2009). This technique allows the formation of different nanoparticle sizes by adjusting the electrolytes, current density, and solvent. Stabilizers and capping agents are added during this process to ensure an enhanced dispersibility. These stabilizers do hinder the aggregation of the nanoparticles during the formation process.

3.2.2.2 Photochemical Technique

The photochemical technique is a reduction method that utilizes light as the source of irradiation. When light is present, the precursors will undergo an excitation to generate free radicals responsible for reducing A^+ to A^0. For the sustainable approach, many researchers use sunlight as the light source. The light source intensity, wavelength, and irradiation time play an essential role in determining the particles' size (Zaarour et al. 2014). For example, increasing the irradiation time and intensity will result in an improved reduction process. This technique enables a yield of a high dispersible nanoparticle instantaneously. This technique also requires simple equipment and does not use any harmful reducing agents.

3.2.2.3 Microwave-Assisted Technique

The microwave-assisted technique involves the rapid heating of a precursor with microwave irradiation to activate the nuclei generation. The concentration of the precursor is important because it will influence the formation of the nanoparticles. In addition, the type of stabilizer, together with the irradiation time, will determine the phase and the nanoparticles' size. Due to the high dielectric losses, alcohol and water are the perfect media as stabilizers (Pal et al. 2009). This technique has the benefit of short irradiation time, high-energy conversion efficacy, and large-scale production.

3.2.2.4 Sonochemical Technique

The sonochemical technique denotes the cavitation effect resulting from the ultrasonic irradiation; this means that it generates a hot spot for the formation of nanoparticles. The huge pressure from the ultrasonic irradiation enables a uniform mix of solution and produces active bubbles. The gas phase adiabatic compression bubbles will augment the A^+ contact with the reducing agent. This will then reduce the A^+ to A^0 (Darroudi et al. 2012). This technique prevents the agglomeration of nanoparticles, which then causes the formation of smaller sized nanoparticles. The advantages of this technique are that it is simple, cost-effective, and environmentally friendly.

3.2.3 Biological Method

Recently, the biological-based synthesis technique has drawn significant attention in developing nanoparticles for antiviral purposes. This method can be done by applying either a microorganism or plant-based mediated technique. The presence of microorganisms can grow metal tolerance genes and decrease the cytotoxicity of the metal that resulted in the forming of metal nanoparticles. The plant-based mediated technique uses plant-organic functional clusters such as O–H and =C–H to reduce the A^+ to A^0. Therefore, these biological components serve as the reducing agent for the precursor to form nanoparticles. Compared to the physical and chemical techniques, the biological technique can be performed at any average temperature and pressure and avoids the toxic additives usage.

3.2.3.1 Bacteria Mediated Technique

The bacteria mediated technique in synthesizing nanoparticles is either intracellular or extracellular. This technique can be performed using cell-free extracts, derivative bacteria components, supernatant, and biomass. However, the extracellular method is beneficial compared to the intracellular method because of the simple recovery of the nanoparticles. The mechanism of each biosynthesis is different due to the different organic substances used as the reducing agent during the synthesis process. Various organic substances are used as reducing agents, such as reductase, peptides, c-type cytochromes, exopolysaccharide, and cofactors (Xu et al. 2020). Besides this technique, various enzymes have also been incorporated to synthesize nanoparticles, such as lactate dehydrogenase and nitrate reductase and peptides with special amino acid (Ali et al. 2019). Moreover, some organic substances can use a stabilizer and capping agents to prevent the nanoparticle from aggregating (Galvez et al. 2019). The bacteria mediated technique is simple, environmentally friendly, and yields high efficacy.

3.2.3.2 Algae Mediated Technique

The algae mediated technique is one of the most reliable techniques because algae is a renewable living resource. Algae contain diverse biological organic substances such as polysaccharides, proteins, carbohydrates, enzymes, vitamin, pigments, and secondary metabolites (Alassali et al. 2016). Thus, these organic-rich components make algae a promising material for the biosynthesis of nanoparticles as a reducing agent to formulate shape and size-controlled nanoparticles. Furthermore, the biomolecules in algae extracts, such as protein, amino acids, and sulphated polysaccharides, can also serve as capping agent and stabilizer (Aziz et al. 2015). Hence, algae mediated synthesis technique gives rise to a simple, sustainable, and green synthesis method for forming nanoparticles as an antiviral agent.

3.2.3.3 Fungi-Mediated Technique

The fungi-mediated technique can be referred to as intracellular and extracellular using mycelia and fungal cell-free filtrate. However, extracellular synthesis using fungi is ideal mainly due to the benefit of appropriate collection. Fungi such as *Penicillium polonicum*, *Trichoderma harzianum*, *Phomopsis liquidambaris*, and

Fusarium oxysporum are commonly chosen due to their exclusive properties of high acceptance in the metal-rich setting, multiple extracellular enzymes secretion, and their economical feasibility. The various organic substances of fungi contributed to the effective synthesis of nanoparticles by reducing the precursor. Furthermore, the presence of protein does add the advantages of fungi as the capping agents to control the formation of the nanoparticle size. Nevertheless, some fungi, such as *F. oxysporum,* is pathogenic, which is unsuitable for the fabrication of nanoparticles for antiviral purposes. Thus, the fungi-mediated technique is a cost-effective, energy-saving, and facile technique.

3.2.3.4 Plant-Mediated Technique

The plant-based technique is proven to be a promising technique in the amalgamation of nanoparticles. This is because the extracts of the diverse parts of the plants can be used as the substance for the biosynthesis of nanoparticles in different sizes and shapes (Kuppusamy et al. 2016). The extract from the different parts of the plant is rich in organic compounds, such as flavonoids, alcohols, alkaloids, quinines, enzymes, oils, phenolic group, and terpenoids (Rajeshkumar and Bharath 2017; Ovais et al. 2018). This resulted in the various functional groups present in these organic compounds, such as carbonyl, hydroxyl, and amidogen that activated the reduction of A^+ to A^0. However, the parameters that will influence the synthesis process are reaction time, temperature, pH, precursor, and concentration of the plant extract. Therefore, it is important to monitor these parameters to be able to synthesis the proper size and shape of the nanoparticles. Hence, this technique that uses plant extract possesses high economic benefit, simplicity, and nontoxicity in synthesizing the nanoparticles for antiviral application.

3.2.4 Other Synthesis Methods

3.2.4.1 Microemulsions

Microemulsions are a surfactant-based nanoparticles synthesis route designed for thermodynamically stable systems for antivirals agents. This technique uses the bottom-up approach where the colloidal is mixed with suitable forms and quantity of oil, water, and surfactant, and the appropriate conditions, for example, ionic strength, pH, and temperature. During this synthesis process, a designated quantity of energy was required to activate and overwhelm the kinetic energy barriers within the individual colloidal structures. For example, in the oil-in-water, microemulsions consist of trivial surfactant-rich nanoparticles disseminated within the water. These surfactants are organized to ensure that the hydrophilic head groups are successfully interact with water.

Meanwhile, a nonpolar core in the particles' interior will be formed by the hydrophobic tails. Furthermore, this synthesis route incorporates hydrophobic antivirals into their cores while the amphiphilic antimicrobials are easily formed into their interfacial sections. Next, as for the hydrophilic antivirals' purposes, it can be encapsulated using the water-in-oil type microemulsions where the surfactant-based nanoparticles are disseminated in oil. Therefore, the hydrophilic head groups are forming the core while the hydrophobic tails form the shell.

Many studies have proved the ability to use the microemulsions technique to produce nanoparticles as an antiviral agent. Shishu and Kamalpreet (2009) reported using a water-oil microemulsion technique containing acyclovir as a hydrophilic antiviral agent formed from Tween 20, Span 20. This acted as a surfactant, together with isopropyl myristate. The medium-chain triglycerides acted as oil while mixing with water/dimethylsulfoxide. The formulated nanoparticles show that optimized microemulsions can suppress mice skin lesions triggered by herpes simplex virus I infection (Shishu and Kamalpreet 2009). In addition, the report found that using olive leaves extract as an antiviral agent was able to tackle the herpes simplex virus by capturing it within oil-water based microemulsions. This was possible due to the ability of olive leaves' extract to hinder viral attachment to the host cell receptors (Khattab et al. 2016).

3.2.4.2 Nanoliposomes

Nanoliposomes have been extensively used as an antiviral agent for many applications. The small size nanoparticles synthesis can be divided into two categories resulting from the characteristics of their internal structures: single-ring vesicles (unilamellar) and onion-ring structure (multilamellar). The inner region is formed by the bilayer of phospholipid molecules in each ring and their nonpolar tails, while the outer layers are formed by the hydrophilic head group (Mozafari et al. 2008). Nanoliposomes were first prepared by dissolving the phospholipids in organic solvent and drying them to produce phospholipid bilayers and form nanoliposomes. It can also be incorporated with mechanical dispersion methods, which are appropriate for the large-scale creation of nanoliposomes. Nanoliposomes can distribute both lipophilic and hydrophilic antivirals, mainly due to containing nonpolar and polar areas within the assemblies of the bilayer. The report has found that using the nanoliposome method can overcome the common antiviral agent limitation, such as low permeability thru cell membranes and weak cell internalization.

Similarly, PEGylated nanoliposomes were synthesized in targeting in transfer of interferon alpha-2b (IFN α-2b) to prevent the human papillomavirus (HPV). This study exposed that (IFN α-2b) encumbered with nanoliposomes enhanced the dissemination thru sheep vaginal tissue. This phenomenon was attributed to PEG molecules that hinder the adhesive connections of the nanoliposomes with mucus (Joraholmen et al. 2017).

3.2.4.3 Nanoemulsions

Nanoemulsions differ from microemulsions, which have thermodynamically uneven colloidal distributions. These are designed from water, oil, and emulsifier together with stabilizers. It can be formulated in the same way as oil-in-water (O/W) and water-in-oil (W/O) types. Furthermore, the additional complicated approach includes oil-in-water-in-oil (O/W/O) or water-in-oil-in-water (W/O/W) nanoemulsions. Nanoemulsions are classically formed using mechanical strategies such as homogenizers, high-pressure valve homogenizers, sonicators, and microfluidizers. These devices' function is to activate the intense, disruptive forces that can easily break up the mixture to form tiny droplets. This technique proved its ability by showing an improved efficiency as a newsworthy antiviral invention. The antiviral activity of coumestrol, a compound in soybeans and alfalfa, is used against Herpes Simplex Virus. This formulation comprised coumestrol-incoporated with oil droplets disseminated inside a hydrogel matrix (Argenta et al. 2018). Similarly, the encapsulation of curcumin through the O/W nanoemulsions technique improved its efficacy for interesting treatment of lesions, which causes human papillomavirus impurity (Bonfim et al. 2020). Another study proved the ability to adopt an O/W nanoemulsion technique to deliver antiviral agents that improved the efficacy by incorporating the lipoproteins that convey hydrophobic substances within the bloodstream (Prabhakar et al. 2013).

3.3 CHARACTERIZATION

3.3.1 Transmission Electron Microscopy

Transmission electron microscopy (TEM) is a vital analysis method to apprehend the fine details of the morphology structure of the nanoparticles. This method can be performed to the smallest size, nanosized particles. Therefore, TEM is an important tool to portray the virucidal conditions with vibrant images on the binding between the viruses and the nanoparticles (Kennedy et al. 2016). The TEM analysis can provide a clear visual interpretation of the infested cells and reveal the viricidal capability of the nanoparticles. Besides that, the unstained specimens can be observed through Cryo-TEM, which indicates the actual nature of the specimens.

3.3.2 Confocal Imaging

Confocal imaging characterization is used to detect the host cells' deviations when infested by viruses (Alonas et al. 2014). This powerful imaging technique helps find out the development of the viral contamination in a host cell. Similarly, this imaging can perform the indirect immunofluorescence analysis to show the changes of the virus protein structure before and after the treatment by using nanoparticles.

3.3.3 Real-Time Polymerase Chain Reaction

This is a quantitative polymerase chain reaction (PCR) characterization mainly based on the molecular biology knowledge on a polymerase chain reaction. It is used to display the intensification of a besieged DNA in the polymerase chain reaction process. This characterization can also be used to distinguish the virus DNA after contacting the nanoparticles (Gall et al. 2015). The PCR characterization can evidently show the evolution and the digestion status of the host cells before and subsequent to the treatment process on the virus infection. This was done by detecting gene appearance of host cells as a clear gauge of viral contagion. Hence, PCR analysis strongly indicate the antiviral ability when in contact with nanoparticles.

3.3.4 Plaque Assay

This characterization is used to detect the virion absorption ability and monitor the outcome of nanoparticles on the virion (Rigotto et al. 2011). The plaque assessment calculates the infectivity of the virus and the nanoparticle's ability as an antiviral agent. As for a typical plaque assessment, the virus is diluted 10-fold, and 0.1 mL aliquot diluted solution is injected onto the monolayers of vulnerable cells. Then the virus will be attached to the cells after the incubation period, where the monolayers are enclosed with a nutrient medium jelly. Subsequent to a designated incubation period, the original nutrient cells will proclaim the virus progeny, and this gel will restrict the spread to the neighbouring cells. Therefore, it will create a zone of infested cells called the plague, and it will grow more prominent in size, noticeable to the bare eye under normal temperature environments. Hence, the titre of a virus standard can be estimated from the plague forming units (PFU) per millilitre that indicate the actual viral infectivity value. The PFU values give rise to the viricidal ability before and after the binding with binding nanoparticles.

3.3.5 β-Galactosidase

This assessment is to detect the virus entry into any specimen. It can hydrolyze the β-glycosidic bond between the galactose and its organic portion can be hydrolyze thru exoglycosidase. This is vital for the organisms as it provides energy and carbon basis through the interruption of lactose into galactose and glucose (Yin et al. 2017). Therefore, it uses as a reporter gene to evaluate the expression level in the host cell, where it delivers indirect indication for the capability of nanoparticles treatment in treating the virus infection.

3.3.6 Flow Cytometry

This is a unique technique in sensing the physical and chemical appearances of the cell particles' population. It can be utilized to sort the different cell sizes and simultaneously monitor their properties and functions. In this investigation, massive quantities of cells will be counted and scrutinized, with the computer gathering and processing its data. Hence, the computer will count the virus-infected cells and evaluate the removal efficacy of the nanoparticles.

3.3.7 Western Blot

This is an individual virus qualitative detection together with its modification. It can classify the degradation percentage of the virus due to the nanoparticle treatment. First, the sample will undergo protein denaturation before proceeding with the electrophoresis analysis and forming a primary antibody. This was produced via an artificial or animal-derived path. Then, the primary antibody solution is used to wash the electrophoresis membrane before adding the secondary antibody. This antibody identifies and binds the primary antibody. The preparation of the secondary antibody can be imagined through immunofluorescence, staining, and permitting indirect discovery of a targeted protein. This will show the variations in the protein structure due to the antiviral effect of the presence of nanoparticles.

3.3.8 Computer Simulation

The advancement of technology has created a new pathway to computer simulation study to reproduce chemical and biological behaviour. As for antiviral application, it is essential to estimate the dissimilarities in a definite DNA and protein site thru the interaction between the nanoparticles and the virus. This will provide strong justification and evidence on predicting the virus transformation and host cell after the treatment with nanoparticles (Capuzzi et al. 2018).

3.4 CONCLUSION AND PROSPECTS

In summary, different characteristics of nanoparticles for the antiviral application can be synthesis thru physical, chemical, biological, and other methods. Numerous aspects should be considered in selecting the appropriate synthesis method, including the anticipated size and shape and the external energy contribution like heat, electricity, microwave, and light. In addition, the type of additives and precursor salts as a reducing agent, capping agent, and stabilizer are essential parameters that will distress the synthesis product outcome in achieving a maximum performance of the nanoparticles for antiviral application. Besides that, synthesis constraints like temperature, pH, reaction time, and precursor concentration play a part in the synthesis process. The way forward, the biological synthesis route that uses fungi, bacteria, and plant extract verifies to a modest and green environment approach that promotes the green synthesis concept where it does not severely harm the environment associated with the physical and chemical amalgamation route. It is cost-effective and does not require expensive equipment and massive energy sources, which makes it a more reliable synthesis method for large-scale production. This chapter also discussed the different types of characterization technique that can be applied to elucidate the synthesized nanoparticle interaction behaviour with the viruses. These techniques will be able to justify the performance efficacy of the nanoparticles as an antiviral agent. However, some important factors need to be considered for future approach, especially on the toxicity level of the nanoparticles. It is understood that the low amount of nanoparticle usage is safe, but it is worth conducting a further study on the toxicity level. Besides that, the biocompatibility can be further enhanced and improved as the current study reveals that nanoparticles' dissolubility declines the discharging efficiency of the antiviral drugs. Next, further reserve the nanoparticles antiviral activity during the viruses binding. Finally, an in-depth study should be performed to expand the virus mutation target range and improve the structure of the nanoparticles to strive with any form of new virus or mutation. Hence, this will be a novel breakthrough for utilizing nanoparticles technology to combat the active formation of new viruses.

REFERENCES

Alassali, A., Cybulska, I., Brudecki, G.P., Farzanah, R. and Thomsen, M.H. 2016. Methods for upstream extraction and chemical characterization of secondary metabolites from algae biomass. *Advanced Techniques in Biology and Medicine* 4:1–16. doi: 10.4172/2379-1764.1000163.

Ali, J., Ali, N., Wang, L., Waseem, H. and Pan, G. 2019. Revisiting the mechanistic pathways for bacterial mediated synthesis of noble metal nanoparticles. *Journal of Microbiological Methods* 159:18–25. doi:10.1016/j.mimet.2019.02.010

Alonas, E., Lifland, A.W., Gudheti, M., et al. 2014. Combining single RNA sensitive probes with subdiffraction-limited and live-cell imaging enables the characterization of virus dynamics in cells. *ACS Nano* 8:302–315. doi:10.1021/nn405998v

Amendola, V. and Meneghetti, M. 2009. Laser ablation synthesis in solution and size manipulation of noble metal nanoparticles. *Physical Chemistry Chemical Physics* 11: 3805–3821. doi:10.1039/B900654K

Argenta, D.F., Bidone, J., Koester, L.S., Bassani, V.L., Simoes, C.M.O. and Teixeira, H.F. 2018. Topical delivery of coumestrol from lipid nanoemulsions thickened with hydroxyethylcellulose for antiherpes treatment. *AAPS PharmSciTech* 19:192–200. doi: 10.1208/s12249-017-0828-8

Aziz, N., Faraz, M., Pandey, R., et al. 2015. Facile algae-derived route to biogenic silver nanoparticles: Synthesis, antibacterial, and photocatalytic properties. *Langmuir* 31: 11605–11612. doi: 10.1021/acs.langmuir.5b03081

Bonfim, C.M., Monteleoni, L.F., Calmon, M.F., Candido, N.M., Provazzi, P.J.S. and Lino, V.S. 2020. Antiviral activity of curcumin-nanoemulsion associated with photodynamic therapy in vulvar cell lines transducing different variants of HPV-16. *Artificial Cells, Nanomedicine, and Biotechnology* 48:515–524. doi:10.1080/21691401.2020.1725023

Capuzzi, S.J., Sun, W., Muratov, E.N., et al. 2018. Computer-aided discovery and characterization of novel Ebola virus inhibitors. *Journal of Medicinal Chemistry* 61: 3582–3594. doi:10.1021/acs.jmedchem.8b00035

Chaudhuri, S., Symons, J.A. and Deval, J. 2018. Innovation and trends in the development and approval of antivirus medicines: 1987-2017 and beyond. *Antiviral Research* 155:76–88. doi:10.1016/j.antiviral.2018.05.005

Chen, L., Zhang, X.W., Zhou, G.H., et al. 2012. Simultaneous determination of human enterovirus 71 and coxsackievirus B3 by dual-color quantum dots and homogeneous immunoassay. *Analytical Chemistry* 84:3200–3207. doi:10.1021/ac203172x

Chung, D.S., Kim, H., Ko, J., et al. 2018. Microwave synthesis of silver nanoparticles using different pentose carbohydrates as reducing agents. *Journal of Chemistry and Chemical Engineering* 12:1–10. doi:10.17265/1934-7375/2018.01.001

Darroudi, M., Zak, A.K., Muhamad, M., Huang, N. and Hakimi, M. 2012. Green synthesis of colloidal silver nanoparticles by sonochemical method. *Materials Letters* 66:117–120. doi:10.1016/j.matlet.2011.08.016

Delshadi, R., Bahrami, A., McClements, D.V., Moore, M.D. and Williams, L. 2021. Development of nanoparticle-delivery systems for antiviral agents: A review. *Journal of Controlled Release* 331:30–44. doi:10.1016/j.jconrel.2021.01.017

Diaz-Nunez, P., Gonzalez-Izquierdo, J., Gonzalez-Rubio, G., et al. 2017. Effect of organic stabilizers on silver nanoparticles fabricated by femtosecond pulsed laser ablation. *Applied Sciences* 7:793. doi:10.3390/app7080793

Gall, A.M., Shister, J.L. and Marinas, B.J. 2015. Analysis of the viral replication cycle of adenovirus serotype 2 after inactivation by free chlorine. *Environmental Science Technology* 49:4584–4590. doi:10.1021/acs.est.5b00301

Galvez, A.M., Ramos, K.M., Teja, A.J. and Baculi, R. 2019. Bacterial exopolysaccharide mediated synthesis of silver nanoparticles and their application on bacterial biofilms. *Journal of Microbiology, Biotechnology and Food Sciences* 9:970–978. doi:10.15414/jmbfs.2019.8.4.970-978

Joraholmen, M.W., Basnet, P., Acharya, G. and Skalko-Basnet, N. 2017. PEGylated liposomes for topical vaginal therapy improve delivery of interferon alpha. *European Journal of Pharmaceutics and Biopharmaceutics* 113:132–139. doi:10.1016/j.ejpb.2016.12.029

Ju-Nam, Y. and Lead, J.R. 2008. Manufactured nanoparticles: An overview of their chemistry, interactions and potential environmental implications. *Science of the Total Environment* 400:396–414. doi:10.1016/j.scitotenv.2008.06.042

Kennedy, E., Nelson, E.M., Tanaka, T., Damiano, J. and Timp, G. 2016. Live bacterial physiology visualized with 5 mm resolution using scanning transmission electron microscopy. *ACS Nano* 10:2669–2677. doi:10.1021/acsnano.5b07697

Khattab, R.A., Hosny, A.E.D.S., Abdelkawy, M.A., Fahmy, R.H. and Elmenoufy, N.A. 2016. Anti-HSV type-1 activity of olive leaves extract crude form acting as a microemulsion dosage form. *African Journal of Microbiology Research* 10(22):820–828. doi:10.5897/AJMR2016.7972.

Khayati, G. and Janghorban, K. 2012. The nanostructure evolution of Ag powder synthesized by high energy ball milling. *Advanced Powder Technology* 23:393–397. doi:10.1016/j.apt.2011.05.005

Khaydarov, R.A., Khaydarov, R.R., Gapurova, O., Estrin, Y. and Scheper, T. 2009. Electrochemical method for the synthesis of silver nanoparticles. *Journal of Nanoparticle Research* 11:1193–1200. doi:10.1007/s11051-008-9513-x

Kuntyi, O., Kytsya, A., Mertsalo, I., et al. 2019. Electrochemical synthesis of silver nanoparticles by reversible current in solutions of sodium polyacrylate. *Colloid and Polymer Science* 297:689–695. doi:10.1007/s00396-019-04488-4

Kuppusamy, P., Yusoff, M.M., Maniam, G.P. and Govindan, N. 2016. Biosynthesis of metallic nanoparticles using plant derivatives and their new avenues in pharmacological applications - An updated report. *Saudi Pharmaceutical Journal* 24:473–484. doi:10.1016/j.jsps.2014.11.013

Li, Y., Xiao, Y., Chen, Y. and Huang, K. 2021. Nano-based approaches in the development of antiviral agents and vaccines. *Life Sciences* 265:118761. doi:10.1016/j.lfs.2020.118761.

Luo, K., Jung, S., Park, K.H. and Kim, Y.R. 2018. Microbial biosynthesis of silver nanoparticles in different culture media. *Journal Agricultural and Food Chemistry* 66:957–962. doi:10.1021/acs.jafc.7b05092

Mozafari, M.R., Khosravi-Darani, K., Borazan, G.G., Cui, J., Pardakhty, A. and Yurdugul, S. 2008. Encapsulation of food ingredients using nanoliposome technology. *International Journal of Food Properties* 11(4):833–844. doi:10.1080/10942910701648115

Ovais, M., Khalil, A.T., Islam, N.U., et al. 2018. Role of plant phytochemicals and microbial enzymes in biosynthesis of metallic nanoparticles. *Applied Microbiology and Biotechnology* 102:6799–6814. doi:10.1007/s00253-018-9146-7

Pal, A., Shah, S. and Devi, S. 2009. Microwave-assisted synthesis of silver nanoparticles using ethanol as a reducing agent. *Materials Chemistry and Physics* 114:530–532. doi:10.1016/j.matchemphys.2008.11.056

Prabhakar, K., Afzal, S.M., Surender, G. and Kishan, V. 2013. Tween 80 containing lipid nanoemulsions for delivery of indinavir to brain. *Acta Pharmaceutica Sinica B* 3:345–353. doi:10.1016/j.apsb.2013.08.001

Rajeshkumar, S. and Bharath, L. 2017. Mechanism of plant-mediated synthesis of silver nanoparticles-A review on biomolecules involved, characterisation and antibacterial activity. *Chemico-Biological Interactions* 273:219–227. doi:10.1016/j.cbi.2017.06.019

Rigotto, C., Hanley, K., Rochelle, P.A., Leon, R.D., Barardi, C.R.M. and Yates, M.V. 2011. Survival of adenovirus types 2 and 41 in surface and ground waters measured by a plaque assay. *Environmental Science Technology* 45:4145–4150. doi:10.1021/es103922r

Shishu, Rajan, S. and Kamalpreet. 2009. Development of novel microemulsion-based topical formulations of acyclovir for the treatment of cutaneous herpetic infections. *AAPS PharmSciTech* 10(2):559–565. doi:10.1208/s12249-009-9242-1

Song, X., Bayati, P., Gupta, M., Elahinia, M. and Haghshenas, M. 2021. Fracture of magnesium matrix nanocomposites - A review. *International Journal of Lightweight Materials and Manufacture* 4:67–98. doi:10.1016/j.ijlmm.2020.07.002

Tien, D.-C., Tseng, K.-H., Liao, C.-Y., Huang, J.-C. and Tsung, T.-T. 2008. Discovery of ionic silver in silver nanoparticles suspension fabricated by arc discharge method. *Journal of Alloys and Compounds* 463:408–411. doi:10.1016/j.jallcom.2007.09.048

Xu, L., Wang, Y.Y., Huang, J., Chen, C.Y., Wang, Z.X. and Xie, H. 2020. Silver nanoparticles: Synthesis, medical applications and biosafety. *Theranostics* 10(20): 8996–9031. doi:/10.7150/thno.45413.

Yin, H.F., Pijning, T., Meng, X.F., Dijkhuizen, L. and Leeuwen, S.S.V. 2017. Engineering of the *Bacillus circulans* β-galactosidase product specificity. *Biochemistry* 56:704–711. doi:10.1021/acs.biochem.7b00032

Zaarour, M., El Roz, M., Dong, B., et al. 2014. Photochemical preparation of silver nanoparticles supported on zeolite crystals. *Langmuir* 30:6250–6256. doi:10.1021/la5006743

Zheng, X., Peng, Y., Lombardi, J.R., Cui, X. and Zheng, W. 2016. Photochemical growth of silver nanoparticles with mixed-light irradiation. *Colloid and Polymer Science* 294:911–916. doi:10.1007/s00396-016-3846-2

4 Nanoparticle Synthesis and Administration Routes for Antiviral Uses

João Augusto Oshiro-Júnior,
Kammila Martins Nicolau Costa, Isadora Frigieri,
and Bruna Galdorfini Chiari-Andréo

CONTENTS

4.1 INTRODUCTION

The study of nanomaterials, elements whose particle sizes are found in the nanometre range, gained significant importance in the late 20th Century, leading to the emergence and consolidation of an entire area of knowledge, now recognized as "Nanoscience and Nanotechnology." These sciences are based on the fact that the properties of materials are strongly dependent on their particle size (Zarbin 2007).

Based on this observation, it is possible to understand that all material properties (optical, electrical, magnetic, transport, catalytic, etc.), manifest themselves from a certain size, called critical. When the particles of some materials are below this critical size, this property becomes differentiated, as in the case of nanomaterials (Zarbin 2007).

For the same material, the critical size is often different for each of its properties. This means that one property could be obtained in a determined critical size, for example, 20 nm, which is different from the critical size to obtain another property,

DOI: 10.1201/9781003136644-5

for example 100 nm. On the other hand, for the same property, the critical size is different for different materials. Also, the properties of the material depend on the shape of the particles; that is, spherical nanoparticles with a diameter of 5 nm have different properties from those found for the same material, but with stick-shaped nanoparticles 5 nm in length (Zarbin 2007). Thus, the particle size and shape should be adjusted according to the aims or needs.

The development of nanomaterials has been used lately as a tool in several areas of science, including biological applications, such as in the pharmaceutical and medical fields. Nanomaterials can help in the elimination or softening of difficulties encountered in the development of new drugs and medicines, as insufficient drug concentration due to unsuitable absorption, rapid metabolism, and elimination; distribution of drugs to tissues different from the target tissue; high drug toxicity; low drug solubility; high variation in plasma levels, and inactivation after oral administration. To overcome these obstacles to high safety and efficiency, the most diverse nanomaterials can be used, including zero-dimensional nanomaterials, as is the case of nanoparticles; one-dimensional nanomaterials, as is the case of nanotubes; and, also, the two-dimensional ones, as the nanofilms (Mehnert and Mäder 2012; Abd Elkodous et al. 2019; Mendonça 2020). In addition to the previously mentioned difficulties, all of these properties of nanomaterials can be employed to improve antiviral activity of molecules, a current subject of wide interest, considering the current pandemic situation.

Antivirals were developed basically to inhibit the virus genome replication within infected cells, thus preventing the production of new virus particles. The use of nanomaterials has proven to be useful for ameliorating the pharmacological properties of inhibitors of viral genome replication, including improvement in the circulation time of these substances impacting the effectiveness time. Therefore, the use of nanomaterials is advantageous in helping to treat existing viral infections and to prevent these infections, being finally a subject to be explored (Gelman and Glenn 2011; Jackman et al. 2016, 2020).

These nanomaterials for inhibiting viral replication act, more specifically, inhibiting the multiplication of viral particles; promoting the attachment of the virus to the nanoparticles, thus cancelling infectivity in some way; preventing infection of cells by means of cell membrane baits that bind to virus particles; by means of extracorporeal blood filters that can remove virus particles circulating from the bloodstream; producing biomimetic nanoparticle vaccines that mimic the multivalent presentation of antigens on the surfaces of viral particles, inducing improved immune responses for greater protection against future exposure to the virus (Jackman et al. 2020).

Between the nanomaterials, nanoparticles are a kind of nanosystem that have shown relevant properties such as the controlled and localized action due to the ability of targeting the release of active substances (Mehnert and Mäder 2012; Abd Elkodous et al. 2019; Mendonça 2020). Thus, this chapter discusses the nanoparticles employment, including synthesis and characterization of them, to improve biological active substances' properties, especially drugs for antiviral use.

4.2 OVERVIEW ABOUT NANOPARTICLES

Considering the aim of this discussion, it's helpful to first remember that the action against the virus can be done by using two mechanisms: virostatic and/or virucidal substances. A valid strategy in the use of nanoscale systems is the combination of both properties. Considering the nanoscale, a variety of systems have been used, and some of them are briefly described below.

A first example are the Cyclodextrins, nanoparticles that, according to the literature, are composed by D-(1)-glycopyranose units linked by α-1,4 bonds. Naturally formed cyclodextrins can be obtained by the bacterial enzymatic degradation of starch, being classified according to the units of glycopyranose, which composes its cavities (Andreaus et al. 2010; Chiari-Andréo et al. 2019). The advantages of theses nanoparticles are based on the protection of the active substance on its cavities, biocompatibility and alteration of solubility properties (Moya-Ortega et al. 2012).

Dendrimers, another example, are spherical hyperbranched nanoparticles with cavities in which a drug can be inserted (Manaia et al. 2017; Chiari-Andréo et al. 2020).

Liposomes are reservoir systems, characterized by self-forming vesicles with an aqueous content surrounded by one or more phospholipid bilayers. Phosphatidylcholine and phosphatidylserine are examples of phospholipids that can be employed. The drug, when aqueous, could be incorporated in the core, however, when of lipid nature, can be inserted on the phospholipids' membrane (Manaia et al. 2017; Chiari-Andréo et al. 2020).

A protease inhibitor used for the treatment of HIV, indinavir, was incorporated into liposomes for the purpose of vectoring the drug to lymphoid organs, and the results showed that they were not only able to accomplish this, but they also maintained their activity by conferring protection for the molecule (Das Neves et al. 2010). Zidovudine has also been the subject of studies involving liposomes, where an increase in plasma concentrations of about 12 was observed when compared to the drug in its commonly used dosage form (Das Neves et al. 2010).

Micelles are also a self-assembling reservoir system, similar to liposomes; however, they have a hydrophobic content delimited by a membrane composed of phospholipids and surfactants (Chiari-Andréo et al. 2020).

Nanofilms are nanometre-scale films that offer various properties to the surfaces where they are applied, such as anticoagulant and antioxidant. The molecules used to form the nanofilms must have nanostructural characteristics that allow their use (Decher et al. 1992a, b, 1994; Faria-Tischer and Tischer, 2012; Medeiros et al. 2013).

Polymeric nanoparticles include drug-carrier systems called nanocapsules and nanospheres. Nanocapsules are systems responsible for the delivery of drugs found inside an oily or aqueous cavity, surrounded by a polymeric membrane, or adsorbed on that membrane (forming like a shell, i.e. reservoir). The nanospheres are composed of a polymeric matrix, where the drugs are dispersed or adsorbed (matricial system, without a reservoir). These two types of carriers differ from each other, therefore, due to their composition and structural organization (Schaffazick et al. 2003; Durán et al. 2006; Santos and Fialho 2007; Oliveira 2009; Melo et al. 2010).

In the case of nanocapsules, it can allow, when it is necessary, to carry different incompatible drugs, keeping one of them in the core and the another in the shell (Chiari-Andréo et al. 2019).

Nanocapsules and microcapsules receive different denominations based on their size. This is because nanocapsules are not able to reach the size of micrometres, ideally between 100 and 500 nm (Daudt et al. 2013).

Solid lipid nanoparticles are used for the incorporation of lipophilic substances. They are a solid hydrophobic core of biocompatible lipids that are solid at room and body temperature. The drugs to be carried by these nanoparticles are dissolved or dispersed in the lipophilic matrix (Manaia et al. 2017; Chiari-Andréo et al. 2019). Lipid nanoparticles are incorporated with lopinavir, tenofovir, and ritonavir for the purpose of a synergistic activity. Varying the formulation components, an increase in activity against HIV of 3 and 30 times greater was observed than when compared with its free forms (Walvekar et al. 2019).

4.3 SYNTHESIS OF DIFFERENT NANOPARTICLES

Considering the synthesis of organic nanoparticles, structures ranging from 10–1000 nm can be formed. The choice of the suitable technique and parameters, such as stirring, temperature, concentration of components, are able to directly affect the nanoparticles' features and biological activity. In the development, the evaluation of several conditions is indispensable to reach the most suitable state.

In general, the colloidal solution of nanoparticles is formed by an organic phase where the bioactive compound is dispersed or dissolved together with the matrix, which can be polymeric or lipid and will be chosen based on the physicochemical characteristics of the compound to be encapsulated. This organic phase will be added in an aqueous phase (FA) composed of water and surfactant, giving rise to an oil-in-water emulsion (Crucho and Barros 2017; Li et al. 2017). Being more specific, in the case of cyclodextrins, they can be synthesized using enzymes, glycosyltransferases, or acting in polysaccharides such as starch. The polysaccharides act as precursors of cyclic oligosaccharides (Venturini et al. 2008).

Kulkarni et al. (2011) describes the liposomes' obtaining based on hydration of the phospholipid, followed by the design of the liposome, finalized by the removal of the drug not incorporated. The design, as previously mentioned, occurs by the self-assembling of phospholipids in contact with a hydrophilic medium. Beyond the conventional method, based on rehydration in an aqueous solvent, of a thin lipid film formed by evaporating the organic solvent used to solubilize the lipid, Manaia et al. (2017) list other techniques: the reverse-phase evaporation, the injection of phospholipids dissolved in an organic phase into a drug-containing aqueous phase, the detergent dialysis, microfluidic hydrodynamic focusing (MHF), supercritical reverse-phase evaporation (SRPE), electroformation, and microfluid liposome formation.

The production of micelles is based on the critical micellar concentration (CMC), which should be reached by phospholipids and polymers in water media to the formation of the hydrophobic core and the hydrophilic shell (Manaia et al. 2017). To produce nanofilms, the Layer-by-Layer (LbL) methodology stands out, created by Decher (1997). Basically, this methodology consists of immersing a substrate in cationic and anionic solutions, which causes an electrostatic self-assembly. For this purpose, polyelectrolytes, polymers that contain ionic groups in their repetitive units, are used. These ionic groups in aqueous solution are partially ionized, and thus, different charge densities in the polymer are obtained (Decher et al. 1992a,b, 1994; Faria-Tischer and Tischer 2012; Medeiros et al. 2013).

Polymeric nanoparticles (nanocapsules and nanospheres), in turn, can be prepared by *in situ* polymerization, nanoprecipitation of preformed polymers, emulsification-diffusion, and emulsification-solvent evaporation methods (Chiari-Andréo et al. 2019). According to Guterres et al. (2007), the difference between the obtaining of nanocapsules and nanospheres is based in the use of an oily substance, which is used for nanoparticles' preparation, but is not in the nanospheres preparation.

A variety of polymers are available for its use, including synthetic, semisynthetic, and natural ones. Important examples that can be cited are: chitosan, cellulose derivatives, polyglycolic acid, and poly(lactic-co-glycolic acid) (PLGA) (Pillai and Panchagnula 2001; Schaffazick et al. 2003).

In the production of solid lipid nanoparticles, the lipids that will be chosen for the production should be a raw material solid at room and body temperature; such lipids include monoglycerides, diglycerides, triglycerides, fatty acids, steroids, and waxes (Li et al. 2013; Tapeinos et al. 2017; Gordillo-Galeano and Mora-Huertas 2018). The techniques that describe the synthesis are hot and cold homogenization, high-pressure homogenization, microemulsion, and nanoprecipitation methods (Manaia et al. 2017).

4.4 ADMINISTRATION ROUTES OF NANOPARTICLES FOR ANTIVIRAL USE

As with other forms of medicines presentation, the use of nanoparticles addresses advantages and challenges that are currently known related to the stages of release, absorption, distribution, metabolism, and excretion.

The oral route of administration is the most commonly used in the ingestion of medications because it is a more comfortable means for the patient; it's versatile, with no pain, and possible to be administered by the patient himself (self-administered) (Sastry et al. 2000; Fuhrmann and Fuhrmann 2017; Fan et al. 2018). However, it is contraindicated for patients who have difficulty or are unable to swallow (dysphagia) the pharmaceutical form (Sastry et al. 2000), in addition to those who have any inconvenience present in the digestive tract and may develop nausea and vomiting. In addition, the presence of food can interfere with the speed of absorption of the drug (Fleisher et al. 1999). Aiming at these disadvantages, it is understood that drugs administered orally are not applicable for emergencies due to the numerous barriers that must be crossed and, consequently, the drugs' slow absorption (Shreya et al. 2018; Homayun et al. 2019).

Every drug administered orally goes through a series of pathways until it is completely absorbed. The drug is initially administered through the mouth, and it ultimately reaches the mesenteric veins and then the hepatic portal vein; then it is transported directly to the liver, which normally metabolizes a large part of the medication, launching the active drug to reach systemic circulation, but with a reduced concentration range. This process is called first-pass or pre-systemic metabolism, where a small portion of the medication is eliminated by the organ responsible for its elimination. This process determines that the oral route has a lower bioavailability when compared to an intravenous administration (Ahmad et al. 1984).

Other factors that interfere with the absorption of the drug are its degree of ionization and solubility and, also, the disintegration time related to the excipients; thus, many high-potency and important drugs, such as insulin, are not able to pass through the gastrointestinal tract. Currently, many drugs have a low degree of solubility in the intestine, causing the elimination of a large portion of the drug and compromising its bioavailability; this could result in developing a need for the application of high doses to the patient (Palacio et al. 2016; Nielsen et al. 2018; Shreya et al. 2018).

Considering the disadvantages presented by compounds and formulations orally administered, nanotechnology could help circumvent these obstacles. For oral administration, liposomes, cyclodextrins, solid lipid nanoparticles, nanostructured lipid carriers, polymeric micelles, nanoparticles and nanocrystals could ameliorate the properties.

For liposomes, the oral route can be considered a preferred route of administration since they have a tendency to accumulate in the liver tissue, representing a promising drug-delivery system with a liver target, such as drugs for the treatment of hepatitis C (Lembo et al. 2018). Regarding cyclodextrins, which exhibit a high range of biological and physical-chemical properties, they are being studied and manipulated to fit the therapeutic needs attributed to oral administration. Studies also indicate an increased ability to incorporate high drug content, a controlled and targeted drug-release mechanism, gastro-stability and low toxicity (Adeoye and Cabral-Marques 2017).

Lamivudine, an oral agent indicated for the treatment of infection caused by the human immunodeficiency virus (HIV), presented drug resistance associated with the mutation of the RT gene, thus limiting its clinical application. The study carried out by Grande et al. (2019) demonstrated that several nanocarriers synthesized with biodegradable networks; dendritic networks or nanogels were able to direct the delivery of the drug to its site of action, demonstrating increased effectiveness in inhibiting HIV and requiring a low drug concentration, consequently decreasing the toxicity of the molecule and adverse reactions.

In the same study, the authors cited that PLGA nanoparticles coated with bovine serum albumin were considered a promising delivery system. In addition, tenofovir, used for the treatment of chronic hepatitis B (HBV) and also for the treatment and prevention of HIV, was dispersed in lipid nanoparticles in order to establish a long-term release of the drug and improve bioavailability (Grande et al. 2019).

Two other widely used antivirals that act as HIV protease inhibitors are ritonavir and saquinavir. Both could be incorporated into self-emulsifying drug-delivery systems (SNEDDS) to promote several advantages to the drugs, such as reduction in the frequency of dosage and dose reduction, since the incorporation in the innovative system demonstrated an increase in its bioavailability and better stability (Lembo et al. 2018).

Acyclovir, used mainly for the treatment of initial or recurrent infections caused by the Herpes simplex virus (HSV) on the skin and mucous membranes, was analyzed when incorporated into niosomes, analogous to liposomes, but formulated with non-ionic surfactants as opposed to lipids used in liposomes. The studies revealed that when incorporated into niosomes, acyclovir developed a better performance than liposomes in relation to controlled release, in addition to presenting oral bioavailability index twice higher in relation to the free drug (Lembo and Cavalli 2010).

Another possible administration route, including for nanoparticles, is the vaginal route; during the last few years, this route has been gaining more attention. The vagina is characterized as a suitable route for local and systemic administrations due to its high local vascularization, exhibiting an extensive network of blood and lymph vessels. In addition, the vaginal route is able to bypass first-pass metabolism and prevents enzymatic degradation and gastric irritation.

Thus, vaginal application results in higher concentrations and improved efficacy due to the mucosa being highly permeable and easy to administer. Drug absorption can occur through passive diffusion (through epithelial cells), diffusion between cells through narrow junctions (paracellular route) and by means of vesicular/receptor mediated drug transport. However, some disadvantages are presented regarding the acceptability of this pathway, including possible systemic adverse effects, affection of the drug absorption due to the vaginal microbiota, as well as oestrogen levels that can interfere with the permeability characteristics of the vagina. In addition, the acid vaginal pH (4.0–5.0) may limit the stability of the drug (Vanić and Škalko-Basnet 2014; Ferreira et al. 2017; Leyva-Gómez et al. 2018, 2019; De Araújo et al. 2019; Lalan et al. 2021).

Several nanosystems have been developed to improve and circumvent the disadvantages presented by mucous membranes as routes of administration, including cyclodextrins, liposomes, niosomes, nanostructured lipid carriers, and polymeric nanoparticles for the modulation and adaptation of the pharmacokinetics of the incorporated drug (Vanić and Škalko-Basnet 2014). In a system containing lamivudine incorporated in polymeric nanoparticles for vaginal administration, atoxicity, high drug entrapment rate, and improved stability in relation to the free drug were observed (Grande et al. 2019).

Tenofovir and maraviroc were analyzed when incorporated into dendrimers, and the results demonstrated a more pronounced anti-HIV activity than when compared to free drugs. In the dendrimers, these molecules still presented synergistic interactions, not showing irritation to the vaginal mucosa attributed to a reduced dose of both drugs (Macchione et al. 2020).

In another study about maraviroc and tenofovir, this time developed by Zhou et al. (2021), the drugs were incorporated into hydrogel nanoparticles with lipid layers (nanolipogels) that presented an inactivity index of the HIV virus more effective, when compared with free drugs. In addition, presented an expressive retention rate in the mucosa, reducing toxic systemic effects of drugs (Zhou et al. 2021).

Other studies have shown that inert inorganic nanoparticles have antiviral action (Akbarzadeh et al. 2018) and when loaded with acyclovir, the HSV inhibition has undergone a considerable increase, in addition to an increase in its bioavailability. Considering the polymeric nanoparticles, an improvement of permeability in the vaginal mucosa was verified, as well as in the distribution of the drug when compared to the free active molecule (Singh et al. 2017). Acyclovir has also been incorporated into liposomes for administration to the vaginal mucosa to the local treatment of genital herpes. The system containing liposomes involved in a bioadhesive hydrogel showed a sustained release of the drug (Lembo and Cavalli 2010).

Rectal route of administration was a route used in the 16th Century and is still extensively explored today. Despite being used for the purpose of local actions, this route can also be used for patients such as babies, the elderly, or those who have restrictions with the oral route, such as problems with the gastrointestinal tract. The use of this route prevents the drug from undergoing first-pass metabolism, which, like the vaginal route, results in higher plasma levels of drug and with rapid absorption, although in some cases, irregular or incomplete. Disadvantages, on the other hand, include discomfort at the application site, mucosal irritation, and uncomfortable position for the patient (Nawata et al. 2016; Melo et al. 2018).

Medicines for administration through the rectal route have followed the same growth parameter as those mentioned for the vaginally administrated ones, considering the treatment of viral diseases. The use of nanosystems can provide several advantages such as increased solubility for fat-soluble drugs, controlled drug release, in addition to the protection of the molecule (Mesquita et al. 2019).

Efavirenz incorporated into polymeric nanoparticles was analyzed and presented satisfactory results in improving the pharmacokinetics of the molecule, ameliorating its colorectal bioavailability compared to the free drug (Macchione et al. 2020). When coated with polyethylene glycol (PEG), polymeric nanoparticles established a longer residence time in contact with the mucosa, thus generating a prophylactic action against long-term HIV transmission (Grande et al. 2019).

Already aiming immunization against hepatitis B, liposomes were loaded on their surface with hepatitis B antigen and adjuvants. This formulation showed stability and increased mucosal immune response when compared to a product already available on the market, this being a vaccine for intramuscular application (Sahu and Pandey 2016).

The use of intravenous route is based on the use of solutions that must be clear, nongreasy, transparent, and without visible crystals in suspension. This route is used for quick effect and guarantees complete bioavailability, being one of the most used when high plasma concentrations are necessary. However, it establishes risks for greater toxicity when compared to other routes, such as the oral route. The half-life of drugs administered by this route is generally shortened. The doses injected intravenously do not undergo the absorption stage because, in general, they first enter the central compartment of the organism, characterizing the blood, heart, and lungs, and then they are transported to the peripheral compartments, such as the liver, kidneys, and brain, just like the whole body (Dhanasekaran et al. 2018; Kang et al. 2018).

Among the nanosystems that are allowed to be safely administered intravenously are polymeric and metallic nanoparticles, nanostructured lipid carriers, solid lipid nanoparticles, as well as liposomes. However, these are at risk of being captured by the liver and spleen, thus interfering in the bioavailability of the compound. Non-nucleoside reverse transcriptase (NNRTI) inhibitors loaded on polymeric nanoparticles showed a higher degree of cell uptake, strong activity against HIV, in addition to a synergistic activity observed in the incorporated drugs (Walvekar et al. 2019).

Nevirapine was incorporated into nanocrystals coated on its surface with PEG, polysaccharides or albumin in order to vectorize the drug after administration (Lembo et al. 2018). In polymeric nanoparticles coated with PEG, Zidovudine suffered a delayed release at pH 7.4, when compared with the free drug, generating advantages such as greater stability, higher plasma circulation time and, consequently, greater bioavailability (Joshy et al. 2018).

The atazanavir liposome-encapsulated was analyzed by Lembo et al. (2018) to direct the drug to macrophages, specifically to the lectin receptors present on them. The results showed atoxicity, in addition to a better distribution of the drug when compared to its free form. Stavudine, a drug that exhibits several adverse reactions and instability, was dispersed in solid lipid nanoparticles and demonstrated stability control, greater cell uptake, and prolonged therapeutic action close to the therapeutic target (Grande et al. 2019).

The intramuscular route is currently the commonly used way to administer vaccines. It is also used for drugs that could be deactivated by the first-pass mechanism or enzymes and is possible thanks to the vascularization of the target tissue. Comparing with other administration routes, a faster absorption than the oral route and a slower absorption than the intravenous route were observed. The more developed the muscle of administration, the safer is the application of the drug, since the risk of reaching large-calibre vessels decreases considerably. When compared to the subcutaneous route, it is preferable to use the intramuscular route if the application is of a larger volume of medication (5 ml); in addition, it is also allowed to apply oily medications (Shatsky 2009; Jin et al. 2015).

Polymeric nanoparticles have been used for the development of vaccines targeting the immunization against HIV, containing antigenic peptides from the virus. In this case, chitosan was used, being observed absence of skin toxicity in addition to good distribution with fast elimination of the action site (Macchione et al. 2020). Others already on the market are Epaxal®, Inflexal® V (Bulbake et al. 2017) and Influvac® Plus used for immunization against hepatitis A virus (HAV) and influenza, respectively, incorporated into virosomes, thus presenting reduced toxicity, greater tolerability at the application site and superior immune response (Cojocaru et al. 2020).

Due to the seriousness of the pandemic, originated from SARS-CoV-2 infection, several vaccines are under development and are being analyzed with care. Some of them are still under study, and others are already being administered. The vaccines consist of a liposomal envelope surrounding the antigen, with the purpose of protecting and preventing degradation of the mRNA molecule, as well as assisting in its cell permeation stage (Kohyama et al. 2009; Ohno et al. 2009; Huang et al. 2020).

Administration by the subcutaneous route occurs through the hypodermic adipose tissue and depends on the blood capillaries present in this site; it ends up becoming a more comfortable and accessible route than the intravenous route. However, a lower speed and lower volume of infusion (1.5 ml) is observed, being that higher volumes end up associated with local irritation, pain, and adverse reactions. It is an accessible route for protein delivery, self-administration, frequent and long-term treatments, and is becoming the standard route of administration for patients undergoing treatment for diabetes and rheumatoid arthritis (Mathaes et al. 2016; Bittner et al. 2018; Turner and Balu-Iyer 2018; Viola et al. 2018).

The nasal route is characterized by the absorption of the drug through the nasal mucosa to establish both local and systemic actions. It provides several advantages, such as ease of application and a painless and non-invasive process when compared to intravenous and intramuscular routes. Furthermore, drugs administered by this route do not undergo a first-pass mechanism, avoiding hepatic metabolism, increasing the bioavailability of the drug and a quick onset of action when compared to drugs orally administered. However, disadvantages interfere with the choice of this route. Due to the mechanism of mucociliary clearance, there is a shorter absorption time of the drug, as well as a difficult permeation through the mucous barrier, mainly for hydrophilic drugs, and a limited volume for application (Costa et al. 2019; Akel et al. 2020).

A delivery called nose-brain has been well explored among researchers because it has the advantage of transposing the brain blood barrier, as well as avoiding first-pass metabolism. However, a high molecular weight of the drug can result in a lower degree of permeability through the nasal mucosa, as well as causing local irritation and being susceptible to enzymatic metabolism at the administration site (Costa et al. 2019; Akel et al. 2020).

In this context, the nasal route presents itself as a promising alternative to improve the targeting of the drug, as well as the reduction of metabolism through nanosystems such as liposomes, nano and microemulsions, as well as solid lipid nanoparticles (SLN) and nanostructured lipid carriers (NLC). In addition, the nasal mucosa is also able to act as a gateway for some types of viruses, such as influenza, where the administration of antiviral nanosystems can be considered very promising (Hu et al. 2018).

Another promising nanosystem for this route of administration are nanosponges, suitable for the administration of antiviral drugs through the nasal mucosa, in addition to providing selective and targeted delivery to viruses that infect the lower respiratory tract; these include influenza and rhinovirus, also HSV, HIV, and HBV. Acyclovir dispersed in β-cyclodextrin nanosponges demonstrated a higher rate of antiviral efficacy (against HSV) when compared to free drug, in addition to a sustained in vitro release profile. This β-cyclodextrin nanosponges acts in cell surface receptors used by the virus to carry out cell adhesion (Pawar et al. 2019).

Tenofovir was incorporated into solid lipid nanoparticles that showed better flow of the drug in the nasal mucosa, in addition to a low toxicity rate (Grande et al. 2019). An oil-in-water nanoemulsion loaded with a group of HIV recombinant gp120 glycoproteins for the purpose of promoting a vaccine was synthesized and a significant immune response was observed. Also, using gp120, vaccines based on liposomes were able to induce immune response (Das Neves et al. 2010).

When drugs are intended for administration to the skin, the topical (cutaneous) route is used. The purpose can be the local action, maximizing the concentration of the drug at the application site; however, the penetration though the skin could be a possibility. The transport of substances through the skin can occur around the cells (using the extracellular compartment), directly through cells and through the skin's attachments.

It is considered an alternative for those patients who are not in favour of the oral administration route since, through the topical route, the drug does not present first-pass metabolism, in addition to having a lower index of systemic adverse reactions. The topical route is an important option when the therapeutic target is the skin itself. Drugs can be oily or hydrophilic in nature, in the form of emulsions, ointments, creams, and/or gels, for example. The absorption of the drug is directly linked to the physiological conditions in which the organ is found (Hagen and Baker 2017; De Matos et al. 2019).

Cyclodextrins, solid lipid nanoparticles, nanostructured lipid carriers, nanogels, polymeric films, and nanoemulsions are among the nanosystems employed for the purpose of topical or cutaneous application. Dendrimers, used to incorporate and attach various molecules to their branched structure, have been widely used to improve the degree of solubility of drugs, sharpeners of penetration through the skin, as well as vectoring (Papakostas et al. 2011). Etosomes, vesicles similar to liposomes but containing ethanol, phospholipids, and water in their composition, are also being extensively analyzed due to fusing with the skin's lipids and, consequently, causing a more efficient distribution through the organ (Lembo et al. 2018). Inorganic nanoparticles have also been showing a satisfactory degree of viral inhibition, including for HIV and HBV (Lembo and Cavalli 2010).

Liposomes loaded with Zidovudine demonstrated that the incorporation in the nanosystem increased the permeability of the drug at the administration site by up to 12 times, increasing its plasma concentrations when compared to the drug being in its conventional formulation (Das Neves et al. 2010). Nanospheres composed of chitosan containing acyclovir were tested against strains of HSV, and when compared to the commercial product resulted in an improved index of about 5 times higher penetration into the skin (Cojocaru et al. 2020).

Penciclovir and valacyclovir were also analyzed when incorporated into nanosystems. In this case, solid lipid nanoparticles were used to improve their delivery to the therapeutic target and release profile (Lembo et al. 2018). Coumestrol, an isoflavonoid-like compound that has a capacity to inhibit HSV replication, was dispersed in nanoemulsions and showed promising results in terms of delivery of the compound to the site of action, in addition to an increase in the inhibitory activity of HSV (Cojocaru et al. 2020).

The ophthalmic route is employed to treat eye disorders, requiring more specific production patterns, such as sterility. In addition, several natural barriers affect the degree of bioavailability of the drugs administrated by this route, including the short contact time of the formulations with the action site due to the tear fluid turnover.

Other barriers that could be cited are hemato-ocular barrier, corneal epithelium, and nasolacrimal drainage where only about 5–10% of the drug is able to overcome barriers (Agarwal et al. 2016; Dubald et al. 2018; Imperiale et al. 2018; Xu et al. 2018; Maharjan et al. 2019; Wu et al. 2019).

Compared to the oral route, the ophthalmic administration still provides a more satisfactory ocular bioavailability. Also, it is considered comfortable for patients since its administration is relatively easy and there are fewer adverse reactions when compared to systemic administration. Conventional ophthalmic formulations exhibit satisfactory rate of pathologies combat when in the anterior ocular portion; however, when associated with the posterior portion, this index declines due to the presence of the previously cited physiological barriers (Agarwal et al. 2016; Dubald et al. 2018; Imperiale et al. 2018; Xu et al. 2018; Maharjan et al. 2019; Wu et al. 2019).

Cyclodextrins, liposomes, polymeric micelles, solid lipid nanoparticles, and nanostructured lipid carriers are some of the nanosystems that are characterized as safe systems for administration by the ophthalmic route. Polymeric nanoparticles were developed to optimize the delivery of acyclovir to the ocular surface. The results demonstrate a satisfactory, safe, and efficient delivery, when compared to the drug in its free form (Alkholief et al. 2019).

Since liposomes have the ability to increase the penetration of the drug into the cornea, studies have been carried out to compare the drug ganciclovir incorporated in liposomes and its free form. For those who received the formulation containing liposomes, the pathology was eradicated, in addition to the therapeutic levels of the drug remaining days after administration (Pathak et al. 2016).

4.5 CHARACTERIZATION OF THE NANOPARTICLES

There are several techniques described for the characterization of nanoparticles in terms of shape, size, and stability. They are presented below.

4.5.1 Microscopy Techniques

Transmission electron microscopy (TEM), atomic force microscopy (AFM), and scanning electron microscopy (SEM) are some examples of widely used techniques for characterizing these systems. Both the TEM technique and the SEM use electron beams as a source of illumination to obtain images with a resolution higher than that generated by optical microscopy. In the case of SEM, the information collected in detail refers to the morphology and size of the nanoscale particles, generally in greater quantity, however, generally linked to the two surfaces. When referring to TEM, it provides information about the particle morphology and degree of crystallinity of nanoparticles (Taheri et al. 2016; Franken et al. 2017; Tang and Yang 2017; Callahan et al. 2018; Fan et al. 2019). AFM allows, as well as the other techniques mentioned, the generation of images; however, in a three-dimensional format, the interactions present in the compound can even, in some cases, be quantified through its manipulation, in addition to providing details about its mechanisms of action (Alsteens et al. 2017; Dufrêne et al. 2017; Pavlicek and Gross 2017; Collins et al. 2019).

The structure of polymeric nanoparticles containing Zidovudine was observed using the TEM technique and provided images that demonstrated a nucleus containing the compound encased in a nanoparticle envelope. In addition, information was provided on the size of the particles that differed from those provided by the dynamic light scattering (DLS) technique presented by the same group, later explained by a possible shrinkage of the particles when preparing the TEM technique. Both techniques were able to provide information about nonexistent aggregations of the particles (Joshy et al. 2018). The TEM technique also provides the analysis of images that show the surface of nanoparticles, determining whether or not there was the removal of all the solvent used on its synthesis (Sahu and Pandey 2016).

The SEM technique, in addition to providing information about size and morphology, is able to determine whether the drug or compound was trapped inside nanoparticles by means of image contrast, as in the case of ganciclovir used in formulations for ophthalmic administration (Yang et al. 2016). For Martins et al. (2019), the SEM helped in the evaluation of the cell membrane integrity and the erythrocyte morphology when they come into contact with polymeric nanoparticles containing efavirenz for the treatment of HIV; where the conservation of the cell structure was observed, without significant alteration in its morphology, being hemocompatible for administration by the intravenous route (Martins et al. 2019).

Lima et al. (2018) used AFM to determine the shape and surface of the drug, as well as of the nanoparticles for the treatment of HSV. They used images generated by the device in 2D and 3D, where appropriate characteristics for the nanosystem were observed, but diverging in some aspects with DLS. The authors attribute explanations due to the sample preparation procedure (Lima et al. 2018).

Also through the AFM technique, Joshy et al. (2018) were able to analyze whether or not there was aggregation of the nanoparticles containing Zidovudine developed by the group. It is important to note that the association of microscopy techniques brings more reliable and detailed information about nanoparticles, allowing to observe behaviours such as aggregation, shrinkage, etc.

4.5.2 Dynamic Light Scattering

Dynamic light scattering is a technique that aims to determine the diffusion coefficients of macromolecules in solution, allowing to obtain information such as aggregation and the average size of the particles. This information is obtained in a simpler and easier way than using microscopy techniques; however, the results are less accurate (Fischer and Schmidt 2016; Stetefeld et al. 2016; Malm and Corbett 2019; Nepomnyashchaya 2019).

Huang et al. (2020) demonstrated in their research that the characterization through DLS was able to demonstrate the size and shape of the analyzed liposomes for the purposes of formulating vaccines. This technique also helped the researchers in providing the stability of the liposomes. Following the characteristics by DLS, they observed no significant difference between the data collected on day 1 and one week after (Huang et al. 2020). Joshy et al. (2018) also used DLS. They evaluated Zidovudine in different concentrations incorporated into nanoparticles and the results showed that the size of the particles was directly linked to the concentration of the drug (Joshy et al. 2018). For Martins et al. (2019), some data provided by DLS may present discrepancies when compared to TEM. It happens since TEM evaluates only a limited number of nanoparticles, whereas DLS analyzes the average of thousand nanoparticles (Martins et al. 2019).

4.5.3 Zeta Potential

The zeta potential (ZP) demonstrates the charges present on the surfaces of the analyzed particles, revealing their electrostatic potential when such particles in solid state come into contact with a fluid, usually aqueous, establishing information about their physical stability and their aggregation potential. Analyzing the results, values of about ±30 mV indicate stability when present in a colloidal system, which can be influenced by the polymer present in the formulation (Pathak et al. 2016; Předota et al. 2016; Smith et al. 2017; Biriukov et al. 2020).

According to Joshy et al. (2018), the zeta potential is essential for determining the stability of suspended nanosystems, with negative values of ZP, the nanoparticles loaded with Zidovudine guarantee dispersion stability, justified by repelling the components present in the blood medium with negative charge and, consequently, reducing the rate of coagulation and/or cell damage.

With negative values of −28.2 ± 1.7 mV, nanosponges containing acyclovir were also considered very stable for application through the nasal route (Pawar et al. 2019). The ZP of aciclovir incorporated in polymeric nanoparticles were also analyzed, aiming the comprehension of the stability. The measures were carried on day 0, after a week, after a month and three months later, subjecting the samples to changes on the temperature. Change in the stability of the nanoparticles were observed due to exposure to changes in temperature and, also, aggregation during the storage period (Alkholief et al. 2019).

4.5.4 Thermal Analysis

Thermal analysis, composed by differential scanning calorimetry (DSC), differential thermal analysis (DTA), and thermogravimetric analysis (TGA), are able to analyze several physical (crystallization or fusion) and chemical (oxidation or decomposition) aspects in relation to the compounds present in the formulations in function of the temperature. DSC determines the difference in thermal energy absorption between the sample to be analyzed and some standard compound or reference under the influence of a constant increase in temperature at a constant rate (Ibarra-Molero et al. 2016; Knopp et al. 2016; Durowoju et al. 2017). DTA and TGA informs loss of mass, phase transitions, as well as degradation of the compounds present in the formulation (Juarez-Perez et al. 2016; Cai et al. 2018; Kaya et al. 2020).

Considering the use of polymers in the formulation, a glass transition temperature is observed in DSC, as well as a decomposition temperature of the drug present in the nanosystem (Yang et al. 2016a). DSC was used as a characterization technique for nanoparticles containing both embedded lamivudine and efavirenz, showing results of molecular interaction between the drug and the polymer used in the formulation and chemical stability (Cojocaru et al. 2020).

Yang et al. (2016b) studied the use of curcumin in silver nanoparticles to inhibit infections in the respiratory tract by the respiratory syncytial virus. TGA made on pure curcumin and in the curcumin present in nanoparticles confirmed the purity of the natural compound, differing significantly from the nanoparticles, due to its composition. Ghaffari et al. (2019) used TGA to confirm the presence of PEG on the surface of zinc oxide nanoparticles used to treat the H1N1 influenza virus, characterized by loss of mass at a specific temperature related to PEG.

4.5.5 Spectroscopy Techniques

The spectroscopy techniques can be used to determine/identify structures and evaluate the quality of compounds and degree of crystallinity, among other characteristics related to chromophore groups; the techniques of ultraviolet-visible spectroscopy (UV-Vis) and fluorescence spectroscopy stand out for being widely used.

These techniques are capable of quantitatively measuring the properties of the compounds using their frequency and wavelength (UV-Vis) or are based on the presentation of light emission when their molecule is excited to a higher quantum state and, later, returns to its fundamental state (emitting fluorescence) (Kondo et al. 2017; Bobadilla et al. 2018; Rocha et al. 2018; Power et al. 2019).

More sensitive chromatographic techniques, such as high performance liquid chromatography (HPLC) and nuclear magnetic resonance (NMR) spectroscopy, are more expensive than the techniques previously mentioned; however, they have been widely used for identification of structures, degree of purity, properties, and conformation of certain atomic nuclei present in the analyzed compound (Reuhs 2017; Sugiki et al. 2017; Abdul Jameel et al. 2018; Eills et al. 2019; Žuvela et al. 2019).

Using a specific absorption range for the drug or compound to be observed, UV-Vis can be used to quantify the release of the drug from the nanoparticle. It was previously applied to Zidovudine, incorporated in nanoparticles, as well as to evaluate the haemolysis promoted by this compound, with the purpose of checking toxicity to blood cells (Joshy et al. 2018).

As stated, like UV-Vis, HPLC is also able to determine the efficiency of entrapment or encapsulation, and the loading of the drug in nanoparticles, but in a more selective way (Yang et al. 2016a). Martins et al. (2019) used this technique to quantify the encapsulation of efavirenz, in addition to observing the release of the drug incorporated into polymeric nanoparticles (Martins et al. 2019).

X-ray diffraction technique (XRD) provides information about the degree of drug crystallinity, as well as the components present in the formulation, as pointed out by Yang et al. (2016a). These authors showed the XRD of ganciclovir incorporated in nanoparticles, where it was possible to observe the non-crystalline nature of the drug before its incorporation, as well as the amorphous nature of the nanoparticles, suggesting the encapsulation of the drug.

Through NMR, Yang et al. (2016a) also proved the formation of a new molecule indispensable for its formulation. In order to investigate the interaction of the link between the flavonoids (flavones) hesperetin and hesperidin with the M2-1 protein, a target for the inhibition of viral replication of the human respiratory syncytial virus, fluorescence spectroscopy was applied. It generated promising results, with a satisfactory interaction between the molecules (Piva et al. 2020).

4.6 CONCLUSION

This chapter demonstrates the possibility of synthesis of several nanoparticles for antiviral effect, which can be used by the most varied routes of administration, such as oral, parenteral, as well as through the use of skin and mucous membranes, with more pronounced effects regarding stability, bioavailability, and maintenance of plasma concentrations than the free drugs or incorporated drugs in conventional dosage forms. The techniques described, including microscopies and spectroscopies, among others, guarantee the reliable study of the systems, allowing understanding the physical and chemical characteristics of the nanoparticles.

Despite the great advances available in science regarding the use of nanoparticles with an antiviral effect, either for the control of established infections or for their prevention, there are still great possibilities and needs in the medical fields in this regard, especially in the current pandemic moment, in which the use of nanoparticles can propitiate the development of innovative and effective formulations for quick service to the entire world population.

REFERENCES

Abd Elkodous, M., El-Sayyad, G., Abdelrahman, I. et al. 2019. Therapeutic and diagnostic potential of nanomaterials for enhanced biomedical applications. *Colloids Surfaces B Biointerfaces* 180:411–428. doi: 10.1016/j.colsurfb.2019.05.008.

Abdul Jameel, A.G., Van Oudenhoven, V., Emwas, A.H. and Sarathy, S.M. 2018. Predicting octane number using nuclear magnetic resonance spectroscopy and artificial neural networks. *Energy and Fuels* 32(5):6309–6329. doi: 10.1021/acs.energyfuels.8b00556.

Adeoye, O. and Cabral-Marques, H. 2017. Cyclodextrin nanosystems in oral drug delivery: A mini review. *International Journal of Pharmaceutics* 531(2):521–531. doi: 10.1016/j.ijpharm.2017.04.050.

Agarwal, R., Iezhitsa, I., Agarwal, P., et al. 2016. Liposomes in topical ophthalmic drug delivery: An update. *Drug Delivery* 23(4):1075–1091. doi: 10.3109/10717544.2014.943336.

Ahmad, A.B., Bennett, P.N. and Rowland, M. 1984. Influence of route of hepatic administration on drug availability. *Journal of Pharmacology and Experimental Therapeutics* 230(3):718–725.

Akbarzadeh, A., Kafshdooz, L., Razban, Z., et al. 2018. An overview application of silver nanoparticles in inhibition of herpes simplex virus. *Artificial Cells, Nanomedicine and Biotechnology* 46(2):263–267. doi: 10.1080/21691401.2017.1307208.

Akel, H., Ismail, R. and Csóka, I. 2020. Progress and perspectives of brain-targeting lipid-based nanosystems via the nasal route in Alzheimer's disease. *European Journal of Pharmaceutics and Biopharmaceutics* 148:38–53. 10.1016/j.ejpb.2019.12.014

Alkholief, M., Albasit, H., Alhowyan, A. et al. 2019. Employing a PLGA-TPGS based nanoparticle to improve the ocular delivery of acyclovir. *Saudi Pharmaceutical Journal* 27(2):293–302. doi: 10.1016/j.jsps.2018.11.011.

Alsteens, D., Gaub, H.E., Pfreundschuh, M. et al. 2017. Atomic force microscopy-based characterization and design of biointerfaces. *Nature Reviews Materials* 2(5):1–16. doi: 10.1038/natrevmats.2017.8.

Andreaus, J., Dalmolin, M.C., de Oliveira Junior, I.B. and Barcellos, I.O. 2010. Aplicação de ciclodextrinas em processos têxteis. *Química Nova* 33(4):929–937. Doi: 10.1590/S0100-40422010000400031.

Biriukov, D., Fibich, P. and Předota, M. 2020. Zeta potential determination from molecular simulations. *Journal of Physical Chemistry C* 124(5):3159–3170. doi: 10.1021/acs.jpcc.9b11371.

Bittner, B., Richter, W. and Schmidt, J. 2018. Subcutaneous administration of biotherapeutics: An overview of current challenges and opportunities. *BioDrugs* 32(5): 425–440. doi: 10.1007/S40259-018-0295-0.

Bobadilla, L.F., Santos, J.L., Ivanova, S., Odriozola, J.A. and Urakawa, A. 2018. Unravelling the role of oxygen vacancies in the mechanism of the reverse water-gas shift reaction by operando DRIFTS and ultraviolet-visible spectroscopy. *ACS Catalysis* 8(8):7455–7467. doi: 10.1021/acscatal.8b02121.

Bulbake, U., Doppalapudi, S., Kommineni, N. and Khan, W. 2017. Liposomal formulations in clinical use: An updated review. *Pharmaceutics* 9(2):1–33. doi: 10.3390/pharmaceutics9020012.

Cai, J., Xu, D., Dong, Z. et al. 2018. Processing thermogravimetric analysis data for isoconversional kinetic analysis of lignocellulosic biomass pyrolysis: Case study of corn stalk. *Renewable and Sustainable Energy Reviews* 82:2705–2715. doi: 10.1016/j.rser.2017.09.113.

Callahan, P.G., Stinville, J.C., Yao, E.R. et al. 2018. Transmission scanning electron microscopy: Defect observations and image simulations. *Ultramicroscopy* 186:49–61. doi: 10.1016/j.ultramic.2017.11.004.

Chiari-Andréo, B.G., Abuçafy, M.P., Manaia, E.B. et al. 2020. Drug delivery using theranostics: An overview of its use, advantages and safety assessment. *Current Nanoscience* 16:3–14. doi: 10.2174/1573413715666190618162321.

Chiari-Andréo, B.G., de Almeida-Cincotto, M.G.J., Oshiro, J.A., Taniguchi, C.Y.Y., Chiavacci, L.A. and Isaac, V.L.B. 2019. Nanoparticles for cosmetic use and its application. In *Nanoparticles in Pharmacotherapy*, ed. A.M. Grumezescu, 113–146. New York: William Andrew Publishing. doi: 10.1016/B978-0-12-816504-1.00013-2.

Cojocaru, F.D., Botezat, D., Gardikiotis, I. et al. 2020. Nanomaterials designed for antiviral drug delivery transport across biological barriers. *Pharmaceutics* 12(2): 1–34. doi: 10.3390/pharmaceutics12020171.

Collins, L., Liu, Y., Ovchinnikova, O.S. and Proksch, R. 2019. Quantitative electromechanical atomic force microscopy. *ACS Nano* 13(7):8055–8066. doi: 10.1021/acsnano.9b02883.

Costa, C., Moreira, J.N., Amaral, M.H., Sousa Lobo, J.M. and Silva, A.C. 2019. Nose-to-brain delivery of lipid-based nanosystems for epileptic seizures and anxiety crisis. *Journal of Controlled Release* 295(2019):187–200. doi: 10.1016/j.jconrel.2018.12.049.

Crucho, C.I.C. and Barros, M.T. 2017. Polymeric nanoparticles: A study on the preparation variables and characterization methods. *Materials Science and Engineering: C* 80: 771–784. doi: 10.1016/j.msec.2017.06.004

Das Neves, J., Amiji, M.M., Bahia, M.F. and Sarmento, B. 2010. Nanotechnology-based systems for the treatment and prevention of HIV/AIDS. *Advanced Drug Delivery Reviews* 62(4–5):458–477. doi: 10.1016/j.addr.2009.11.017.

Daudt, R.M., Emanuelli, J., Külkamp-Guerreiro, I.C., Pohlmann, A.R. and Guterres, S.S. 2013. A nanotecnologia como estratégia para o desenvolvimento de cosméticos. *Ciência e Cultura* 65:28–31. doi: 10.21800/S0009-67252013000300011.

De Araújo, P.R., Calixto, G.M.F., da Silva, I.C. et al. 2019. Mucoadhesive *in situ* gelling liquid crystalline precursor system to improve the vaginal administration of drugs. *AAPS PharmSciTech* 20(6):225–237. doi: 10.1208/s12249-019-1439-3.

De Matos, S.P., Teixeira, H.F., de Lima, Á.A.N., Veiga-Junior, V.F. and Koester, L.S. 2019. Essential oils and isolated terpenes in nanosystems designed for topical administration: A review. *Biomolecules* 9(4):138–157. doi: 10.3390/biom9040138.

Decher, G. 1997. Fuzzy nanoassemblies: Toward layered polymeric multicomposites. *Science* 277(5330):1232–1237. doi: 10.1126/science.277.5330.1232.

Decher, G., Hong, J.-D. and Schmitt, J. 1992a. Buildup of ultrathin multilayer films by a self-assembly process: III. Consecutively alternating adsorption of anionic and cationic polyeletrolytes on charged surfaces. *Thin Solid Films* 210/211:831–835. doi: 10.1016/0040-6090(92)90417-A.

Decher, G. and Schmitt, J. 1992b. Fine – Tuning of the film thickness of ultrathin multilayer films composed of consecutively alternating layers of anionic and cationic polyelectrolytes. *Progress in Colloid and Polymer Science* 89:160–164. doi: 10.1007/BFb0116302.

Decher, G., Lehr, B., Lowack, K., Lvov, Y. and Schmitt, J. 1994. New nanocomposite films for biosensors: Layer by-layer absorbed films of polyelectrolytes. Proteins or DNA. *Biosensors and Bioelectronics* 9:677–684. doi: 10.1016/0956-5663(94)80065-0.

Dhanasekaran, H.R., Sharma, C.P. and Haridoss, P. 2018. Drug delivery nanosystems - An introduction. *Micro and Nano Technologies* 2018:1–12. doi: 10.1016/B978-0-323-50922-0.00001-8.

Dubald, M., Bourgeois, S., Andrieu, V. and Fessi, H. 2018. Ophthalmic drug delivery systems for antibiotherapy - A review. *Pharmaceutics* 10(1):1–31. doi: 10.3390/pharmaceutics10010010.

Dufrêne, Y.F., Ando, T., Garcia, R., et al. 2017. Imaging modes of atomic force microscopy for application in molecular and cell biology. *Nature Nanotechnology* 12(4):295–307. doi: 10.1038/nnano.2017.45.

Durán et al. 2006. *Nanotecnologia introdução, preparação e caracterização de nanomateriais e exemplos de aplicação*, 208. São Paulo: Artiliber.

Durowoju, I.B., Bhandal, K.S., Hu, J., Carpick, B. and Kirkitadze, M. 2017. Differential scanning calorimetry - A method for assessing the thermal stability and conformation of protein antigen. *Journal of Visualized Experiments* 2017(121):1–8. doi: 10.3791/55262.

Eills, J., Hale, W., Sharma, M., Rossetto, M., Levitt, M.H. and Utz, M. 2019. High-resolution nuclear magnetic resonance spectroscopy with picomole sensitivity by hyperpolarization on a chip. *Journal of the American Chemical Society* 141(25):9955–9963. doi: 10.1021/jacs.9b03507.

Fan, J.X., Li, Z.-H., Liu, X.-H., Zheng, D.-W., Chen, Y. and Zhang, X.-Z. 2018. Bacteria-mediated tumor therapy utilizing photothermally-controlled TNF-α expression via oral administration. *Nano Letters* 18(4):2373–2380. doi: 10.1021/acs.nanolett.7b05323.

Fan, Z., Zhang, L., Baumann, D. et al. 2019. *In situ* transmission electron microscopy for energy materials and devices. *Advanced Materials* 31(33):1–22. doi: 10.1002/adma.201900608.

Faria-Tischer, P.C.S. and Tischer, C.A. 2012. Nanobiotechnology: Platform technology for biomaterials and biological applications the nanostructures. *BBR - Biochemistry and Biotechnology Reports* 1(1):32–53. doi: 10.5433/2316-5200.2012v1n1p32.

Ferreira N.N., Perez, T.A., Pedreiro, L.N. et al. 2017. A novel pH-responsive hydrogel-based on calcium alginate engineered by the previous formation of polyelectrolyte complexes (PECs) intended to vaginal administration. *Drug Development and Industrial Pharmacy* 43(10):1656–1668. doi: 10.1080/03639045.2017.1328434.

Fischer, K. and Schmidt, M. 2016. Pitfalls and novel applications of particle sizing by dynamic light scattering. *Biomaterials* 98:79–91. doi: 10.1016/J.Biomaterials.2016.05.003.

Fleisher, D., Li, C., Zhou, Y., Pao, L.H. and Karim, A. 1999. Drug, meal and formulation interactions influencing drug absorption after oral administration. Clinical implications. *Clinical Pharmacokinetics* 36(3):233–254. doi: 10.2165/00003088-19993603 0-00004.

Franken, L.E., Boekema, E.J. and Stuart, M.C.A. 2017. Transmission electron microscopy as a tool for the characterization of soft materials: Application and interpretation. *Advanced Science* 4(5):1–9. doi: 10.1002/advs.201600476.

Fuhrmann, K. and Fuhrmann, G. 2017. Recent advances in oral delivery of macromolecular drugs and benefits of polymer conjugation. *Current Opinion in Colloid and Interface Science* 31:67–74. doi: 10.1016/j.cocis.2017.07.002.

Gelman, M.A. and Glenn, J.S. 2011. Mixing the right hepatitis C inhibitor cocktail. *Trends in Molecular Medicine* 17(1):34–36. doi: 10.1016/j.molmed.2010.10.005.

Ghaffari, H., Tavakoli, A., Moradi, A. et al. 2019. Inhibition of H1N1 influenza virus infection by zinc oxide nanoparticles: Another emerging application of nanomedicine. *Journal of Biomedical Science* 26(1):1–10. doi: 10.1186/s12929-019-05 63-4.

Gordillo-Galeano, P. and Mora-Huertas C.E. 2018. Solid lipid nanoparticles and nanostructured lipid carriers: A review emphasizing on particle structure and drug release. European *Journal of Pharmaceutics and Biopharmaceutics* 133:285–308. doi: 10.101 6/J.Ejpb.2018.10.017.

Grande, F., Ioele, G., Occhiuzzi, M.A. et al. 2019. Reverse transcriptase inhibitors nanosystems designed for drug stability and controlled delivery. *Pharmaceutics* 11(5):1–26. doi: 10.3390/pharmaceutics11050197.

Guterres, S.S., Alves, M.P. and Pohlmann, A.R. 2007. Polymeric nanoparticles, nanospheres and nanocapsules for cutaneous apllications. *Drug Target Insights* 2:148–157. doi: 10.1177/117739280700200002.

Hagen, M. and Baker, M. 2017. Skin penetration and tissue permeation after topical administration of diclofenac. *Current Medical Research and Opinion* 33(9):1623–1634. doi: 10.1080/03007995.2017.1352497.

Homayun, B., Lin, X. and Choi, H.J. 2019. Challenges and recent progress in oral drug delivery systems for biopharmaceuticals. *Pharmaceutics* 11(3):129–158. doi: 10.3390/ pharmaceutics11030129.

Hu, C.M.J., Chen,Y.-T., Fang, Z.-S., Chang, W.-S. and Chen, H.-W. 2018. Antiviral efficacy of nanoparticulate vacuolar ATPase inhibitors against Influenza virus infection. *International Journal of Nanomedicine* 13:8579–8593. doi: 10.2147/IJN. S185806.

Huang, W.C., Zhou, S., He, X. et al. 2020. SARS-CoV-2 RBD neutralizing antibody induction is enhanced by particulate vaccination. *Advanced Materials* 2005637:1–11. doi: 10.1002/adma.202005637.

Ibarra-Molero, B., Naganathan, A.N., Sanchez-Ruiz, J.M. and Muñoz, V. 2016. Modern analysis of protein folding by differential scanning calorimetry. *Methods in Enzymology* 567:281–318. doi: 10.1016/bs.mie.2015.08.027.

Imperiale, J.C., Acosta, G.B. and Sosnik, A. 2018. Polymer-based carriers for ophthalmic drug delivery. *Journal of Controlled Release* 285:106–141. doi: 10.1016/j.jconrel.201 8.06.031.

Jackman, J.A., Lee, J. and Cho, N.J. 2016. Nanomedicine for infectious disease applications: Innovation towards broad-spectrum treatment of viral infections. *Small* 12(9):1133–1139. doi: 10.1002/smll.201500854.

Jackman, J.A., Yoon, B.K., Ouyang, L. et al. 2020. Biomimetic nanomaterial strategies for virus targeting: Antiviral therapies and vaccines. *Advanced Functional Materials* 31(12):1–25. doi: 10.1002/adfm.202008352.

Jin, J.F., Zhu, L.-L., Chen, M. et al. 2015. The optimal choice of medication administration route regarding intravenous, intramuscular, and subcutaneous injection. *Patient Preference and Adherence* 9:923–942. doi: 10.2147/PPA.S87271.

Joshy, K.S.Babu, S.S., Joseph, B. et al. 2018. Novel core-shell dextran hybrid nanosystem for anti-viral drug delivery. *Materials Science and Engineering C* 93:864–872. doi: 10.1016/j.msec.2018.08.015.

Juarez-Perez, E.J., Hawash, Z., Raga, S.R., Ono, L. and Qi, Y. 2016. Thermal degradation of $CH_3NH_3PbI_3$ perovskite into NH_3 and CH_3I gases observed by coupled thermogravimetry-mass spectrometry analysis. *Energy and Environmental Science* 9(11):3406–3410. doi: 10.1039/c6ee02016j.

Kang, H., Hu, S., Hong, S.H., et al. 2018. Theranostic nanosystems for targeted cancer therapy. *Nano Today* 23:59–72. doi: 10.1016/j.nantod.2018.11.001.

Kaya, O., Kose, C., Donderalp, V., Gecim, T. & Taskın, T. 2020. Last updates on cell death point, bud death time and exothermic characteristics of flower buds for deciduous fruit species by using differential thermal analysis. *Scientia Horticulturae* 270(2020): 109403. doi: 10.1016/j.scienta.2020.109403.

Knopp, M.M., Löbmann, K., Elder, D.P., Rades, T. and Holm, R. 2016. Recent advances and potential applications of modulated differential scanning calorimetry (mDSC) in drug development. *European Journal of Pharmaceutical Sciences* 87:164–173. doi: 10.1016/j.ejps.2015.12.024.

Kohyama, S., Ohno, S., Suda, T. et al. 2009. Efficient induction of cytotoxic T lymphocytes specific for severe acute respiratory syndrome (SARS)-associated coronavirus by immunization with surface-linked liposomal peptides derived from a non-structural polyprotein 1a. *Antiviral Research* 84(2):168–177. doi: 10.1016/j.antiviral.2009.09.004.

Kondo, T., Chen, W.J., Schlau-Cohen, G.S. 2017. Single-molecule fluorescence spectroscopy of photosynthetic systems. *Chemical Reviews* 117(2):860–898. doi: 10.1021/acs.chemrev.6b00195.

Kulkarni, P.R., Yadav, J.D. and Vaidya, K.A. 2011. Lipossomes: A novel drug delivery system. *International Journal of Current Pharmaceutical Research* 3:11–18.

Lalan, M.S., Patel, V.N. and Misra, A. 2021. Polymers in vaginal drug delivery: Recent advancements. In *Applications of polymers in drug delivery*, eds. A. Misra and A. Shahiwala, 281–303. Elsevier. doi: 10.1016/b978-0-12-819659-5.00010-0.

Lembo, D. and Cavalli, R. 2010. Nanoparticulate delivery systems for antiviral drugs. *Antiviral Chemistry and Chemotherapy* 21(2):53–70. doi: 10.3851/IMP1684.

Lembo, D., Donalisio, M., Civra, A., Argenziano, M. and Cavalli, R. 2018. Nanomedicine formulations for the delivery of antiviral drugs: A promising solution for the treatment of viral infections. *Expert Opinion on Drug Delivery* 15(1):93–114. doi: 10.1080/17425247.2017.1360863.

Leyva-Gómez, G., Piñón-Segundo, E., Mendoza-Muñoz, N., Zambrano-Zaragoza, M.L., Mendoza-Elvira, S. and Quintanar-Guerrero, D. 2018. Approaches in polymeric nanoparticles for vaginal drug delivery: A review of the state of the art. *International Journal of Molecular Sciences* 19(6):1–19. doi: 10.3390/ijms19061549.

Li, F., Zhu, A., Song, X., Ji, L. and Wang, J. 2013. The internalization of fluorescence-labeled PLA nanoparticles by macrophages. *International Journal of Pharmaceutics* 453(2):506–513. doi: 10.1016/j.ijpharm.2013.06.033.

Li, Q., Cai, T., Huang, Y., Xia, X., Cole, S.P.C. and Cai, Y. 2017. A review of the structure, preparation, and application of NLCs, PNPs, and PLNs. *Nanomaterials* 7(6):122–147. doi: 10.3390/nano7060122.

Lima, T.L.C., de Carvalho Feitosa, R., Dos Santos-Silva, E., et al. 2018. Improving encapsulation of hydrophilic chloroquine diphosphate into biodegradable nanoparticles: A promising approach against herpes virus simplex-1 infection. *Pharmaceutics* 10(4):1–18. doi: 10.3390/pharmaceutics10040255.

Macchione, M.A., Bedoya, D.A., Figueroa, F.N., Muñoz-Fernández, M.Á., Strumia, M.C. 2020. Nanosystems applied to HIV infection: Prevention and treatments. *International Journal of Molecular Sciences* 21(22):1–29. doi: 10.3390/ijms21228647.

Maharjan, P., Cho, K.Y., Maharjan, A., Shin, M.C., MoonC., Min, K.A. 2019. Pharmaceutical challenges and perspectives in developing ophthalmic drug formulations. *Journal of Pharmaceutical Investigation* 49(2):215–228. doi: 10.1007/s40005-018-0404-6.

Malm, A.V. and Corbett, J.C.W. 2019. Improved dynamic light scattering using an adaptive and statistically driven time resolved treatment of correlation data. *Scientific Reports* 9(1):1–11. doi: 10.1038/s41598-019-50077-4.

Manaia, E.B., Abuçafy, M.P., Chiari-Andréo, B.G., Silva, B.L., Oshiro Junior, J.A. and Chiavacci, L.A. 2017. Physicochemical characterization of drug nanocarriers. *International Journal of Nanomedicine* 12:4991–5011. doi: 10.2147/IJN.S133832.

Martins, C., Araújo, F., Gomes, M.J. et al. 2019. Using microfluidic platforms to develop CNS-targeted polymeric nanoparticles for HIV therapy. *European Journal of Pharmaceutics and Biopharmaceutics* 138:111–124. doi: 10.1016/j.ejpb.2018.01.014.

Mathaes, R., Koulov, A., Joerg, S., Mahler, H.C. 2016. Subcutaneous injection volume of biopharmaceuticals - Pushing the boundaries. *Journal of Pharmaceutical Sciences* 105(8):2255–2259. doi: 10.1016/j.xphs.2016.05.029.

Medeiros, B.G.S. et al. 2013. *Heparina e lisozima: características nanoestruturais que permitem à construção de nanofilmes para revestimento de superfícies.* https://www.editorarealize.com.br/artigo/visualizar/2233

Mehnert, W. and Mäder, K. 2012. Solid lipid nanoparticles: Production, characterization and applications. *Advanced Drug Delivery Reviews* 64:83–101. doi: 10.1016/j.addr.2012.09.021.

Melo F.S., Grillo, R., Rosa, A.H., et al. 2010. Desenvolvimento e caracterização de nanocápsulas de poli (L-lactídeo) contendo benzocaíne. *Química Nova* 33(1):65–69. doi: 10.1590/S0100-40422010000100013.

Melo, M., Sarmento, B., Nunes, R. and das Neves, J. 2018. Rectal administration of nanosystems: From drug delivery to diagnostics. *Materials Today Chemistry* 10(2018):128–141. doi: 10.1016/j.mtchem.2018.09.001.

Mendonça, G.A. 2020. *Avaliação da atividade antiviral do fitol contra herpesvírus humano.* http://dspace.sti.ufcg.edu.br:8080/xmlui/handle/riufcg/14280.

Mesquita, L., Galante, J., Nunes, R., Sarmento, B. and das Neves, J. 2019. Pharmaceutical vehicles for vaginal and rectal administration of anti-hivmicrobicide nanosystems. *Pharmaceutics* 11(3). doi: 10.3390/pharmaceutics11030145.

Moya-Ortega, M.D., Alvarez-Lorenzo, C., Concheiro, A., Loftsson, T. 2012. Cyclodextrin-based nanogels for pharmaceutical and biomedical applications. *International Journal of Pharmaceutics* 428(1–2):152–163. doi: 10.1016/j.ijpharm.2012.02.038.

Nawata, S., Kohyama, N., Uchida, N., 2016. The pharmacokinetics of mianserin suppositories for rectal administration in dogs and healthy volunteers: A pilot study. *Journal of Pharmaceutical Health Care and Sciences* 2(1):1–7. doi: 10.1186/s40780-016-0046-7.

Nepomnyashchaya, E.K. 2019. On approximation and experimental accuracy in dynamic light scattering. *Journal of Physics: Conference Series* 1236(2019):012041–012046. doi: 10.1088/1742-6596/1236/1/012041.

Nielsen, L.H., Keller, S.S., Boisen, A. 2018. Microfabricated devices for oral drug delivery. *Lab On A Chip* 18(16):2348–2358. doi: 10.1039/c8lc00408k.

Ohno, S., Kohyama, S., Taneichi, M., et al. 2009. Synthetic peptides coupled to the surface of liposomes effectively induce SARS coronavirus-specific cytotoxic T lymphocytes and viral clearance in HLA-A*0201 transgenic mice. *Vaccine* 27(29):3912–3920. doi: 10.1016/j.vaccine.2009.04.001.

Oliveira, S.S.A.M. 2009. *Preparação e caracterização in vitro de micropartículas de heparina fracionada potencialmente aplicáveis ao tratamento da trombose 74 venosa profunda.* https://www.teses.usp.br/teses/disponiveis/60/60137/tde-30042009-143701/publico/Dissertacao_de_Mestrado_USP_Samantha_abril_2009.pdf.

Palacio, J., Agudelo, N.A., Lopez, B.L. 2016. PEGylation of PLA nanoparticles to improve mucus-penetration and colloidal stability for oral delivery systems. *Current Opinion in Chemical Engineering* 11:14–19. doi: 10.1016/j.coche.2015.11.006.

Papakostas, D., Rancan, F., Sterry, W.Blume-Peytavi, U., Vogt, A. 2011. Nanoparticles in dermatology. *Archives of Dermatological Research* 303(8):533–550. doi: 10.1007/s00403-011-1163-7.

Pathak, Y., Sutariya, V. and Hirani, A.A. 2016. *Nano-biomaterials for ophthalmic drug delivery*, 627. Switzerland: Springer Nature.

Pavlicek, N. and Gross, L. 2017. Generation, manipulation and characterization of molecules by atomic force microscopy. *Nature Reviews Chemistry* 1:0005. doi: 10.1038/s41570-016-0005.

Pawar, S., Shende, P. and Trotta, F. 2019. Diversity of β-cyclodextrin-based nanosponges for transformation of actives. *International Journal of Pharmaceutics* 565:333–350. doi: 10.1016/j.ijpharm.2019.05.015.

Pillai, O. and Panchagnula, R. 2001. Polymers in drug delivery. *Current Opinion in Chemical Biology* 5:447–451. doi: 10.1016/S1367-5931(00)00227-1.

Piva, H.M.R., Sá, J.M., Miranda, A.S., et al. 2020. Insights into interactions of flavanones with target human respiratory syncytial virus M2-1 protein from STD-NMR, fluorescence spectroscopy, and computational simulations. *International Journal of Molecular Sciences* 21(6)-2241–2260. doi: 10.3390/ijms21062241.

Power, A.C., Chapman, J., Chandra, S. and Cozzolino, D. 2019. Ultraviolet-visible spectroscopy for food quality analysis. In *Woodhead Publishing Series in Food Science, Technology and Nutrition*, eds. J. Zhong and X. Wang, 91–104. Woodhead Publishing.

Předota, M., Machesky, M.L. and Wesolowski, D.J. 2016. Molecular origins of the zeta potential. *Langmuir* 32(40):10189–10198. doi: 10.1021/acs.langmuir.6b02493.

Reuhs, B.L. 2017. High-performance liquid chromatography. In *Food analysis*, ed. S.S. Nielsen, 213–226. Switzerland: Springer Nature. doi: 10.1007/978-3-319-45776-5.

Rocha, F.S., Gomes, A.J., Lunardi, C.N., Kaliaguine, S. and Patience, G.S. 2018. Experimental methods in chemical engineering: Ultraviolet visible spectroscopy—UV-Vis. *Canadian Journal of Chemical Engineering* 96(12):2512–2517. doi: 10.1002/cjce.23344.

Sahu, K.K. and Pandey, R.S. 2016. Immunological evaluation of colonic delivered hepatitis B surface antigen loaded TLR-4 agonist modified solid fat nanoparticles. *International Immunopharmacology* 39:343–352. doi: 10.1016/j.intimp.2016.08.007.

Santos, R.M.M. and Fialho, S.L. 2007. Nanopartículas. *Biotecnologia Ciência e Desenvolvimento* 10(37):53–59.

Sastry, S. V., Nyshadham, J. R. and Fix, J. A. 2000. Recent technological advances in oral drug delivery - A review. *Pharmaceutical Science and Technology Today* 3(4):138–145. doi: 10.1016/S1461-5347(00)00247-9.

Schaffazick, S.R., Stanisçuaski, G.S., de Lucca, F.L. and Pohlmann, A.R. 2003. Caracterização e estabilidade físico-química de sistemas poliméricos nanoparticulados para administração de fármacos. *Química Nova* 26(5):726–737. doi: 10.1590/S0100-4 042200300050001 7.

Shatsky, M. 2009. Evidence for the use of intramuscular injections in outpatient practice. *American Family Physician* 79(4):297–300.

Shreya, A.B., Raut, S.Y., Managuli, R.S., Udupa, N. and Mutalik, S. 2018. Active targeting of drugs and bioactive molecules via oral administration by ligand-conjugated lipidic nanocarriers: recent advances. *AAPS PharmSciTech* 20(1):15. doi: 10.1208/s12249-018-1262-2.

Singh, L., Kruger, H.G., Maguire, G.E.M., Govender, T. and Parboosing, R. 2017. The role of nanotechnology in the treatment of viral infections. *Therapeutic Advances in Infectious Disease* 4(4):105–131. doi: 10.1177/2049936117713593.

Smith, M.C., Crist, R.M., Clogston, J.D. and McNeil, S.E. 2017. Zeta potential: A case study of cationic, anionic, and neutral liposomes. *Analytical and Bioanalytical Chemistry* 409(24):5779–5787. doi: 10.1007/s00216-017-0527-z.

Stetefeld, J., Mckenna, S.A. and Patel, T.R. 2016. Dynamic light scattering: A practical guide and applications in biomedical sciences. *Biophysical Reviews* 8(4):409–427. doi: 10.1007/s12551-016-0218-6.

Sugiki, T., Kobayashi, N. and Fujiwara, T. 2017. Modern technologies of solution nuclear magnetic resonance spectroscopy for three-dimensional structure determination of proteins open avenues for life scientists. *Computational and Structural Biotechnology Journal* 13(15):328–339. doi: 10.1016/j.csbj.2017.04.001.

Taheri, M.L., Stach, E.A., Arslan, I., et al. 2016. Current status and future directions for in situ transmission electron microscopy. *Ultramicroscopy* 170:86–95. doi: 10.1016/j.ultramic.2016.08.007.

Tang, C.Y. and Yang, Z. 2017. Transmission electron microscopy (TEM). In *Membrane characterization*, eds. N. Hilal, A.F. Ismail, T. Matsuura, and D. Oatley-Radcliffe, 145–159. Elsevier B.V. doi: 10.1016/B978-0-444-63776-5.00008-5.

Tapeinos, C., Battaglini, M. and Ciofani, G. 2017. Advances in the design of solid lipid nanoparticles and nanostructured lipid carriers for targeting brain diseases. *Journal of Controlled Release* 264:306–332. doi: 10.1016/j.jconrel.2017.08.033.

Turner, M.R. and Balu-Iyer, S.V. 2018. Challenges and opportunities for the subcutaneous delivery of therapeutic proteins. *Journal of Pharmaceutical Sciences* 107(5):1247–1260. doi: 10.1016/j.xphs.2018.01.007.

Vanić, Ž. and Škalko-Basnet N. 2014. Mucosal nanosystems for improved topical drug delivery: Vaginal route of administration. *Journal of Drug Delivery Science and Technology* 24(5):435–444. doi: 10.1016/S1773-2247(14)50085-8.

Venturini, C.G., Nicolini, J., Machado, C. and Machado, V.G. 2008. Propriedades e aplicações recentes das ciclodextrinas. *Química Nova* 31(2):360–368. doi: 10.1590/S0100-40422008000200032.

Viola, M., Sequeira, J., Seiça, R., et al. 2018. Subcutaneous delivery of monoclonal antibodies: how do we get there? *Journal of Controlled Release* 286:301–314. doi: 10.1016/j.jconrel.2018.08.001.

Walvekar, P., Gannimani, R. and Govender, T. 2019. Combination drug therapy via nano-carriers against infectious diseases. *European Journal of Pharmaceutical Sciences* 127:121–141. doi: 10.1016/J.Ejps.2018.10.017.

Wu, Y., Liu, Y., Li, X., et al. 2019. Research progress of in-situ gelling ophthalmic drug delivery system. *Asian Journal of Pharmaceutical Sciences* 14(1):1–15. doi: 10.1016/j.ajps.2018.04.008.

Xu, J., Xue, Y., Hu, G., et al. 2018. A comprehensive review on contact lens for ophthalmic drug delivery. *Journal of Controlled Release* 281:97–118. doi: 10.1016/j.jconrel.2018.05.020.

Yang, X.X., Li, C.M. and Huang, C.Z. 2016a. Curcumin modified silver nanoparticles for highly efficient inhibition of respiratory syncytial virus infection. *Nanoscale* 8: 3040–3048. doi: 10.1039/C5NR07918G.

Yang, X., Shah, S.J., Wang, Z., Agrahari, V., Pal, D. and Mitra, A.K. 2016b. Nanoparticle-based topical ophthalmic formulation for sustained release of stereoisomeric dipeptide prodrugs of ganciclovir. *Drug Delivery* 23(7):2399–2409. doi: 10.3109/10717544.2014.996833.

Zarbin, A.J.G. 2007. (Nano) materials chemistry. *Química Nova* 30(6):1469–1479. doi: 10.1590/S0100-40422007000600016.

Zhou, J., Krishnan, N., Jiang, Y., Fang, R.H. and Zhang, L. 2021. Nanotechnology for virus treatment. *Nano Today* 36:101031. doi: 10.1016/j.nantod.2020.101031.

Žuvela, P., Skoczylas, M., Liu, J.J., et al. 2019. Column characterization and selection systems in reversed-phase high-performance liquid chromatography. *Chemical Reviews* 119(6):3674–3729. doi: 10.1021/acs.chemrev.8b00246.

5 Role of Nanostructures in Inhibition and Treatment of Viral Infections

Pavani Sanapala and Sudhakar Pola

CONTENTS

DOI: 10.1201/9781003136644-6

5.1 INTRODUCTION

Microorganisms, namely bacteria, viruses, fungi and parasites, play an important role as infectious pathogens distressing the world with around 15 million fatalities during the past decades (Panáček et al. 2009). The major outbreaks of viral infections were COVID-19, acute respiratory infections by Influenza viruses i.e. H5N1 (Avian flu) and Swine flu H1N1 (Spanish flu) and H3N2 (Hong-Kong flu), Nipah, Zika, Ebola, Dengue, Chikungunya diseases, Human immunodeficiency virus (HIV) and many more (Qasim et al. 2014; Gurunathan et al. 2020).

Viruses constitute a significant challenge to the body's immune system as they penetrate the cells, making it hard for antibodies to reach them. As opposed to bacterial contaminations, which are generally treated utilizing antitoxins, vaccination against viral diseases is not generally conceivable. Viral infections have a major impact on both civilization and individual health (Mehendale et al. 2013; Antunes et al. 2020). Treating viral infections is a challenging task for scientists due to the continuous emergence of novel viruses having a huge inpact on public health and society, and the nonavailability of antiviral therapeutics and vaccines that are the potent tools for treating and preventing the onset of viral infection (Gerhard et al. 2006; Krammer 2015; Szunerits et al. 2015). Though vaccines are available for few viral infections, such as measles, mumps, hepatitis A and B diseases, some viruses mutate internally while infecting from one person to the next, making immunization a complicated mission since the viruses change by the time a vaccine is accessible (Szunerits et al. 2015). Nevertheless, the efficiency of antiviral therapies is known to be reduced as the viral strains resist the antiviral drugs due to prolonged usage, which poses a global challenge. One such example is the novel Coronavirus disease (COVID-19), the existing pandemic that affected the whole world claiming 2.27 lakhs causalities as of now and, lowering the world economy.

Over the past 30 years, scientists have been establishing antiviral agents to control the spread of infections. A new model for enhancing the efficiency of bioactive modules, i.e. the use of nanomedicine (nanotherapeutics and nano-drug delivery system) has been of great interest. Nanotechnology has captivated both the scientific and medical fields in developing drugs and in basic research (Milroy and Featherstone 2002). The extensive appliance of nanoparticles (NPs) synthesis and expansion in current advanced biological and material sciences are due to its substantial scientific interest because they have valuable properties that compute and influence physical substances from atomic to molecular point of view compared with voluminous materials (Sanapala and Polo 2020). Generally, NPs such as silver nanoparticles (AgNPs), gold NPs, carbon and quantum dots, silicon materials, dendrimers, polymeric NPS, liposomes, and graphene oxide reported significant antiviral and antimicrobial activities (Qasim et al. 2014; Gurunathan et al. 2018, 2019; Jeyaraj et al. 2019; Cojocaru et al. 2020). Thus, it is essential to study nanomaterials that have exclusively antiviral and drug-delivery properties. This chapter deals with modern applications, roles, and characteristics of nanostructure in inhibiting and treating viral infections.

5.1.1 Viral Infection and Mechanism of Pathogenesis

The majority of viral infections are subclinical, where the host immune system arrests the infection before clinical symptoms occur. These infections are of great interest concerning epidemiological implications because they infect the population through spreading. Pathogenesis of viral disease includes the following stages: (1) anchoring of virus onto the target entry point, (2) penetration into the host cell, (3) uncoating of the virus at the nuclear pore or in the cytoplasm, (4) synthesis of virus-specific proteins through replication, (5) assemblage of the capsid, and, finally, (6) Virions released, causing the massive spread of infection in the host (Chinchar 1999). Significant factors transforming pathogenesis mechanisms are accessibility of tissue to the causative virus, the vulnerability of the cell to virus replication, and virus resistance to the host defense mechanism. Factors affecting the affinity of virus infection include the occurrence of virus-specific receptors on the host cell and transcription factors that facilitate the cell to identify viral promoter and enhancer sequences, confined temperature, pH, and the existence of enzymes that may inactivate the virus (Baron et al. 1996). The general mechanism espoused by the virus to impair the host cell is by blocking the synthesis of macromolecules. Another indirect process of cell damage is the assimilation of the viral genome with the host genome leading to mutation. Host infection can be detected by the virus type, virulence, viral genes, virus reliant factors, the quantity of sample or the inoculum, rate of replication occurring, and the magnitude of infection spread (Manjarrez-Zavala et al. 2013). The major concern with the viral study is due to its hardship in analyzing the process on how the host resistance would relate with the virus, whether it prevents the growth of the virus or may rouse immunological response in the affected tissue (Rouse and Sehrawat 2010).

5.1.2 Virus and Immune System

Once the virus crosses the mechanical chemical barrier of the cell, it competes for its existence by either innate or adaptive mechanisms. Immune response to viruses initially shows up during the primary infection of a vulnerable, nonimmune host and enhances during the second reencounter of an immune host. However, the initiation of unusual immune functions, the period, and the enormity of the immune response rely on the interaction between the host cells and the virus (i.e. cytolytic, steady-state, latent and/or integrated infection) and also the manner of spreading, such as either confined, primary and secondary hematogenous, or through the nervous system. As a result, viral antigens tend to accumulate in various body parts based on the way of spread and infection phase. Immune mechanisms that are efficient against the viruses are (a) cell-mediated immunity relating T cells (lymphocytes and cytokine effectors), (b) antibody by and devoid its interaction with complement and antibody-dependent cell-mediated cytotoxicity, (c) natural killer cells, (d) macrophages and (e) lymphokines. However, few mechanisms may relate, usually synergistically with nonimmune defense machinery (Klimpel 1996). However, sometimes viruses escape from the site of an immune response, masking the immune mechanism. To surmount this occurrence, the body system makes use of protein molecules, namely, class I major histocompatibility complex (MHC class I) to show off even small pieces of proteins from inside the cell ahead of the cell surface. Viruses in the process of cell infection fragment these portions of proteins.

In humoral immunity, the virus or the infected host cell stimulates B lymphocytes to produce antigen-specific antibodies. The three antibody types, IgM, IgG, and IgE, show to wield antiviral property. Neutralization of antibodies is most effective in larger fluids, such as serum, or on damp surfaces of the body, such as the gastrointestinal tract and the respiratory system. The general mechanisms of antibodies neutralizing the virus are either by hindering the interaction of the virus and host or by identifying viral antigens on the infected cells that direct antibody-dependent cytotoxic cells or complement intervened lysis.

5.1.3 Overview of Nanochemistry

The technology referring development and applications of nanoparticles is known as nanotechnology (Nalwa 1999; Parboosing et al. 2012). Nanoparticles (NPs) are miniaturized particles ranging in size from 10^{-9} to 1–100 nm. The study of nanoscience in correlation with biological methods is referred to as nanobiotechnology (Scheller et al. 1995; Mirkin and Niemeyer 2004), while the relation of nanoscience and medical systems, namely disease diagnosis, treatment, and prevention, is known as nanomedicine. Nanoparticle biosynthesis adjusts to the sustainable path using organic substances; either plants or animals are used for the formation and stabilization of NPs as a substitute for other components that might be harmful and toxic (Santos et al. 2019). The biosynthesis of nanoparticles begins with a diminution process within the ions, followed by the growth and nucleation phase of NPs to create a colloidal matrix. From these properties, it is assumed that nanomaterials have antimicrobial, anti-inflammatory, and antioxidant properties (Thakkar et al. 2010; Oliveira and De 2013).

Nanosystems in medicine were exercised initially to enhance the effectiveness of existing, hitherto dose-limiting and inadequately bioaccessible drugs (Schütz et al. 2013). NPs wield antiviral activities through varied mechanisms. Treatment of NPs for viral infections is due to their distinctive properties; that is, particle size that eases drug deliverance to restricted sites (Kumar et al. 2012; Parboosing et al. 2012), the ratio of volume to large surface area, and surface charge-assisting cellular entry covering the cellular membrane charged negatively (Caron et al. 2010; Petros and DeSimone 2010). Another significant role of NPs is that they contain biomimetic properties that result in inherent antiviral properties (Bowman et al. 2008; Sanvicens and Marco 2008; Gagliardi 2017). Well-known examples with these properties are silver NPs (Sun et al. 2005; Lara et al. 2010) and dendrimers (Wang et al. 2006; Mallipeddi and Rohan 2010). Also, there is the prospect of drug encapsulation (Chiappetta et al. 2011a; Kumar et al. 2012) functionalization by constructing established structures (Goldberg et al. 2007); or alteration (Alexis et al. 2008) will be able to direct optimized drug dosing and produce better drug deliverance by escalating stability (Santos-Martinez et al. 2014) and the drug-withholding period (Gabizon et al. 1994).

5.1.4 Nanosystem and Antiviral Mechanism

The discovery of nanotechnology has made it likely to figure out the cellular mechanism of the living cells and to build up significant technologies to enable early prognosis and curing of various viral infections (Mendes 2013; Villanueva-Flores et al. 2020). Few applications relating nanosystem and antiviral mechanisms are drug and gene delivery, exploitation of fluorescent-labeled dye, identification and analysis of proteins, tumors and pathogens, separation and purification of biological molecules and cells, and pharmacokinetic and tissue engineering studies (Joo et al. 2008; Saini et al. 2010; Cojocaru et al. 2020). Nanoformulations on their own may act as antiviral agents by various mechanisms besides abetting in overcoming the issues raised related to drug solubility and bioavailability. Nanoformulation on conventional drugs is advantageous because of the following reasons.

a. Their smaller size increases the surface-to-volume ratio and increases bioavailability, where conventional drugs have limited solubility in an aqueous medium.
b. Usage of a minimal dose of the drug is crucial; targeted drug delivery proves to be promising compared to traditional antiviral drugs that may have high toxic effects.
c. Patient acquiescence due to lessening dose frequency where they show nonconformity toward conventional drugs.
d. Nanoformulation has distinctive characteristics that increase the therapeutic efficiency and can control serious disease that seems deficient in traditional antiviral drugs.
e. Conventional antiviral drugs have a short half-life time, high rate of degradation activity, and slow uptake of the body, whereas nanocarriers have decreased the degradation rate of the drug by site-specific action.
f. Resistance to drug development is lessened by the coating and filtration process, which is not possible in traditional drugs.

Thus, nanotechnology has been of great interest to researchers and scientists all over the world to treat viral infections traditionally with less or no harmful substances than are by using nanoparticles (Blecher et al. 2011).

5.2 CHARACTERISTICS OF NANOCARRIERS AND THEIR APPLICATION IN VIRAL THERAPEUTICS

Nanocarriers are mainly organic based. For establishing a proficient nanocarrier, certain criteria ought to be considered: (1) biologically compatibility, (2) biodegradability, and (3) ability to bio-absorbable with the drug. Furthermore, NP must be modified to optimize pharmacokinetics characterization, in vivo bio-circulation, and viral interaction (Lembo et al. 2018). However, overcoming a few negative aspects is required to optimize for NPs; to prevail over the confines of antiviral therapy, a few of them are listed in Table 5.1. Some of the nanocarriers with nanoformulations for antiviral are described below:

TABLE 5.1
Lists of the Strategies by Nanosystems to Overcome Challenges in Antiviral Therapeutics

Challenges to Overcome in Antiviral Therapeutics	Strategies Essential by Nanosystems to Limit Antiviral Drugs
Surmounting blood brain barrier (BBB)	• Use of polymeric NPs that have biodegradable activity • Transportation of more drugs to the specific site • Stimuli responsive of nanosystems (NSs)
Target of lymph nodes, lymphoid tissues, macrophages	• Enhance targeting selectively associated ligands to NSs
Liver function	• Develop targeting selectively associated ligands to NSs, especially galactose
Gastric degradation, Less solubility in water, Low mucosal permeability, short dwelling time	• Gastro absorptive NSs • Polymeric NSs with extreme drug load and great bioavailability • NS with persistent drug release • Mucoadhesive and muco penetrating NSs
Improvement of vaginal microbicides for prevention and also therapy of STDs	• Polymeric NSs mainly chitosan and PLGA • Optimizing targeting specificity • Formulation of vehicles (gel creams, vaignal rings) • Electrospun polymeric fibers

5.2.1 Liposomes

Liposomes are smaller spherical vesicles, sizes ranging from 15–1000 nm. Liposomes consist of phospholipid bilayers that envelope an aqueous core where the drug is circulated (Zazo et al. 2016). These vesicles are capable of trapping and delivering a broad range of hydrophilic and hydrophobic drugs by enhancing the solubility of compounds through compartmentalizing and dissolving. Liposomes have unique characteristics such as biocompatibility, biodegradability, and less-lethal activity, which makes this nanocarrier extremely interesting as a drug-delivery vector (Lembo et al. 2018). Though the vesicles have shown effectiveness as nano-delivery systems, they are constrained for therapeutical application due to their reduced encapsulation efficacy, physical instability during storage and administration, and high production rate (He et al. 2019). Many methods are available, such as detergent removal method, supercritical fluid and freeze-drying technology, passive and active loading technique, solvent and mechanical (especially sonication) dispersion methods, dual-asymmetric centrifugation, cross-flow filtration, and membrane contactor technologies (Huang et al. 2014). Liposomes have a normal inclination for hepatic accumulation and thus act as the best drug-delivery system for the curing of liver diseases besides hepatitis C virus infections (Ripoli et al. 2016).

5.2.2 Niosomes and Ethosome

Niosomes are derived from surfactant monomers on hydration through self-assemblage. They are referred to as vesicular delivery systems and are advantageous over liposomes with regard to the stability of the compound, the low rate of chemicals, and the presence of a vast number of surfactants accessible for the design of the vesicular system (Mehendale et al. 2013). Niosomes are structurally nonionic, they encapsulate both hydrophilic molecules (within vascular core present in the aqueous state) and lipophilic molecules (encapsulation is accomplished by separating into the lipophilic realm of the bilayers) (Lembo et al. 2018).

Ethosomes are lipid multilamellar vesicles made up of ethanol, water, and phospholipids. Due to their capability to blend with skin lipids, ethosomes distribute relevant agents onto the skin more competently than liposomes. Based on the composition of ethanol compounds, ethosomes are categorized into three types, namely: classical, binary, and trans ethosomes. Binary ethosomes are composed of an additional alcohol group, either propylene glycol or isopropyl alcohol to the classical ethosome. Transethosomes are developed by extra compounds that act as an enhancer or an edge activator, besides the fundamental components of classical ethosomes. Of the three groups, classical ethosomes are more likely to penetrate the skin (Abdulbaqi et al. 2016).

5.2.3 Polymeric Nanoparticles

Basic nanocarriers, namely polymeric NPs, are produced by a mixture of natural or synthetic polymers using preparative techniques such as emulsion-diffusion evaporation, salting out method, polymer dispersion, dialysis, coacervation, phase-inversion temperature method, nanoprecipitation, spray drying, supercritical fluid technology and interfacial polymerization (Nagavarma et al. 2012). Polymeric NPs are further classified into nanocapsules and nanospheres based on drug incorporation (detailed in sections 5.2.3.1 and 5.2.3.2). The major benefits of polymeric NPs as nanocarriers are their controlled liberation, enhanced cellular intake, ability to shield the drug from depriving, site-specific delivery, low toxicity, and their capacity to be used as theragnostic (Qian et al. 2017; Ferrari et al. 2018). The characteristics of polymeric NPs have been extensively studied for increasing the efficiency of anti-retroviral drugs since polymeric NPs have a high ability to target the monocytes and macrophages of the brain and lymphatic system, which acts as the reservoir for viral dissemination at some stage during HIV infectivity (Edagwa et al. 2014).

5.2.3.1 Nanocapsules

Nanocapsules are structurally hollow spheres composed of a core and a shell; they range in size from 50 to 300 nm. The polymeric shell with a confined inner core encapsulates the drug. Nanocapsules are usually made from the nontoxic polymer by coating, layer-by-layer method, precipitation, emulsion-diffusion, emulsion coacervation, and evaporation (Mayer 2005). Special characteristics of nanocapsules that make them a significant nanocarrier are their high drug-loading capacity, site-specific delivery, low density, and controlled release.

5.2.3.2 Nanospheres

Nanospheres are colloidal systems, also defined as homogenous matrix systems, where the drug is dissolved or dispersed physically or consistently, either on the surface or ensnared within the polymeric matrix; sizes vary from 100 to 200 nm (Soppimath et al. 2001; Letchford and Burt 2007). Studies demonstrated the efficiency of nanospheres in treating Hepatitis (Zeng et al. 2011), Herpes (Baram-Pinto et al. 2012), and influenza (Dehghan et al. 2013).

5.2.4 Micelles

Micelles are polymer forms of nanocarriers that have structural sizes ranging from 10–100 nm (Letchford and Burt 2007). They are composed of both hydrophobic (inner core) and hydrophilic (outer shell) block copolymers per each unimer (Kataoka et al. 2001; Miyata et al. 2011; Cabral et al. 2018). The hydrophobic inner core aids in the absorption of poorly water-soluble drugs, and the hydrophilic outer shell acts as a barrier protecting the drug from external influences and reduces the nonspecific interactions with other molecules. Hydrophilic shells have an activity that enhances circulation time and, as a result, improves accumulation and therapeutic efficiency (Ochekpe et al. 2009; Gill et al. 2015; Mandal et al. 2017). Encapsulation of the drug might be accomplished by conjoining the drug chemically with the hydrophobic part of the block copolymer, allowing higher drug lading and limiting premature discharge (Trivedi and Kompella 2010). Compelling the micellar surface with specific ligands or monoclonal antibodies has shown the significance of targeted drug delivery using polymeric micelles (Jhaveri and Torchilin 2014). The best example of polymeric micelles is polyethylene glycol, which is used as a shield-forming block of the surfactant. Shells acuity of pH and temperature are intended by two derivatives, namely poly N-isopropyl acrylamide and polymethacrylic acid. To formulate polymeric micelles, certain techniques, such as solvent extraction, solution casting, and dialysis methods, have been in use (Kumari et al. 2014).

5.2.5 Dendrimers

Dendrimers are symmetrical, macromolecule, and highly branched diffuse from the central core through connectors and branching units. Dendrimers are structurally composed of three domains, i.e. the inner shell, central core, and outer shell with terminal functionalities (Caminade et al. 2005; Tomalia 2005). Dendrimers are usually 2–10 nm in diameter size; they are also known as synthetic nanoarchitectures. The unoccupied and pliable spaces in the dendrimers ease the allurement of the foreign molecules, and the functional groups on the outer surface permit interaction with the preferred target; they can transmit the drug by either electrostatic interaction or chemical conjugation (Tripathy and Das 2013). Sequential adding of molecules (building blocks) to the initiator via convergent or contrary approach results in producing dendrimers (Sebestik et al. 2017). The use of dendrimers is of great significance because they are easily assimilated by cells, have longer circulation times, and have increased stability and solubility and selected delivery (Mayer 2005). The most traded dendrimers are polyamidoamine, poly L-lysine (PLL), and polypropylene imine (PPI) (Garg et al. 2011). The functionality of dendrimers is regulated with three major biological molecules, namely proteins, carbohydrates, and anionic groups. Excitingly, dendrimers have an inbuilt function through antimicrobial or antiviral activities. Dendrimers decked with anionic groups intermingle with the V3 loop of glycoprotein 120, and they hinder viral replication at the initial stage. Dendrimers PPI and PLL with sulfated galactose and cellobiose residue, respectively, have shown more activity against viruses (Kensinger et al. 2004). A study by Kannan et al. (2014) demonstrated the significant effect of glycodendropeptide (having 16 peptides and 9 mannose units) in optimization of the immune response toward antiviral activity (Kannan et al. 2014).

5.2.6 Solid Lipid NPs

Solid lipid nanoparticles (SLNs) are colloidal matrix systems with sizes ranging from about 10 to 1000 nm. The main SLNs are lipids and their derivatives such as triglycerides, partial glycerides, fatty acids, steroids, and waxes. The main purpose of SLNs production is to attain the controlled release of the drug, enhance stability, and improve the loading competence and site-specific drug-delivery system (Zhao et al. 2014). Industrially SLNs are ascendible. They are used for the delivery of various antiviral drugs, such as darunavir, zidovudine, efavirenz, ritonavir, lopinavir, and maraviroc in combination with lipids like Compritol 888ATO and Gelucire 44/14, which are water-soluble excipient. Techniques such as ultrasonication, emulsification-diffusion, supercritical-assisted injection in a liquid antisolvent (SAILA), spray-drying, high-pressure homogenization, solvent emulsification, or evaporation and extraction of fluid emulsions are preferred (Baram-Pinto et al. 2012).

Nanostructured lipid carriers (NLCs) are précised as second-generation SLNs. NLCs use lipids in a liquid state, which allows better loading capacity as well as improved stability and controlled drug release, compared to SLN. To attain target specificity, NLCs are modified on the external surface. The utilization of SLNs intends to join the advantages of conventional nanocarriers while evading few limitations. One such example is the production of polymeric NPs at a large scale that confines their efficacy in drug delivery. Alternatively, the production of SLNs can be accomplished in two easy and cost-effective ways i.e. by homogenization technique at high pressure and microemulsion method (Müller et al. 2000).

5.2.7 Cyclodextrin Derivatives

Cyclodextrin derivatives are nanometric materials composed of alpha-1–4 cyclic linkage glucopyranose oligomers. Cyclodextrin is generally of two types: the α and γ-cyclodextrin constituting 6 to 8 glucopyranose units. The distinctive feature of cyclodextrin derivatives is their unique form with a toroidal structure composed of multiple glucose units linked between each other, thereby producing a hydrophilic exterior and a hydrophobic cavity. The making of cyclodextrin drug complexes has been extensively studied for enhancing the stability, solubility, bioavailability, and dissolution rate of the therapeutic agents. Cyclodextrin-based nanosponges are hyper cross-linked; they have cyclodextrin units as building blocks, which mark the capability to encapsulate an enormous range of substances (Trotta et al. 2012).

5.2.8 Nano Emulsions and Suspensions

Nanoemulsions are biphasic thermodynamically stable systems with a magnitude range of 20–500 nm. Nanoemulsions (NE) constitute water, oil, and emulsifier as surfactants and cosurfactants. The significance of adding an emulsifier is crucial for the formation of small droplets as it decreases the interfacial tension (Helgeson 2016). Techniques such as microfluidization, homogenization at higher pressure and ultrasonication (phase inversion temperature), or low-energy emulsification method are used for NE preparations (Gupta et al. 2019). Oils, namely castor oil, soyabean oil, and peanut oil, are used as lipids in NE preparation (Hobson et al. 2018). For creating a small droplet to lessen the interfacial tension within the oil and water phases of the emulsion, the main prerequisite is adding an emulsifier (Helgeson 2016). The advantages of NEs are their capability to increase solubility in an aqueous medium, high loading capacity, comparatively high dwelling time in GIT, increased absorption and bioavailability (Sutradhar and Amin 2013). Few examples of NEs developed are nevirapine (proven to show low drug efflux, improved permeability, and profile liberation) (Manyarara et al. 2018), saquinavir (Vyas et al. 2008), and indinavir (Prabhakar et al. 2013); saquinavir and indinavir both demonstrate improved brain-targeting potential.

Nanosuspensions, also referred to as nanocrystals, are nanoformulations that are stabilized by polymers or surfactants. They are produced industrially through pharmaceutical technologies by either top-down method or bottom-up method, or a combination of both, especially in liquid media. The purpose of nanosuspensions is to increase the rate of solubility and dissolution, and the biological availability of the drug through different administration paths. To date, around nine nanocrystals are derived in the market.

5.3 NANOTHERAPEUTICS AGAINST SPECIFIC VIRAL INFECTIONS

Many nanomedicines have been approved and are being investigated for the treatment of various viral infections. A few of the investigations are illustrated in the below sections.

5.3.1 HIV

Human immunodeficiency virus (HIV) is the source organism causing acquired immunodeficiency syndrome (AIDS), which is the main public health issue and cause of many deaths worldwide (Fauci et al. 2013). Yet there is no effective vaccine against HIV, due to its atypical pathogenesis. Studies for antiretroviral drugs against the infection were of great concern; currently, only 29 drugs have been approved based on their replication phases. However, due to their toxicity, systemic side effects, efficient concentrations in the viral pool, drug-drug interaction, poor patient acquiescence, and poor bioavailability, antiretroviral drugs have not been used as potential actions. So, to prevail over these problems, nanoformulation approaches have been developed in recent decades; a few are listed in Table 5.2. Some other nanoformulations are polyanionic carbosilane dendrimers (PCDs), used as topical microbicides; their mode of action is that they inhibit entry of the virus by fusion process (Telwatte et al. 2011). Tenofovir and maraviroc, in mixture with PCDs, are shown to have high efficacy as monotherapy; the combination decreases side effects and the number of doses needed, and also reduced the appearance of multidrug-resistant mutants of HIV (Sepulveda-Crespo et al. 2015).

TABLE 5.2
List of Antiviral Drugs with Nanoformulations (Lembo et al. 2018)

Therapeutic Agents	Nanomaterials	Reference
Enhancing drug aqueous solubility		
Nevirapine	Nanocrystals	Shegokar and Singh (2011)
Efavirenz	Nanodispersion, Polymeric NPs, Polymeric micelles	Chiappetta et al. (2010); Gaur et al. (2014); Tshweu et al. (2014)
Nelfinavir mesylate, Darunavir, Atazanavir	Solid lipid NPs	Patel et al. (2014)
Saquinavir	Cyclodextrins-complexes	Sathigari et al. (2009)
Brain targeting		
Stavudine, AZT, Lamivudine, Delavirdine, Ritonavir,	Polymeric NPs	Kuo and Chen (2006); Borgmann et al. (2011)
Nevirapine	Quantum rods	Kuo et al. (2011)
Saquinavir	Polymeric NPs, Dendrimers, Solid lipid NPs	Vyas et al. (2008); Mahajan et al. (2014)
Zidovudine, Nelfinavir, Lamivudine, Ritonavir	Magnetoliposomes	Kuo and Chen (2006); Saiyed et al. (2010)
Macrophage uptake and drug release		
Saquinavir, Ritonavir, Lopinavir, Efavirenz, Didanosine, AZT-TP	Polymeric NPs	Hillaireau et al. (2006); Hillaireau et al. (2007); Destache et al. (2010); Pattnaik et al. (2012); Giacalone et al. (2013)
AZT-TP	Magnetic NPs	Saiyed et al. (2009)
Atazanavir, Efavirenz, Ritonavir, Indinavir	Nanocrystals	Nowacek et al. (2011)
Saquinavir	PEGylated liposomes	Ramana et al. (2012)

(Continued)

TABLE 5.2 (Continued)
List of Antiviral Drugs with Nanoformulations (Lembo et al. 2018)

Therapeutic Agents	Nanomaterials	Reference
Vaginal delivery		
Saquinavir, Tenofovir, Efavirenz, Lopinavir, Etravirine, Maraviroc, Raltegravir, Rilpivirine	Polymeric nanoparticles	Ramana et al. (2012)
Maraviroc, Tenofovir, Etravirine	Electrospun polymeric fiber	Huang et al. (2012)
Macrophage targeting		
Efavirenz	Dendrimers	Dutta et al. (2007)
AZT, Stavudine	Liposomes	Garg and Jain (2006)
Ritonavir	Polymeric NP	Borgmann et al. (2011)
Lymph node targeting		
Indinavir; Lopinavir, Ritonavir, Tenofovir, Efavirenz	Lipid NPs	Endsley and Ho (2012); Freeling et al. (2014); Roy et al. (2015)
AZT	Liposomes	Kaur et al. (2008)
Long-acting injectable delivery		
Rilpivirine, Atazanavir, Ritonavir, Efavirenz	Nanosuspension	Williams et al. (2015)
Atazanavir, Darunavir, Ritonavir, Tenofovir	Lipid nanoparticles	Duan et al. (2014)
Rilpivirine, Cabotegravir	Nanocrystals	Andrews and Heneine (2015)
Cell-based drug-delivery system		
Ritonavir, Indinavir, Efavirenz, Atazanavir	Nanoparticles into Macrophages	Martinez-Skinner et al. (2013)
Improving drug mucosal permeability		
Tenofovir	Proliposomes	Zidan et al. (2013)
Lopinavir	Solid lipid NPs	Negi et al. (2013)

5.3.2 HSV

Herpes simplex virus (HSV) infection is caused by the herpes virus belonging to the Herpesviridae family. HSV is distinguished into three different types based on the subfamilies (α, β, and γ). All types of viruses of the family exhibit life-long latency based on different cell types and tissues, where later reactivation results in symptomatic or asymptomatic infections. A range of nanomedicines is at present under study for the management of herpetic infections. Infection caused by HSV is not curable; however, studies targeted the inhibition of the replication process of virus followed by damaging the epithelial cells by therapeutic agents with modifications through nanoformulations. The therapeutic agents that are frequently in use for HSV are acyclovir, ganciclovir, and famciclovir.

Acyclovir has a short half-life, is less soluble in water, and absorption is slower; due to these reasons, various nanoparticles based on cyclodextrin have been developed to enhance the physico-chemical factors of ACV (Bencini et al. 2008; Cavalli et al. 2009; Ghera et al. 2009). A study by Kamel and his colleagues developed sneaky PLGA NPs laden with ACV to lengthen the drug plasma half-life after intravenous administration (Kamel et al. 2009). For the bioavailability of ACV, nanoformulations with nanomaterials especially galactosylated NPs and solid lipid NPs, played a key role in the drug-delivery system (Gupta et al. 2013; Seyfoddin and Al-Kassas 2013). Solid lipid NPs can be administrated in both ocular and dermal delivery systems (Jain et al. 2011; Chetoni et al. 2016). Another class of nanomaterials has been developed for dermal delivery of ACV: that is, nanoemulsions by the phase inversion temperature method (Schwarz et al. 2012).

Similarly, to improve drug bioavailability of ganciclovir (GCV), nanocarriers such as nanoemulsions, micellar NPs, solid lipid NPs, and nanoniosomes played a key role in nanoformulation of GCV drug (Akhter et al. 2012; Ren et al. 2013; Sawdon and Peng 2015). Mucoadhesive nanoemulsion formulation with GCV reported nontoxic and nonirritant for ocular drug-delivery system (Akhter et al. 2011).

In another study with ammonium terminated amphiphilic Janus, dendrimers were shown to self-accumulate in water to form micelles capable of transporting the antiviral drug camptothecin. These drug-loaded dendrimers were established to be highly effective against replicating HCV (Lancelot et al. 2017).

5.3.3 Influenza Virus

Orthomyxoviruses, which are seasonal, cause human influenza infection that affects nearly 5 to 10% of the population worldwide annually. So far, very few drugs have been approved i.e. adamantanes (an ion channel matrix protein inhibitor), analogs of sialic acid, and lastly, viral RNA polymerase that are inhibitors of neuraminidase (example, zanamivir and oseltamivir). However, these drugs do not show strong bioavailability; hence, using the double emulsion solvent evaporation method, zanamivir is loaded with solid lipid NPs for oral bioavailability (Cao et al. 2011) and dendrimer cyclodextrin conjugation with amantadine for drug-loading ability (Wang et al. 2012).

5.3.4 HBV Infection

The organism hepatitis B virus (HBV) causes Hepatitis B (liver infection). Besides having a prophylactic vaccine, deaths due to HBV have been on the rise every year. Drugs, namely adefovir, dipivoxil, and lamivudine, are more in use. Nevertheless, adefovir dipivoxil is formulated with solid lipid NPs and nanosuspension nanocarriers due to its low bioavailability; later on, nanoformulation increased the efficiency of bioavailability by 52% and 78% (Dodiya et al. 2013).

5.4 CONCLUSION

Nanotechnology has modernized globally through innovative techniques toward multiple issues allied with the health-care system. Despite numerous studies conducted in producing nanomedicines yet, research should be directed to accomplish numerous functions (i.e. multifunctionalization) of nanomaterials to overcome the drawbacks of the therapeutic agents, which seem toxic and show low bioavailability against infectious disease. Boosting immune response and usage of nanocarriers as adjuvants toward antiviral vaccines is promising in preventing and managing viral infections. Due to emerging insights in the fields of sciences with advanced technologies, it is optimistic that controlling and preventing the new emerging viral infections is inevitable.

ACKNOWLEDGEMENTS

The authors Pavanai Sanapala and Sudhakar Pola are grateful to authorities of Andhra University Visakhapatnam, India for providing necessary facilities to carry out the research work and for extending constant support.

REFERENCES

Abdulbaqi, I.M., Darwis, Y., Khan, N.A.K., Abou Assi, R. and Khan, A.A. 2016. Ethosomal nanocarriers: The impact of constituents and formulation techniques on ethosomal properties, in vivo studies, and clinical trials. *International Journal of Nanomedicine* 11:2279.

Akhter, S., Kushwaha, S., Warsi, M.H., et al. 2012. Development and evaluation of nanosized niosomal dispersion for oral delivery of Ganciclovir. *Drug Development and Industrial Pharmacy* 38(1):84–92.

Akhter, S., Talegaonkar, S., Khan, Z.I., Jain, G.K., Khar, R.K. and Ahmad, F.J., 2011. Assessment of ocular pharmacokinetics and safety of Ganciclovir loaded nanoformulations. *Journal of Biomedical Nanotechnology* 7(1):144–145.

Alexis, F., Pridgen, E., Molnar, L.K. and Farokhzad, O.C. 2008. Factors affecting the clearance and biodistribution of polymeric nanoparticles. *Molecular Pharmaceutics* 5(4):505–515.

Andrews, C.D. and Heneine, W. 2015. Cabotegravir long-acting for HIV-1 prevention. *Current Opinion in HIV and AIDS* 10(4):258–263.

Antunes Filho, S., dos Santos, O.A.L., dos Santos, M.S. and Backx, B.P. 2020. Exploiting nanotechnology to target viruses. *Journal of Nanotechnology and Nanomaterials* 1(1):11–15.

Baram-Pinto, D., Shukla, S., Richman, M., Gedanken, A., Rahimipour, S. and Sarid, R. 2012. Surface-modified protein nanospheres as potential antiviral agents. *Chemical Communications* 48(67):8359–8361.

Baron, S., Fons, M. and Albrecht, T. 1996. Viral pathogenesis. In *Medical Microbiology*, ed. S. Baron. Galveston (TX): University of Texas Medical Branch at Galveston.

Bencini, M., Ranucci, E., Ferruti, P., et al. 2008. Preparation and in vitro evaluation of the antiviral activity of the acyclovir complex of a β-cyclodextrin/poly (amidoamine) copolymer. *Journal of Controlled Release* 126(1):17–25.

Blecher, K., Nasir, A. and Friedman, A. 2011. The growing role of nanotechnology in combating infectious disease. *Virulence* 2(5):395–401.

Borgmann, K., Rao, K.S., Labhasetwar, V. and Ghorpade, A. 2011. Efficacy of Tat-conjugated ritonavir-loaded nanoparticles in reducing HIV-1 replication in monocyte-derived macrophages and cytocompatibility with macrophages and human neurons. *AIDS Research and Human Retroviruses* 27(8):853–862.

Bowman, M.C., Ballard, T.E., Ackerson, C.J., et al. 2008. Inhibition of HIV fusion with multivalent gold nanoparticles. *Journal of the American Chemical Society*, 130(22): 6896–6897.

Cabral, H., Miyata, K., Osada, K. and Kataoka, K. 2018. Block copolymer micelles in nanomedicine applications. *Chemical Reviews* 118(14):6844–6892.

Caminade, A.M., Laurent, R. and Majoral, J.P. 2005. Characterization of dendrimers. *Advanced Drug Delivery Reviews* 57(15):2130–2146.

Cao, Q., Wu, H., Zhu, L., et al. 2011. Preparation and evaluation of zanamivir-loaded solid lipid nanoparticles. *Journal of Controlled Release* 152(1):e2–e4.

Caron, J., Reddy, L.H., Lepêtre-Mouelhi, S., et al. 2010. Squalenoyl nucleoside monophosphate nanoassemblies: New prodrug strategy for the delivery of nucleotide analogues. *Bioorganic and Medicinal Chemistry Letters* 20(9):2761–2764.

Cavalli, R., Donalisio, M., Civra, A., et al. 2009. Enhanced antiviral activity of Acyclovir loaded into β-cyclodextrin-poly (4-acryloylmorpholine) conjugate nanoparticles. *Journal of Controlled Release* 137(2):116–122.

Chetoni, P., Burgalassi, S., Monti, D., et al. 2016. Solid lipid nanoparticles as promising tool for intraocular tobramycin delivery: Pharmacokinetic studies on rabbits. *European Journal of Pharmaceutics and Biopharmaceutics* 109:214–223.

Chiappetta, D.A., Facorro, G., de Celis, E.R. and Sosnik, A. 2011a. Synergistic encapsulation of the anti-HIV agent efavirenz within mixed poloxamine/poloxamer polymeric micelles. *Nanomedicine: Nanotechnology, Biology and Medicine* 7(5):624–637.

Chiappetta, D.A., Hocht, C., Taira, C. and Sosnik, A. 2010. Efavirenz-loaded polymeric micelles for pediatric anti-HIV pharmacotherapy with significantly higher oral bioavailability. *Nanomedicine* 5(1):11–23.

Chinchar, V.G. 1999. Replication of viruses. *Encyclopedia of Virology* 1471–1478.

Cojocaru, F.D., Botezat, D., Gardikiotis, I., et al. 2020. Nanomaterials designed for antiviral drug delivery transport across biological barriers. *Pharmaceutics* 12(2):171.

Dehghan, S., Kheiri, M.T., Tabatabaiean, M., Darzi, S. and Tafaghodi, M. 2013. Dry-powder form of chitosan nanospheres containing influenza virus and adjuvants for nasal immunization. *Arch Pharm* Res. 36(8):981–992.

Destache, C.J., Belgum, T., Goede, M., Shibata, A. and Belshan, M.A. 2010. Antiretroviral release from poly (DL-lactide-co-glycolide) nanoparticles in mice. *Journal of Antimicrobial Chemotherapy* 65(10):2183–2187.

Dodiya, S., Chavhan, S., Korde, A. and Sawant, K.K. 2013. Solid lipid nanoparticles and nanosuspension of adefovir dipivoxil for bioavailability improvement: Formulation, characterization, pharmacokinetic and biodistribution studies. *Drug Development and Industrial Pharmacy* 39(5):733–743.

Duan, J., Freeling, J.P., Koehn, J., Shu, C., and Ho, R.J. 2014. Evaluation of atazanavir and darunavir interactions with lipids for developing pH-responsive anti-HIV drug combination nanoparticles. *Journal of Pharmaceutical Sciences* 103(8): 2520–2529.

Dutta, T., Agashe, H.B., Garg, M., Balasubramanium, P., Kabra, M. and Jain, N.K. 2007. Poly (propyleneimine) dendrimer based nanocontainers for targeting of efavirenz to human monocytes/macrophages in vitro. *Journal of Drug Targeting* 15(1):89–98.

Edagwa, B., Zhou, T., M McMillan, J., Liu, X.M. and Gendelman, H.E. 2014. Development of HIV reservoir targeted long acting nanoformulated antiretroviral therapies. *Current Medicinal Chemistry* 21(36):4186–4198.

Endsley, A.N. and Ho, R.J. 2012. Design and characterization of novel peptide-coated lipid nanoparticles for targeting anti-HIV drug to CD4 expressing cells. *The AAPS Journal* 14(2):225–235.

Fauci, A.S., Folkers, G.K. and Dieffenbach, C.W., 2013. HIV-AIDS: Much accomplished, much to do. *Nature Immunology* 14(11):1104–1107.

Ferrari, R., Sponchioni, M., Morbidelli, M. and Moscatelli, D. 2018. Polymer nanoparticles for the intravenous delivery of anticancer drugs: The checkpoints on the road from the synthesis to clinical translation. *Nanoscale* 10(48):22701–22719.

Freeling, J.P., Koehn, J., Shu, C., Sun, J. and Ho, R.J. 2014. Long-acting three-drug combination anti-HIV nanoparticles enhance drug exposure in primate plasma and cells within lymph nodes and blood. *AIDS* 28(17):2625.

Gabizon, A., Catane, R., Uziely, B., et al. 1994. Prolonged circulation time and enhanced accumulation in malignant exudates of doxorubicin encapsulated in polyethylene-glycol coated liposomes. *Cancer Research* 54(4):987–992.

Gagliardi, M. 2017. Biomimetic and bioinspired nanoparticles for targeted drug delivery. *Therapeutic Delivery* 8(5):289–299.

Garg, M. and Jain, N.K. 2006. Reduced hematopoietic toxicity, enhanced cellular uptake and altered pharmacokinetics of azidothymidine loaded galactosylated liposomes. *Journal of Drug Targeting* 14(1):1–11.

Garg, T., Singh, O., Arora, S. and Murthy, R. 2011. Dendrimer - A novel scaffold for drug delivery. *International Journal of Pharmaceutical Sciences Review and Research* 7(2):211–220.

Gaur, P.K., Mishra, S., Bajpai, M. and Mishra, A. 2014. Enhanced oral bioavailability of efavirenz by solid lipid nanoparticles: in vitro drug release and pharmacokinetics studies. *BioMed Research International* 2014:1–9.

Gerhard, W., Mozdzanowska, K. and Zharikova, D. 2006. Prospects for universal influenza virus vaccine. *Emerging Infectious Diseases* 12(4):569–574.

Ghera, B.B., Perret, F., Chevalier, Y. and Parrot-Lopez, H. 2009. Novel nanoparticles made from amphiphilic perfluoroalkyl α-cyclodextrin derivatives: Preparation, characterization and application to the transport of acyclovir. *International Journal of Pharmaceutics* 375(1-2):155–162.

Giacalone, G., Bochot, A., Fattal, E. and Hillaireau, H. 2013. Drug-induced nanocarrier assembly as a strategy for the cellular delivery of nucleotides and nucleotide analogues. *Biomacromolecules* 14(3):737–742.

Gill, K.K., Kaddoumi, A. and Nazzal, S. 2015. PEG–lipid micelles as drug carriers: Physiochemical attributes, formulation principles and biological implication. *Journal of Drug Targeting* 23(3):222–231.

Goldberg, M., Langer, R. and Jia, X. 2007. Nanostructured materials for applications in drug delivery and tissue engineering. *Journal of Biomaterials Science, Polymer Edition* 18(3):241–268.

Gupta, P.K., Bhandari, N., Shah, H., et al. 2019. An update on nanoemulsions using nanosized liquid in liquid colloidal systems. *Nanoemulsions-Properties, Fabrications and Applications*. Kai Seng Koh and Voon Loong Wong, IntechOpen. https://www.intechopen.com/books/nanoemulsions-properties-fabrications-and-applications/an-update-on-nanoemulsions-using-nanosized-liquid-in-liquid-colloidal-systems.

Gupta, S., Agarwal, A., Gupta, N.K., Saraogi, G., Agrawal, H. and Agrawal, G.P. 2013. Galactose decorated PLGA nanoparticles for hepatic delivery of acyclovir. *Drug Development and Industrial Pharmacy* 39(12):1866–1873.

Gurunathan, S., Qasim, M., Choi, Y., et al. 2020. Antiviral potential of nanoparticles - Can nanoparticles fight against coronaviruses?. *Nanomaterials* 10(9):1645.

Gurunathan, S., Qasim, M., Park, C.H., et al. 2019. Cytotoxicity and transcriptomic analyses of biogenic palladium nanoparticles in human ovarian cancer cells (SKOV3). *Nanomaterials* 9(5):787.

Gurunathan, S., Qasim, M., Park, C., Yoo, H., Kim, J.H. and Hong, K. 2018. Cytotoxic potential and molecular pathway analysis of silver nanoparticles in human colon cancer cells HCT116. *International Journal of Molecular Sciences* 19(8):2269.

He, H., Lu, Y., Qi, J., Zhu, Q., Chen, Z. and Wu, W. 2019. Adapting liposomes for oral drug delivery. *Acta Pharmaceuticasinica B* 9(1):36–48.

Helgeson, M.E. 2016. Colloidal behavior of nanoemulsions: Interactions, structure, and rheology. *Current Opinion in Colloid and Interface Science* 25:39–50.

Hillaireau, H., Le Doan, T., Appel, M. and Couvreur, P. 2006. Hybrid polymer nanocapsules enhance in vitro delivery of azidothymidine-triphosphate to macrophages. *Journal of Controlled Release* 116(3):346–352.

Hillaireau, H., Le Doan, T., Chacun, H., Janin, J. and Couvreur, P. 2007. Encapsulation of mono-and oligo-nucleotides into aqueous-core nanocapsules in presence of various water-soluble polymers. *International Journal of Pharmaceutics* 331(2): 148–152.

Hobson, J.J., Edwards, S., Slater, R.A., Martin, P., Owen, A. and Rannard, S.P., 2018. Branched copolymer-stabilised nanoemulsions as new candidate oral drug delivery systems. *RSC Advances* 8(23):12984–12991.

Huang, C., Soenen, S.J., van Gulck, E., et al. 2012. Electrospun cellulose acetate phthalate fibers for semen induced anti-HIV vaginal drug delivery. *Biomaterials* 33(3):962–969.

Huang, Z., Li, X., Zhang, T., et al. 2014. Progress involving new techniques for liposome preparation. *asian journal of Pharmaceutical Sciences* 9(4):176–182.

Jain, S., Mistry, M.A. and Swarnakar, N.K. 2011. Enhanced dermal delivery of acyclovir using solid lipid nanoparticles. *Drug Delivery and Translational Research* 1(5): 395–406.

Jeyaraj, M., Gurunathan, S., Qasim, M., Kang, M.H. and Kim, J.H. 2019. A comprehensive review on the synthesis, characterization, and biomedical application of platinum nanoparticles. *Nanomaterials* 9(12):1719.

Jhaveri, A.M. and Torchilin, V.P. 2014. Multifunctional polymeric micelles for delivery of drugs and siRNA. *Frontiers in pharmacology* 5:77.

Joo, K.I., Lei Y., Lee C.L., et al. 2008. Site-specific labeling of enveloped viruses with quantum dots for single virus tracking. *ACS Nano* 2:1553–1562.

Kamel, A.O., Awad, G.A., Geneidi, A.S. and Mortada, N.D. 2009. Preparation of intravenous stealthy acyclovir nanoparticles with increased mean residence time. *AAPS Pharmscitech* 10(4):1427–1436.

Kannan, R.M., Nance, E., Kannan, S. and Tomalia, D.A. 2014. Emerging concepts in dendrimer-based nanomedicine: From design principles to clinical applications. *Journal of Internal Medicine* 276(6):579–617.

Kataoka, K., Harada, A. and Nagasaki, Y. 2001. Block copolymer micelles for drug delivery: Design, characterization and biological significance. *Advanced Drug Delivery Reviews* 47(1):113–131.

Kaur, C.D., Nahar, M. and Jain, N.K. 2008. Lymphatic targeting of zidovudine using surface-engineered liposomes. *Journal of Drug Targeting* 16(10):798–805.

Kensinger, R.D., Catalone, B.J., Krebs, F.C., Wigdahl, B. and Schengrund, C.L. 2004. Novel polysulfated galactose-derivatized dendrimers as binding antagonists of human immunodeficiency virus type 1 infection. *Antimicrobial Agents and Chemotherapy* 48(5): 1614.

Klimpel, G.R. 1996. Immune defenses. In *Medical Microbiology*, ed. S. Baron, Galveston (TX): University of Texas Medical Branch at Galveston.

Krammer, F. 2015. The quest for a universal flu vaccine: Headless HA 2.0. *Cell Host and Microbe* 18(4):395–397.

Kumar, A., Ma, H., Zhang, X., et al. 2012. Gold nanoparticles functionalized with therapeutic and targeted peptides for cancer treatment. *Biomaterials* 33(4):1180–1189.

Kumari, A., Singla, R., Guliani, A. and Yadav, S.K. 2014. Nanoencapsulation for drug delivery. *EXCLI Journal* 13:265.

Kuo, Y.C. and Chen, H.H. 2006. Effect of nanoparticulate polybutylcyanoacrylate and methylmethacrylate–sulfopropylmethacrylate on the permeability of zidovudine and lamivudine across the in vitro blood–brain barrier. *International Journal of Pharmaceutics* 327(1-2):160–169.

Kuo, Y.C., Lin, P.I. and Wang, C.C. 2011. Targeting nevirapine delivery across human brain microvascular endothelial cells using transferrin-grafted poly (lactide-co-glycolide) nanoparticles. *Nanomedicine* 6(6):1011–1026.

Lancelot, A., Clavería-Gimeno, R., Velázquez-Campoy, A., Abian, O., Serrano, J.L. and Sierra, T. 2017. Nanostructures based on ammonium-terminated amphiphilic Janus dendrimers as camptothecin carriers with antiviral activity. *European Polymer Journal* 90:136–149.

Lara, H.H., Ayala-Nuñez, N.V., Ixtepan-Turrent, L. and Rodriguez-Padilla, C. 2010. Mode of antiviral action of silver nanoparticles against HIV-1. *Journal of Nanobiotechnology* 8(1):1–10.

Lembo, D., Donalisio, M., Civra, A., Argenziano, M. and Cavalli, R. 2018. Nanomedicine formulations for the delivery of antiviral drugs: A promising solution for the treatment of viral infections. *Expert Opinion on Drug Delivery* 15(1):93–114.

Letchford, K. and Burt, H. 2007. A review of the formation and classification of amphiphilic block copolymer nanoparticulate structures: Micelles, nanospheres, nanocapsules and polymersomes. *European Journal of Pharmaceutics and Biopharmaceutics* 65(3): 259–269.

Mahajan, H.S., Mahajan, M.S., Nerkar, P.P. and Agrawal, A. 2014. Nanoemulsion-based intranasal drug delivery system of saquinavir mesylate for brain targeting. *Drug Delivery* 21(2):148–154.

Mallipeddi, R. and Rohan, L.C. 2010. Progress in antiretroviral drug delivery using nanotechnology. *International Journal of Nanomedicine* 5:533.

Mandal, A., Bisht, R., Rupenthal, I.D. and Mitra, A.K. 2017. Polymeric micelles for ocular drug delivery: From structural frameworks to recent preclinical studies. *Journal of Controlled Release* 248:96–116.

Manjarrez-Zavala, M.E., Rosete-Olvera, D.P., Gutiérrez-González, L.H., Ocadiz-Delgado, R. and Cabello-Gutiérrez, C. 2013. Pathogenesis of viral respiratory infection. *Respiratory Disease and Infection: A New Insight* 1:3–32.

Manyarara, T.E., Khoza, S., Dube, A. and Maponga, C.C. 2018. Formulation and characterization of a paediatricnanoemulsion dosage form with modified oral drug delivery system for improved dissolution rate of nevirapine. *MRS Advances* 3(37):2203–2219.

Martinez-Skinner, A.L., Veerubhotla, R.S., Liu, H., et al. 2013. Functional proteome of macrophage carried nanoformulated antiretroviral therapy demonstrates enhanced particle carrying capacity. *Journal of Proteome Research* 12(5):2282–2294.

Mayer, C. 2005. Nanocapsules as drug delivery systems. *The International Journal of Artificial Organs* 28(11):1163–1171.

McNeil, S.E. 2011. Unique benefits of nanotechnology to drug delivery and diagnostics. In *Characterization of Nanoparticles Intended for Drug Delivery*, 3–8.

Mehendale, R., Joshi, M. and Patravale, V.B. 2013. Nanomedicines for treatment of viral diseases. *Critical Reviews in Therapeutic Drug Carrier Systems* 30(1):1–49.

Mendes, P.M. 2013. Cellular nanotechnology: Making biological interfaces smarter. *Chemical Society Reviews* 42(24):9207–9218.

Milroy, D. and Featherstone, J. 2002. Antiviral market overview. *Nature Reviews Drug Discovery* 1(1):11–12.

Mirkin, C.A. and Niemeyer, C.M. 2004. *Nanobiotechnology: Concepts, Applications, and Perspectives*. Wiley-VCH.

Miyata, K., Christie, R.J. and Kataoka, K. 2011. Polymeric micelles for nano-scale drug delivery. *Reactive and Functional Polymers* 71(3):227–234.

Müller, R. H., Mäder, K., and Gohla, S. 2000. Solid lipid nanoparticles (SLN) for controlled drug delivery – A review of the state of the art. *European Journal of Pharmaceutics and Biopharmaceutics* 50(1):161–177.

Nagavarma, B.V.N., Yadav, H.K., Ayaz, A.V.L.S., Vasudha, L.S. and Shivakumar, H.G. 2012. Different techniques for preparation of polymeric nanoparticles - A review. *Asian Journal of Pharmceutical Clincal Resarch* 5(3):16–23.

Nalwa, H.S. 1999. *Handbook of Nanostructured Materials and Nanotechnology, Vol. 1.* San Diego, 1–3. Cambridge: Academic Press.

Negi, J.S., Chattopadhyay, P., Sharma, A.K. and Ram, V. 2013. Development of solid lipid nanoparticles (SLNs) of lopinavir using hot self nano-emulsification (SNE) technique. *European Journal of Pharmaceutical Sciences* 48(1-2):231–239.

Nowacek, A.S., Balkundi, S., McMillan, J., et al. 2011. Analyses of nanoformulated antiretroviral drug charge, size, shape and content for uptake, drug release and antiviral activities in human monocyte-derived macrophages. *Journal of Controlled Release* 150(2):204–211.

Ochekpe, N.A., Olorunfemi, P.O. and Ngwuluka, N.C. 2009. Nanotechnology and drug delivery part 2: Nanostructures for drug delivery. *Tropical Journal of Pharmaceutical Research* 8(3):275–287.

Oliveira, K.A. and De, M. 2013. Antimicrobial activity and quantification of total flavonoids and phenols in different extracts of propolis. *Seminar: Biological and Health Sciences* 211–222.

Panáček, A., Kolář, M., Večeřová, R., et al. 2009. Antifungal activity of silver nanoparticles against *Candida* spp. *Biomaterials* 30(31):6333–6340.

Parboosing, R., Maguire, G. E., Govender, P., et al. 2012. Nanotechnology and the treatment of HIV infection. *Viruses* 4(4):488–520.

Patel, A., Shelat, P. and Lalwani, A. 2014. Development and optimization of solid self-nanoemulsifying drug delivery system (S-SNEDDS) using Scheffe's design for improvement of oral bioavailability of nelfinavir mesylate. *Drug Delivery and Translational Research* 4(2):171–186.

Pattnaik, G., Sinha, B., Mukherjee, B., et al. 2012. Submicron-size biodegradable polymer-based didanosine particles for treating HIV at early stage: an in vitro study. *Journal of Microencapsulation* 29(7):666–676.

Petros, R.A. and DeSimone, J.M. 2010. Strategies in the design of nanoparticles for therapeutic applications. *Nature Reviews Drug Discovery* 9(8):615–627.

Prabhakar, K., Afzal, S.M., Surender, G. and Kishan, V. 2013. Tween 80 containing lipid nanoemulsions for delivery of indinavir to brain. *Acta Pharmaceutica Sinica B* 3(5):345–353.

Qasim, M., Baipaywad, P., Udomluck, N., Na, D. and Park, H. 2014. Enhanced therapeutic efficacy of lipophilic amphotericin B against *Candida albicans* with amphiphilic poly (N-isopropylacrylamide) nanogels. *Macromolecular Research* 22(10):1125–1131.

Qasim, M., Lim, D.J., Park, H. and Na, D. 2014. Nanotechnology for diagnosis and treatment of infectious diseases. *Journal of Nanoscience and Nanotechnology* 14(10):7374–7387.

Qian, C.G., Chen, Y.L., Feng, P.J., et al. 2017. Conjugated polymer nanomaterials for theranostics. *Acta Pharmacologica Sinica* 38(6):764–781.

Ramana, L.N., Sharma, S., Sethuraman, S., Ranga, U. and Krishnan, U.M. 2012. Investigation on the stability of saquinavir loaded liposomes: Implication on stealth, release characteristics and cytotoxicity. *International Journal of Pharmaceutics* 431(1-2):120–129.

Ren, J., Zou, M., Gao, P., Wang, Y. and Cheng, G. 2013. Tissue distribution of borneol-modified ganciclovir-loaded solid lipid nanoparticles in mice after intravenous administration. *European Journal of Pharmaceutics and Biopharmaceutics* 83(2):141–148.

Ripoli, M., Angelico, R., Sacco, P., Ceglie, A. and Mangia, A. 2016. Phytoliposome-based silibinin delivery system as a promising strategy to prevent hepatitis C virus infection. *Journal of Biomedical Nanotechnology* 12(4):770–780.

Rouse, B.T. and Sehrawat, S. 2010. Immunity and immunopathology to viruses: What decides the outcome?. *Nature Reviews Immunology* 10(7):514–526.

Roy, U., Ding, H., Pilakka-Kanthikeel, S., et al. 2015. Preparation and characterization of anti-HIV nanodrug targeted to microfold cell of gut-associated lymphoid tissue. *International Journal of Nanomedicine* 10:819.

Saini, R., Saini, S. and Sharma, S. 2010. Nanotechnology: The future medicine. *The Journal of Cutaneous and Aesthetic Surgery* 3(1):32–33.

Saiyed, Z.M., Gandhi, N.H. and Nair, M.P. 2009. AZT 5′-triphosphate nanoformulation suppresses human immunodeficiency virus type 1 replication in peripheral blood mononuclear cells. *Journal of Neurovirology* 15(4):343–347.

Saiyed, Z.M., Gandhi, N.H. and Nair, M.P. 2010. Magnetic nanoformulation of azidothymidine 5'-triphosphate for targeted delivery across the blood–brain barrier. *International Journal of Nanomedicine* 5:157.

Sanapala, P. and Pola, S. 2020. Toxicological evaluation of nanoparticles using prokaryotic model organisms. In *Model Organisms to Study Biological Activities and Toxicity of Nanoparticles*, 277–296. Springer: Singapore.

Santos, D. S., Santos Filho, D. L. and de Souza, M. F. 2019. Can green synthesis of nanoparticles be efficient all year long. *Nanomaterial Chemistry and Technology* 1(1):32–35.

Santos-Martinez, M.J., Rahme, K., Corbalan, J.J., et al. 2014. Pegylation increases platelet biocompatibility of gold nanoparticles. *Journal of Biomedical Nanotechnology* 10(6):1004–1015.

Sanvicens, N. and Marco, M.P. 2008. Multifunctional nanoparticles – Properties and prospects for their use in human medicine. *Trends in Biotechnology 26*(8):425–433.

Sathigari, S., Chadha, G., Lee, Y.P., et al. 2009. Physicochemical characterization of efavirenz–cyclodextrin inclusion complexes. *AAPS Pharmscitech* 10(1):81–87.

Sawdon, A.J. and Peng, C.A. 2015. Ring-opening polymerization of ε-caprolactone initiated by ganciclovir (GCV) for the preparation of GCV-tagged polymeric micelles. *Molecules* 20(2):2857–2867.

Scheller, F.W., Bier, F.F. and Pfeiffer, D. 1995. Biosensoren: Grundlagen und anwendungen. *Tm-Technisches Messen*, 62(JG): 213–219.

Schütz, C.A., Juillerat-Jeanneret, L., Mueller, H., et al. 2013. Therapeutic nanoparticles in clinics and under clinical evaluation. *Nanomedicine* 8(3):449–467.

Schwarz, J.C., Klang, V., Karall, S., Mahrhauser, D., Resch, G.P. and Valenta, C. 2012. Optimisation of multiple W/O/W nanoemulsions for dermal delivery of aciclovir. *International Journal of Pharmaceutics* 435(1):69–75.

Šebestík, J., Reiniš, M. and Ježek, J. 2012. Synthesis of dendrimers: Convergent and divergent approaches. In *Biomedical Applications of Peptide-, Glyco- and Glycopeptide Dendrimers, and Analogous Dendrimeric Structures*, 55–81. Springer, Vienna.

Sepulveda-Crespo, D., Gomez, R., De La Mata, F.J., Jimenez, J.L. and Muñoz-Fernández, M.Á. 2015. Polyanionic carbosilane dendrimer-conjugated antiviral drugs as efficient microbicides: Recent trends and developments in HIV treatment/therapy. *Nanomedicine: Nanotechnology, Biology and Medicine* 11(6):1481–1498.

Seyfoddin, A. and Al-Kassas, R. 2013. Development of solid lipid nanoparticles and nanostructured lipid carriers for improving ocular delivery of acyclovir. *Drug Development and Industrial Pharmacy* 39(4):508–519.

Shegokar, R. and Singh, K.K. 2011. Nevirapine nanosuspensions for HIV reservoir targeting. *Die Pharmazie - An International Journal of Pharmaceutical Sciences* 66(6): 408–415.

Soppimath, K.S., Aminabhavi, T.M., Kulkarni, A.R. and Rudzinski, W.E. 2001. Biodegradable polymeric nanoparticles as drug delivery devices. *Journal of Controlled Release* 70(1-2):1–20.

Sun, R.W.Y., Chen, R., Chung, N.P.Y., et al. 2005. Silver nanoparticles fabricated in Hepes buffer exhibit cytoprotective activities toward HIV-1 infected cells. *Chemical Communications* 40:5059–5061.

Sutradhar, K.B. and Amin, M.L. 2013. Nanoemulsions: Increasing possibilities in drug delivery. *European Journal of Nanomedicine* 5(2):97–110.

Szunerits, S., Barras, A., Khanal, M., Pagneux, Q. and Boukherroub, R. 2015. Nanostructures for the inhibition of viral infections. *Molecules* 20(8):14051–14081.

Telwatte, S., Moore, K., Johnson, A., et al. 2011. Virucidal activity of the dendrimer microbicide SPL7013 against HIV-1. *Antiviral Research* 90(3):195–199.

Thakkar, K.N., Mhatre, S.S. and Parikh, R.Y. 2010. Biological synthesis of metallic nanoparticles. *Nanomedicine: Nanotechnology, Biology and Medicine* 6(2):257–262.

Tomalia, D.A. 2005. The dendritic state. *Materials Today* 8(3):34–46.

Tripathy, S. and Das, M.K. 2013. Dendrimers and their applications as novel drug delivery carriers. *Journal of Applied Pharmaceutical Sciences* 3(9): 142–149.

Trivedi, R. and Kompella, U.B. 2010. Nanomicellar formulations for sustained drug delivery: Strategies and underlying principles. *Nanomedicine* 5(3):485–505.

Trotta, F., Zanctti, M. and Cavalli, R. 2012. Cyclodextrin-based nanosponges as drug carriers. *Beilstein Journal of Organic Chemistry* 8(1):2091–2099.

Tshweu, L., Katata, L., Kalombo, L., et al. 2014. Enhanced oral bioavailability of the antiretroviral efavirenz encapsulated in poly (epsilon-caprolactone) nanoparticles by a spray-drying method. *Nanomedicine* 9(12):1821–1833.

Villanueva-Flores, F., Castro-Lugo, A., Ramírez, O.T. and Palomares, L.A. 2020. Understanding cellular interactions with nanomaterials: Towards a rational design of medical nanodevices. *Nanotechnology* 31(13):132002.

Vyas, T.K., Shahiwala, A. and Amiji, M.M. 2008. Improved oral bioavailability and brain transport of Saquinavir upon administration in novel nanoemulsion formulations. *International Journal of Pharmaceutics* 347(1-2):93–101.

Wang, H., Shao, N., Qiao, S. and Cheng, Y. 2012. Host–guest chemistry of dendrimer–cyclodextrin conjugates: selective encapsulations of guests within dendrimer or cyclodextrin cavities revealed by NOE NMR techniques. *The Journal of Physical Chemistry B* 116(36):11217–11224.

Wang, W., Guo, Z., Chen, Y., Liu, T. and Jiang, L. 2006. Influence of generation 2–5 of PAMAM dendrimer on the inhibition of Tat peptide/TAR RNA binding in HIV-1 transcription. *Chemical Biology and Drug Design* 68(6):314–318.

Williams, P.E., Crauwels, H.M. and Basstanie, E.D. 2015. Formulation and pharmacology of long-acting rilpivirine. *Current Opinion in HIV and AIDS* 10(4):233–238.

Zazo, H., Colino, C.I. and Lanao, J.M. 2016. Current applications of nanoparticles in infectious diseases. *Journal of Controlled Release* 224:86–102.

Zeng, P., Xu, Y., Zeng, C., Ren, H. and Peng, M. 2011. Chitosan-modified poly (D,L-lactide-co-glycolide) nanospheres for plasmid DNA delivery and HBV gene-silencing. *International Journal of Pharmaceutics* 415(1-2):259–266.

Zhao, L., Seth, A., Wibowo, N., et al. 2014. Nanoparticle vaccines. *Vaccine* 32(3):327–337.

Zidan, A.S., Spinks, C.B., Habib, M.J. and Khan, M.A. 2013. Formulation and transport properties of tenofovir loaded liposomes through Caco-2 cell model. *Journal of Liposome Research* 23(4):318–326.

Section II

Viral Nanomaterials: Biomedical Applications

6 Photocatalytic Inactivation of Pathogenic Viruses Using Metal Oxide and Carbon-Based Nanoparticles

Lan Ching Sim, Wei Qing Wee, Shien Yoong Siow, Kah Hon Leong, Jit Jang Ng, and Pichiah Saravanan

CONTENTS

6.1 INTRODUCTION

The world is currently facing a significant threat from Coronavirus disease 2019 (COVID-19); it is caused by a Coronavirus called severe acute respiratory syndrome Coronavirus 2 (SARS-CoV-2). The transmission route of COVID-19 includes direct contact, fomite contact, and airborne transmission (Ge et al. 2020; Guan et al. 2020; Van Doremalen et al. 2020). At the same time, the world is still facing the continual threat of the evolving influenza viruses, which have similar disease presentation as COVID-19. Different methods, such as microwave, ozone,

DOI: 10.1201/9781003136644-8

ultraviolet, plasma, filtration, and photocatalysis, have been used to control the airborne transmission of viruses (Yu et al. 2009; Xu et al. 2011; Li et al. 2021). Szeto et al. (2020) used vacuum-UV (VUV) lamps with ozone production to inactivate influenza A viruses H1N1 and H3N2. The viral load was reduced by 4-log10 folds after 20 minutes of VUV illumination. Wu et al. (2015) found that the release of reactive oxygen species (ROS) from atmospheric-pressure cold plasma (APCP) could damage the virus surface proteins and RNA genes, thus leading to 95% inactivation of MS2 virus after exposure at 28 W. According to Wu and Yao (2014), more than 90% inactivation efficiency was achieved when airborne MS2 viruses were exposed to microwave irradiation at 700 W for less than 2 minutes.

Woo et al. (2012) designed a microwave-irradiation-assisted filtration system to inactivate viral aerosols, which can reach a high inactivation efficiency of around 5-log. Xia et al. (2019) demonstrated the efficacy of nonthermal plasma (NTP) against MS2 virus using a packed bed nonthermal plasma reactor. Approximately 2.3-log virus was reduced in which ~2-log of the MS2 inactivated and ~0.35-log physically removed in the packed bed. Wigginton et al. (2012) used five different disinfectants, such as free chlorine (FC), heat, UV irradiation, singlet oxygen, and chlorine dioxide (ClO_2), to inactivate MS2 virus. They found that each treatment method resulted in a unique inactivation mechanism. For example, ClO_2 or heat treatments may be suitable for inactivating double-stranded DNA viruses with genome repair mechanisms. UV treatment was more effective for inactivating single-stranded RNA viruses without genome repair mechanisms. Nonthermal plasma (NTP) produces chemically active species, such as atomic oxygen, hydroxyl radicals, and ozone, to remove bioaerosol. However, the production of toxic byproduct, such as CO, O_3, NO_X and aerosol particles, restricts its application (Yu et al. 2009). Other conventional methods like UV irradiation, thermal treatment, and microwave irradiation are not practical because they require high energy consumption. Therefore, heterogeneous photocatalysis has recently emerged as an alternative technology to the current viral inactivation since the foremost discovery by Sjogren and Sierka (1994).

Heterogeneous photocatalysis is initiated by the absorption of photon with energy equal or larger than the band gap energy of the semiconductor (Yu et al. 2016). Metal oxide nanoparticles (NPs), such as titanium dioxide (TiO_2), zinc oxide (ZnO), and iron (III) oxide (Fe_2O_3), have been favoured as semiconductor photocatalysts due to their ability to transport charge carriers and absorb light (Le Bahers et al. 2014). The electron will be excited from the valence band (VB) to the conduction band (CB), leading to the production of electron (e^-) and hole (h^+) pairs. The photogenerated charges will reduce oxygen (O_2) to form superoxide radicals anion ($^{\bullet}O_2^-$) and the holes will oxidise water (H_2O) to form hydroxyl radicals ($^{\bullet}OH$) (Ahmed et al. 2010). TiO_2 are most widely used owing to its outstanding properties, such as economical viability, environment friendly, nontoxicity, high photocatalytic capability, and high chemical stability (Pookmanee and Phanichphant 2009).

Nakano et al. (2012) demonstrated the antiviral effect of TiO_2 against influenza virus and found that these reactive oxidising species (ROS) could digest the protein of virus and attack the influenza virus RNA. Kim and Jang (2018) investigated VUV photocatalysis for the inactivation of MS2 viruses and degradation of ozone by using Pd-TiO_2 catalyst with short irradiation time (0.004–0.125 s). More than 90% of inactivation efficiency was achieved at irradiation time of 0.009 s. The residual ozone was degraded to a safe level of 35 ppb. Khaiboullina et al. (2021) recently explored the antiviral efficacy of TiO_2 against human Coronavirus using a close genetic relative of SAR-CoV-2, HCoV-NL63. Their results showed that the coating of TiO_2 increased the viral inactivation efficiency by reducing the viral genomic RNA stability and virus infectivity at short exposure of UV. It is proven that the self-cleaning property of TiO_2 is helpful in controlling the transmission of the virus via fomites.

This chapter deals with the application of metal oxide NPs and carbon-based NPs for viral inactivation. Their antiviral efficacy and inactivation mechanism are summarised. The challenges and future perspective in this field are listed as well.

6.2 METAL OXIDE NANOPARTICLES

Various metal oxide nanoparticles have been studied to inactivate viruses because of their superior photocatalytic activity and nontoxicity nature. The interaction between the metal oxides and the viruses brings about several chemical and physical changes to the viruses' cells and leads to such viruses' death. The commonly used metal oxides for the inactivation of viruses are TiO_2, ZnO, Fe_2O_3, CuO, and WO_3 (Takehara et al. 2010; Yamaguchi et al. 2016; Esfahani et al. 2020; Habibi-Yangjeh et al. 2020) With the utilisation of the photocatalysis theory, where the metal oxide nanoparticles are irradiated with light intensity higher than or equal to the band energy of the metal oxide, electron excitation will take place. These electrons' mobility will then ensue to the generation of reactive oxygen species (ROS), which can oxidise the absorbed viruses resulting from their protein and genes damage (Ahmed and Haider 2018; Gupta and Modak 2020). Therefore, photocatalysis plays a vital role in the inactivation of viruses mainly contributed by its extensive antivirus range, high efficacy, and abiding (Zhang et al. 2019). Such radicals are believed to give rise to physical damage to the virus formation due to the sharp edges on the nanostructure surface, while their presence is toxic to the virus due to the metal ions.

Therefore, there's been a tremendous evolution in using metal oxide nanoparticles for the inactivation of viruses where Sjogren and Sierka (1994) reported the first study, where they use the TiO_2 photocatalyst for inactivation MS2 virus. Their study's promising outcome had inspired the further development of such based photocatalyst to inactivate waterborne and airborne viruses. Nakano et al. (2012) reported that using TiO_2 photocatalysis could inactive the influenza virus, even with low light intensity. Besides that, it was found that the addition of TiO_2 under the UV light irradiation enhanced the inactivation of MS2 efficacy compared to the absence of TiO_2 nanoparticles (Lee and Ko 2013). This proves the ability of the TiO_2 to generate active radicals during the photocatalysis process and damages the virus components. In addition, Zhang's group also found that removing phage F2 using a nano-TiO_2 membrane coupled system to yield a better removal efficacy compared to the pure membrane. However, the limitation of pure TiO_2 of its wide band gap energy and rapid recombination of electrons and holes retarded the overall photocatalysis efficiency (Zheng et al. 2013). Correspondingly, Kim et al. (2017) found that the internalised MNV-1 are able to reduce by $2.9\log_{10}$ and further reduce to $5.5\log_{10}$ with the incorporation of high hydrostatic pressure during the treatment process. Recently, Tong et al. (2021) testified that TiO_2 nanoparticles are able to inactivate pathogenic viruses through viral genome attack under light illumination.

Besides the common TiO_2 as a promising metal oxide nanoparticle for viruses' inactivation, other metal oxides nanoparticles could be the next alternative to achieve the same outcome as TiO_2. Mazurkow et al. (2020) reported that the incorporation of Cu and CuO with dried alumina granules created an effective adsorbent against viruses. Their findings show that it achieved a 99.9% removal of MS2 bacteriophages. Similarly, Tavakoli and Hashemzadeh (2020) found that using CuO nanoparticles showed a high inhibition rate (83.3%) and demonstrated a significant antiviral potency against HSV-1. Meanwhile, Gutierrez et al. (2009) coated the glass fibre with hematite Fe_2O_3 nanoparticles to increase the surface area (80.75 m^2/g), and that achieved an enhanced removal efficacy of both rotavirus and MS2 bacteriophage. Bradley et al. (2011) fabricated iron oxide nanoparticles with a biosand filter to remove viruses. They attained an MS2 bacteriophage removal efficiency of 5-log compared to 0.5-log with the absence of iron oxide nanoparticles. This enhancement is due to the strong interaction through the electrostatic of positively charged iron oxides with the negatively charged MS2 particles. Furthermore, Ryan et al. (2002) reported that attaching iron oxide with quartz sand had enhanced the inactivation of PRD1 and MS2 viruses by 3 times faster than the absence of iron oxide through the surface inactivation. Not only that, Michen et al. (2013) used magnesium oxyhydroxide (MgO) to improve the virus's removal from water. This MgO showed a significant removal efficiency of approximately 4-log PhiX174 and MS2 bacteriophages. Giannakis et al. (2017) used iron oxide and H_2O_2 to improve the inactivation process of MS2 phage to achieve improved efficiency of 5-log in 30 minutes under simulated solar energy.

6.2.1 Modification to Metal Oxide Nanoparticles

Therefore, various modifications to pure TiO_2 have been performed to maximise the virus's inactivation in the broader light spectrum. Loading nano metal onto TiO_2 nanoparticles is a compelling way to boost the inactivation of the virus. Moongraksathum et al. (2019) successfully doped silver nanoparticles on TiO_2 surface and coated them on a glass substrate. It was found that the developed composite shows a complete inactivation of the H1N1 virus under visible light irradiation. Zheng et al. (2018) also reported similar findings where they developed Cu-TiO_2 nanofiber, which completely inactivates bacteriophage F2 under visible light irradiation. Also, Kozlova et al. (2010) coated Pt with acidic TiO_2 on glass plates to inactivate the influenza A virus with the presence of light source. Subsequent to 30 minutes of light illumination, the activation amount of Pt/TiO_2 to H3N2 was 99.8%, which is higher than that of pure TiO_2. Similarly, Kim and Jang (2018) successfully developed Pd-TiO_2 that can easily generate reactive oxygen radicals that lead to the high inactivation efficiency of 90% of the MS2 virus within a short period. Meanwhile, Horovitz et al. (2018) synthesised N-doped TiO_2 loaded photocatalytic membrane reactor to enhance MS2 bacteriophage inactivation. This hybrid process achieved a high efficacy of removal (4.9 ± 0.1 log (>99.99%) in natural surface water. In addition, Miyauchi et al. (2020) reported using Cu_xO/TiO_2 as a photocatalyst for antiviral functions under visible light irradiation. It was also reported that this photocatalyst was used to minimise the virus contagion risk through an antiviral coating material. Liga et al. (2013) reported a 37-fold boosted viral removal efficacy by doping 5% of SiO_2 with TiO_2 nanoparticles. It was recommended that such an elimination approach would benefit the outcome of utilising a low-cost green material. Yamaguchi et al. (2017) developed a composite combining Rh with $SrTiO_3$ to inactivate phage Qβ under the presence of visible light. It achieved 5-log removal in 120 minutes.

6.2.2 Mechanism

Figure 6.1 shows the schematic of the comprehensive mechanism of using metal oxide nanoparticles for virus inactivation. The photocatalytic inactivation of viruses principally involves the redox reaction due to the generation of ROS. The generation of ROS could only happen when the metal oxide nanoparticles are irradiated with a light source, which is higher than the respective metal oxide nanoparticles' band gap energy. Due to this excitation phenomena, a portion of the light-generated electrons and holes will transfer to the material's surface due to the internal electric field effect. The inactivation of viruses involved mainly two distinct processes: the decomposition of the cytoplasmic membrane and cell wall. The metal oxide nanoparticles will generate ROS when their nanoparticles are being excited with the light sources. It will then excite the electrons and holes from their respective band. The electrons will then transfer from the valence band to the metal oxide's

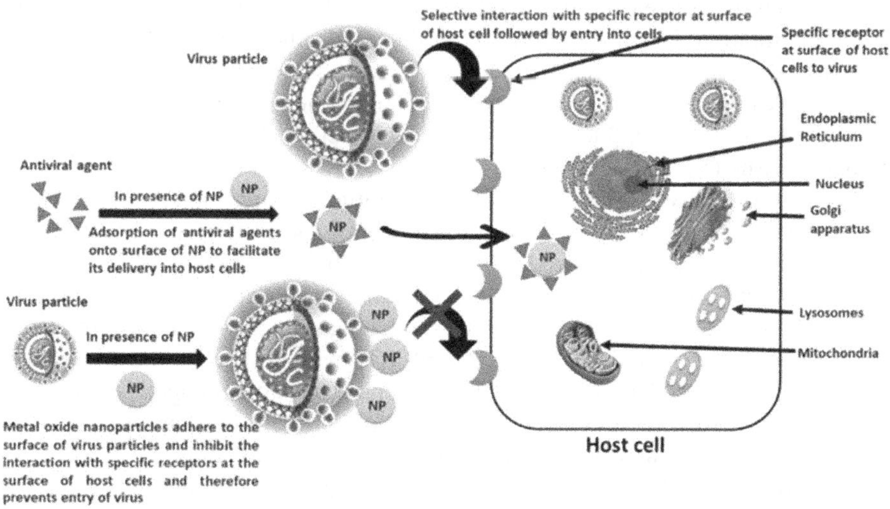

FIGURE 6.1 Schematic mechanism of virus inactivation using metal oxide nanoparticles (Abo-zeid and Williams 2020).

conduction band, while the holes will remain at the valence band. Thus, the electron transfer will then undergo a redox reaction in which the electrons will react with oxygen (O_2) to generate highly reactive radicals such as ($^{\bullet}O_2^-$, $^{\bullet}OH$ and $^{\bullet}OOH$). Meanwhile, the remaining holes at the valence band will react with water (H_2O) to generate hydroxyl radicals ($^{\bullet}OH$) (Abo-zeid and Williams 2020; Habibi-Yangjeh et al. 2020). Therefore, these ROS will then oxide the viruses and damage their protein and genes that attach to metal oxide nanoparticles' surface. This ROS has a strong ability to damage the virus's protein and genes (Li et al. 2021).

6.3 CARBON-BASED NANOPARTICLES

Various studies conducted in the past decades has shown that carbon-based nanoparticles (CN) possess excellent antimicrobial activity (Dizaj et al. 2015). Attention has turned towards the antiviral properties of CN due to its ability to produces reactive oxygen species (ROS) such as singlet oxygen ($^1O_2$) and superoxide ($^{\bullet}O_2^-$). ROS has been shown to oxidise and deforms the capsid protein (Lu and Liang 2020), effectively deactivates the target virus. The virus capsid protects the genomes from the environment. With the capsid protein structure compromised it could reduce the ability of viral attachment, stopping the release of viral genomes to host cell (Matharu et al. 2020). This section discusses the interest and potential of CN in combating viral infections.

6.3.1 Fullerenes

Fullerene is a cage-like structure formed by connecting carbon atom with one another through single bond or double bond (Georgakilas et al. 2015). Since the discovery of fullerene, it has been widely studied and used due to its flexibility. This caged structure with electronic cloud allows the compound to act as a charge-transfer complex (Gacem et al. 2020). Studies have shown that fullerene and its derivatives possess antiviral activity toward various viruses, such as cytomegalovirus (CMV) (Klimova et al. 2020), human immunodeficiency virus-1 (HIV-1) (Yasuno et al. 2021), influenza virus (Kraevaya et al. 2020a, b), and vesicular stomatitis virus (VSV) (Donskyi et al. 2018).

The low solubility of fullerene in water is limiting the application in biology. As such, soluble alternatives have been studied deeply. Donskyi et al. (2018) synthesised one of these fullerene derivatives, known as fullerene-polyglycerol sulphates (FPS), with different fullerene weight ratio to target VSV with baby hamster kidney cells as the host cell. The combination of sulphate group and fullerene core of FPS was found to be preventing the interaction of VSV with host cells. They concluded, based on the finding, that FPS have a high potential for antiviral applications. They possesses strong binding to viral and leucocyte proteins at low concentration while having good water solubility and biocompatibility.

Kraevaya et al. (2020a, b) synthesised water soluble chlorofullerene with different unprotected carboxylic acids through one-step Friedel-Crafts arylation method to target HIV-1 NL4.3(X4), HIV-1 BaL (R5), and HIV-2 ROD. The previously inaccessible water-soluble derivatives were found to inhibit the replication of strain R5 at low dosage, close to commercial drug (tenofovir). Dosage required for HIV-2 strains, on the other hand, is slightly higher than tenofovir. With cellular toxicity of higher than 100 μM, the synthesised compounds were found to be effective against strain X4 at EC_{50} near 1 μM.

6.3.2 Carbon Dots

Carbon dots (CDs) are small carbon nanoparticles with sizes smaller than 10 nm. They were first discovered unintentionally by Xu et al. (2004) during the purification of single-walled carbon nanotubes. Since the discovery of CDs, the applications in various fields have been explored heavily, including in biomedicine due to no toxicity in vitro and in vivo (Zhi et al. 2018). Studies of CDs on antiviral properties are very recent as the discovery itself is very recent. Nevertheless, they are a promising alternative to other CNs.

Dong et al. (2017) have recently accessed the antiviral activity of CDs to human norovirus (NoV) virus-like-particles (VLPs), GI.1 and GII.4 VLPs with different surface passivation molecules, namely 2,2-(ethylenedioxy)bis(ethylamine) (EDA)-CDs and 3-ethoxypropylamine (EPA)-CDs. The result shows that both EDA-CDs and EPA-CDs were able to achieve high inhibition (85–100%) on both strains of VLPs. The derivative of CDs has shown to inhibit VLPs binding to histo-blood group antigens (HBGA) receptors, which are the main entry pathway for VLPs. With the absent of effective vaccine for NoV, CDs may be a potential solution to prevent NoV from spreading.

Lin et al. (2019) and Tong et al. (2020) have reported the studies of CDs derived from Chinese medicine, namely from licorice root and curcumin respectively. Glycyrrhizin (glycyrrhizin acid) is the main active constituent of licorice root. Glycyrrhizin (Fu et al. 2005) and curcumin (Gera et al. 2017) have both shown antiviral, anti-oxidation, anti-inflammation, and antimicrobial properties. However, cytotoxicity and water solubility limit the application of both compounds. To overcome the disadvantages, they have demonstrated that by converting glycyrrhizin and curcumin to CDs (Gly-CDs and Cur-CDs respectively), the CDs were able to show high biocompatibility. Gly-CDs showed excellent inactivation of porcine reproductive and respiratory syndrome virus (PRRSV). It was found to inhibit PRRSV invasion and replication, as well as ROS production induced by PRRSV infection. Cur-CDs, on the other hand, showed excellent inactivation of enterovirus 71 (EV71). Cur-CDs were found to inhibit viral attachment to host cell thus suppressing the EV71-induced ROS (Lin et al. 2019; Tong et al. 2020).

6.3.3 Carbon Nanotubes

Carbon nanotubes (CNTs) are a hollow cylindrical structure made of carbon. They are often synthesised by winding single layers or multiple layers of graphene sheets. Based on the number of layers wound, CNTs are classified into two categories: single-walled (SW) CNTs, which use single layers of graphene sheets, and multi-walled (MW) CNTs, which use more than one layer of graphene sheets. CNTs have the unique properties among other carbon nanoparticles to transport drugs or biomolecules to various types of targeted cells, such as cancer cells and T cells (Date and Destache 2013). Due to the conjugation and complexation of CNTs, it is possible to insert more than one type of functional group to the surface of CNTs. In theory, it is also possible to transport molecules through the internal cavity of CNTs (Bianco 2004). Pristine CNTs are known to be harmful to cell due to its properties of hydrophobicity and tendency to aggregate due to strong van der Waals force (Zhou et al. 2017). Methods have been studied to reduce the cytotoxicity by reducing the length of CNTs, such as introducing hydrophilic groups to the surface of CNTs to allow ease of transport through body and introducing structural defects to allow degradation by oxidative enzyme (Russier et al. 2011).

Iannazzo et al. (2015) explored the antiviral properties of carboxylated multi-walled CNTs (ox-MWCNT) and their derivatives toward HIV-1. They showed that by introducing carboxylic group onto the surface, the IC_{50} of ox-MWCNT is half of pristine MWCNT, EC_{50} of ox-MWCNT, which is seven times lower than that of pristine MWCNT and the cytotoxicity has improved by a third. With selectivity of 10 (cytotoxicity to EC_{50}), the obtained result could be a promising start for the use of CNT-based nanomaterial for HIV infection treatment. By introducing carboxylic group onto CNTs, the hydrophilicity has improved compared to pristine CNTs, which induce better water dispersion than is necessary for biological applications (Bianco 2004).

Severe acute respiratory symptoms Coronavirus 2 (SARS-CoV-2) pandemic has been rampaging around the world since December of 2019. The virus is spread through respiratory droplets or fomites. Studies have shown that the smaller droplets were able to remain suspended in air and travel up to 8 m away (Bourouiba 2016). Current personal protection gears (PPE) such as masks are mostly dependent on physical filtration using small pore size. However, most fabric masks are not able to block all the expulsion of CoV due to size difference being significant (30–140 nm compared to >1 micron), nor does the fabric inactivate the virus (Leung et al. 2020). Aasi et al. (2020) have investigated the effectiveness of metal decorated SWCNTs to act as antiviral surfaces. Hydrogen peroxide is being physiosorbed onto metal decorated SWCNTs. This ROS promoting molecules has proved to be detrimental to many viruses. Using this method, it was noted that rhodium decorated and ruthenium decorated SWCNTs show high potential to act as filter for virus removal and inactivation with long shelf life. By heating at 100 °C under UV radiation, the regeneration rate of peroxides modelled was in milliseconds.

6.3.4 Graphene Oxides

Graphene oxide (GO) is the oxide of a two-dimensional structure consisting of single layer of carbon atoms arranged in honeycomb lattice. Application of graphene has been studied intensely in fields such as electronics, sensing, photonics, and antibacterial (Innocenzi and Stagi 2020). GO and reduced GO (rGO) has shown excellent antibacterial activity toward both Escherichia coli and Staphylococcus aureus. It has been observed that the GO will fold over the bacteria by attaching to the bacterial membrane's functional group while penetrating with doped metal, effectively killing the cell (Vi et al. 2020). The rGO, on the other hand, will embed the cell into large aggregates. The antibacterial activities of GO and rGO has brought up the possibility of applying the same mechanism toward antiviral action. The main mechanisms of antimicrobial activities can be classified as either through direct physical contact with the target cell or through ROS produced from GO or rGO acting as photocatalyst (Singh et al. 2019), with the antibacterial example above falling under the former category.

Performance of GO against pseudorabies virus (PRV) and porcine epidemic diarrhea virus (PEDV) have been conducted (Ye et al. 2015). By conjugating GO with polyvinylpyrrolidone (nonionic polymer), the infection rate of PRV and PEDV has been suppressed by 99%. It was found that GO possess close to zero antiviral activity when conjugated with poly(diallyldimethylammonium chloride) (PDDA). The nanosheets structure of GO is also proved to be important for antiviral activity. Based on the studies, GO inactivate both PRV and PEDV by deforming the capsid protein prior to viral entry.

Respiratory syncytial virus (RSV) has contributed to severe lung disease in infants. Yang et al. (2017) developed a β-cyclodextrin (CD) functionalised GO composite. By incorporating curcumin functional group onto CD, the infection rate of RSV was reduced by 99.99% based on viral titers. CD-GO has shown to have no obvious cytotoxic effect when tested with cell counting kit-8 assay. The mechanisms of GO in antiviral activity of this study agree with Ye et al. (2015), that CD-GO prevent RSV infection by inactivating the virus and inhibit functional group attachment.

6.4 CONCLUSION AND FUTURE PERSPECTIVES

The use of a photocatalyst is environmental-friendly and sustainable because of the employment of renewable energy as the energy source of the photocatalytic viral inactivation. This chapter provides a review on the development of photocatalytic inactivation of viruses using metal oxide NPs and carbon-based NPs. Despite recent advances, some potential research is yet to be explored to overcome new challenges provoked by viral pandemic cases. Future work should explore the coating of photocatalysts on face masks or other surfaces to control the viral spread of Coronavirus 2 (SARS-CoV-2) via fomite and aerosol. Further, molecular imprinting could be used to selectively adsorb viruses and concentrate them near photocatalytic sites for inactivation. It has been proven to be effective in waterborne viral inactivation. There is lack of investigation on the viral inactivation mechanism using a combination of quantitative analytical tools since the foremost discovery by Wigginton et al. (2012). This approach helps to identify and quantify the extent of modifications in the virus genome or proteins by measuring the damage at well-defined levels of inactivation (Wigginton et al. 2012). Most of the viral inactivation methods are reported by using UVC light at 254 nm, which is harmful to skin and eyes. Hence, it is suggested to use far-UVC light (207–222 nm) or visible light (390–750 nm) to reduce the negative effects of UVC light. And for that, it will be necessary to develop a photocatalyst material that could harvest more visible light.

ACKNOWLEDGEMENTS

The research was supported by the Ministry of Higher Education (MoHE) Malaysia through Fundamental Research Grant Scheme project FRGS/1/2019/TK10/UTAR/02/5.

REFERENCES

Aasi, A., Aghaei, S.M., Moore, M.D. and Panchapakesan, B. 2020. Pt-, Rh-, Ru-, and Cu-single-wall carbon nanotubes are exceptional candidates for design of anti-viral surfaces: A theoretical study. *International Journal of Molecular Science* 21(15):5211.

Abo-zeid, Y. and Williams, G.R. 2020. The potential anti-infective applications of metal oxide nanoparticles: A systematic review. *Wiley Interdisciplinary Reviews: Nanomedicine and Nanobiotechnology* 12(2):e1592.

Ahmed, S., Rasul, M.G., Martens, W.N., Brown, R. and Hashib, M.A. 2010. Heterogeneous photocatalytic degradation of phenols in wastewater: A review on current status and developments. *Desalination* 261(1):3–18.

Ahmed, S.N. and Haider, W. 2018. Heterogeneous photocatalysis and its potential applications in water and wastewater treatment: A review. *Nanotechnology* 29(34):342001.

Bianco, A. 2004. Carbon nanotubes for the delivery of therapeutic molecules. *Expert Opinion on Drug Delivery* 1(1):57–65.

Bourouiba, L. 2016. A sneeze. *The New England Journal of Medicine* 375(8):e15.

Bradley, I., Straub, A., Maraccini, P., Markazi, S. and Nguyen, T.H. 2011. Iron oxide amended biosand filters for virus removal. *Water Research* 45(15):4501–4510.

Date, A. and Destache, C.J. 2013. A review of nanotechnological approaches for the prophylaxis of HIV/AIDS. *Biomaterials* 34(26):6202–6208.

Dizaj, S.M., Mennati, A., Jafari, S., Khezri, K. and Adibkia, K. 2015. Antimicrobial activity of carbon-based nanoparticles. *Advanced Pharmaceutical Bulletin* 5(1):19–23.

Dong, X., Moyer, M.M., Yang, F., Sun, Y. and Yang, L. 2017. Carbon dots' antiviral functions against noroviruses. *Scientific Reports* 7(1):1–10.

Donskyi, I., Drüke, M., Silberreis, K., et al. 2018. Interactions of fullerene-polyglycerol sulfates at viral and cellular interfaces. *Small* 14(17):1800189.

Esfahani, A.R., Batelaan, O., Hutson, J.L. and Fallowfield, H.J. 2020. Role of biofilm on virus inactivation in limestone aquifers: Implications for managed aquifer recharge. *Journal of Environmental Health Science and Engineering* 18(1):21–34.

Fu, B., Liu, J., Li, H., Li, L., Lee, F. S.C. and Wang, X. 2005. The application of macroporous resins in the separation of licorice flavonoids and glycyrrhizic acid. *Journal of Chromatography A* 1089(1-2):18–24.

Gacem, M.A., Gacem, H. and Aminata, O. 2020. Nanocarbons: Antibacterial, antifungal, and antiviral activity and the underlying mechanism. In *Carbon Nanomaterials for Agri-Food and Environmental Applications*, ed. A.A. Kamel, 505–533. Netherlands: Elsevier.

Ge, Z.Y., Yang, L.M., Xia, J.J., Fu, X.H. and Zhang, Y.Z. 2020. Possible aerosol transmission of COVID-19 and special precautions in dentistry. *Journal of Zhejiang University-Science B* 21(5):361–368.

Georgakilas, V., Perman, J.A., Tucek, J. and Zboril, R. 2015. Broad family of carbon nanoallotropes: Classification, chemistry, and applications of fullerenes, carbon dots, nanotubes, graphene, nanodiamonds, and combined superstructures. *Chemical Reviews* 115(11):4744–4822.

Gera, M., Sharma, N., Ghosh, M., Huynh, D.L., Lee, S.J. and Min, T. 2017. Nanoformulations of curcumin: An emerging paradigm for improved remedial application. *Oncotarget* 8(39):66680–66698.

Giannakis, S., Liu, S., Carratala, A., Rtimi, S., Bensimon, M. and Pulgarin, C. 2017. Iron-oxide-mediated semiconductor photocatalysis vs. heterogeneous photo-Fenton treatment of viruses in wastewater. Impact of the oxide particle size. *Journal of Hazardous Materials* 339:223–231.

Guan, W.J., Ni, Z.Y., Hu, Y., et al. 2020. Clinical characteristics of coronavirus disease 2019 in China. *New England Journal of Medicine* 382(18):1708–1720.

Gupta, R. and Modak, J. 2020. Bacterial lysis via photocatalysis - A critical mechanistic review. *ChemCatChem* 12(8):2148–2170.

Gutierrez, L., Li, X., Wang, J., et al. 2009. Adsorption of rotavirus and bacteriophage MS2 using glass fiber coated with hematite nanoparticles. *Water Research* 43(20):5198–5208.

Habibi-Yangjeh, A., Asadzadeh-Khaneghah, S., Feizpoor, S. and Rouhi, A. 2020. Review on heterogeneous photocatalytic disinfection of waterborne, airborne, and foodborne viruses: Can we win against pathogenic viruses? *Journal of Colloid and Interface Science* 580:503–514.

Horovitz, I., Avisar, D., Luster, E., Lozzi, L., Luxbacher, T. and Mamane, H. 2018. MS2 bacteriophage inactivation a N-doped TiO_2-coated photocatalytic membrane reactor: Influence of water-quality parameters. *Chemical Engineering Journal* 354:995–1006.

Iannazzo, D., Pistone, A., Galvagno, S., et al. 2015. Synthesis and anti-HIV activity of carboxylated and drug-conjugated multi-walled carbon nanotubes. *Carbon* 82:548–561.

Innocenzi, P. and Stagi, L. 2020. Carbon-based antiviral nanomaterials: Graphene, C-dots, and fullerenes. A perspective. *Chemical Science* 11(26):6606–6622.

Khaiboullina, S., Uppal, T., Dhabarde, N., Subramanian, V.R. and Verma, S.C. 2021. Inactivation of human coronavirus by titania nanoparticle coatings and UVC radiation: Throwing light on SARS-CoV-2. *Viruses* 13(1):19.

Kim, J. and Jang, J. 2018. Inactivation of airborne viruses using vacuum ultraviolet photocatalysis for a flow-through indoor air purifier with short irradiation time. *Aerosol Science and Technology* 52(5):557–566.

Kim, S-H., Shahbaz, H.M., Park, D., et al. 2017. A combined treatment of UV-assisted TiO_2 photocatalysis and high hydrostatic pressure to inactivate internalized murine norovirus. *Innovative Food Science and Emerging Technologies* 39:188–196.

Klimova, R., Andreev, S., Momotyuk, E., et al. 2020. Aqueous fullerene C_{60} solution suppresses herpes simplex virus and cytomegalovirus infections. *Fullerenes, Nanotubes and Carbon Nanostructures* 28(6):487–499.

Kozlova, E.A., Safatov, A.S., Kiselev, S.A., et al. 2010. Inactivation and mineralization of aerosol deposited model pathogenic microorganisms over TiO_2 and Pt/TiO_2. *Environmental Science and Technology* 44(13):5121–5126.

Kraevaya, O.A., Peregudov, A.S., Fedorova, N.E., et al. 2020. Thiophene-based water-soluble fullerene derivatives as highly potent antiherpetic pharmaceuticals. *Organic and Biomolecular Chemistry* 18(42):8702–8708.

Kraevaya, O.A., Peregudov, A.S., Godovikov, I.A., et al. 2020. Direct arylation of $C_{60}Cl_6$ and $C_{70}Cl_8$ with carboxylic acids: A synthetic avenue to water-soluble fullerene derivatives with promising antiviral activity. *Chemical Communications* 56(8):1179–1182.

Le Bahers, T., Rerat, M. and Sautet, P. 2014. Semiconductors used in photovoltaic and photocatalytic devices: Assessing fundamental properties from DFT. *The Journal of Physical Chemistry C* 118(12):5997–6008.

Leung, N.H., Chu, D.K., Shiu, E.Y., et al. 2020. Respiratory virus shedding in exhaled breath and efficacy of face masks. *Brief Communications* 26(5):676–680.

Lee, J.E. and Ko, G. 2013. Norovirus and MS_2 inactivation kinetics of UV-A and UV-B with and without TiO_2. *Water Research* 47(15):5607–5613.

Liga, M.V., Maguire-Boyle, S.J., Jafry, H.R., Barron, A.R. and Li, Q. 2013. Silica decorated TiO_2 for virus inactivation in drinking water simple synthesis method and mechanisms of enhanced inactivation kinetics. *Environmental Science and Technology* 47(12): 6463–6470.

Li, R., Cui, L., Chen, M. and Huang, Y. 2021. Nanomaterials for airborne virus inactivation: A short review. *Aerosol Science and Engineering* 5:1–11.

Lin, C.J., Chang, L., Chu, H.W., et al. 2019. High amplification of the antiviral activity of curcumin through transformation into carbon quantum dots. *Small* 15(41): 1902641.

Lu, C. and Liang, J. 2020. An overview of functional nanoparticles as novel emerging antiviral therapeutic agents. *Materials Science and Engineering:C* 112:110924.

Matharu, R.K., Porwal, H., Chen, B., Ciric, L. and Edirisinghe, M. 2020. Viral filtration using carbon-based materials. *Medical Devices and Sensors* 3(4):e10107.

Mazurkow, J.M., Yuzbasi, N.S., Domagala, K.W., Pfeiffer, S., Kata, D. and Graule, T. 2020. Nano-sized copper (Oxide) on alumina granules for water filtration: Effect of copper oxidation state on virus removal performance. *Environmental Science and Technology* 54(2):1214–1222.

Michen, B., Meder, F., Rust, A., Fritsch, J., Aneziris, C. and Graule, T. 2013. Improved virus removal in ceramic depth filters based on diatomaceous earth. *Environmental Science and Technology* 47(3):1526–1533.

Miyauchi, M., Sunada, K. and Hashimoto, K. 2020. Antiviral effect of visible light-sensitive Cu_xO/TiO_2 photocatalysts. *Catalysts* 10(9):1093.

Moongraksathum, B., Chien, M-Y. and Chen, Y-W. 2019. Antiviral and antibacterial effects of silver-doped TiO_2 prepared by the peroxo sol-gel method. *Journal of Nanoscience and Nanotechnology* 19(11):7356–7362.

Nakano, R., Ishiguro, H., Yao, Y., et al. 2012. Photocatalytic inactivation of influenza virus by titanium dioxide thin film. *Photochemical and Photobiological Sciences* 11(8):1293–1298.

Pookmanee, P. and Phanichphant, S. 2009. Titanium dioxide powder prepared by a sol-gel method. *Journal of Ceramic Processing Research* 10(2):167–170.

Russier, J., Ménard-Moyon, C., Venturelli, E., et al. 2011. Oxidative biodegradation of single- and multi-walled carbon nanotubes. *Nanoscale* 3(3):893–896.

Ryan, J.N., Harvey, R.W., Metge, D., Elimelech, M., Navigato, T. and Pieper, A.P. 2002. Field and laboratory investigations of inactivation of viruses (PRD1 and MS2) attached to iron oxide-coated quartz sand. *Environmental Science and Technology* 36(11):2403–2413.

Singh, K., Mishra, A., Sharma, D. and Singh, K. 2019. Antiviral and antimicrobial potentiality of nano drugs. In *Applications of Targeted Nano Drugs and Delivery Systems*, eds. S.S. Mohapatra, S. Ranjan, N. Dasgupta, R.K. Mishra, and S. Thomas, 343–356. Netherlands: Elsevier.

Sjogren, J.C. and Sierka, R.A. 1994. Inactivation of phage MS2 by iron-aided titanium dioxide photocatalysis. *Applied and Environmental Microbiology* 60(1):344–347.

Szeto, W., Yam, W. C., Huang, H. and Leung, D. Y. 2020. The efficacy of vacuum-ultraviolet light disinfection of some common environmental pathogens. *BMC Infectious Diseases* 20(1):1–9.

Takehara, K., Yamazaki, K., Miyazaki, M., et al. 2010. Inactivation of avian influenza virus H1N1 by photocatalyst under visible light irradiation. *Virus Research* 151(1): 102–103.

Tavakoli, A. and Hashemzadeh, M.S. 2020. Inhibition of herpes simplex virus type 1 by copper oxide nanoparticles. *Journal of Virological Methods* 275:113688.

Tong, T., Hu, H., Zhou, J., et al. 2020. Glycyrrhizic-acid-based carbon dots with high antiviral activity by multisite inhibition mechanisms. *Small* 16(13):1906206

Tong, Y., Shi, G., Hu, G., et al. 2021. Photo-catalyzed TiO_2 inactivates pathogenic viruses by attacking viral genome. *Chemical Engineering Journal* 414:128788.

Van Doremalen, N., Bushmaker, T., Morris, D.H., et al. 2020. Aerosol and surface stability of SARS-CoV-2 as compared with SARS-CoV-1. *New England Journal of Medicine* 382(16):1564–1567.

Vi, T.T.T., Kumar, S.R., Pang, J.S., Liu, Y., Chen, D.W. and Lue, S.J. 2020. Synergistic antibacterial activity of silver-loaded graphene oxide towards *Staphylococcus aureus* and *Escherichia coli. Nanomaterials* 10(2):366.

Wigginton, K.R., Pecson, B.M., Sigstam, T., Bosshard, F. and Kohn, T. 2012. Virus inactivation mechanisms: Impact of disinfectants on virus function and structural integrity. *Environmental Science and Technology* 46(21):12069–12078.

Woo, M.H., Grippin, A., Wu, C.Y. and Wander, J. 2012. Microwave-irradiation-assisted HVAC filtration for inactivation of viral aerosols. *Aerosol and Air Quality Research* 12(3):295–303.

Wu, Y., Liang, Y., Wei, K., et al. 2015. MS2 virus inactivation by atmospheric-pressure cold plasma using different gas carriers and power levels. *Applied and Environmental Microbiology* 81(3):996–1002.

Wu, Y. and Yao, M. 2014. In situ airborne virus inactivation by microwave irradiation. *Chinese Science Bulletin* 59(13):1438–1445.

Xia, T., Kleinheksel, A., Lee, E.M., Qiao, Z., Wigginton, K.R. and Clack, H.L. 2019. Inactivation of airborne viruses using a packed bed non-thermal plasma reactor. *Journal of Physics D: Applied Physics* 52(25):255201.

Xu, X., Ray, R., Gu, Y., et al. 2004. Electrophoretic analysis and purification of fluorescent single-walled carbon nanotube fragments. *Journal of the American Chemical Society* 126(40):12736–12737.

Xu, Z., Wu, Y., Shen, F., Chen, Q., Tan, M. and Yao, M. 2011. Bioaerosol science, technology, and engineering: Past, present, and future. *Aerosol Science and Technology* 45(11):1337–1349.

Yamaguchi, Y., Shimodo, T., Usuki, S., et al. 2016. Different hollow and spherical TiO_2 morphologies have distinct activities for the photocatalytic inactivation of chemical and biological agents. *Photochemical and Photobiological Sciences* 15(8):988–994.

Yamaguchi, Y., Usuki, S., Kanai, Y., et al. 2017. Selective inactivation of bacteriophage in the presence of bacteria by use of ground Rh-Doped $SrTiO_3$ photocatalyst and visible light. *Applied Materials and Interfaces* 9(37):31393–31400.

Yang, X.X., Li, C.M., Li, Y.F., Wang, J. and Huang, C.Z. 2017. Synergistic antiviral effect of curcumin functionalized graphene oxide against respiratory syncytial virus infection. *Nanoscale* 9(41):16086–16092.

Yasuno, T., Ohe, T., Kataoka, H., et al. 2021. Fullerene derivatives as dual inhibitors of HIV-1 reverse transcriptase and protease. *Bioorganic and Medicinal Chemistry Letters* 31:127675.

Ye, S., Shao, K., Li, Z., et al. 2015. Antiviral activity of graphene oxide: How sharp edged structure and charge matter. *Applied Materials and Interfaces* 7(38):21571–21579.

Yu, B.F., Hu, Z.B., Liu, M., Yang, H.L., Kong, Q.X. and Liu, Y.H. 2009. Review of research on air-conditioning systems and indoor air quality control for human health. *International Journal of Refrigeration* 32(1):3–20.

Yu, L., Wang, C., Chen, F., Zhang, J., Ruan, Y. and Xu, J. 2016. Investigating the synergistic effects in tourmaline/TiO_2-based heterogeneous photocatalysis: Underlying mechanism insights. *Journal of Molecular Catalysis A: Chemical* 411:1–8.

Zhang, C., Li, Y., Shuai, D.M., Shen, Y. and Wang, D.W. 2019. Progress and challenges in photocatalytic disinfection of waterborne viruses: A review to fill current knowledge gaps. *Chemical Engineering Journal* 355:399–415.

Zheng, X., Chen, D., Wang, Z., Lei, Y. and Cheng, R. 2013. Nano-TiO_2 membrane adsorption reactor (MAR) for virus removal in drinking water. *Chemical Engineering Journal* 230:180–187.

Zheng, X., Shen, Z.-P., Cheng, C., Shi, L., Cheng, R. and Yuan, D.-H. 2018. Photocatalytic disinfection performance in virus and virus/bacteria system by Cu-TiO_2 nanofibers under visible light. *Environmental Pollution* 237:452–459.

Zhi, B., Gallagher, M.J., Frank, B.P., et al. 2018. Investigation of phosphorous doping effects on polymeric carbon dots: Fluorescence, photostability, and environmental impact. *Carbon* 129:438–449.

Zhou, L., Forman, H.J., Ge, Y. and Lunec, J. 2017. Multi-walled carbon nanotubes: A cytotoxicity study in relation to functionalization, dose and dispersion. *Toxicology in Vitro* 42:292–298.

7 Non-Photocatalytic and Photocatalytic Inactivation of Viruses Using Antiviral Assays and Antiviral Nanomaterials

Suman Tahir, Noor Tahir, Tajamal Hussain, Zubera Naseem, Muhammad Zahid, and Ghulam Mustafa

CONTENTS

DOI: 10.1201/9781003136644-9

7.1 INTRODUCTION

Viruses are smaller obligate intracellular parasites that generally range from 20 to 30 nm and comprise either single or double-stranded RNA or DNA surrounded in a protein coat termed as capsid (Sharma et al. 2012). Viruses are contagious disease-causing agents that can damage tissues, cells, and organs, and even may cause death in animals and human beings (Gerba and Betancourt 2017). It is estimated that viruses cause 2,000,000 human demises yearly (Colpitts et al. 2015). Viruses are unfortunately challenging to be inactivated completely because of their smaller size and higher resistance to strict environmental circumstances. Up to now, the ideal method to inhibit viral diseases is the vaccination approach. There are no adequate and useful vaccines against several viruses, along with their variations, proposing the alternative to develop extremely efficient drugs to inhibit viral entrance into host cells for repetition. Presently, numerous chemical drugs are present for the medical cure of certain viral ailments. For instance, a famous antiviral drug, acyclovir, is used for treating hepatitis B virus (HBV) and hepatitis C virus (HCV) (Chaudhry et al. 2015), and for treating varicella and other contagions in children, ganciclovir is used (Dibrov et al. 2014). Investigation on the efficiency of the chemical medicines has shown their inevitable adverse impacts on healthy humanoid cells, particularly in the case of children (Nitsche et al. 2014), hence restraining their extensive usage in medical therapies and implying the requirement to develop novel efficient and safe antiviral remedies. Amid several antiviral approaches, nanotechnology has displayed the potent capability to deal with this matter, and the developing nanoparticles (NPs) have been investigated and show remarkable efficiency against viral infection and replication (Chaudhuri et al. 2018).

The beginning of nanotechnology and nanoscience has provided incredible development in the research work, not owing to the structural design of the NPs, though their characteristics are also investigated (Seil and Webster 2012). Because of the size-dependent characteristics of NPs, nanotechnology is a significant field of nanomedicine. Additionally, NPs diameter is far less than that of typical bacterial size (0.5–2.0 μm). So, NPs can enter into microbes easily. The growth of various microorganisms is inhibited by nanostructured materials (NMs). NMs can aid as an efficient and active vehicular carrier for many biomedical purposes, for instance, targeted drug delivery, immunomodulation, vaccines, and antimicrobials (Salata 2004). The antimicrobial activities of metal oxides (TiO_2, CuO, CaO, Fe_2O_3, Ag_2O, SiO_2, ZnO, and MgO) and inorganic NPs (metals such as Au, Pt, Ag, and Cu) have been examined for practical integration of nanotechnology into pharmaceuticals (Dizaj et al. 2014). Initially, nanotechnology-based probes have extensively been employed in viral recognition, directing to the fabrication of numerous bioelectronics and biosensors based on innovative NPs (Chen et al. 2012). During recent decades, NPs have been broadly utilised in antiviral therapy because of the advancement of surface functionalization approaches (Szunerits et al. 2015). For instance, Au (Rafiei et al. 2015), Ag (Lu et al. 2008), SiO_2 (Botequim et al. 2012), $CuCl_2$ (Sucipto et al. 2017), TiO_2 (Levina et al. 2016) and CeO_2 (Lysenko et al. 2018) NPs have been used against various viruses, including H3N2 and H1N1 (Mazurkova et al. 2010), HBV (Lu et al. 2008), herpes simplex virus (HSV) (Hu et al. 2014), HIV-1 (Lara et al. 2010), foot-and-mouth disease (Rafiei et al. 2015), dengue virus type-2 (Sucipto et al. 2017), and vesicular stomatitis (Lokshyn et al. 2015).

7.2 CLASSIFICATION AND INFECTION MECHANISM OF VIRUSES

According to the Baltimore classification system, viruses are categorised into families depending on their method of replication and the type of genome is given in Table 7.1 (Sharma et al. 2012).

For self-reproduction, viruses use their host cells. Viral contagion mainly includes six stages: attachment, entrance, uncoating, repetition, assembly, and release (Oswald et al. 2017). Through binding proteins in capsid surrounded in the enveloped virus, viruses bind to particular receptor positions on the cell membrane of the host. This connection specificity establishes host cells that may be affected through a specific viral type. The capsid remains exterior to the cell; only bacteriophage nucleic acid enters into the host cells. Viruses of animals and

TABLE 7.1
Viruses Classification Based on Replication Method

Class	Replication Method	Examples
I	dsDNA viruses	Herpesviruses, Adenoviruses, Poxviruses
II	ssDNA viruses	Parvoviruses
III	dsRNA viruses	Reoviruses
IV	(+) sense ssRNA viruses	Picornaviruses, Togaviruses
V	(-) sense ssRNA viruses	Rhabdoviruses, Orthomyxoviruses
VI	(+) sense ssRNA viruses that duplicate using DNA intermediate	Retroviruses
VII	dsDNA viruses that duplicate using RNA intermediate (single-stranded)	Hepadnaviruses

certain florae can penetrate the cell by endocytosis, i.e. the whole virus being enclosed and engulfed through the cell membrane. Once the virus envelope combines with cell membrane, enveloped viruses will penetrate into the host. Within the host cell, when virus capsid is damaged, a discharge of virus nucleic acid results, which afterward can be used for transcription and multiplication. Mechanism of repetition is dependent on the genome of virus. Generally, DNA viruses consume proteins and enzymes of host cells to generate extra DNA, which is transcribed to messenger RNA (mRNA), and later utilised for the formation of protein. Typically, RNA viruses consume core of RNA for the production of mRNA, along with virus genomic RNA. The last step of viral multiplication is a discharge of new virions made in the host cells, permitting contagion of neighboring cells and duplication of self-multiplication cycles. In host cells, virus repetition cycle can make severe biochemical and structural variations and cause harm to them (Yang et al. 2017).

7.3 ANTIVIRAL ASSAY METHODS FOR NANOPARTICLES

7.3.1 Confocal Imaging Assay

For observing the variations in host cells once affected through viruses, virion adsorption assay is used (Alonas et al. 2014). This effective method for demonstrating the process of virus contagion in host cells is broadly employed in the antiviral assessment of NPs. Using confocal microscopy imaging, indirect immunofluorescence analysis can also be operated to exhibit variations in the viral proteins' structure earlier, and afterward treatment through NPs. The assay is an imagining method to show the antiviral consequences of NPs.

7.3.2 Transmission Electron Microscopy

Transmission electron microscopy (TEM) is the main technique in biological, chemical, and physical disciplines, which can attain delicate features even as tiny as an individual atom columns, which is thousands-folds lesser than the resolvable entity observed in the light microscope. So, TEM can be utilised to reveal viricidal states with clear images regarding the connection between NPs and viruses (Kennedy et al. 2016). TEM also offers visual representation concerning the affected cells and establishes the viricidal capability of NPs. Transmission electron cryo-microscopy, another important TEM technology, can examine the unfixed or unstained specimen and show them in the native environment. In an antiviral investigation, it can also be employed for characterisation of NPs effects on virus-infected cells.

7.3.3 Real-Time Polymerase Chain Reaction

Quantitative polymerase chain reaction (qRT-PCR) is another term of RT-PCR and is mostly utilised to observe the targeted DNA amplification in the course of the PCR process. The method can be utilised to identify viral DNA after its connection with NPs (Gall et al. 2015). For example, it can identify gene expression of host cells as a sign of virus contagion. The results of PCR can evidently display the metabolism status and growth of host cells after viral contagion and therapy with NPs. So, the PCR method offers an indirect indication of antiviral capability of NPs.

7.3.4 Flow Cytometry Assay

The assay for identifying chemical and physical features of the cells' population can be utilised to classify various cells and count a huge number of cells, in addition to identifying their features and functions. In this assay, huge quantities of cells can be rapidly observed, with statistics being collected and converted through a computer. Hence, the assay can be employed to count the viral affected cells and establish the destroying efficacy of NPs (Oakey et al. 2010).

7.3.5 Plaque Assay

Plaque assay, the process typically utilised for virion absorption examination, can identify the influence of NPs on virion (Rigotto et al. 2011). Using plaque assay, virus infectivity can be measured and the antiviral capability of functional NPs can be assessed. In the usual assay, the stock of virus is diluted 10 times, and 0.1 mL of aliquot dilution is inoculated on monolayers of vulnerable cells. After the incubation phase, a virus is enabled to bind to the cells, and nutrient medium agar is coated on the monolayers. After a longer incubation phase, the original affected cells discharge viral progeny, and the existence of the gel limits their spreading to adjacent cells; this results in the development of the zone of affected cells termed as plaque, which become huge, sufficient and noticeable to bare eye in room temperature circumstances. Titer of stock virus can be estimated in PFU (plaque forming units), which is accurate value of virus contagion. Virucidal capability of NPs can precisely be established through analysing PFU value of viral previously and afterward attaching with antiviral NPs.

7.3.6 Tissue Culture Infectious Dose Assay

$TCID_{50}$ assay is the usual viral quantification analysis for measuring contagious virus titer that can induce cytopathic impact in tissue culture for approximately 5–20 days though cells that persist in feasible circumstances. In the usual method, dilution stages are necessary to measure the quantity of virus for destroying 50% of affected host cells or inducing a cytopathic impact in 50% of inoculated tissue cultured cells. The method is generally utilised in the medical investigation for quantitation of deadly dosage of viruses. Once utilised in tissue culturing, host cells are plated, and sequential viral dilutions are also included. Afterward, incubation ratio of cell decease is evaluated and noted for every viral dilution, and outcomes are gathered as the result of $TCID_{50}$ (Bibby et al. 2015).

7.3.7 β-Galactosidase Assay

The procedure is established by Food and Drug Administration (FDA) of the United States to determine host indications after vaccination, which is currently adapted to be particular for identifying viral entrance (Yin et al. 2017). It can be utilised in genetics, life sciences, and molecular biology. In the antiviral characterisation of NPs, it's normally employed as a reporter gene to observe expression level in host cells, offering indirect indication for capability of NPs to cure virus contagions through displaying the physiological condition of the affected host cells.

7.3.8 Western Blot Assay

In biochemistry, this assay is the widely employed process for qualitative identification of specific proteins and can recognise protein denaturation of viruses when treated with NPs (Thadani et al. 2018). In the usual experiment, sample experiences with protein degradation succeeded after gel electrophoresis, and the primary antibody formed through the man-made or animal derivative method for identifying and connecting to a particular target. Afterward, the electrophoresis membrane was washed two times in the solution comprising the primary antibody prior to an additional antibody being washed away with the addition of a secondary antibody to identify and attach to the primary antibody. Various methods can visualise the secondary antibody, for instance, immunofluorescence, staining, and radioactivity, allowing indirect recognition of the particular targeted protein. In the antiviral characterisation of NPs, this can exhibit the changes in viral protein structures after connection with NPs.

7.3.9 In Vivo Analysis

In vivo analysis states to test the impact of numerous biological facts on entire living cells or beings, involving humans, plants, and animals. The study is meant to determine the influence of drugs or chemicals on the biological system. In the antiviral investigation, the goal was to shield lives through chemicals and vaccines, so an emphasis was placed on safety of chemical drugs and vaccines for humans, plants, and animals. Pardi and Hogan coworkers established that the single less-dosage intradermal immunisation with the lipid-NP-encapsulated nucleoside-functionalized mRNA (mRNA-LNP) coding pre-membrane and enveloped glycoproteins form Zika virus (ZIKV) strain (Pardi et al. 2017). By vaccinating a living rat, a single dosage of 50 μg was realised to be sufficient for shielding nonhuman primates contrary to the task of five weeks afterward vaccination. The effort displays that the nucleoside-functionalized mRNA-LNP might be a probable nominee as an anti-ZIKV vaccine. This in vivo mice data established that animal vaccine analysis is an obligatory and valuable process for studying antiviral action of NPs.

7.3.10 Computer Simulation

This method represents the reproduction of biological or chemical actions through the computer for representing the results of numerical modeling. In biochemistry, computer simulation has become a valuable approach for numerical modeling of several natural systems. In an antiviral investigation, this is an auxiliary way to analyse variants in particular protein and DNA position in the course of interaction between NPs and virus. It offers needed indirect indication for foreseeing the variations in viral and host cells after the interference of NPs (Capuzzi et al. 2018).

7.4 NON-PHOTOCATALYTIC INACTIVATION OF VIRUSES

7.4.1 Silver Nanoparticles

Ag NPs with exceptional electronic, optical, and antibacterial properties have been broadly utilised in photonics (Hu and Chan 2004), electronics (Alshehri et al. 2012), biosensing (Habouti et al. 2010), and antimicrobial (Chen et al. 2014) applications. The physicochemical properties of Ag NPs, which involve shape, surface charge, size (surface area), and agglomeration, are specifically significant for establishing their biological connections. The smaller particle has a greater surface area and hence has the higher toxic capability (Johnston et al. 2010). The shape of Ag NPs can drastically affect their chemical and physical properties. The commonly employed Ag nanostructures in the biomedical discipline involve Ag spherical NPs, nanorods, nanocubes, nanowires, and nanoplates (Rycenga et al. 2011). Because of the peculiar physicochemical characteristics and exceptional biological functions, silver NPs are utilised as antibacterial, antiviral, anti-angiogenesis, anti-inflammatory, antifungal, anticancer, and antiplatelet agents (Gurunathan et al. 2018). The Ag NPs characterisation is significant to estimate functional characteristics of prepared particles and can be characterised through various analytical techniques, including X-ray diffractometry (XRD), dynamic light scattering (DLS), Fourier transform infrared spectroscopy (FTIR), atomic force microscopy (AFM), UV-vis spectroscopy, X-ray photoelectron spectroscopy (XPS), scanning electron microscopy (SEM), and transmission electron microscopy (TEM) (Zhang et al. 2016). The characterisation of Ag NPs through various techniques has been shown in Figure 7.1.

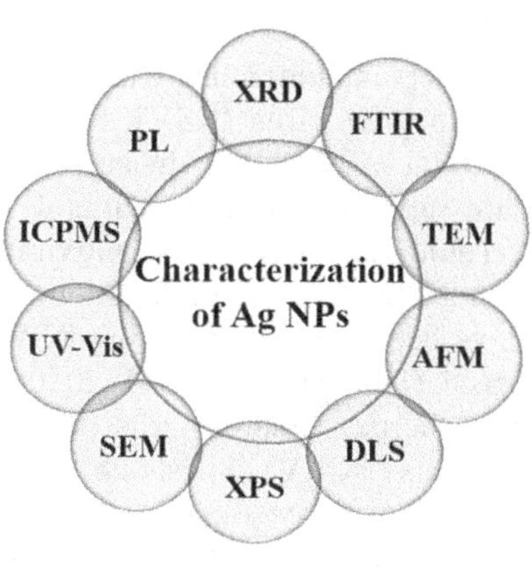

FIGURE 7.1 Characterisation of Ag NPs through various techniques.

7.4.1.1 Synthesis of Silver Nanoparticles

Various biological, chemical, and physical routes are utilised for the Ag NPs synthesis. Chemical approaches can be subdivided into the irradiation-assisted chemical process (Sotiriou and Pratsinis 2010), chemical reduction (Zhang et al. 2011), pyrolysis (Sotiriou et al. 2011), and electrochemical techniques (Roldán et al. 2013). Ag NPs preparation in solution generally comprises three chief constituents: metal precursor, stabilising/capping agents, and reducing agents. The broadly employed reducing agents comprise sodium citrate, alcohol, borohydride, ascorbic acid, and hydrazine compounds. Ag/Si NPs with relatively limited size distribution were prepared by flame spray pyrolysis (Sotiriou et al. 2011). On the other hand, the physical method does not include harmful chemicals, generally has rapid processing time, and involves energy through the ball milling process (Kosmala et al. 2011), direct current (DC) magnetron sputtering (Asanithi et al. 2012), vapour condensation (Abou El-Nour et al. 2010), and arc-discharge (Tien et al. 2008). Narrow size distribution of the prepared Ag NPs is another advantage of the physical methods (Asanithi et al. 2012). However, high energy consumption is its major drawback. In biological synthesis, harmful stabilisers and reducing agents are substituted by harmless molecules, such as antioxidants, carbohydrates, and proteins made by living beings and involving fungi, bacteria, yeast, and florae. The biological approaches are based on microbes, for instance, fungi (Li et al. 2012), yeast (Mourato et al. 2011), and bacteria (Shivaji et al. 2011), which have been extensively investigated. The inexpensive plants, for instance, aloe vera, alfalfa, neem, safeda, tulsi, mustard, lemongrass, seaweed, tea, and lotus, have been studied for Ag NPs synthesis. The probable mechanism of biological synthesis involves enzymatic (i.e. NADPH reductase) and nonenzymatic reduction (Ge et al. 2014). Generally, Ag NPs synthesis through plant extract is mainly employed as an environmentally friendly process. Because of their easy preparation, Ag NPs have fascinated attention in medical and biological applications. For example, Ag NPs have been utilised in antibacterial lotion and as biomedical therapeutic agents; for instance, long-term burn-care products and wound dressings comprising Ag NPs are commercially available. Figure 7.2 shows the different synthesis route for synthesis of Ag NPs.

7.4.1.2 Silver Nanoparticles as Antiviral Agents

Ag NPs as antiviral candidates have been studied versus numerous viruses; for example, monkeypox virus (MPV), respiratory syncytial virus (RSV), human immunodeficiency virus (HIV), HBV, and HSV (Rai et al. 2016). In antiviral treatments, Ag NPs are novel agents. Ag NPs are supposed to inhibit virus development as they target all parts of the virus. Table 7.2 provides the antiviral activities of Ag NPs for numerous viruses. For understanding antiviral action of Ag NPs contrary to viruses, important efforts have been offered involving RSV (Sun et al. 2008), human immunodeficiency virus-1 (HIV-1) (Lara et al. 2010), MPV (Rogers et al. 2008), HBV (Lu et al. 2008), herpes simplex virus type-1 (HSV-1) (Baram-Pinto et al. 2010), and tacaribe virus (TCRV) (Speshock et al. 2010). The utilisation of Ag NPs contrary to HIV-1 was investigated and observed that Ag NPs exhibited dimension-based connections with the HIV-1

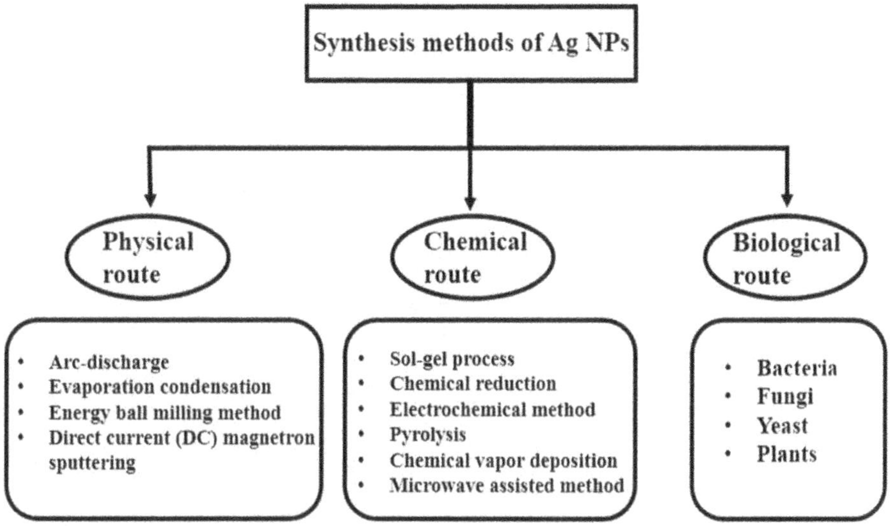

FIGURE 7.2 Different synthesis methods of Ag NPs.

(Elechiguerra et al. 2005). Similarly, chitosan (Ch) based Ag NPs composites were synthesised to investigate their antiviral action on the influenza A virus H1N1 (Mori et al. 2013). Antiviral action of Ag/Ch composite was estimated through comparison with $TCID_{50}$ value of the influenza A virus H1N1 with and deprived of Ch-based Ag NPs. Higher content of Ag NPs exhibited efficient antiviral action; however, with chitosan, no antiviral action was noticed. A novel Ag NPs-modified protocol for the assessment of cytotoxic and antiviral characteristics of Ag NPs was developed versus HIV-1 (Trefry and Wooley 2012). The three various sizes (13, 33, and 46 nm) of Ag NPs were synthesised and tannic acid was used as a capping and reducing agent (Orlowski et al. 2014). The prepared Ag NPs were efficient on the mice model of HSV-2 contagion. Ag NPs utilising curcumin as capping and reducing agents were prepared to examine their antiviral action on RSV (Yang et al. 2016). Curcumin-based Ag NPs repressed RSV before entering to cells. The studies have shown that Ag NPs are active antiviral candidates. Though, as stated by some Ag NPs antiviral investigations, an antiviral mechanism can be elucidated, as explained below.

7.4.1.3 Antiviral Mechanism of Silver Nanoparticles

Ag NPs displayed various kinds of antiviral mechanisms contrary to numerous viruses (Narayanan and Park 2014). When the Ag NPs performed versus HIV-1, the multiplication of HIV-1 was repressed. Also, Ag NPs were observed to inhibit HBV RNA through repressing formation of the extracellular virions in vitro. Furthermore, cytotoxicity and antiviral activity of Ag NPs were observed through haemagglutination assay (Narayanan and Park 2014). Consequently, Ag NPs were proved to possess the robust capability to destroy glycoproteins of the viral membrane. The mercaptoethane sulfonate (MES) shielded Ag NPs have been also

TABLE 7.2
Antiviral Properties of Ag Nanoparticles

Nanoparticles	Virus Name	Virus Family	Mechanism of Action	References
Tannic acid-based Ag NPs	Herpes simplex virus type 2 (HSV-2)	*Herpesviridae*	Interact with viral glycoproteins hence inhibit with cell attachment	Orlowski et al. (2014)
Magnetic hybrid colloid Ag NPs	Bacteriophage MS2	*Leviviridae*	Damage proteins of the viral coat	Park et al. (2014)
Mycosynthsized Ag NPs	HSV-1 and HSV-2	*Herpesviridae*	Block interaction of virus and cells	Gaikwad et al. (2013)
PVP-based Ag NPs	HIV	*Retroviridae*	Repress the interaction between gp120 and cell membrane receptors	Lara et al. (2010)
Chitosan-coated Ag NPs	H1N1 influenza A	*Orthomyxoviridae*	Prevent the virus connection with host cells and interaction of Ag NPs with virus glycoproteins	Mori et al. (2013)
Pure Ag NPs	Poliovirus		Binding with viral, therefore, inhibit connection with host receptor cells and suppression of virus proteins	Huy et al. (2017)
PVP-based Ag NPs	RSV	*Paramyxoviridae*	Interfere with viral binding through attachment with glycoproteins	Morris et al. (2019)
Ag NPs	HBV	*Hepadnaviridae*	Decrease the HBV DNA formation by attachment with HBV dsDNA and virion	Lu et al. (2008)
Ag NPs	Adenovirus type 3	*Adenoviridae*	Destruction of the virus and binds to virus DNA	Chen et al. (2013)
Naked and polysaccharide coating	TCRV	*Arenaviridae*	Prior to entrance, inactivation of virus	Speshock et al. (2010)
PVP, RF-412, and BSA-based Ag NPs	RSV	*Paramyxoviridae*	Interfere with the viral binding	Sun et al. (2008)
Naked and polysaccharide coating	MPV	*Poxviridae*	Blockage of virus-host cell attachment and entrance	Rogers et al. (2008)
Curcumin-based Ag NPs	RSV	*Paramyxoviridae*	Prevent the virus multiplication	Yang et al. (2016)

utilised as possible antiviral candidates for HSV-1 (Galdiero et al. 2011). However, it was examined that MES-based Ag NPs played as binding agents to attach with viral host cells; they repressed HSV-1 contagions and obstructed viral entrance into host cells. To investigate probable utilisation of capping agent on Ag NPs surfaces, three capping agents such as poly N-vinyl-2-pyrrolidone (PVP), bovine serum albumin (BSA), and recombinant F protein (RF 412) from RSV were fabricated on Ag NPs to evaluate the antiviral action on RSV in Hep-2 cell culture (Elgorban et al. 2016). RF-412 and BSA-Ag NPs were noticed to efficiently connect with the RSV without particular binding, while PVP-based Ag NPs connected with G proteins that occur on enveloped of RSV virus. Exploration of antiviral action of Ag NPs on MPV show that Ag NPs have possibly inhibited viral entry into the host cells through either preventing viral host cell connections or distorting intracellular multiplication of viral (Holt and Bard 2005). Contagion of TCRV reduced through the connection of Ag NPs with the glycoproteins of virus membrane, thus repressing internalisation of virus (Speshock et al. 2010). Consequently, TCRVRNA-dependent RNA polymerase displayed an important breakup to cease viral multiplication. The possible Ag NPs antiviral mechanism is shown in Figure 7.3. In conclusion, Ag NPs are established as probable antiviral candidates contrary to several viruses. Hence, for effectively treating a widespread spectrum of viruses, Ag NPs functionalized antiviral drugs can be established.

7.4.1.4 Toxicity of Silver Nanoparticles

Ag NPs are progressively employed owing to their intrinsic antimicrobial action and numerous useful characteristics for the improvement of humanity (Ip et al. 2006). On the other hand, their increased utilisation leads to numerous concerns associated with environmental and health effects.

7.4.1.4.1 Cellular Toxicity

It has been studied that at a lower concentration, Ag NPs exhibit exceptional antimicrobial action; however, at greater concentration, they can cause harmful impact to organisms (Han et al. 2012). Hence, with increasing research in the discipline of nanotechnology, there has been escalating threat to NPs exposure. Ag NPs toxicity is estimated through carrying out various biochemical analysis in vivo and in vitro. The toxicity estimation generally involves studies of effects, for instance, metabolic activity in the course of exposure, generation of ROS, measuring variations in cellular morphology, and expression of numerous stress markers. Now, several investigations are presented which show numerous impacts and mechanism of Ag NPs toxic activity. It is suggested that beyond a particular limit, Ag NPs cause harmful impacts both in vitro and in vivo. Also, it is recognised that human beings are subjected to NPs through inhalation, ingestion, and penetration in the skin. The in vitro evaluation of the harmful impact of Ag NPs generally depends on the utilisation of cell cultures. However, such reports do not represent the actual state as that of in vivo state, only providing a vision of possible risky impact up to a specific level. Thus, in vivo investigations are more significant to acquire a closer view of exposure impacts. In rats, inhalation of Ag NPs for the prolonged-time period can result in lung accumulation, which leads to toxic impact. Ji and his colleagues

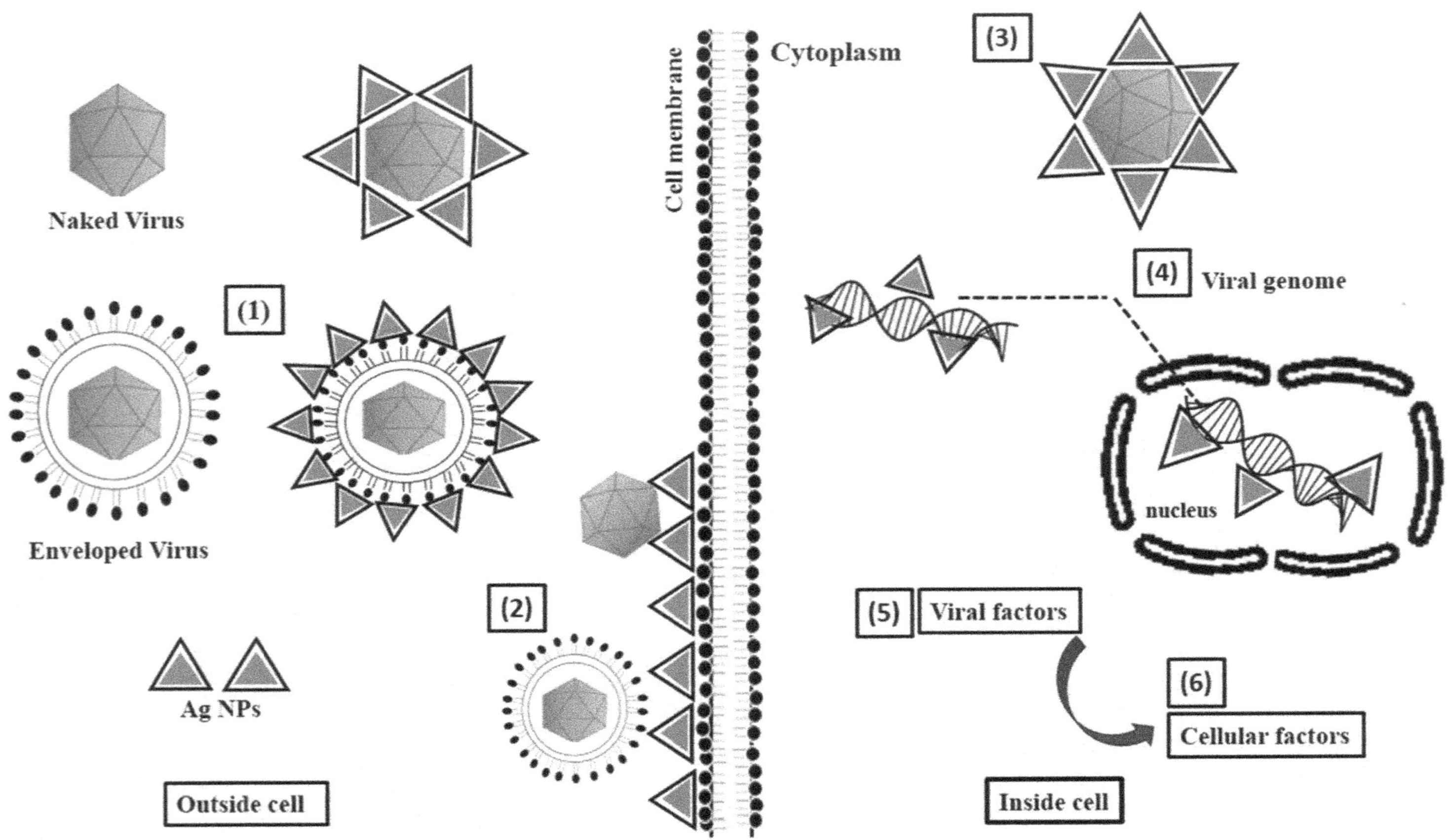

FIGURE 7.3 Proposed Ag NPs antiviral mechanism for numerous viruses: (1) connection of Ag NPs with virus enveloped surface proteins; (2) connection of Ag NPs with cell membrane and succeeding blockage of virus entry; (3) hindering of cellular pathways of virus entrance through Ag NPs; (4) connection of Ag NPs with virus genome; (5) Ag NPs connection in the course of viral repetition; (6) Ag NPs connection with cellular features compulsory for productive virus multiplication (Reproduced, adapted and modified from Rai et al. 2016).

assessed inhalation influence of Ag NPs for four weeks (Ji et al. 2007). Their experiment showed the deposits and transportation of Ag NPs from lungs to gastrointestinal region and liver. Moreover, consumption of Ag colloids was also observed to affect kidney and liver destruction. Increasing toxicological in vivo and in vitro data has much importance for designing biocompatible Ag NPs to describe the borderline beyond which Ag NPs would be harmful and within which they would be harmless for their utilisation in several biomedical and other applications.

7.4.1.4.2 Environmental Risk

Evaluation of environmental risk needs the inquiry of environmental release, bioavailability, mobility, and toxic action of NPs. Discharge of Ag NPs into the environment happens from point resources, such as from origination, in the course of synthesis and from non-point resources; these include accumulation in the atmosphere and products comprising NPs. Because of nano-dimension, NPs may have a substantial influence on the biotic constituent and bioavailability of contaminants or nutrient. Probable connection of NPs with harmful organic compounds can intensify or lessen their toxic action. Hence, NPs and organic contaminants can establish a static system for the living beings, in which harmful candidates are adsorbed using NPs, lessening their free content and decreasing toxicity. Since Ag NPs are robust antimicrobial agent, ecologists believe that if NPs are discharged into the environment, they may cause severe impact on natural systems, namely soil and water communities. In bulk state, Ag is harmful to algae, fish, some water florae, zebrafish, crustaceans, soil-forming bacteria, and some fungi. Ag NPs have also certain lethal effects on useful bacteria, including ammonifying bacteria, chemolithotrophic bacteria, and nitrogen-fixing bacteria. These microbes show significant roles in the biodegradation of organic matter and nitrogen fixation. Particular bacteria have also a symbiotic association with leguminous plants, offering nitrogen to plants. Consequently, the ultimate toxic action of NPs demands to be addressed carefully, especially concerning possible harm to eukaryotic cells; however, the worldwide point of view regarding environmental influence demands appropriate attention.

7.4.2 Gold Nanoparticles

Au NPs have attracted great concern in nanomedicine and industry because of their exceptional electrical, mechanical, optical, and biological properties (Gupta et al. 2016). Au NPs are utilised to identify proteins, DNA sequences, viruses, and bacteria and are commonly used in cancer investigations and as antibacterial and antiviral agents. One of the earliest prepared nanomaterials, Au NPs have earned consideration in nanomedical and nanodiagnostic implementations owing to their attractive surface impact, quantum size impact, and macroscopic quantum tunneling effect. Highly monodispersed quasi-spherical Au NPs were synthesised and utilised to repress the HSV (Halder et al. 2018). The hyaluronic acid (HA) based Au NPs/interferon (IFN) complex have been employed for the cure of HCV contagion (Lee et al. 2012). Compared with acyclovir drug, Au NPs are safe and don't cause any drug-resistant virus strains, and they show outstanding viricidal characteristics

(Halder et al. 2018). Modified Au NPs inhibit HIV, influenza virus, and HSV. Au NPs probably enhance antiviral impact by multivalent connections; dendronized Au NPs restrain HIV with more efficiently than dendrons only (Andresen et al. 2014). The sialic acid-based Au NPs prevent influenza virus contagion through multivalent connections, comparatively well compared to the few synthetic medical drugs, for instance, oseltamivir and zanamivir, which are susceptible to resistance developed by influenza virus (Papp et al. 2010). Gold nanoclusters (Au NCs), comprised of tens and hundreds of Au atoms with typical dimension of 1–2 nm, perform significant functions in biomedical applications. The entrance of PRRSV (porcine reproductive and respiratory syndrome virus) into the host cell was investigated, and it was observed that it can be repressed by Au NCs (Bai et al. 2018). Au NCs selectively repressed proliferation and protein expression of PRRSV. In antiviral activity of Au NCs, surface modification acts a significant role; hence, investigators altered surface of Au NCs with two diverse compounds such as histidine (His) and MES. Amid them, His-stabilised Au NCs exhibited robust prevention of proliferation of pseudorabies virus (PRV) (Bai et al. 2018). Consequently, Au NPs modified with diverse molecules play vital roles in antiviral action.

7.5 PHOTOCATALYTIC INACTIVATION OF VIRUSES

7.5.1 Zinc Oxide Nanoparticles

ZnO owns a wide bandgap of 3.3 eV in the UV range and holds exceptional chemical sensing, optical, electrical conductivity, piezoelectric, and semiconducting properties (Sirelkhatim et al. 2015). Commonly, ZnO NPs are added in paints, coatings, and sunscreens to absorb UV light, and they play a significant part in several industries, for instance, food, rubber, and pharmaceuticals. ZnO NPs are incorporated as antimicrobials into surface coatings, textiles, cellulose fibers, and cosmetics to prevent microbial growth (Dimapilis et al. 2018). ZnO NPs are considered as a suitable antibacterial agent since they are stable in severe processing circumstances and are recognised as harmless materials for animals and humans. Moreover, ZnO NPs are employed as an effective drug-delivery system. Along with its exceptional antifungal and antibacterial features, ZnO NPs own higher photochemical and catalytic activities. Also, ZnO NPs display vital microbial action versus numerous kinds of microorganisms involving viruses. ZnO NPs based drugs for the suppression of H1N1 influenza viral contagion were examined, and it was observed that PEGylated-based ZnO NPs might be an innovative, efficient, and auspicious antiviral candidate (Ghaffari et al. 2019).

7.5.2 Titanium Dioxide Nanoparticles

The most investigated semiconductor, TiO_2, is also used for viral photocatalytic inactivation in aqueous media (Lee and Ko 2013). However, it works only under ultraviolet (UV) light that credits ~4% of the solar range. Up to now, TiO_2 has been widely tailored by surface engineering or molecular doping, for instance, Mn/Co-TiO_2, Ag/TiO_2, and TiON/PdO to increase its consumption in solar light for viral inactivation (Venieri et al. 2015). Antibacterial features of TiO_2 have been well recognised and are ascribed for the formation of ROSs, particularly hydrogen peroxide (H_2O_2) and hydroxyl radicals (.OH). However, some reports have examined antiviral characteristics of TiO_2, and its capability for inactivating viruses has been established. But, the rate of inactivation attained in many of these investigations was very low. Cho and coworkers established that merely ~1 log elimination of MS2 virus afterward 2 hours of radiation using 1 g/L of P25 TiO_2 (Cho et al. 2005). TiO_2 inactivates influenza virus H3N2 through directly damaging viral particles (Mazurkova et al. 2010). In another study, Ag-doped TiO_2 NPs were examined for the ability of bacteriophage MS2 viral inactivation in aqueous solution (Liga et al. 2011). Using a photochemical reduction method, nano-dimensioned Ag deposits were synthesised using two commercial TiO_2 nanopowders. The MS2 virus inactivation rate was improved more than 5 times, and deactivation efficacy improved with the increasing Ag concentration. The enlarged generation of (.OH) was found to be accountable for improved viral inactivation.

7.5.3 Graphene-Based Nanomaterials

GO is a single atom thick and 2D carbon-based material that is organised in the hexagonal matrix. GO and its derivative are of huge concern to the investigators because of their extraordinary mechanical, thermal, and electronic properties (Ghosal and Sarkar 2018). GO is broadly utilised as an anticancer and antibacterial agent (Gurunathan et al. 2019). GO perform as an antiviral agent by inactivating the pathogenic agent of endemic gastrointestinal avian influenza A virus H9N2 and hand-foot-and-mouth ailment EV71 (Song et al. 2015). The nanocomposites comprising of GO and partly reduced GO sulfonated composite (rGO/SO_3) displayed antiviral action contrary to HSV-1 through restraining HSV-1 from binding to the host cells (Sametband et al. 2014). Antiviral action of rGO and GO was estimated contrary to PEDV (RNA virus) and PRV (DNA virus), showing that GO considerably repressed contagion of PEDV and PRV by 2-log decline in viral titer at non-cytotoxic concentration. GO prevented the virus entrance into host cells through structural damage (Ye et al. 2015). Nanocomposites comprising of GO and Ag NPs exhibited probable antiviral action versus non-enveloped and enveloped viruses, such as infectious bursal disease virus (IBDV) and feline coronavirus (FCoV). The virus inhibition assays determined that GO-Ag NPs repressed IBDV infection by 23% and FCoV by 25%; GO only repressed FCoV contagion 16%, however displayed no antiviral action versus IBDV contagion. So, combination of Ag NPs and GO showed superior antiviral action compared to either Ag or GO alone (Chen et al. 2016). To improve antiviral efficacy of the GO, it is typically modified with more antiviral agents to create the synergistic viral inhibition agent versus viral destructive attack. Huang and his colleagues established the curcumin-based GO to oppose RSV by adding a huge quantity of curcumin on cyclodextrin functionalized GO composite (Yang et al. 2017). It was established that curcumin-loaded GO has a considerable inhibitory impact on the RSV contagion and higher biocompatibility with the host cell. The probable antiviral mechanism of GO includes the following three modes: curcumin-modified GO can (i) directly inactivate the virus, (ii) restrain the virus from binding to host cells, and, (iii) lastly, disrupt the viral multiplication. Multifunctional GO-modified antiviral candidates deliver the new perception into nanomedicine assembly.

7.5.4 Quantum Dots

QDs are the semiconductor material with a typical dimension of 2–10 nm and are extensively utilised for image tracking, virus and cell labelling, and detection because of their renowned luminescent characteristics; these characteristics include narrow and bright emission spectroscopy, broad excitation spectroscopy, size-dependent emission wavelengths, and long fluorescence lifetime (Chen et al. 2018). However, limited reports are available on the utilisation of QDs as an antiviral candidate. The glutathione (GSH)-based cadmium telluride (CdTe) QDs were investigated; it was established that they varied PRV surface proteins structure, preventing the viral entrance into host cells (Du et al. 2015). Moreover, attaching of CdTe QDs to cell membrane itself lessened the virus count. The antiviral impact of carbon dots (CDots) with less toxic action were investigated with PRRSV and PRV as trial models of RNA and DNA viruses. The CDots meaningfully suppressed replication of both PRRSV and PRV.

7.5.5 Graphitic Carbon Nitride

g-C_3N_4 possesses 2-D structure with a reasonable 2.7 eV band gap, appropriate visible light absorption (approximately at 450–460 nm), and high chemical stability which results in exceptional optoelectronic characteristics (Zhang et al. 2019). Initially, g-C_3N_4 was described as the photocatalyst for the splitting of water in the visible light range (Wang et al. 2009) and appeals to huge consideration for the environmental implementations (Liu et al. 2015). g-C_3N_4 can be synthesised directly from inexpensive nitrogen-rich precursors, i.e. by melamine heating (Yan et al. 2009). g-C_3N_4 has been resistant to photo-corrosion, biocompatible with minor toxicity, and remains chemically stable in solvents such as diluted acids and bases (Cao et al. 2015). Earlier reports have mentioned that g-C_3N_4 is an encouraging candidate for the inactivation of bacteria (Huang et al. 2014), though no efficient work has been achieved for the virus inactivation using g-C_3N_4. Metal-free robust photocatalyst, g-C_3N_4 was investigated to inactivate bacteriophage MS2 virus in visible light radiation. In visible light radiation for 6 h, MS2 virus with initial content of 1×10^8 PFU/mL was entirely deactivated using g-C_3N_4 of 150 mg/L loading capacity (Li et al. 2016). No decline in g-C_3N_4 antiviral performance was noticed in five runs of consecutive MS2 photocatalytic inactivation. The reactive oxygen species (ROSs) were determined through a variety of scavengers, and photogenerated electrons and ($.O_2^-$) were observed to be a foremost provider for the inactivation of the virus. The images of TEM showed that after photocatalysis, the viral shape was deformed and the capsid covering was broken. Through photocatalytic oxidation, virus surface proteins, specifically replicase proteins, were destroyed. The protein damage would cause the escape and quick damage of inner constituents (four basic kinds of RNA genes), which ultimately result in virus decease without regrowth (Li et al. 2016).

7.5.5.1 Viral Inactivation Mechanism of g-C_3N_4

For understanding the mechanism of g-C_3N_4 induced viral inactivation, successions of scavengers were employed separately for comparison. Sodium oxalate, Fe(II)-EDTA, Cr(VI), isopropanol, L-histidine, and TEMPOL were employed to quench reactions of MS2 with OH, O_2^-, e^-, H_2O_2, h^+, and $^1O_2$, respectively. Preliminary trials showed that the existence of scavengers at chosen concentration did not deactivate the viruses (Li et al. 2016). Adding the isopropanol or TEMPOL, photocatalytic inactivation efficacy for MS2 was repressed in comparison with no addition of scavenger. It is proposed that OH and O_2^- were major ROSs which performed the essential task in the photocatalytic viral inactivation method. The major virucidal impact of $\cdot$ O_2^- rather than $\cdot$ OH are predictable, deeming the reality that e^- formed from conduction band (CB) of g-C_3N_4 can directly reduce O_2 to O_2^- however, h^+ formed from the valence band (VB) of g-C_3N_4 cannot oxidise directly H_2O/ OH to OH (Dong et al. 2014). On the other hand, identified OH must be originated from the conversion of $\cdot$ O_2^- by e^- in reduction site of g-C_3N_4. Consequently, feeding pristine O_2 might improve virus inactivation, which is probably through encouraging the generation of O_2^- and $\cdot$ OH. Thus, photogenerated electron and its derivative ROSs, O_2^- are thought to be efficient for virus inactivation in g-C_3N_4 photocatalytic system.

The virus inactivation through photocatalysis has been designated as the non-selective reaction followed from photogenerated e^- and its derivative ROSs (chiefly OH and O_2^-) are shown in Figure 7.4. Damage and oxidative destruction of viral surface proteins lead to escape and quick damage to inner constituents, specifically RNA, ultimately result in virus demise without regrowth. g-C_3N_4-based photocatalysts with improved antiviral characteristics, for instance, g-C_3N_4/TiO_2, have been described with an exceptional functioning for both bacterial inactivation (Li et al. 2015) and pollutant degradation, which is essential to be established and validated in more investigation.

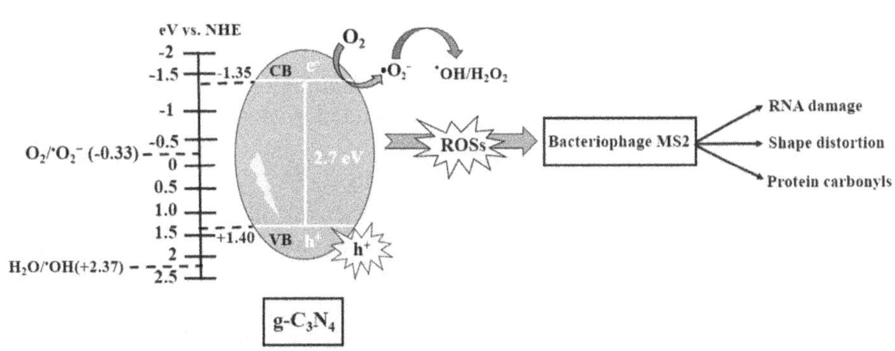

FIGURE 7.4 Proposed mechanism of virus inactivation using g-C_3N_4 in visible light radiation. (Reproduced, adapted and modified from Li et al. 2016).

7.5.6 Other Nanoparticles as Antiviral Agents

With the advancement of nanotechnology, apart from the above-mentioned NPs, numerous kinds of nanomaterials with antiviral action have been developed. Fluorescent NPs Gd_2O_3:Tb^{3+}/Er^{3+} as microcarriers were synthesised for ZIKV antigenic peptide (Ortega-Berlanga et al. 2020). Spray-dried Al_2O_3 granules fabricated with CuO NPs were reported and evaluated, showing that Cu oxidation state influence viral elimination capability (Mazurkow et al. 2019). The influence of CuO NPs on HSV-1 contagion was examined and realised that CuO NPs have a higher inhibitory impact on HSV-1 (Tavakoli and Hashemzadeh 2020). Exposing the HSV-1 to the CuO NPs at a greater nontoxic intensity (100 µg/mL) led to 2.8 log10 of $TCID_{50}$ decreases in contagious viral titer against viral control. Curcumin-modified (cur) nanoemulsions were reported to enhance solubility and cell uptake efficiency of cur and attained an exceptional dengue virus (DENV) inhibition outcome from nanoemulsion (Nabila et al. 2020). Nickel oxide nanostructures (NiO NS) were synthesised and assessed their direct antiviral action contrary to cucumber mosaic virus (CMV) (Derbalah and Elsharkawy 2019). Results displayed that when cucumber plant interacted with NiO NS, it might powerfully inhibit CMV contagion against the nontreated plant. Zirconia (ZrO_2) NPs antiviral impact on highly pathogenic avian influenza virus H5N1 was observed and established that ZrO_2 NPs can shield mice versus H5N1 infection without any adverse effect (Huo et al. 2020). NPs have been also employed as antiviral remedy carriers through self-assembling structures for inhibition of ZIKV spread (Surnar et al. 2019). In short, NPs have displayed exceptional ability in virus targeting and inhibition, also lesser toxic action than chemical drugs in vivo. At present, investigations on the nanomaterials-modified antiviral candidates is on arise, and in future, further auspicious materials will be utilised for the antiviral treatment. Table 7.3 offers the antiviral activities of different NPs.

7.6 CONCERNING PROBLEMS WITH NANOPARTICLES AS ANTIVIRAL AGENTS

There are a few critical problems to be resolved for NPs:

i. It is mandatory to further enhance the biocompatibility of NPs. Presently, NPs dissolubility declines the releasing efficiency of several antiviral drugs, and an efficient preparation method is one of the ways to improve biocompatibility of NPs-modified antiviral drugs.
ii. Targeting the particular virus is an extremely demanding effort, and enhancing the specificity of NPs-functionalized antiviral drugs is crucial to resolve this issue.
iii. It is vital to maintain the NPs antiviral action, once attaching with viruses, and the structural design of nano-drug and assembly procedure is significant to resolve this concern.

TABLE 7.3
Antiviral Activities of Various Nanoparticles

Nanoparticles	Virus Name	Mechanism of Action	References
GO	RSV	Inactivate the virus directly and repress attachment	Yang et al. (2017)
GO	Herpesvirus	Attachment inhibition	Sametband et al. (2014)
ZrO_2 NPs	H5N1 influenza virus	Stimulate the cytokines expression	Huo et al. (2020)
Gd_2O_3:Tb^{3+}/Er^{3+} NPs	ZIKV	As the antigen micro-carriers for Zk2 peptide of ZIKV	Ortega-Berlanga et al. (2020)
Au NPs	HSV	Inhibit viral binding and entrance	Halder et al. (2018)
Sialic acid based Au NPs	Influenza virus	Repression of viral attachment to plasma membrane	Papp et al. (2010)
MES-based Au NPs	HSV-1	Competition for attachment of virus to the cell	Baram-Pinto et al. (2010)
CuO-NPs	HSV-1	Viral proteins oxidation and breakdown of the virus genome	Tavakoli and Hashemzadeh (2020)
NiO NS	CMV	Enhance expression of the pod, pr1, and pal1 genes	Derbalah and Elsharkawy (2019)
ZnO NPs	Influenza virus H1N1	Prevent virus only afterward virus entrance into host cells	Ghaffari et al. (2019)
TiO_2 NPs	H3N2 Influenza virus	Destruction of the virus envelope	Mazurkova et al. (2010)
Ag-doped TiO_2 NPs	Bacteriophage MS2	–	Liga et al. (2011)
g-C_3N_4	MS2	Damage to the viral surface proteins, specifically replicase proteins through photocatalytic oxidation.	Li et al. (2016)

iv. Up to now, the viricidal mechanism of NPs is still mainly unknown, particularly for the repression of drug-resistant viruses, regardless of numerous investigations concerning the inhibition impact of NPs on drug-resistant viral strains.

v. Finally, an important issue about the toxicity of NPs cannot be ignored; hence, degradation and stability topics must be a continuous investigation subject together with the improvement of NPs. Though using less concentration is perhaps nontoxic and safe to the active cells, it is still compulsory to examine the concentration restrictions of antiviral NPs to host cells, particularly in human medical experiments. As shown above, toxicity of many NPs is dosage-dependent and antiviral action works in a dosage-dependent way also. Hence, optimisation should be operated for each antiviral candidate to achieve its ideal concentration.

7.7 CONCLUSIONS AND PERSPECTIVES

Infectious ailments count for more than 20% of deaths worldwide, and viruses are accountable for approximately one-third of these demises. Researchers are rapidly struggling to identify and develop appropriate nano-vaccines and nano-based drugs. In both viral disease diagnosis and therapeutics, nanotechnology plays a significant role. NPs exhibit immense capability in biomedical implementations, particularly in those patients who revert after completion of traditional antiviral treatment. Moreover, nano-based methods are nontoxic, cost-effective, feasible, biocompatible, and a suitable approach for dealing with numerous kinds of viral contagions. With the benefits of size-dependent electrical and optical properties, higher surface reaction activity, and larger surface-to-volume ratio, NPs are being widely investigated in the domain of biomedicines and biosensors, and their viral activity has been systematically explored as well. The photocatalytic and non-photocatalytic inactivation of viruses have been examined through several nanoparticles; these include silver, titanium dioxide, zinc oxide, gold, and carbon-based materials, for instance, quantum dots, graphene oxide, and graphitic carbon nitride, along with several other emerging nanoparticles. In the future, these NPs may offer an innovative platform for the assembly of bio-safe and efficient drugs for the nanoscale cure of viral contagious ailments.

REFERENCES

Abou El-Nour, K.M., Eftaiha, A.A., Al-Warthan, A. and Ammar, R.A. 2010. Synthesis and applications of silver nanoparticles. *Arabian Journal of Chemistry* 3(3): 135–140.

Alonas, E., Lifland, A.W., Gudheti, M., et al. 2014. Combining single RNA sensitive probes with subdiffraction-limited and live-cell imaging enables the characterization of virus dynamics in cells. *ACS Nano* 8(1):302–315.

Alshehri, A.H., Jakubowska, M., Młożniak, A., et al. 2012. Enhanced electrical conductivity of silver nanoparticles for high frequency electronic applications. *ACS applied Materials and Interfaces* 4(12):7007–7010.

Andresen, H., Mager, M., Grießner, M., et al. 2014. Single-step homogeneous immunoassays utilizing epitope-tagged gold nanoparticles: On the mechanism, feasibility, and limitations. *Chemistry of Materials* 26(16):4696–4704.

Asanithi, P., Chaiyakun, S. and Limsuwan, P. 2012. Growth of silver nanoparticles by DC magnetron sputtering. *Journal of Nanomaterials* 2012:1–8.

Bai, Y., Zhou, Y., Liu, H., Fang, L., Liang, J. and Xiao, S. 2018. Glutathione-stabilized fluorescent gold nanoclusters vary in their influences on the proliferation of pseudorabies virus and porcine reproductive and respiratory syndrome virus. *ACS Applied Nano Materials* 1(2):969–976.

Baram-Pinto, D., Shukla, S., Gedanken, A. and Sarid, R. 2010. Inhibition of HSV-1 attachment, entry, and cell-to-cell spread by functionalized multivalent gold nanoparticles. *Small* 6(9):1044–1050.

Bibby, K., Fischer, R.J., Casson, L.W., Stachler, E., Haas, C.N. and Munster, V.J. 2015. Persistence of Ebola virus in sterilized wastewater. *Environmental Science and Technology Letters* 2(9):245–249.

Botequim, D., Maia, J., Lino, M., et al. 2012. Nanoparticles and surfaces presenting antifungal, antibacterial and antiviral properties. *Langmuir* 28(20):7646–7656.

Cao, S., Low, J., Yu, J. and Jaroniec, M. 2015. Polymeric photocatalysts based on graphitic carbon nitride. *Advanced Materials* 27(13):2150–2176.

Capuzzi, S.J., Sun, W., Muratov, E.N., et al. 2018. Computer-aided discovery and characterization of novel Ebola virus inhibitors. *Journal of Medicinal Chemistry* 61(8): 3582–3594.

Chaudhry, R.M., Nelson, K.L. and Drewes, J.R.E. 2015. Mechanisms of pathogenic virus removal in a full-scale membrane bioreactor. *Environmental Science and Technology* 49(5):2815–2822.

Chaudhuri, S., Symons, J.A. and Deval, J. 2018. Innovation and trends in the development and approval of antiviral medicines: 1987–2017 and beyond. *Antiviral Research* 155: 76–88.

Chen, F., Yao, Y., Lin, H., et al. 2018. Synthesis of CuInZnS quantum dots for cell labelling applications. *Ceramics International* 44:S34–S37.

Chen, G., Lu, J., Lam, C. and Yu, Y. 2014. A novel green synthesis approach for polymer nanocomposites decorated with silver nanoparticles and their antibacterial activity. *Analyst* 139(22):5793–5799.

Chen, L., Zhang, X., Zhou, G., et al. 2012. Simultaneous determination of human enterovirus 71 and coxsackievirus B3 by dual-color quantum dots and homogeneous immunoassay. *Analytical Chemistry* 84(7):3200–3207.

Chen, N., Zheng, Y., Yin, J., Li, X. and Zheng, C. 2013. Inhibitory effects of silver nanoparticles against adenovirus type 3 in vitro. *Journal of Virological Methods* 193(2): 470–477.

Chen, Y.-N., Hsueh, Y.-H., Hsieh, C.-T., Tzou, D.-Y. and Chang, P.-L. 2016. Antiviral activity of graphene–silver nanocomposites against non-enveloped and enveloped viruses. *International Journal of Environmental Research and Public Health* 13(4): 430–441.

Cho, M., Chung, H., Choi, W. and Yoon, J. 2005. Different inactivation behaviors of MS-2 phage and *Escherichia coli* in TiO_2 photocatalytic disinfection. *Applied and Environmental Microbiology* 71(1):270–275.

Colpitts, C.C., Verrier, E.R. and Baumert, T.F. 2015. Targeting viral entry for treatment of hepatitis B and C virus infections. *ACS Infectious Diseases* 1(9):420–427.

Derbalah, A.S.H. and Elsharkawy, M.M. 2019. A new strategy to control Cucumber mosaic virus using fabricated NiO-nanostructures. *Journal of Biotechnology* 306:134–141.

Dibrov, S.M., Parsons, J., Carnevali, M., et al. 2014. Hepatitis C virus translation inhibitors targeting the internal ribosomal entry site: Miniperspective. *Journal of Medicinal Chemistry* 57(5):1694–1707.

Dimapilis, E.A.S., Hsu, C.-S., Mendoza, R.M.O. and Lu, M.-C. 2018. Zinc oxide nanoparticles for water disinfection. *Sustainable Environment Research* 28(2):47–56.

Dizaj, S.M., Lotfipour, F., Barzegar-Jalali, M., Zarrintan, M.H. and Adibkia, K. 2014. Antimicrobial activity of the metals and metal oxide nanoparticles. *Materials Science and Engineering: C* 44:278–284.

Dong, F., Wang, Z., Li, Y., Ho, W.-K. and Lee, S. 2014. Immobilization of polymeric g-C_3N_4 on structured ceramic foam for efficient visible light photocatalytic air purification with real indoor illumination. *Environmental Science and Technology* 48(17):10345–10353.

Du, T., Cai, K., Han, H., Fang, L., Liang, J. and Xiao, S. 2015. Probing the interactions of CdTe quantum dots with pseudorabies virus. *Scientific Reports* 5(1):1–10.

Elechiguerra, J.L., Burt, J.L., Morones, J.R., et al. 2005. Interaction of silver nanoparticles with HIV-1. *Journal of Nanobiotechnology* 3(1):1–10.

Elgorban, A.M., El-Samawaty, A.E.-R.M., Yassin, M.A., et al. 2016. Antifungal silver nanoparticles: Synthesis, characterization and biological evaluation. *Biotechnology and Biotechnological Equipment* 30(1):56–62.

Gaikwad, S., Ingle, A., Gade, A., et al. 2013. Antiviral activity of mycosynthesized silver nanoparticles against herpes simplex virus and human parainfluenza virus type 3. *International Journal of Nanomedicine* 8:4303–4314.

Galdiero, S., Falanga, A., Vitiello, M., Cantisani, M., Marra, V. and Galdiero, M. 2011. Silver nanoparticles as potential antiviral agents. *Molecules* 16(10):8894–8918.

Gall, A.M., Shisler, J.L. and Mariñas, B.J. 2015. Analysis of the viral replication cycle of adenovirus serotype 2 after inactivation by free chlorine. *Environmental Science and Technology* 49(7):4584–4590.

Ge, L., Li, Q., Wang, M., Ouyang, J., Li, X. and Xing, M.M. 2014. Nanosilver particles in medical applications: Synthesis, performance, and toxicity. *International Journal of Nanomedicine* 9:2399–2407.

Gerba, C.P. and Betancourt, W.Q. 2017. Viral aggregation: Impact on virus behavior in the environment. *Environmental Science and Technology* 51(13):7318–7325.

Ghaffari, H., Tavakoli, A., Moradi, A., et al. 2019. Inhibition of H1N1 influenza virus infection by zinc oxide nanoparticles: Another emerging application of nanomedicine. *Journal of Biomedical Science* 26(1):1–10.

Ghosal, K. and Sarkar, K. 2018. Biomedical applications of graphene nanomaterials and beyond. *ACS Biomaterials Science and Engineering* 4(8):2653–2703.

Gupta, A., Moyano, D.F., Parnsubsakul, A., et al. 2016. Ultrastable and biofunctionalizable gold nanoparticles. *ACS Applied Materials and Interfaces* 8(22):14096–14101.

Gurunathan, S., Arsalan Iqbal, M., Qasim M., et al. 2019. Evaluation of graphene oxide induced cellular toxicity and transcriptome analysis in human embryonic kidney cells. *Nanomaterials* 9(7):969–990.

Gurunathan, S., Qasim, M., Park, C., et al. 2018. Cytotoxicity and transcriptomic analysis of silver nanoparticles in mouse embryonic fibroblast cells. *International Journal of Molecular Sciences* 19(11):3618–3639.

Habouti, S., Solterbeck, C.-H. and Es-Souni, M. 2010. Synthesis of silver nano-fir-twigs and application to single molecules detection. *Journal of Materials Chemistry* 20(25): 5215–5219.

Halder, A., Das, S., Ojha, D., Chattopadhyay, D. and Mukherjee, A. 2018. Highly monodispersed gold nanoparticles synthesis and inhibition of herpes simplex virus infections. *Materials Science and Engineering: C* 89:413–421.

Han, D.-W., Woo, Y.I., Lee, M.H., Lee, J.H., Lee, J. and Park, J.-C. 2012. In-vivo and in-vitro biocompatibility evaluations of silver nanoparticles with antimicrobial activity. *Journal of Nanoscience and Nanotechnology* 12(7):5205–5209.

Holt, K.B. and Bard, A.J. 2005. Interaction of silver (I) ions with the respiratory chain of *Escherichia coli*: An electrochemical and scanning electrochemical microscopy study of the antimicrobial mechanism of micromolar Ag^+. *Biochemistry* 44(39): 13214–13223.

Hu, R., Li, S., Kong, F., Hou, R., Guan, X. and Guo, F. 2014. Inhibition effect of silver nanoparticles on herpes simplex virus 2. *Genet Mol Res* 13(3):7022–7028.

Hu, X. and Chan, C. 2004. Photonic crystals with silver nanowires as a near-infrared superlens. *Applied Physics Letters* 85(9):1520–1522.

Huang, J., Ho, W. and Wang, X. 2014. Metal-free disinfection effects induced by graphitic carbon nitride polymers under visible light illumination. *Chemical Communications* 50(33):4338–4340.

Huo, C., Xiao, J., Xiao, K., et al. 2020. Pre-treatment with zirconia nanoparticles reduces inflammation induced by the pathogenic H5N1 influenza virus. *International Journal of Nanomedicine* 15:661–674.

Huy, T.Q., Thanh, N.T.H., Thuy, N.T., et al. 2017. Cytotoxicity and antiviral activity of electrochemical–synthesized silver nanoparticles against poliovirus. *Journal of Virological Methods* 241:52–57.

Ip, M., Lui, S.L., Poon, V.K., Lung, I. and Burd, A. 2006. Antimicrobial activities of silver dressings: An in vitro comparison. *Journal of Medical Microbiology* 55(1): 59–63.

Ji, J.H., Jung, J.H., Kim, S.S., et al. 2007. Twenty-eight-day inhalation toxicity study of silver nanoparticles in Sprague-Dawley rats. *Inhalation Toxicology* 19(10):857–871.

Johnston, H.J., Hutchison, G., Christensen, F.M., Peters, S., Hankin, S. and Stone, V. 2010. A review of the in vivo and in vitro toxicity of silver and gold particulates: Particle attributes and biological mechanisms responsible for the observed toxicity. *Critical Reviews in Toxicology* 40(4):328–346.

Kennedy, E., Nelson, E.M., Tanaka, T., Damiano, J. and Timp, G. 2016. Live bacterial physiology visualized with 5 nm resolution using scanning transmission electron microscopy. *ACS Nano* 10(2):2669–2677.

Kosmala, A., Wright, R., Zhang, Q. and Kirby, P. 2011. Synthesis of silver nano particles and fabrication of aqueous Ag inks for inkjet printing. *Materials Chemistry and Physics* 129(3):1075–1080.

Lara, H.H., Ayala-Nuñez, N.V., Ixtepan-Turrent, L. and Rodriguez-Padilla, C. 2010. Mode of antiviral action of silver nanoparticles against HIV-1. *Journal of Nanobiotechnology* 8(1):1–10.

Lara, H.H., Ixtepan-Turrent, L., Garza-Treviño, E.N. and Rodriguez-Padilla, C. 2010. PVP-coated silver nanoparticles block the transmission of cell-free and cell-associated HIV-1 in human cervical culture. *Journal of Nanobiotechnology* 8(1):1–11.

Lee, J.E. and Ko, G. 2013. Norovirus and MS2 inactivation kinetics of UV-A and UV-B with and without TiO_2. *Water Research* 47(15):5607–5613.

Lee, M.-Y., Yang, J.-A., Jung, H.S., et al. (2012). Hyaluronic acid–gold nanoparticle/interferon α complex for targeted treatment of hepatitis C virus infection. *ACS Nano* 6(11):9522–9531.

Levina, A.S., Repkova, M.N., Mazurkova, N.A. and Zarytova, V.F. 2016. Nanoparticle-mediated nonviral DNA delivery for effective inhibition of influenza A viruses in cells. *IEEE Transactions on Nanotechnology* 15(2):248–254.

Li, G., He, D., Qian, Y., et al. 2012. Fungus-mediated green synthesis of silver nanoparticles using Aspergillus terreus. *International Journal of Molecular Sciences* 13(1): 466–476.

Li, G., Nie, X., Chen, J., et al. 2015. Enhanced visible-light-driven photocatalytic inactivation of *Escherichia coli* using g-C_3N_4/TiO_2 hybrid photocatalyst synthesized using a hydrothermal-calcination approach. *Water Research* 86:17–24.

Li, Y., Zhang, C., Shuai, D., Naraginti, S., Wang, D. and Zhang, W. 2016. Visible-light-driven photocatalytic inactivation of MS2 by metal-free g-C_3N_4: Virucidal performance and mechanism. *Water Research* 106:249–258.

Liga, M.V., Bryant, E.L., Colvin, V.L. and Li, Q. 2011. Virus inactivation by silver doped titanium dioxide nanoparticles for drinking water treatment. *Water Research* 45(2): 535–544.

Liu, J., Liu, Y., Liu, N., et al. 2015. Metal-free efficient photocatalyst for stable visible water splitting via a two-electron pathway. *Science* 347(6225):970–974.

Lokshyn, M., Lozovski, V., Lysenko, V., et al. 2015. Purification of bioliquids from viruses by surface plasmon-polaritons. *Journal of Bionanoscience* 9(6):431–438.

Lu, L., Sun, R., Chen, R., et al. 2008. Silver nanoparticles inhibit hepatitis B virus replication. *Antiviral Therapy* 13(2):253.

Lu, L., Sun, R.W.-Y., Chen, R., et al. 2008. Silver nanoparticles inhibit hepatitis B virus replication. *Antiviral Therapy* 13:253–262.

Lysenko, V., Lozovski, V., Lokshyn, M., et al. 2018. Nanoparticles as antiviral agents against adenoviruses. *Advances in Natural Sciences: Nanoscience and Nanotechnology* 9(2):025021–025028.

Mazurkova, N., Spitsyna, Y.E., Shikina, N., Ismagilov, Z., Zagrebel'Nyi, S. and Ryabchikova, E. 2010. Interaction of titanium dioxide nanoparticles with influenza virus. *Nanotechnologies in Russia* 5(5):417–420.

Mazurkow, J.M., Yüzbasi, N.S., Domagala, K.W., Pfeiffer, S., Kata, D. and Graule, T. 2019. Nano-sized copper (Oxide) on alumina granules for water filtration: Effect of copper oxidation state on virus removal performance. *Environmental Science and Technology* 54(2):1214–1222.

Mori, Y., Ono, T., Miyahira, Y., Nguyen, V.Q., Matsui, T. and Ishihara, M. 2013. Antiviral activity of silver nanoparticle/chitosan composites against H1N1 influenza A virus. *Nanoscale Research Letters* 8(1):1–6.

Morris, D., Ansar, M., Speshock, J., et al. 2019. Antiviral and immunomodulatory activity of silver nanoparticles in experimental RSV infection. *Viruses* 11(8):732.

Mourato, A., Gadanho, M., Lino, A.R. and Tenreiro, R. 2011. Biosynthesis of crystalline silver and gold nanoparticles by extremophilic yeasts. *Bioinorganic Chemistry and Applications* 2011:546074–546081.

Nabila, N., Suada, N.K., Denis, D., et al. 2020. Antiviral action of curcumin encapsulated in nanoemulsion against four serotypes of dengue virus. *Pharmaceutical Nanotechnology* 8(1):54–62.

Narayanan, K.B. and Park, H.H. 2014. Antifungal activity of silver nanoparticles synthesized using turnip leaf extract (*Brassica rapa* L.) against wood rotting pathogens. *European Journal of Plant Pathology* 140(2):185–192.

Nitsche, C., Holloway, S., Schirmeister, T. and Klein, C.D. 2014. Biochemistry and medicinal chemistry of the dengue virus protease. *Chemical Reviews* 114(22):11348–11381.

Oakey, J., Applegate Jr, R.W., Arellano, E., Carlo, D.D., Graves, S.W. and Toner, M. 2010. Particle focusing in staged inertial microfluidic devices for flow cytometry. *Analytical Chemistry* 82(9):3862–3867.

Orlowski, P., Tomaszewska, E., Gniadek, M., et al. 2014. Tannic acid modified silver nanoparticles show antiviral activity in herpes simplex virus type 2 infection. *PLoS One* 9(8):104113–104127.

Ortega-Berlanga, B., Hernández-Adame, L., del Angel-Olarte, C., Aguilar, F., Rosales-Mendoza, S. and Palestino, G. 2020. Optical and biological evaluation of upconverting Gd_2O_3: Tb^{3+}/Er^{3+} particles as microcarriers of a Zika virus antigenic peptide. *Chemical Engineering Journal* 385:123414–123424.

Oswald, M., Geissler, S. and Goepferich, A. 2017. Targeting the central nervous system (CNS): A review of rabies virus-targeting strategies. *Molecular Pharmaceutics* 14(7):2177–2196.

Papp, I., Sieben, C., Ludwig, K., et al. 2010. Inhibition of influenza virus infection by multivalent sialic-acid-functionalized gold nanoparticles. *Small* 6(24):2900–2906.

Pardi, N., Hogan, M.J., Pelc, R.S., et al. 2017. Zika virus protection by a single low-dose nucleoside-modified mRNA vaccination. *Nature* 543(7644):248–251.

Park, S., Park, H.H., Kim, S.Y., Kim, S.J., Woo, K. and Ko, G. 2014. Antiviral properties of silver nanoparticles on a magnetic hybrid colloid. *Applied and Environmental Microbiology* 80(8):2343–2350.

Rafiei, S., Rezatofighi, S.E., Ardakani, M.R. and Rastegarzadeh, S. 2015. Gold nanoparticles impair foot-and-mouth disease virus replication. *IEEE Transactions on Nanobioscience* 15(1):34–40.

Rai, M., Deshmukh, S.D., Ingle, A.P., Gupta, I.R., Galdiero, M. and Galdiero, S. 2016. Metal nanoparticles: The protective nanoshield against virus infection. *Critical Reviews in Microbiology* 42(1):46–56.

Rigotto, C., Hanley, K., Rochelle, P., De Leon, R., Barardi, C. and Yates, M. 2011. Survival of adenovirus types 2 and 41 in surface and ground waters measured by a plaque assay. *Environmental Science and Technology* 45(9):4145–4150.

Rogers, J.V., Parkinson, C.V., Choi, Y.W., Speshock, J.L. and Hussain, S.M. 2008. A preliminary assessment of silver nanoparticle inhibition of monkeypox virus plaque formation. *Nanoscale Research Letters* 3(4):129–133.

Roldán, M., Pellegri, N. and de Sanctis, O. 2013. Electrochemical method for Ag-PEG nanoparticles synthesis. *Journal of Nanoparticles* 2013:524150–524156.

Rycenga, M., Cobley, C.M., Zeng, J., et al. 2011. Controlling the synthesis and assembly of silver nanostructures for plasmonic applications. *Chemical Reviews* 111(6):3669–3712.

Salata, O.V. 2004. Applications of nanoparticles in biology and medicine. *Journal of Nanobiotechnology* 2(1):1–6.

Sametband, M., Kalt, I., Gedanken, A. and Sarid, R. 2014. Herpes simplex virus type-1 attachment inhibition by functionalized graphene oxide. *ACS Applied Materials and Interfaces* 6(2):1228–1235.

Seil, J.T. and Webster, T.J. 2012. Antimicrobial applications of nanotechnology: Methods and literature. *International Journal of Nanomedicine* 7:2767–2781.

Sharma, P., Chawla, A., Arora, S. and Pawar, P. 2012. Novel drug delivery approaches on antiviral and antiretroviral agents. *Journal of Advanced Pharmaceutical Technology and Research* 3(3):147–159.

Shivaji, S., Madhu, S. and Singh, S. 2011. Extracellular synthesis of antibacterial silver nanoparticles using psychrophilic bacteria. *Process Biochemistry* 46(9):1800–1807.

Sirelkhatim, A., Mahmud, S., Seeni, A., et al. 2015. Review on zinc oxide nanoparticles: Antibacterial activity and toxicity mechanism. *Nano-Micro Letters* 7(3):219–242.

Song, Z., Wang, X., Zhu, G., et al. 2015. Virus capture and destruction by label-free graphene oxide for detection and disinfection applications. *Small* 11(9-10):1171–1176.

Sotiriou, G.A. and Pratsinis, S.E. 2010. Antibacterial activity of nanosilver ions and particles. *Environmental Science and Technology* 44(14):5649–5654.

Sotiriou, G.A., Teleki, A., Camenzind, A., et al. 2011. Nanosilver on nanostructured silica: Antibacterial activity and Ag surface area. *Chemical Engineering Journal* 170(2–3):547–554.

Speshock, J.L., Murdock, R.C., Braydich-Stolle, L.K., Schrand, A.M. and Hussain, S.M. 2010. Interaction of silver nanoparticles with Tacaribe virus. *Journal of Nanobiotechnology* 8(1):1–9.

Sucipto, T.H., Churrotin, S., Setyawati, H., Kotaki, T., Martak, F. and Soegijanto, S. 2017. Antiviral activity of copper (II) chloride dihydrate against dengue virus type-2 in vero cell. *Indonesian Journal of Tropical and Infectious Disease* 6(4):84–87.

Sun, L., Singh, A.K., Vig, K., Pillai, S.R. and Singh, S.R. 2008. Silver nanoparticles inhibit replication of respiratory syncytial virus. *Journal of Biomedical Nanotechnology* 4(2):149–158.

Surnar, B., Kamran, M.Z., Shah, A.S., et al. 2019. Orally administrable therapeutic synthetic nanoparticle for Zika virus. *ACS Nano* 13(10):11034–11048.

Szunerits, S., Barras, A., Khanal, M., Pagneux, Q. and Boukherroub, R. 2015. Nanostructures for the inhibition of viral infections. *Molecules* 20(8):14051–14081.

Tavakoli, A. and Hashemzadeh, M.S. 2020. Inhibition of herpes simplex virus type 1 by copper oxide nanoparticles. *Journal of Virological Methods* 275:113688–113695.

Thadani, N.N., Dempsey, C., Zhao, J., Vasquez, S.M. and Suh, J. 2018. Reprogramming the activatable peptide display function of adeno-associated virus nanoparticles. *ACS Nano* 12(2):1445–1454.

Tien, D.-C., Tseng, K.-H., Liao, C.-Y., Huang, J.-C. and Tsung, T.-T. 2008. Discovery of ionic silver in silver nanoparticle suspension fabricated by arc discharge method. *Journal of Alloys and Compounds* 463(1-2):408–411.

Trefry, J.C. and Wooley, D.P. 2012. Rapid assessment of antiviral activity and cytotoxicity of silver nanoparticles using a novel application of the tetrazolium-based colorimetric assay. *Journal of Virological Methods* 183(1):19–24.

Venieri, D., Gounaki, I., Binas, V., Zachopoulos, A., Kiriakidis, G. and Mantzavinos, D. 2015. Inactivation of MS2 coliphage in sewage by solar photocatalysis using metal-doped TiO_2. *Applied Catalysis B: Environmental* 178:54–64.

Wang, X., Maeda, K., Thomas, A., et al. 2009. A metal-free polymeric photocatalyst for hydrogen production from water under visible light. *Nature Materials* 8(1):76–80.

Yan, S., Li, Z. and Zou, Z. 2009. Photodegradation performance of g-C_3N_4 fabricated by directly heating melamine. *Langmuir* 25(17):10397–10401.

Yang, M., Sunderland, K. and Mao, C. 2017. Virus-derived peptides for clinical applications. *Chemical Reviews* 117(15):10377–10402.

Yang, X.X., Li, C.M. and Huang, C.Z. 2016. Curcumin modified silver nanoparticles for highly efficient inhibition of respiratory syncytial virus infection. *Nanoscale* 8(5): 3040–3048.

Yang, X.X., Li, C.M., Li, Y.F., Wang, J. and Huang, C.Z. 2017. Synergistic antiviral effect of curcumin functionalized graphene oxide against respiratory syncytial virus infection. *Nanoscale* 9(41):16086–16092.

Ye, S., Shao, K., Li, Z., et al. 2015. Antiviral activity of graphene oxide: How sharp edged structure and charge matter. *ACS Applied Materials and Interfaces* 7(38):21571–21579.

Yin, H., Pijning, T., Meng, X., Dijkhuizen, L. and van Leeuwen, S.S. 2017. Engineering of the *Bacillus circulans* β-galactosidase product specificity. *Biochemistry* 56(5): 704–711.

Zhang, Q., Li, N., Goebl, J., Lu, Z. and Yin, Y. 2011. A systematic study of the synthesis of silver nanoplates: Is citrate a "magic" reagent? *Journal of the American Chemical Society* 133(46):18931–18939.

Zhang, S., Gu, P., Ma, R., et al. 2019. Recent developments in fabrication and structure regulation of visible-light-driven g-C_3N_4-based photocatalysts towards water purification: A critical review. *Catalysis Today* 335:65–77.

Zhang, X.-F., Liu, Z.-G., Shen, W. and Gurunathan, S. 2016. Silver nanoparticles: Synthesis, characterization, properties, applications, and therapeutic approaches. *International Journal of Molecular Sciences* 17(9):1534.

8 Application of Viral Nanomaterials in Medicine

Sudhakar Pola and Dhanalakshmi Padi

CONTENTS

8.1 INTRODUCTION

Nanomaterials are the smallest particles and can be used for many purposes. The word 'nano' is derived from the Greek and it means 'dwarf.' The father of nanotechnology is Richard Feynman, and nanoparticles were discovered by Richard Adolf Zsigmondy. Nanotechnology and nanoscience comprise a multidisciplinary study where science, engineering, and technology are studied at the nanoscale level, and the nanostructure has its dimension less than 100 nanometre (nm).

Nanomaterials have many applications in medicine, and the concept of nanotechnology in medicine is that those tiny nanorobots and their related machines are designed to treat the cellular repairs at the molecular level by releasing the nanomaterials to the site for drug delivery (Feynman 1960). The small structure, multifunctions, surface adaptability, precise delivery of drugs, and the development of new drug products and substances (which refers to nanomedicine) play a critical role in early detection and treatment of various diseases.

DOI: 10.1201/9781003136644-10

8.2 VIRAL NANOMATERIALS

All nanoparticles are nanomaterials, but not all nanomaterials are nanoparticles. Nanomaterials and nanoparticles are slightly differentiated in their size and are naturally produced. Viruses have the natural ability to infect the host and precisely deliver the drug to the targeted sites; therefore, VNPs and VLPs can be used as a vehicle for the site-targeted drug delivery. VLPs are the subpart of VNPs; they are capsid without genome and can be used as vaccines (Garcea and Gissmann 2004). VNPs are obtained from plants and animals that are harmless and noninfectious to humans; they also are biocompatible and biodegradable.

Viral nanotechnology is an emerging interdisciplinary technology in which viral nanoparticles are produced and applied in various areas, such as medicine development and electronics. Also, they are described as programmable because, as per technology, they can be modified (Pokorski and Steinmetz 2011). Physicochemical characteristics and surface bioavailability make NPs to detect the biomarkers in molecular diagnostics techniques (Moffatt 2016). Viral nanomaterials and nanoparticles are the platform for many advanced and next-generation therapeutics and also imaging technology (Steinmetz 2010).

8.2.1 Types of Nanomaterials

Nanomaterials are natural or synthesised particles that are in bound or unbound states with the size dimensions ranging from 1 nm to 100 nm. Nanoparticle clusters composed of two or more particles are called as aggregates or agglomerates. In agglomeration, two or more nanoparticles are agglomerated permanently, changing the size and physicochemical properties of nanoparticles.

Various kinds of nanomaterials are in use; mostly they are synthetic materials and natural bio-nano materials. Natural bio-nanomaterials are viral nanoparticles (VNPs) and virus-like particles (VLPs), whereas synthetic nanomaterials are differentiated according to their chemical and physical properties, such as carbonic, metals, and metal oxides (Yildiz et al. 2011). In cancer treatments and gene therapies, viral nanomaterials are mostly used and have a high success rate when compared to the synthetic nanomaterials. Viral nanomaterials are target specific and can deliver the drugs to the site. Therefore, viral nanotechnology is one of the advanced technologies in treating the cancer.

8.2.2 Engineering of VNPs

The bacteriophages and plant viruses are the major source of engineering VNPs. Engineering of VNPs includes many steps, such as bioconjugation chemistries, encapsulation techniques, biomineralization strategies, and film and hydrogel development (Pokorski and Steinmetz 2011). In the bioconjugation method, viral capsids, which are proteinaceous and have reactive side chains, can effectively conjugate with nucleic acids and other small molecules. These bioconjugation chemistries between biomolecules form complex VNPs, which are efficient in medical technologies for nanotherapeutics.

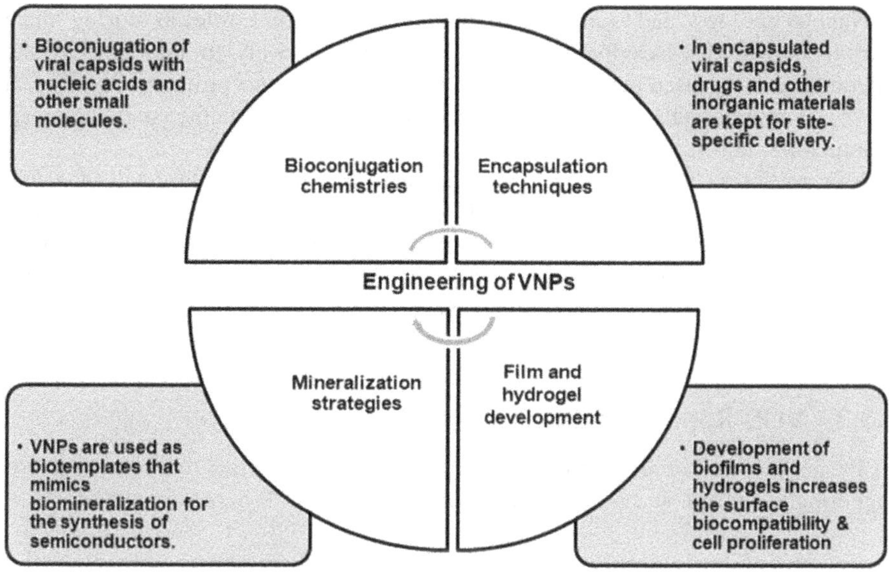

FIGURE 8.1 Different processes of engineering viral nanoparticles (VNPs).

The encapsulation method includes the natural ability of a virus to protect the genetic material by encapsulating it with a proteinaceous layer, a capsid. In addition to protection, it also has the ability to precisely deliver the genome into the host cell for viral replication. Therefore, in medical application, the natural ability of the virus is used for precisely delivering the drugs and inorganic nanomaterials to the site for further treatment.

In biomineralization or metalization strategies, VNPs are used as biotemplates. This means they are exposed to metals and inorganic materials, and they also interact with proteins. The result is the formation of nucleation and synthesis of mineralised semiconducting tubes and wires, which are applied in the medical technologies for imaging and diagnosis.

Hydrogels have many biomedical applications, such as drug delivery, cell culture, and tissue reconstruction during wound healing. The formation of biofilms increased the surface biocompatibility and cellular proliferation effectively; also, hydrogels have the ability to bind the negatively charged particles that are applied to the tissue engineering (Figure 8.1).

8.3 APPLICATION OF NANOTECHNOLOGY IN MEDICINE

8.3.1 Nanomedicine

Nanomaterials' role in medical technology refers to nanomedicine; nanomaterials are used to develop drugs and to diagnosis and treat various human diseases. There are some nanotherapeutic treatments; nanomedicines are approved by FDA for diseases like cancers, autoimmune and cardiovascular diseases, hepatitis, and infectious diseases (Bhaskar et al. 2010; Ventola 2012).

Nanotechnology and nanomedicine play a significant role in curing many chronic human diseases by delivering medicines precisely to the targeted site. Nanomaterials are used in medicine at the nanoscale level, according to Soares et al. (2018). Nanomaterials are used in three types of nanomedicine: nanodiagnosis, nanotherapy, and regenerative medicine.

Nanoparticles can effectively deliver the antiviral drugs to the targeted sites, with controlled release kinetics, enhanced bioavailability, modified pharmacokinetics, and reduced kinetics. Challenges occurring during the production of nano-based drugs are physico-chemical characterisation, biocompatibility, pharmacokinetics, and controlled process (Soares et al. 2018) (Figure 8.2). Therefore, nanotechnology plays a crucial role in antiviral therapeutics (Lembo et al. 2018).

8.3.2 VLPs Role in Immune Responses

VLPs are subviral structures; these contain adjuvant properties that are capable of triggering the innate and adaptive immune responses. VLP-based vaccines, such as

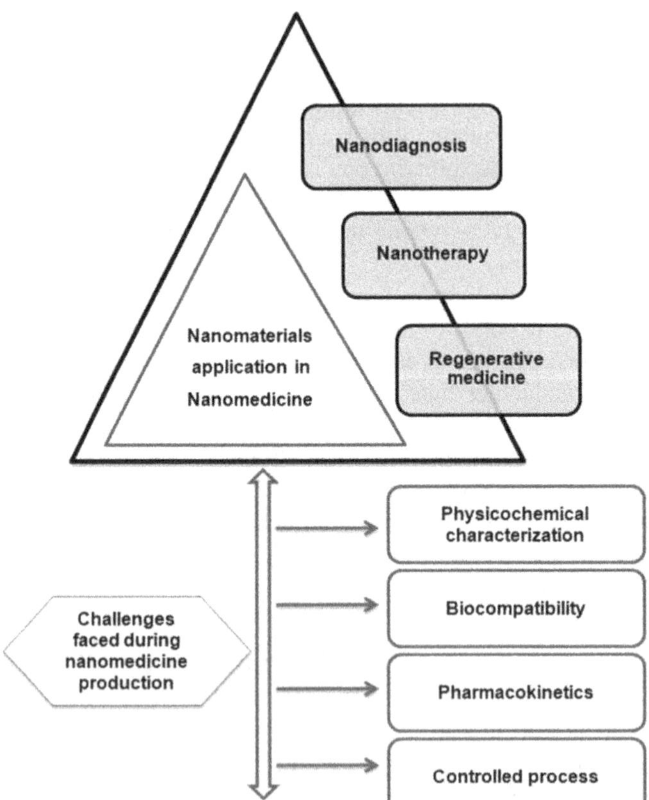

FIGURE 8.2 Applications of nanomaterials in nanomedicine for diagnosis and treatment, and also various challenges faced during the nanomedicine production.

nanovaccines, are used to protect humans from viral infections (Ludwig and Wagner 2007). VLPs are only capsids materials, meaning they lack genetic material; therefore, they are not infectious to humans and can be used as vaccines for infectious diseases.

Due to size, surface geometry, ability to enhance the immune responses, and safer and cheaper yielding, the VLP-based vaccines are available commercially worldwide. Some examples of commercialised VLP-based vaccines are mentioned in the review article by Roldão et al. (2010). They include, against hepatitis B virus (HBV), GlaxoSmithKline's engerix and Merck and Co., Inc's recombivax HB, and against human papilloma virus (HPV), cervarix and gardasil (Roldão et al. 2010; Mohsen et al. 2017). The development of chimeric VLP-based HPV includes many benefits, challenges, and limitations; focusing on these topics is necessary to improve the production of VLP-based vaccines and also to fight against pathogens (Eto et al. 2019).

8.3.3 Nanoparticles' Role against Cancers

Current cancer therapies, such as chemotherapy and radiotherapy, have many side effects and toxicity; however, immunotherapy is an effective clinical treatment for various cancers. Chemotherapeutic drugs inhibit the uncontrollable cell division but also affect the normal cells surrounding the cancerous cells. Therefore, the merger nanotechnology and immunotherapy is significant in treating cancers by allowing the delivery of drugs to the targeted sites using nanoparticles; in turn, it enhances the immune responses to fight against tumours (Pérez-Herrero and Fernández-Medarde 2015). Phagocytes, which play a critical role in both immunosupression and immunostimulation, uptake engineered NPs with immunostimulatory agents and thus induce antitumour immune responses (Sheen et al. 2014).

Gorbet et al. (2020) mentioned the in situ vaccination (ISV), which is injected directly into the targeted tumours; this vaccination has the ability to induce the innate and adaptive immune responses against cancerous tumours. The ISVs have included attenuated microorganisms, NPs, and also recombinant proteins and other small molecules. These induce immunostimulatory tumour microenvironment (TME). Among all, NPs are an advanced and attractive method because they can ably deliver the drugs to the cancerous cells in the TME and also induce antitumour immune responses. Therefore, NP-ISV therapeutics paved a successful way for treating various cancers through an immunotherapy regime (Gorbet et al. 2020; Mao et al. 2020). In upper gastrointestinal (GI) carcinomas, nano-based therapies aid in diagnosis, imaging, and precise delivery of drugs to the cancerous cells (Salapa et al. 2020).

8.3.4 Nanotechnology Role in the COVID-19 Pandemic

The Coronavirus infection disease-19 (COVID-19) pandemic caused a serious concern to the world as it affected 165 million people and killed 3.43 million, according to Johns Hopkins University and Medicine Coronavirus Resource Center. Thus, there is effort to prevent the infection and stop the various mutations of severe

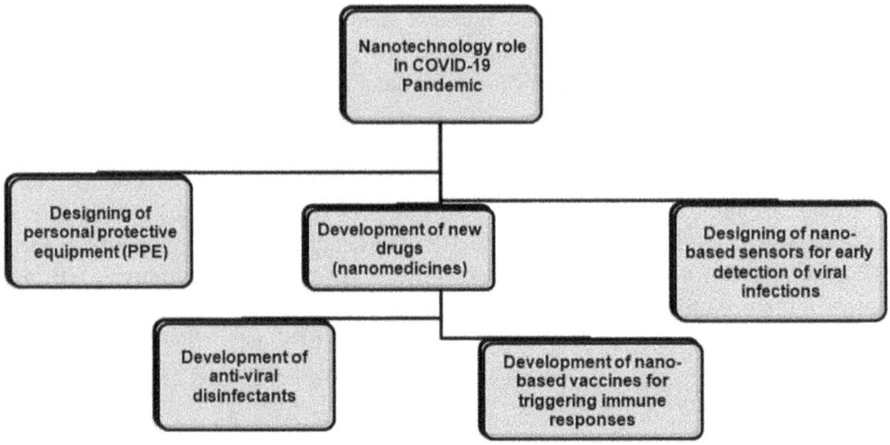

FIGURE 8.3 Schematic representation of various advantages of nanotechnology role in controlling and treating COVID-19 pandemic.

acute respiratory syndrome–Coronavirus-2 (SARS-CoV-2) virus by early tracing and treating techniques. However, the use of nanotechnology aids in ceasing further infections by inhibiting the replication of the viral genome; the design of nontoxic antiviral nanoparticles (NPs) showed an effective treatment for the COVID-19 disease. Therefore, designing an effective viral nanomaterial or nanovaccine for testing and treatment of COVID-19 is required (Tavakol et al. 2021).

The changes in genetic material and the emergence of new mutations and variants of Coronavirus are leading to faster and more dangerous spread of disease. Therefore, the usage of many antiviral drugs, immunotherapies, and other treatments are increased for treatment of COVID-19. The delivery of antiviral drugs at the infected site has been developed amid the COVID-19 pandemic for testing, treating, and controlling the viral diseases (Lembo et al. 2018; Patra et al. 2018).

The nano-based materials that are used for the prevention of infection by SARS-CoV-2 are antiviral disinfectants to cease the viral replication, personal protective equipments (PPE) kits to protect healthcare workers from viral infections, new drug products (nanomedicines), nano-based sensors for early diagnosis and treatment, and nano-based vaccines for enhancing the responses of immune system to produce antibodies against pathogens (Campos et al. 2020; Figure 8.3).

8.4 CONCLUSION

Viral nanomaterials and nanoparticles have a wide application in diagnostics and therapeutics regime. The advent of nanobiotechnology facilitated many inter-related connections and advancements, such as nanodiagnostics, nanomedicines, nanovaccines, and also nano-based sensors for diagnostics, imaging, and treatment. Therefore, various natural or engineered nanomaterials can effectively cure different cancers and infectious diseases.

ACKNOWLEDGEMENTS

The authors Sudhakar Pola and Dhanalakshmi Padi are grateful to authorities of Andhra University, Visakhapatnam, India for extending constant support.

REFERENCES

Bhaskar, S., Tian, F., Stoeger, T., et al. 2010. Multifunctional nanocarriers for diagnostics, drug delivery and targeted treatment across blood-brain barrier: Perspectives on tracking and neuroimaging. *Particle and Fibre Toxicology* 7:3. doi:10.1186/1743-8977-7-3.

Campos, E.V.R., Pereira, A.E.S., de Oliveira J.L., et al. 2020. How can nanotechnology help to combat COVID-19? Opportunities and urgent need. *Journal of Nanobiotechnology* 18(1):125. doi:10.1186/s12951-020-00685-4.

Eto, Y., Saubi, N., Ferrer, P. and Joseph, J. 2019. Designing chimeric virus-like particle-based vaccines for human papillomavirus and HIV: Lessons Learned. *AIDS Reviews* 21(4):218–232. doi:10.24875/AIDSRev.19000114.

Feynman, R.P. 1960. There's plenty of room at the bottom. *Engineering Science (CalTech)* 23:22–36.

Garcea, R.L. and Gissmann, L. 2004. Virus-like particles as vaccines and vessels for the delivery of small molecules. *Current Opinion in Biotechnology* 15(6):513–517. doi:10.1016/j.copbio.2004.10.002.

Gorbet, M.-J., Singh, A., Mao, C., Fiering, S. and Ranjan, A. 2020. Using nanoparticles for *in situ* vaccination against cancer: Mechanisms and immunotherapy benefits. *International Journal of Hyperthermia* 37(3):18–33. doi:10.1080/02656736.2020.1802519.

Lembo, D., Donalisio, M., Civra, A., Argenziano, M. and Cavalli, R. 2018. Nanomedicine formulations for the delivery of antiviral drugs: A promising solution for the treatment of viral infections. *Expert Opinion on Drug Delivery* 15(1):93–114. doi:10.1080/17425247.2017.1360863.

Ludwig, C. and Wagner, R. 2007. Virus-like particles universal molecular toolboxes. *Current Opinion in Biotechnology* 18(6):537–545. doi:10.1016/j.copbio.2007.10.013

Mao, C., Gorbet, M.-J., Singh, A., Ranjan, A. and Fiering, S. 2020. *In situ* vaccination with nanoparticles for cancer immunotherapy: Understanding the immunology. *International Journal of Hyperthermia* 37(3):4–17. doi:10.1080/02656736.2020.1810333.

Moffatt, S. 2016. Nanodiagnostics: A revolution in biomedical nanotechnology. *MOJ Proteomics Bioinform* 3(2):34–36. doi:10.15406/mojpb.2016.03.00080

Mohsen, M.O., Zha, L., Cabral-Miranda, G. and Bachmann, M.F. 2017. Major findings and recent advances in virus-like particle (VLP) - based vaccines. *Seminars in Immunology* 34:123–132. doi:10.1016/j.smim.2017.08.014.

Patra, J.K., Das, G., Fraceto, L.F., et al. 2018. Nano-based drug delivery systems: Recent developments and future prospects. *Journal of Nanobiotechnology* 16(1):71. doi:10.1186/s12951-018-0392-8.

Pérez-Herrero, E. and Fernández-Medarde, A. 2015. Advanced targeted therapies in cancer: Drug nanocarriers, the future of chemotherapy. *European Journal of Pharmaceutics and Biopharmaceutics* 93:52–79. doi:10.1016/j.ejpb.2015.03.018.

Pokorski, J.K. and Steinmetz, N.F. 2011. The art of engineering viral nanoparticles. *Molecular Pharmaceutics* 8(1):29–43. doi:10.1021/mp100225y.

Roldão, A., Mellado, M.C.M., Castilho, L.R., Carrondo, M.J.T. and Alves, P.M. 2010. Virus-like particles in vaccine development. *Expert Review of Vaccines* 9(10):1149–1176. doi:10.1586/erv.10.115.

Salapa, J., Bushman, A., Lowe, K. and Irudayaraj, J. 2020. Nano drug delivery systems in upper gastrointestinal cancer therapy. *Nano Convergence* 7(1):38. doi:10.1186/s40580-020-00247-2.

Sheen, M.R., Lizotte, P.H., Toraya-Brown, S. and Fiering, S. 2014. Stimulating antitumor immunity with nanoparticles. *Wiley Interdisciplinary Reviews: Nanomedicine and Nanobiotechnology* 6(5):496–505. doi:10.1002/wnan.1274.

Soares, S., Sousa, J., Pais, A. and Vitorino, C. 2018. Nanomedicine: Principles, properties, and regulatory issues. *Frontiers in Chemistry* 6:360. doi:10.3389/fchem.2018.00360.

Steinmetz, N.F. 2010. Viral nanoparticles as platforms for next-generation therapeutics and imaging devices. *Nanomedicine* 6(5):634–641. doi:10.1016/j.nano.2010.04.005.

Tavakol, S., Zahmatkeshan, M., Mohammadinejad, R., et al. 2021. The role of nanotechnology in current COVID-19 outbreak. *Heliyon* 7(4):e06841. doi:10.1016/j.heliyon.2021.e06841.

Ventola, C.L. 2012. The nanomedicine revolution, part 2: Current and future clinical applications. *Pharmacy and Therapeutics* 37(10):582–591.

Yildiz, I., Shukla, S., Steinmetz, N.F. 2011. Applications of viral nanoparticles in medicine. *Current Opinion in Biotechnology* 22(6):901–908. doi:10.1016/j.copbio.2011.04.020.

9 Virus-Based Nanocarriers for Targeted Drug Delivery

Semra Akgönüllü, Monireh Bakhshpour, Yeşeren Saylan, and Adil Denizli

CONTENTS

9.1 INTRODUCTION

The use of nanomaterials for treatment and non-invasive diagnosis in nanotechnology medicine is a new and attractive choice. In recent years, it has attracted great attention due to the nanomaterial applications as nanocarriers to deliver drugs to a target tissue. Cancer therapy applications of various nanocarrier platforms have greatly improved. The kinds of nanomaterials utilised for drug delivery systems vary in drug loading capacity, drug stability, targeted delivery ability, drug release rates, toxicity, and environmental impact (Haley and Frenkel 2008; Alemzadeh et al. 2018). New nanocarrier vehicles have shown perfect

DOI: 10.1201/9781003136644-11

potentials in improving targeting efficiencies, drug delivery, and packaging. These platforms present a few advances such as increase the adsorption of the drug in the desired tissues, control over the drug tissue distribution profile, and interacting with its environment, or prevention early drug disruption. The nanocarrier platforms have been performed including dendrimers, liposomes, polymeric nanoparticles, micelles, and virus-based nanoparticles. These nanocarrier systems have advantages and disadvantages according to developed needs in therapeutic applications (Peer et al. 2007).

Nanocarriers typically have a sub-micron nanoparticle size of 500 nanometers (Neubert 2011). Nanocarriers have been broadly researched over the past few years because of their great role in drug-delivery systems (Mishra et al. 2010). Nanocarriers can alter the basic features and drugs' bioactivity owing to their high surface area to volume ratio. Nanocarriers can influence drug-delivery systems through decreased toxicities, improved pharmacokinetics and biodistribution, controlled release, increased solubility and stability, and site-specific delivery of therapeutic agents (How et al. 2013; Din et al. 2017).

The cost-effective design of different nanocarrier materials has been gaining attention over the past few years. Following the discovery of the synthesis and design of small materials, researchers are at present focusing on the direct utilisation of biological structures for the synthesis of natural nanocarrier platforms. Many of these recent advancements are targeted at improving the performance of drug nanocarrier vehicles, displaying reduced toxicity and high differential uptake efficiencies, creating water-soluble, colloidal structures, or making them more biocompatible (Peer et al. 2007; Ma et al. 2012). In this chapter, the employment of viruses as natural drug nanocarriers and between various kinds of nanocarriers in drug-delivery systems is examined in depth. Finally, different virus-based nanomaterials and drug-delivery platforms are summarised in the field of nanotechnology.

9.2 TYPES OF NANOCARRIERS

Nanotechnology has recently attracted increasing interest due to its successful use in diagnosis and treatment (Din et al. 2017). Different nanomaterials may be utilised for the design of nanocarriers (Figure 9.1). The nanocarrier materials that have been broadly investigated contain liposomes, polymeric nanoparticles, micelles, dendrimers, and virus-based particles.

9.1.1 POLYMERIC NANOPARTICLES

Natural polymers, including albumin, heparin, and chitosan, have been used for protein, oligonucleotides, and DNA, as well as drug delivery in medical applications. Many synthetic polymers, including poly-L-glutamic acid, polystyrene-maleic anhydride copolymer, polyethylene glycol, and N-(2-hydroxypropyl)-methacrylamide copolymers, have also been used in drug-delivery applications (Cho et al. 2008). The drug can be conjugated to the polymer with a covalent bond or physically entrapped in the polymer. Both synthetic polymers (Afrooz et al. 2017), as well as natural polymers, have been utilised for nanoparticle preparation.

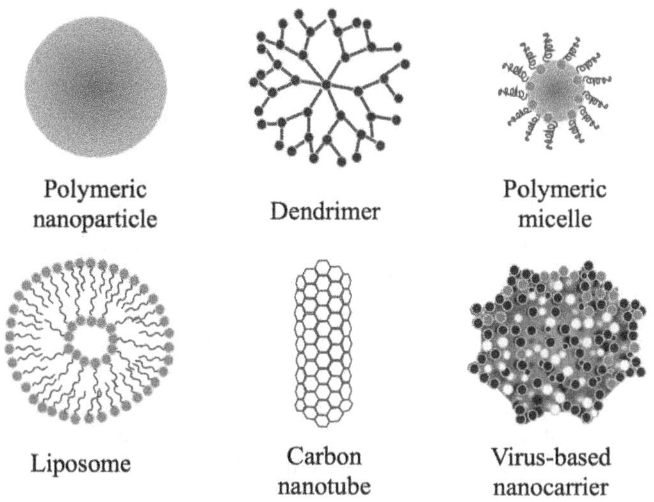

FIGURE 9.1 Types of nanocarriers (polymeric nanoparticle, dendrimer, polymeric micelle, liposome, carbon nanotube, and virus-based nanocarrier) used in drug delivery.

9.1.2 Polymeric Micelles

Micelles come together to create a nanosized core or shell structure in the aqueous environment. The properties of micelles depend on amphiphilic block copolymers. The hydrophobic core part acts as a reservoir for especially hydrophobic drugs. The hydrophilic shell part stabilises the hydrophobic core to make the particle a suitable candidate and make polymers water soluble (Adams et al. 2003). Polymeric micelles obtained by the block copolymers' organisation are mostly studied in drug delivery. Drug loading into polymeric micelles is achieved by physical encapsulation or covalent binding (Cho et al. 2008).

9.1.3 Dendrimers

The dendrimer is a multiple highly branched polymeric nanostructure (Haley and Frenkel 2008). Features related to dendrimers make them attractive as drug-delivery vehicles. Modifiable surface functionality, monodisperse size, available internal cavity, water solubility, and multivalency are some features of dendrimers. The structure of a dendrimer, especially the functional parts on the surface of dendrimer, play a significant role in encapsulating drugs/targets for obtaining a supramolecular construct. The covalent bond can occur between the peripheral functional groups and drugs (Svenson and Tomalia 2012).

9.1.4 Liposomes

Liposomes are self-assembling colloidal structures that form the lipid bilayers. They have spherical shapes in which the outer lipid bilayer envelopes a central aqueous environment (Cho et al. 2008; Haley and Frenkel 2008). The lipid-based system has been applied using different preparation methods for various cancer drugs. Clinical tests are currently evaluating the many liposomal chemotherapeutics. The next step of liposomal drugs may be immunoliposomes, which selectively deliver the drug to the targeted sites of action (Cho et al. 2008).

9.1.5 Viral-Based Nanocarriers

Viruses are tiny particles composed of nucleic acid (DNA, RNA). They are self-assembling to prepare a hollow scaffold-depended packing of the viral nucleic acid. The viral protein outer shell (i.e. the so-called capsid) is considered a nano-template because of its dimensions. Most viruses are limited to a special kind of host. Plant viruses infect some plants. Plant viruses have not caused any side effects for humans. They do not offer a biological hazard (Wellink 1998). Viral capsids are highly robust and stable. The genomic materials are encapsulated in the capsid to replicate within a host. Protecting genetic material is the primary task of the capsid, and it does keep viruses stable under situations such as harsh pH and temperature (Pokorski and Steinmetz 2011; Liu et al. 2012). Recently, viruses have been displayed to be genetically adaptable for chemical reactions as reagents, catalysts, and scaffolds applications (Strable and Finn 2009). Virus-based nanocarriers present the important properties of biocompatibility, uniformity, morphological, and simple functionalization. Additionally, virus-based nanoparticles are found in a diversity of characteristic shapes and sizes (Ma et al. 2012). Viral particles are stable in harsh conditions, such as high temperatures, a broad range of pH, and organic solvent water mixtures. These particles are natural structures, symmetrical, and attractive (Steinmetz and Evans 2007; Ma et al. 2012; Alemzadeh et al. 2018).

Although animal viruses are broadly accepted as a delivery tool, or in gene therapy as a 'vector,' their utilisation as a nanocarrier has stayed comparatively limited owing to security worries. Administration of 1011 plant virus or phage to mice has proven to show no signs of toxicity, but the same dose of animal virus reasoned violent hepatotoxicity (Green et al. 2004; Kaiser et al. 2007; Singh et al. 2007; Vaks and Benhar 2011). Therefore, bacteria and plant-based viruses have met substantially more concern than animal-based viruses in nanomedicine applications. The most researched bacterial-based virus is M13 phage, whereas the cowpea mosaic virus (CMV), tobacco mosaic virus (TMV), and cowpea chlorotic mottle virus (CCMV) are the most broadly examined plant viruses (Flynn et al. 2003). Viruses have sensitive, nanoscale-sized dimensions and structures that are hard to replicate by chemical synthesis or production (Lee et al. 2012). Viruses with different shapes and sizes for the most convenient system for a particular application are given in Figure 9.2. Viruses with higher aspect ratios are more favorable for targeting diseased vessel walls. Flexible rod-shaped viruses have been reported to penetrate tumour tissue better (Shukla et al. 2013; Wen et al. 2015). Besides, the surface features of viruses can be examined from genetic and chemical perspectives without damaging their formational integrity. These properties facilitate spatial control on the position of functional moieties on the surface of a virus and let it design the multifunctional systems (Rong et al. 2011). They can be generated in high quantities via infecting host cells in the

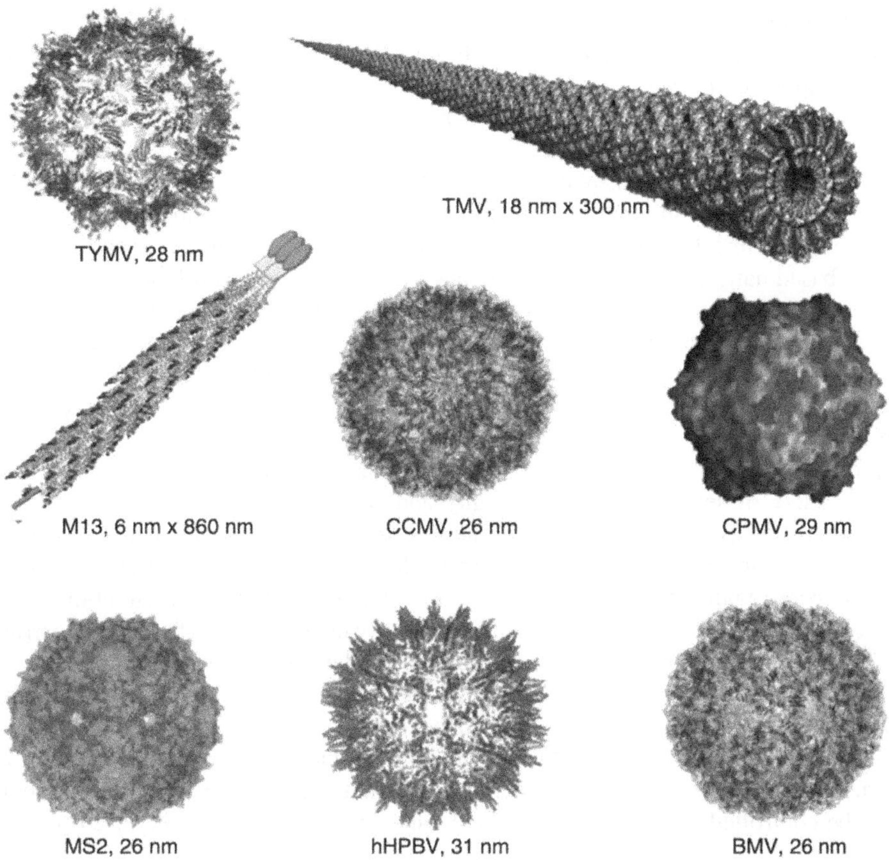

FIGURE 9.2 Structural models of various viral nanoparticles. Turnip yellow mosaic virus (TYMV), M13 bacteriophage (M13), MS2 bacteriophage (MS2), tobacco mosaic virus (TMV), cowpea chlorotic mosaic virus (CCMV), human hepatitis B virus (hHPBV), cowpea mosaic virus (CPMV), and brome mosaic virus (BMV) (Reproduced, adapted and modified from Rong et al. 2011).

nanomaterial synthesis of viruses, so they can be purified inexpensively. Finally, virus-based structures are phenomenon nanomaterials with many distinctive properties compared to synthetic-based nanomaterials (Sokullu et al. 2019).

9.3 VIRUS-BASED DRUG-DELIVERY SYSTEMS

9.3.1 Drug Delivery

The application of virus-based nanocarrier materials has gained attention. These nanocarrier materials have been used in the drug-delivery system for certain tissues. The virus-based nanocarrier materials can be employed for many platforms, especially cancer treatment. The chosen virus-based nanocarrier materials are dependent on the properties of the drug, such as loading capacity, stability, the ability of drug delivery, and release rate (Haley and Frenkel 2008). Many works reported using nanomaterials and biomaterials in drug-delivery system(s) (Bakhshpour et al. 2018; Tamahkar et al. 2019), but recently, the focus has been on virus-based nanocarrier materials, especially plant virus-based, increasing for use in the drug delivery area (Kaiser et al. 2007).

The plant virus-based nanocarrier materials have advantages such as biocompatibility, biodegradability, noninfectious, and readily available used for drug delivery. Chemically-modified and/or genetically-modified viruses, as well as plant virus-based nanocarriers, are used as nanocarriers (Steinmetz and Evans 2007). The virus-based nanocarrier materials commonly are employed in the chemotherapeutic area, and also, these biological materials are used as a specific nanocarrier for the delivery of small drugs to a certain tissue. Some of the plant hormones are used as a prodrug in the virus-based nanocarriers for cancer therapy. Indole-3-acetic acid is a prodrug hormone commonly utilised in prostate cancer treatment (Ma et al. 2012; Mura et al. 2013). The virus-based nanocarriers have a robust potential impact, such as a variety of several forms, uniformity and/or regular geometric size, and high stability in temperature, pH, solvent, and salt. Recently, many types of plant virus-based nanocarriers have been preferred in drug delivery because they are safety nanocarriers against mammalian viruses, noninfectious, and nontoxic/nonbiological risk (Alemzadeh et al. 2018). The other major advantage of virus-based nanocarrier is the multiple sites available on the surface of the virus capsid. This significant advantage makes using a virus attractive for the introduction of foreign ligands/peptides (Gupta et al. 2005). The low concentrations of the target molecule that are used in cell targeting are probable via utilising the targeting characterisation of the virus-based nanocarrier. Unlike lipid/inorganic materials, the level of concentration can be controlled by using a virus-based nanocarrier (Alemzadeh et al. 2017).

9.3.2 The Package Processes of Drugs into Virus-Based Nanocarriers

The viral capsids, as outer protein walls, package the genome of the virus that is used in the delivery of drug molecules. The capsid proteins have governed by the packaging of the viral genome (RNA or DNA). Sequence and structural properties of the capsid protein and the viral RNA(s) play a magnificent role in genome packaging such as disassembly or self-assembly processes (Speir and Johnson 2012). The mechanism of virus packaging plays a significant role in drug delivery enforcement through reengineering capsids or blocking viral propagation. The capsid proteins can be reassembled into null viruses-like nanoparticles. So, this specific coat protein content can be able to utilise in drug delivery systems. These viruses-like nanoparticles have a rod-shaped and/or spherical shape.

9.3.2.1 Using Supramolecular Chemistry for Packaging of Drugs

The straightforward geometric architecture of icosahedral geometry capsids with the easy and basic geometric structure has fewer assembly mechanisms (Zlotnick et al. 2000; Rossmann 2013). After assembly, virus-like nanoparticles prepare inside hole structures that can cover the drug molecules (Figure 9.3a,b) (Gupta et al. 2005). Rod-shaped nanoparticles are separated by the radial site of the subunit,

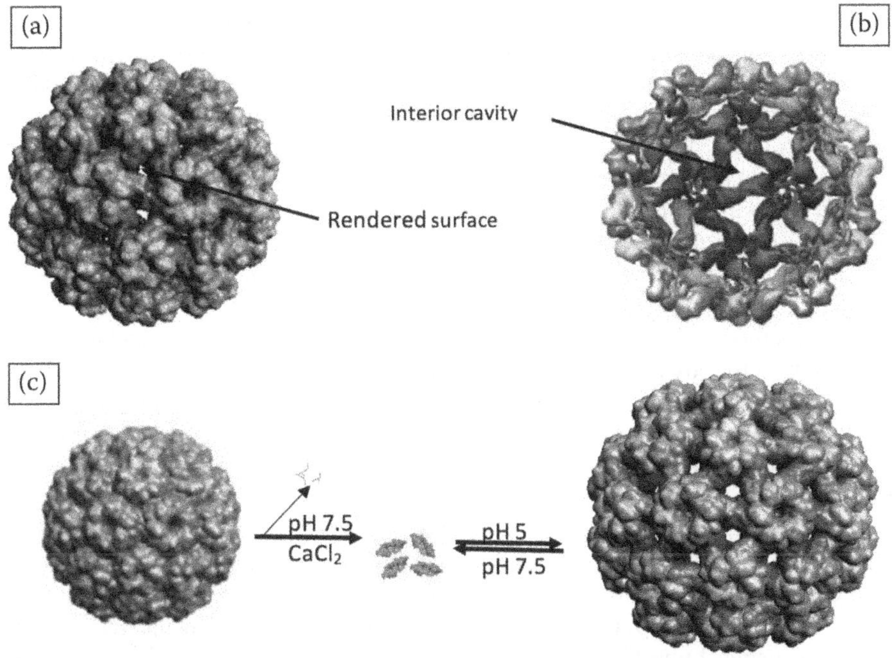

FIGURE 9.3 (a) The rendered surface; (b): Inside view of subunit organisation of CCMV-swollen form and (c) Schematic presentation of the transition process in the structure of CCMV using pH 5.0 and 7.5 (Models were rendered from crystallographic data obtained from VIPERdb (http://viperdb.scripps.edu)).

according to the helix axis. These nanoparticles have 15–30 nm of diameter. These nanoparticles lack enough area for loading; therefore, they are not preferable for use in chemical modification (Koudelka and Manchester 2010).

Lately, many viruses are used to decorate the material in the interior, such as the hibiscus chlorotic ringspot, the brome mosaic, the cowpea chlorotic mottle, the Simian, the human polyomavirus JC, the red clover necrotic mosaic, and the alphaviruses. These viruses can be successfully used for the encapsulation of drugs. Several research groups have used the cowpea chlorotic mottle virus (CCMV) in the encapsulation of therapeutic molecules for varied applications (Douglas and Young 1998). Recent studies showed that different pH values affect the closing/opening of the capsid pores. The inside surface of the native-CCMV capsid has a positive charge. Therefore, this native capsid can encapsulate the negative charge types directly. The pH is a simple and easy method to entrap drugs inside the CCMV; the schematic representation of this process is shown in Figure 9.3c. The CCMV coat assembles at pH 5.0 and disassembles at pH 7.5. This process is used for the encapsulation of targets (drugs) in the presence of a guest species (Comellas-Aragonès et al. 2007).

An icosahedral virus, *Salmonella typhimurium* bacteriophage P22, is a virus-based nanocarrier that is assembled from 420 copies of coat proteins and 300 copies of the scaffolding proteins (Weigele et al. 2005). The structure of P22 capsid could be transformed to the size of an empty procapsid shell from 58 nm to 64 nm diameter in the mature capsid structure, which is started by the packaging of DNA. The transformation could be imitated in vitro via slowly heating at 65°C for 10 minutes. The increasing heating to at least 75°C for at least 20 minutes fabricates a recent form of wiffleball. Therefore, the release of subunits from the mature capsid can occur via a pore at each of the icosahedral vertices. The large pores of wiffleball provide a free molecular change between the exterior and interior environments of the virus-based capsid. The distinctive characteristic of P22 capsids makes them extremely dynamic nano-platforms for synthetic usages (Teschke et al. 2003).

These methods are generally used for the packaging of overall negative charge compounds. The other methods for encapsulation of positive-charge compounds inside CCMV use a leucine zipper motif. Here, an enhanced green fluorescent protein (EGFP), including a negative charge leucine zipper E-coil and a positive charge leucine zipper K-coil, can achieve heterodimerization. Therefore, the EGFP molecules can be encapsulated inside the capsid (Minten et al. 2009).

These days, a general method of loading virus-based nanocarriers has been demonstrated based on CCMV. The polymer DNA amphiphile was loaded into the CCMV capsids. In this method, hydrophobic molecules are loaded into the core of the micelle (Figure 9.4a). Then, the micelles are equipped using hydrophilic molecules using hybridisation for attaching to complementary DNA strands (Figure 9.4b). At the end of the method, using an easy mixing process at neutral pH, CCMV coats protein molecules encapsulate the micelle (Kwak et al. 2010).

Additionally, the inner surface of the CCMV capsid can be altered thanks to the genetic engineering and protein designing process, instead of the positive-charge

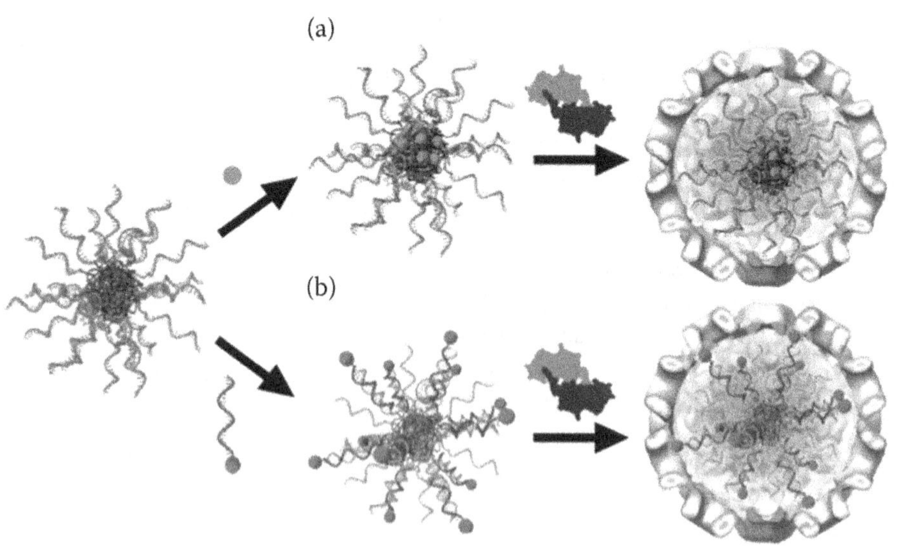

FIGURE 9.4 DNA micelle-template VC formations (Reproduced, adapted, and modified from (Kwak et al. 2010).

inner of the CCMV cavity. Therefore, the drugs with positive-charge properties can be easily packaged inside the virus-based capsid. There are many viral-based nanocarriers that have been investigated to encapsulate the particles/drugs using the basic and easy self-assembly/disassembly procedure (Douglas et al. 2002).

9.3.2.2 Using Chemical Attachment for Loading Drugs

Besides loading drug molecules inside the virus-based nanocarriers, mentioned in the previous section, the covalent attachment of the drug molecules or reactive functional groups on the surface of capsid proteins is usually used in drug cargo. For this reason, there is a chemical technique to attach the drug molecules to the virus (Ma et al. 2012). The chemical modification of the coat proteins of the viruses is the first procedure in this method (Yildiz et al. 2011). The capsid virus has many sites that can be modified with chemical agents, such as cysteine, lysine, and glutamic acid residues.

Functionalization is reached via attracting the peptides, antibodies, oligonucleotides, fluorescent reagents, drug molecules, and proteins to the capsid through various chemical materials, including maleimide, N-hydroxy-succinimidyl ester, carbodiimide, and isothiocyanate. Genetically, a cysteine residue could be let into a subunit that shows a thiol-based group in the assembled surface capsid structure (Wang et al. 2002). The summary of the general bioconjugation techniques used for modification is shown in Figure 9.5 (Alemzadeh et al. 2018). A new way to conduct bioconjugation is through click-chemistry. Magnificent progress in the chemoselective ligation on the surfaces of the virus utilised click-chemistry where the virus-based nanocarriers could be coupled in an orthogonal style without using safety groups. The click-chemistry reaction is chosen for bioconjugation because is quicker, bio-orthogonal, and more easily continues than suitable NHS reactions. The other chemo-selective approach is hydrazone-ligation. All chemical materials in the drug would be immediately bioconjugated to the surface of the virus, which is very important (Pokorski and Steinmetz 2011).

9.4 APPLICATIONS OF VIRUS-BASED NANOCARRIERS IN MEDICINE

The application of nanotechnology to medicine concerns the utilisation of nanocarriers to advance therapeutic and diagnostic modalities. It is an attractive field for various researchers due to the potential alternatives for noninvasive treatment and diagnosis. Nanocarriers have significant features, including ultra-small size, a large ratio of surface area to mass, and high reactivity; these features can be employed to accomplish some of the restrictions found in traditional agents, such as a high capacity for reagents and a high capability for modification of nanomaterials with ligands (Zhang et al. 2008; Yildiz et al. 2011). Moreover, they can be used to conduct a recent generation of reagents rather than current modalities, and also target therapies that are more efficacious and have fewer adverse effects than today's medicine. Virus-based nanocarriers have a broad application range in medicine, including medical imaging, tumour targeting, and vaccination (Alemzadeh et al. 2018). There are different types of virus-based nanocarriers, but plant viruses

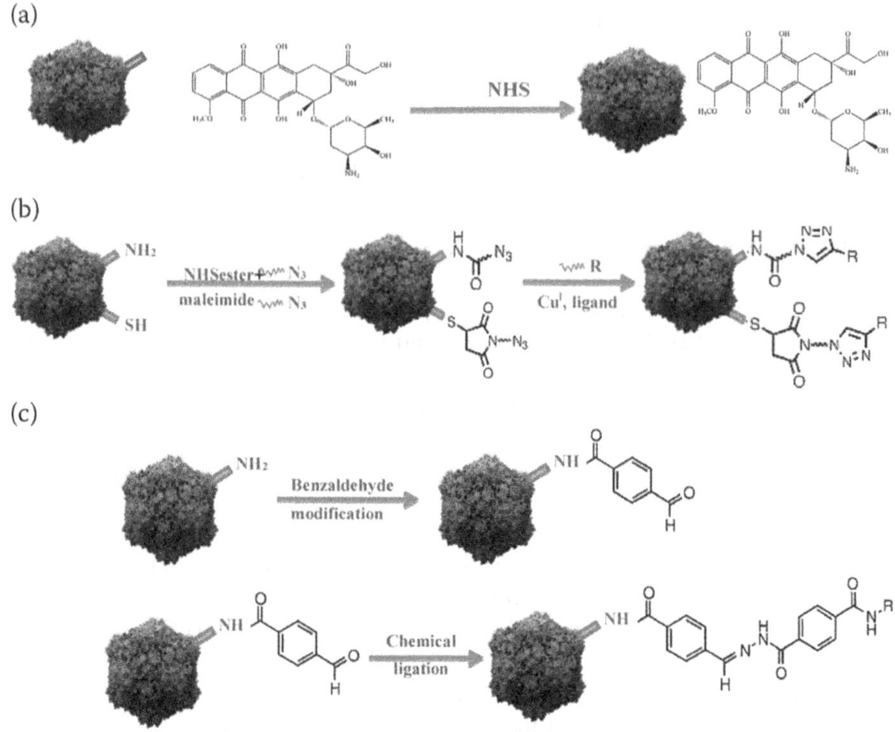

FIGURE 9.5 Chemical modification of virus-based nanocarriers. The coat of the protein modification by (a) NHS, (b) click-chemistry and (c) hydrazone ligation strategy (Reproduced, adapted, and modified from Alemzadeh et al. 2018).

and bacteriophages are generally favourable because they are safer in humans than mammalian viruses. They have also various advantages related to their reproducible and precise modifications. Thus, the identical nanocarriers paired to individual patient profiles can be produced as personalised medicines (Wen et al. 2013).

9.4.1 Medical-Imaging Applications

One of the medical-imaging methods is intravital vascular imaging, a powerful method for noninvasive monitoring and scanning of disease diagnosis. In imaging, nanocarriers are regular agents. But some difficulties may arise in the penetration of the beam and the resolution. The regular agents have also particular utilities such as price, flexibility, toxicity, availability, and stability. Some of these limitations can be accomplished using biological agents (Lewis et al. 2006). Fluorescent dyes as an imaging agent can be employed to covalently attach to the virus-based nanocarriers. The purpose of this attaching is to image intravital vascular, possible due to the biocompatibility properties of these materials. The nanocarriers have spontaneous affinities for vascular endothelium (Voigt et al. 2014). To conjugate, targeting ligands with virus-based nanocarriers could be encouraging to image vasculature during disease states (Manchester and Singh 2006). Another application of medical imaging is magnetic-resonance imaging (MRI), which is also a noninvasive method with a broad range of applications in medicine. MRI has a powerful magnetic field that creates a proton to arrange in the way of the field. Contrast agents can be used to improve the sensitivity and quality of the taken images. They assist to increase protons' relaxation time and also magnetisation time (Liepold et al. 2007). Virus-based nanocarriers are perfect applicants for contrast agents. Gadolinium forms complexes and/or binds the metal-binding sites of the biomolecule (Prasuhn et al. 2007; Steinmetz 2010).

For instance, Datta et al. assessed that contrast agents depended on the conjugation of the gadolinium on the interior and exterior surfaces of MS2 viral capsids. They observed that the magnetic-relaxation profiles of the conjugates indicate a 5-fold rise in relaxivity. The crystal form of the MS2 virus capsid is depicted in Figure 9.6. The amino groups are highlighted on the exterior surface as red (K106), yellow (K113), and orange (the N-terminus), and Y85 is highlighted in green on the interior surface (Datta et al. 2008).

Virus-based nanocarriers have also been employed for other noninvasive imaging techniques, including PET (positron emission tomography). The principle of PET is based on the emission of radiation from a patient, radiopharmaceutical injected, is registered via external detectors positioned at several orientations (Omami et al. 2014). This technology frequently is used for longitudinal follow-up and

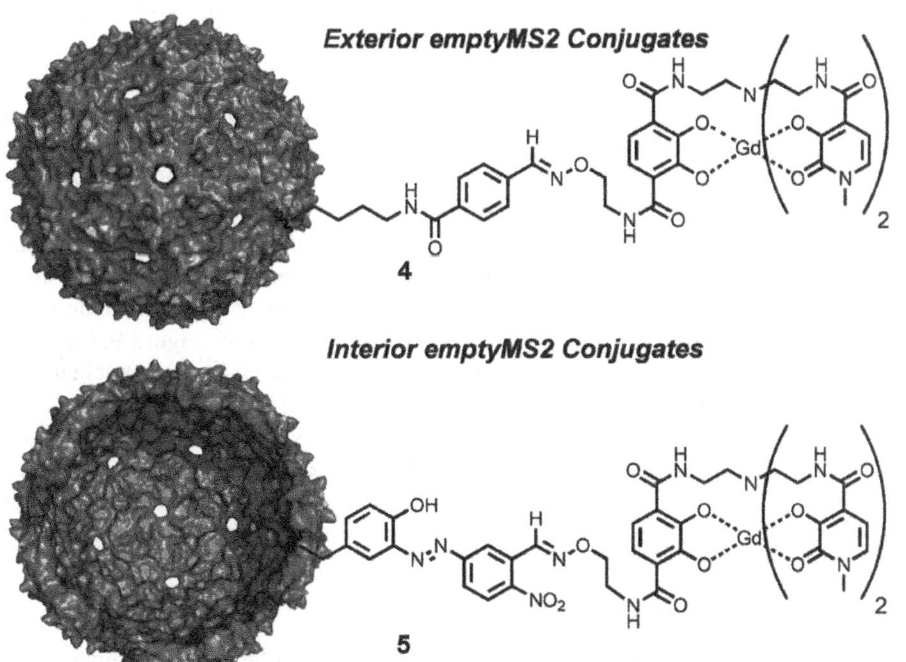

FIGURE 9.6 Externally (4) and internally (5) labeled agents obtained from the conjugates with gadolinium (Reproduced, adapted, and modified from Datta et al. 2008).

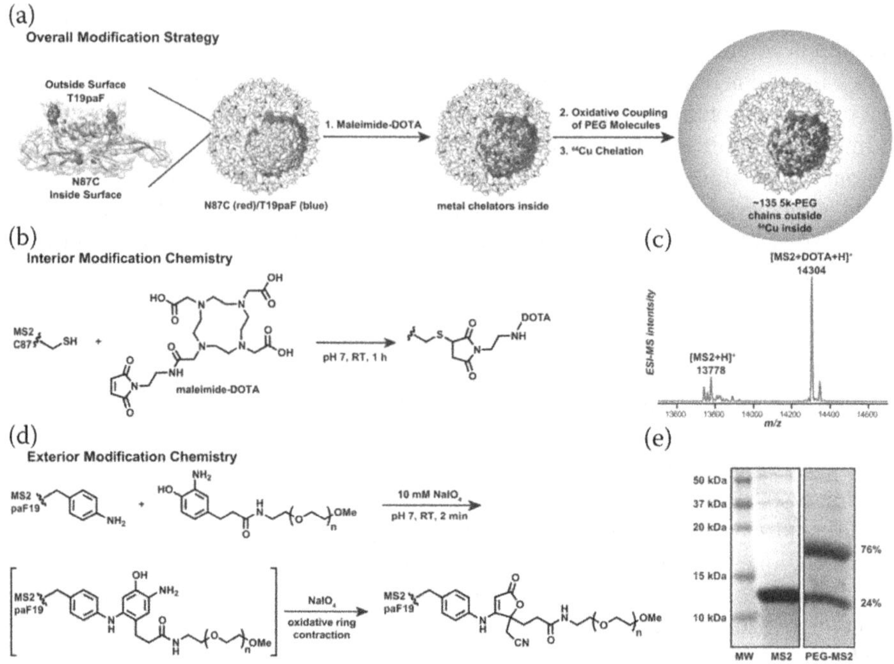

FIGURE 9.7 Surface chemistry steps for genome-free capsids (Reproduced, adapted, and modified from Farkas et al. 2013).

cancer diagnosis. There are many platforms for use in the PET, such as mammalian viruses, bacteriophages, etc. (Shukla and Steinmetz 2015). For example; Farkas et al. modulated the nanoparticle-based agents, which are PEG chains addition. They labeled MS2 and MS2-PEG capsids that have DOTA chelators and injected them into mice that have tumour xenografts. As shown in Figure 9.7a, they used each MS2 coat protein monomer (amino phenylalanine (T19paF) and cysteine mutations) for facing exterior and interior surfaces. They attached maleimide-DOTA to the cysteines to let 64Cu binding for radio labeling (Figure 9.7b). They analysed disassembled proteins with an upper limit of conversion and attached PEG chains applying a fast reaction. They performed SDS-PAGE analysis in a short time to indicate 76% of the proteins (Figures 9.7c-e). They also compared with other capsids and observed that the MS2 agents depicted longer circulation times. Furthermore, the MS2 and MS2-PEG bacteriophages behaved the same, although the latter agent has remarkably less uptake in the spleen (Farkas et al. 2013).

9.4.2 Tumour-Targeting Applications

Some biomolecules on the surface of the tumour cell can be used as biomarkers to separate tumours from regular tissue. Thus, selective destroying the tumour without remarkably impressing regular tissue would be reached employing carrier systems (Allen and Cullis 2004). The spontaneous container-like features of viruses make them appealing for planning targeted drug-delivery systems. Disease diagnosis would be more accurate and offer a more effective signal-to-noise ratio of imaging via these agents. The delivery also develops the impact of drugs and decreases opposite effects on nontarget tissues (de la Zerda et al. 2012). The virus-based nanocarriers have been investigated to deliver therapeutics in neurodegenerative and infectious diseases usage. They are placed principally in the blood-brain barrier endothelium. As an example, Shriver et al. suggested that inflammatory lesions contain macrophages, immunoglobulins, and microglia for detection of CPMV. They performed surface chemistry using N-hydroxysuccinimide ester that acts lysines on the capsid and results in dyes attaching to the surface of the particle (Figure 9.8a). They also carried out the size-exclusion analysis and obtained the peaks for Alexa Fluor 555 (red) coelutes with the viral particles protein content (green) and nucleic acid content (blue) pointing out the labeling of intact CPMV particles (Figure 9.8b) (Shriver et al. 2009).

Steinmetz et al. declared virus-based nanocarriers' synthesis and characterization for targeting gastrin-releasing peptide receptors, which are overexpressed in prostate cancers. They developed an imaging agent containing CPMV–polyethylene glycol–bombesin that was collected in small tumours. They showed that virus-based nanocarriers could be a good alternative to detect prostate cancer and mentioned

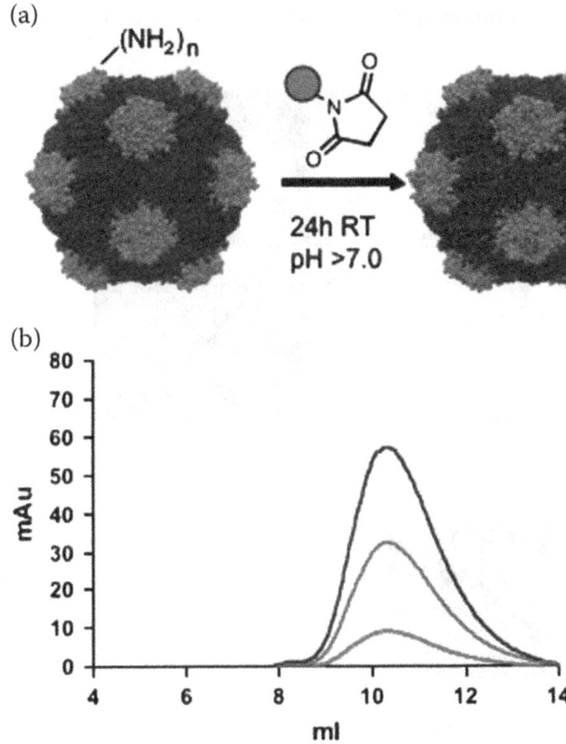

FIGURE 9.8 Alexa Fluor 555 and CPMV attachment (Reproduced, adapted, and modified from Shriver et al. 2009).

that the capability to target specific tumours is an important parameter to be imagined (Steinmetz et al. 2011). Destito et al. studied the CPMV particle bindings to tumour cells by great expression of folate receptors. They demonstrated that CPMV PEGylation is fully extracted binding of the background of the viruses to the tumour cells, and PEG-folate moiety empowered CPMV tumour cells' recognition carrying the folate receptor. They also showed that greater-density loading of ligands on CPMV may not be essential to effectively target tumour cells via performing CPMV formulations with several amounts of the PEG-folate moiety on the surface (Destito et al. 2007). Shukla et al. investigated tumour homing properties to possess architectures of nanocarriers by comparing their counterparts of filamentous such as potato virus X and spherical such as CPMV. The two viruses have different biodistribution profiles and vary in their tumour homing and penetration efficiency (Figure 9.9). Potato virus X has improved tumour homing and tissue penetration, and human tumour xenografts depict higher uptake of PEGylated filamentous Potato virus X against CPMV (Shukla et al. 2013).

9.4.3 Vaccination Applications

Vaccination is an alternative procedure for identifying tumour-related antigens and prompting the immune system toward tumours. The 'in situ vaccination' strategy can accomplish local immunosuppression and then remove residual tumours via

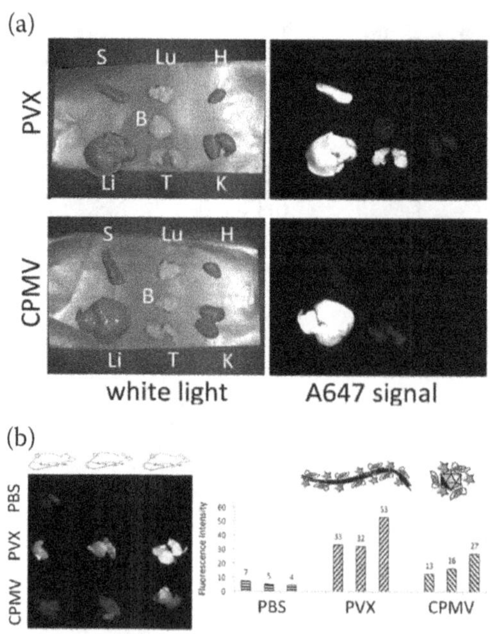

FIGURE 9.9 Tumour homing and biodistribution of virus-based nanocarriers. (a) A647-labeled, PEGylated CPMV, and PVX were directed in HT-29 xenografts and (b) tumours were obtained from three animals and monitored (Reproduced, adapted, and modified from Shukla et al. 2013).

warming antitumour immunity (Lizotte et al. 2016). Especially, plant viruses are immunogenic and ensure a useful growth method for new vaccines (Peruzzi and Chiocca 2016). Plant viruses are effective manufacturing systems for oral vaccines that supply via the high output of a plant virus in comestible plant tissue. They are also durable to gastric medium, easily adapted to the intestine, and absorbed into the circulation (Rae et al. 2005). Oral vaccination can induce both mucosal and systemic immunity, which continues to protect at high levels (Nicholas et al. 2002).

Several studies are based on plant virus-based platforms to own natural immunogenic features that caution antitumour immunity. Le et al. reported that virus-based nanocarriers could act in the role of immunotherapeutic agent as they prompt the local innate immunity system. They mentioned that the actuation may follow in compressing the durability of the cancer microenvironment for treatment. They finalised the study the cancer cell immunity restarts and supplies adaptive immunity (Le et al. 2017).

9.5 CONCLUSION

Nanoparticle technology was recently indicated to hold big hope for applications of drug delivery in nanomedicine because of its positive features, such as bioavailability, better encapsulation, lower toxic effects, and control release. Regardless of the great progress in nanomedicine, there are many limitations before broadly being accepted for medical application. Nearly all types of nanoparticles, such as polymeric micelles, polymeric nanoparticles, nanocrystals, carbon nanotubes, and dendrimers, can be used for medical area applications. One of the latest developments in nanotechnology is the widespread use of new generation nanocarrier drugs that will be used effectively in the treatment of various diseases. Advances in biological-based nanotechnology have caused the improvement of viral nanocarriers applications in imaging and targeted therapy. Viral nanocarriers as novel natural-based nanomaterials show a few advantages against synthetic-based nanoparticles. Biodegradation and the safety of viral nanoparticles derived from plant viruses are the most advantages of these materials. The chemical and/or genetic modification, the high degree of structural control of the ligand, and the possibility of binding with significant flexibility in their structure are the other important advantages. Virus-based nanocarriers have important properties, such as better solubility and stability, targeting, site-specificity, and modified release of medications that can be preferred to use in the drug-release systems. This is beneficial for the correction of the therapeutic behavior of diverse drugs. Evidence indicates that viral-based nanotechnology could be a hopeful interdisciplinary area for the development of therapeutics and smart medical vehicles.

REFERENCES

Adams, M.L., Lavasanifar, A. and Kwon, G.S. 2003. Amphiphilic block copolymers for drug delivery. *Journal of Pharmaceutical Sciences* 92(7):1343–1355.

Afrooz, H., Ahmadi, F., Fallahzadeh, F., Mousavi-Fard, S.H. and Alipour, S. 2017. Design and characterization of paclitaxel-verapamil co-encapsulated PLGA nanoparticles:

Potential system for overcoming P-glycoprotein mediated MDR. *Journal of Drug Delivery Science and Technology* 41:174–181.

Alemzadeh, E., Dehshahri, A., Izadpanah, K. and Ahmadi, F. 2018. Plant virus nanoparticles: Novel and robust nanocarriers for drug delivery and imaging. *Colloids and Surfaces B: Biointerfaces* 167:20–27.

Alemzadeh, E., Izadpanah, K. and Ahmadi, F. 2017. Generation of recombinant protein shells of Johnson grass chlorotic stripe mosaic virus in tobacco plants and their use as drug carrier. *Journal of Virological Methods* 248:148–153.

Allen, T.M. and Cullis, P.R. 2004. Drug delivery systems: Entering the mainstream. *Science* 303(5665):1818–1822.

Bakhshpour, M., Yavuz, H. and Denizli, A. 2018. Controlled release of mitomycin C from PHEMAH–Cu(II) cryogel membranes. *Artificial Cells, Nanomedicine and Biotechnology* 46(1):946–954.

Cho, K., Wang, X., Nie, S., Chen, Z. and Shin, D.M. 2008. Therapeutic nanoparticles for drug delivery in cancer. *Clinical Cancer Research* 14(5):1310–1316.

Comellas-Aragonès, M., Engelkamp, H., Claessen, V.I., et al. 2007. A virus-based single-enzyme nanoreactor. *Nature Nanotechnology* 2(10):635–639.

Datta, A., Hooker, J.M., Botta, M., Francis, M.B., Aime, S. and Raymond, K.N. 2008. High relaxivity gadolinium hydroxypyridonate-viral capsid conjugates: Nanosized MRI contrast agents. *Journal of the American Chemical Society* 130(8): 2546–2552.

Destito, G., Yeh, R., Rae, C.S., Finn, M.G. and Manchester, M. 2007. Folic acid-mediated targeting of cowpea mosaic virus particles to tumor cells. *Chemistry and Biology* 14(10):1152–1162.

de la Zerda, A., Bodapati, S., Teed, R., et al. 2012. Family of enhanced photoacoustic imaging agents for high-sensitivity and multiplexing studies in living mice. *ACS Nano* 6(6):4694–4701.

Din, F., Ud, W., Aman, A., Ullah, I., et al. 2017. Effective use of nanocarriers as drug delivery systems for the treatment of selected tumors. *International Journal of Nanomedicine* 12:7291–7309.

Douglas, T., Strable, E., Willits, D., Aitouchen, A., Libera, M., and Young, M. 2002. Protein engineering of a viral cage for constrained nanomaterials synthesis. *Advanced Materials* 14(6):415–418.

Douglas, T. and Young, M. 1998. Host-guest encapsulation of materials by assembled virus protein cages. *Nature* 393:151–155.

Farkas, M.E., Aanei, I.L., Behrens, C.R., et al. 2013. PET imaging and biodistribution of chemically modified bacteriophage MS2. *Molecular Pharmaceutics* 10(1):69–76.

Flynn, C.E., Lee, S.W., Peelle, B.R. and Belcher, A.M. 2003. Viruses as vehicles for growth, organization and assembly of materials. *Acta Materialia* 51(19):5867–5880.

Green, N.K., Herbert, C.W., Hale, S.J., et al. 2004. Extended plasma circulation time and decreased toxicity of polymer-coated adenovirus. *Gene Therapy* 11(16): 1256–1263.

Gupta, S. Sen, Raja, K.S., Kaltgrad, E., Strahle, E., and Finn, M.G. 2005. Virus-glycopolymer conjugates by copper(I) catalysis of atom transfer radical polymerization and azide-alkyne cycloadditiont. *Chemical Communications* 34:4315–4317.

Haley, B. and Frenkel, E. 2008. Nanoparticles for drug delivery in cancer treatment. *Urologic Oncology: Seminars and Original Investigations* 26(1):57–64.

How, C.W., Rasedee, A., Manickam, S. and Rosli, R. 2013. Tamoxifen-loaded nanostructured lipid carrier as a drug delivery system: Characterization, stability assessment and cytotoxicity. *Colloids and Surfaces B: Biointerfaces* 112:393–399.

Kaiser, C.R., Flenniken, M.L., Gillitzer, E., et al. 2007. Biodistribution studies of protein cage nanoparticles demonstrate broad tissue distribution and rapid clearance in vivo. *International Journal of Nanomedicine* 2(4):715–733.

Koudelka, K.J. and Manchester, M. 2010. Chemically modified viruses: Principles and applications. *Current Opinion in Chemical Biology* 14(6):810–817.

Kwak, M., Minten, I.J., Anaya, D.M., et al. 2010. Virus-like particles templated by DNA micelles: A general method for loading virus nanocarriers. *Journal of the American Chemical Society* 132(23):7834–7835.

Le, D.H.T., Lee, K.L., Shukla, S., Commandeur, U. and Steinmetz, N.F. 2017. Potato virus X, a filamentous plant viral nanoparticle for doxorubicin delivery in cancer therapy. *Nanoscale* 9(6):2348–2357.

Lee, S.Y., Lim, J.S. and Harris, M.T. 2012. Synthesis and application of virus-based hybrid nanomaterials. *Biotechnology and Bioengineering* 109(1):16–30.

Lewis, J.D., Destito, G., Zijlstra, A., Gonzalez, M.J., Quigley, J.P., Manchester, M. and Stuhlmann, H. 2006. Viral nanoparticles as tools for intravital vascular imaging. *Nature Medicine* 12(3):354–360.

Liepold, L., Anderson, S., Willits, D., et al. 2007. Viral capsids as MRI contrast agents. *Magnetic Resonance in Medicine* 58(5):871–879.

Liu, Z., Qiao, J., Niu, Z. and Wang, Q. 2012. Natural supramolecular building blocks: From virus coat proteins to viral nanoparticles. *Chemical Society Reviews* 41(18): 6178–6194.

Lizotte, P.H., Wen, A.M., Sheen, M.R., Fields, J., Rojanasopondist, P., Steinmetz, N.F. and Fiering, S. 2016. In situ vaccination with cowpea mosaic virus nanoparticles suppresses metastatic cancer. *Nature Nanotechnology* 11(3):295–303.

Ma, Y., Nolte, R.J.M. and Cornelissen, J.J.L.M. 2012. Virus-based nanocarriers for drug delivery. *Advanced Drug Delivery Reviews* 64(9):811–825.

Manchester, M. and Singh, P. 2006. Virus-based nanoparticles (VNPs): Platform technologies for diagnostic imaging. *Advanced Drug Delivery Reviews* 58(14):1505–1522.

Minten, I.J., Hendriks, L.J.A., Nolte, R.J.M. and Cornelissen, J.J.L.M. 2009. Controlled encapsulation of multiple proteins in virus capsids. *Journal of the American Chemical Society* 131(49):17771–17773.

Mishra, B., Patel, B.B. and Tiwari, S. 2010. Colloidal nanocarriers: A review on formulation technology, types and applications toward targeted drug delivery. *Nanomedicine: Nanotechnology, Biology, and Medicine* 6(1):9–24.

Mura, S., Nicolas, J. and Couvreur, P. 2013. Stimuli-responsive nanocarriers for drug delivery. *Nature Materials* 12(11):991–1003.

Neubert, R.H.H. 2011. Potentials of new nanocarriers for dermal and transdermal drug delivery. *European Journal of Pharmaceutics and Biopharmaceutics* 77(1):1–2.

Nicholas, B.L., Brennan, F.R., Martinez-Torrecuadrada, J.L., Casal, J.I., Hamilton, W.D. and Wakelin, D. 2002. Characterization of the immune response to canine parvovirus induced by vaccination with chimaeric plant viruses. *Vaccine* 20(21–22):2727–2734.

Omami, G., Tamimi, D. and Branstetter, B.F. 2014. Basic principles and applications of 18F-FDG-PET/CT in oral and maxillofacial imaging: A pictorial essay. *Imaging Science in Dentistry*, 44(4):325–332.

Peer, D., Karp, J.M., Hong, S., Farokhzad, O.C., Margalit, R. and Langer, R. 2007. Nanocarriers as an emerging platform for cancer therapy. *Nature Nanotechnology* 2:751–760.

Peruzzi, P.P. and Chiocca, E.A. 2016. Cancer immunotherapy: A vaccine from plant virus proteins. *Nature Nanotechnology* 11(3):214–215.

Pokorski, J.K. and Steinmetz, N.F. 2011. The art of engineering viral nanoparticles. *Molecular Pharmaceutics*, 8(1):29–43.

Prasuhn, D.E., Yeh, R.M., Obenaus, A., Manchester, M. and Finn, M.G. 2007. Viral MRI contrast agents: Coordination of Gd by native virions and attachment of Gd complexes by azide-alkyne cycloaddition. *Chemical Communications* 12:1269–1271.

Rae, C.S., Wei Khor, I., Wang Q, et al. 2005. Systemic trafficking of plant virus nanoparticles in mice via the oral route. *Virology* 343(2):224–235.

Rong, J., Niu, Z., Lee, L.A. and Wang, Q. 2011. Self-assembly of viral particles. *Current Opinion in Colloid and Interface Science* 16(6):441–450.

Rossmann, M.G. 2013. Structure of viruses: A short history. *Quarterly Reviews of Biophysics* 46(2):133–180.

Shriver, L.P., Koudelka, K.J. and Manchester, M. 2009. Viral nanoparticles associate with regions of inflammation and blood brain barrier disruption during CNS infection. *Journal of Neuroimmunology* 211(1–2):66–72.

Shukla, S., Ablack, A.L., Wen, A.M., Lee, K.L., Lewis, J.D. and Steinmetz, N.F., 2013. Increased tumor homing and tissue penetration of the filamentous plant viral nanoparticle potato virus X. *Molecular Pharmaceutics* 10(1):33–42.

Shukla, S. and Steinmetz, N.F. 2015. Virus-based nanomaterials as positron emission tomography and magnetic resonance contrast agents: From technology development to translational medicine. *Wiley Interdisciplinary Reviews: Nanomedicine and Nanobiotechnology* 7(5):708–721.

Singh, P., Prasuhn, D., Yeh, R.M., et al. 2007. Bio-distribution, toxicity and pathology of cowpea mosaic virus nanoparticles in vivo. *Journal of Controlled Release* 120(1–2): 41–50.

Sokullu, E., Abyaneh, H.S. and Gauthier, M.A. 2019. Plant/bacterial virus-based drug discovery, drug delivery, and therapeutics. *Pharmaceutics* 11(5):211.

Speir, J.A. and Johnson, J.E. 2012. Nucleic acid packaging in viruses. *Current Opinion in Structural Biology* 22(1):65–71.

Steinmetz, N.F. 2010. Viral nanoparticles as platforms for next-generation therapeutics and imaging devices. *Nanomedicine: Nanotechnology, Biology, and Medicine* 6(5):634–641.

Steinmetz, N.F., Ablack, A.L., Hickey, J.L. et al. 2011. Intravital imaging of human prostate cancer using viral nanoparticles targeted to gastrin-releasing peptide receptors. *Small* 7(12):1664–1672.

Steinmetz, N.F. and Evans, D.J. 2007. Utilisation of plant viruses in bionanotechnology. *Organic and Biomolecular Chemistry* 5(18):2891–2902.

Strable, E. and Finn, M.G. 2009. Chemical modification of viruses and virus-like particles. In *Viruses and Nanotechnology*, eds. M. Manchester, and N.F. Steinmetz. 1–21. Springer: Berlin, Heidelberg.

Svenson, S. and Tomalia, D.A. 2012. Dendrimers in biomedical applications-reflections on the field. *Advanced Drug Delivery Reviews* 64:102–115.

Tamahkar, E., Bakhshpour, M. and Denizli, A. 2019. Molecularly imprinted composite bacterial cellulose nanofibers for antibiotic release. *Journal of Biomaterials Science, Polymer Edition* 30(6):450–461.

Teschke, C.M., McGough, A. and Thuman-Commike, P.A. 2003. Penton release from P22 heat-expanded capsids suggests importance of stabilizing penton-hexon interactions during capsid maturation. *Biophysical Journal* 84(4):2595-2592.

Vaks, L. and Benhar, I. 2011. In vivo characteristics of targeted drug-carrying filamentous bacteriophage nanomedicines. *Journal of Nanobiotechnology* 9:1–10.

Voigt, J., Christensen, J., and Shastri, V.P. 2014. Differential uptake of nanoparticles by endothelial cells through polyelectrolytes with affinity for caveolae. *Proceedings of the National Academy of Sciences of the United States of America* 111(8):2942–2947.

Wang, Q., Kaltgrad, E., Lin, T., Johnson, J.E., and Finn, M.G. 2002. Natural supramolecular building blocks: Wild-type Cowpea mosaic virus. *Chemistry and Biology* 9(7): 805–811.

Weigele, P.R., Sampson, L., Winn-Stapley, D., and Casjens, S.R. 2005. Molecular genetics of bacteriophage P22 scaffolding protein's functional domains. *Journal of Molecular Biology* 348(4):841–844.

Wellink, J. 1998. Comovirus isolation and RNA extraction. *Methods in Molecular Biology* 81:205–209.

Wen, A.M., Rambhia, P.H., French, R.H., and Steinmetz, N.F. 2013. Design rules for nanomedical engineering: From physical virology to the applications of virus-based materials in medicine. *Journal of Biological Physics* 39(2):301–325.

Wen, A.M., Wang, Y., Jiang, K., et al. 2015. Shaping bio-inspired nanotechnologies to target thrombosis for dual optical-magnetic resonance imaging. *Journal of Materials Chemistry B* 3(29):6037–6045.

Yildiz, I., Shukla, S., and Steinmetz, N.F. 2011. Applications of viral nanoparticles in medicine. *Current Opinion in Biotechnology* 22(6):901–908.

Zhang, L., Gu, F.X., Chan, J.M., Wang, A.Z., Langer, R.S., and Farokhzad, O.C. 2008. Nanoparticles in medicine: Therapeutic applications and developments. *Clinical Pharmacology and Therapeutics*, 83(5):761–769.

Zlotnick, A., Aldrich, R., Johnson, J.M., Ceres, P., and Young, M.J. 2000. Mechanism of capsid assembly for an icosahedral plant virus. *Virology* 277(2):450–456.

Section III

Antiviral Nanotherapeutics

10 Antiviral Nanomaterials in Therapeutic Interventions

Karan Chaudhary and Dhanraj T. Masram

CONTENTS

10.1 INTRODUCTION

In general, a virus is a particle that is made of a protein shell, called capsid, in which genetic material is enclosed (Cheng and Brooks III 2013). But some viruses further contain an envelope that covers capsid; an example of such a virus is a Coronavirus (Sharma et al. 2020). The virus causes a viral infection that is one of the current global and major challenges toward human health and socioeconomic development (Mehendale et al. 2013). This viral infection disease is one of the contributors to the mortality of humankind (Chakravarty and Vora 2021). In recent times, viruses such as Ebola virus, Nipah virus, Zika virus, Chikungunya virus, Dengue virus, Coronavirus, and influenza virus have caused the occurrence of numerous viral infections (Chakravarty and Vora 2021). And for the treatment of viral infections, several antiviral drugs are used, but there are limitations associated with them. These include side effects due to drug-drug interaction, drug selectivity toward the virus, drug resistance due to extensive use, low bioavailability, frequent uptake of drugs due to short half-life, and limited reach of drugs to viral sites (Gerber 2000; Singh and Lillard 2009; Strasfeld and Chou 2010; Oshikoya et al. 2013; Joshi et al. 2016; Bule et al. 2019). Therefore, it is of prime importance to find effective therapies and antiviral agents for the treatment of emerging viral infections. For the development of antiviral therapies and antiviral agents, it is important to understand the pathogenic mechanism. In this pathogenic mechanism, various stages are (1) virus implantation at point of entry, (2) penetration of the virus into the cell, (3) virus uncoating, (4) replication of the virus, (5) encapsidation, and (6) spread of infection (Chinchar 1999).

DOI: 10.1201/9781003136644-13

In general, nanotechnology can be defined as the development and applications of materials or particles having nanometer dimension(s). Nanoparticles (NPs) or nanomaterials have gained considerable interest and have been proven to be a potential candidate in the field of diagnostics and therapeutics in the past decade (Abd Elkodous et al. 2019). At present time, nanotechnology has emerged as a potential strategy to overcome the limitations associated with antiviral drugs, as well as for the development of effective antiviral agents (Chakravarty and Vora 2021). NPs have unique physical properties, including particle size, structure, large surface area, tunable surface charge, and surface functional groups. These properties of NPs have made them beneficial in therapeutic applications because these properties make them the perfect candidate for drug-delivery systems for antiviral therapies; in addition to this role, they act as antiviral agents (Singh et al. 2017). The drug-encapsulated or loaded NPs as a drug-delivery system has helped to overcome the limitations associated with the treatment of viral infection by drugs in order to improve therapeutic effects (Lin et al. 2017b; Joshy et al. 2018). Further bare NPs inhibit the virus entry into the cell and inhibit viral replication, which proves their potential to act as potential antiviral agents and further enhances the therapeutic effects against viral infections (Li et al., 2021). In this regard, several NPs have been used, such as polymer-based NPs, lipid-based NPs, lipid-polymer-based NPs, carbon NPs, metal and metal oxide NPs to develop antiviral drug carriers, and antiviral agents (Chakravarty and Vora 2021). This chapter presents the development of antiviral nanomaterials for improved or efficient antiviral therapy against viral infections either by intrinsic activity or by loading or encapsulating drugs for enhanced bioavailability and site-specific delivery.

10.2 OVERVIEW OF VIRUSES

Viruses as particles are the smallest parasites that cannot reproduce themselves but can replicate inside a host cell. Mostly, the genetic material of viruses is ribonucleic acid (RNA) or deoxyribonucleic acid (DNA), and they can be single or double-stranded. This genetic material has an outer cover of protein, and sometimes, it is made up of lipid. These viruses are not classified according to the disease they cause but are classified instead based on the nature and structure of the genome and on their replication method (Lodish et al. 2000). Herein, we have briefly discussed the type of genetic material of viruses and the family under which these viruses are classified.

- Porcine reproductive and respiratory syndrome virus (PRRSV) is an RNA virus having a single strand, and it is a positive strand. And as per the classification of viruses, PRRSV has been classified within the *Arteriviridae* family (Du et al. 2018b; Zhou et al. 2020).
- Herpes simplex virus type-1 (HSV-1) is a DNA virus having a double strand and also having an envelop. This HSV-1 belongs to the *Herpesviridae* family (Lima et al. 2018).
- Influenza A virus is a segmented RNA virus having a single strand, and it is a negative strand. The influenza A virus belongs to the *Orthomyxoviridae*

family. This virus is further classified based on the type of surface glycoprotein, whether it is hemagglutinin or neuraminidase (Bouvier and Palese 2008).

- Human immune-deficiency virus (HIV) is an enveloped RNA virus having a single strand, and it is a positive strand. This HIV belongs to the *Retroviridae* family (Smith and Daniel 2006).
- Human T-lymphotropic virus 1 (HTLV-1) is a type C virus that belongs to the *Retroviridae* family. It is an RNA virus having a single strand, and it is a positive strand (Verdonck et al. 2007).
- Enterovirus 71 (EV71) is a nonenveloped RNA virus having a single strand, and it is a positive strand. It belongs to the *Picornaviridae* family (Zhong et al. 2019).
- Zika Virus (ZIKV) is an enveloped RNA virus, nonsegmented, having a single strand, and it is a positive strand. ZIKV belongs to the *Flaviviridae* family (Faye et al. 2014).
- Severe acute respiratory syndrome coronavirus 2 (SARS-CoV-2) is an RNA virus having a single strand, and it is a positive strand. It belongs to the *Coronaviridae* family (Machhi et al. 2020).
- Chikungunya virus (CHIKV) is an enveloped RNA virus, having a single strand, and is a positive strand. It belongs to the *Togaviridae* family (Sharma et al. 2019).
- Porcine epidemic diarrhea virus (PEDV) is an enveloped RNA virus, having a single strand, and is a positive strand. It belongs to *the Coronaviridae* family (Du et al. 2018a).
- Respiratory syncytial virus (RSV) is an enveloped RNA virus having a single strand, and it is a negative strand. It belongs to the *Paramyxoviridae* family (Yang et al. 2017).
- Ebola virus (EBOV) is an enveloped RNA virus, nonsegmented, having a single strand, and it is a negative strand. It belongs to the *Filoviridae* family (Farman et al. 2020).

10.3 NANOMATERIALS FOR ANTIVIRAL THERAPEUTICS

10.3.1 Nanomaterials as Delivery Agents in Antiviral Therapeutics

To overcome the problem of drug bioavailability, solubility, and other associated problems with antiviral drugs for therapies, their encapsulation or loading over nanomaterials have attracted great attention. A variety of nanomaterials have been developed to be used as antiviral drug-delivery agents for improved therapeutic effects of the antiviral drugs. Some of the recently developed nanomaterials have been discussed under this section, and, along with few others, they have been summarised in Table 10.1.

Ramyadevi et al. (2016) developed a drug-delivery system that consists of polymeric NPs of acyclovir drug-loaded *in situ* gelling system that contains drug dose equivalent to 105 mg/day. Through vaginal membrane, this nanoparticle delivery system had improved permeability. Also, for vaginal epithelial cell lines, viability

TABLE 10.1
Nanomaterials as Drug Delivery Agents in Antiviral Therapeutics

Virus	Viral Disease	Drug Loaded/Encapsulated Nanomaterial	Outcome	Reference
HIV	Acquired immune deficiency syndrome	Zidovudine encapsulated amide functionalized alignate NPs	• Slow release in a sustained manner from the AZT-GAAD NPs, NPs internalisation efficiency was significantly higher and NPs were biocompatible	Joshy et al. (2018)
HIV	Acquired immune deficiency syndrome	Poly(lactide-co-glycolide) (PLGA) NPs loaded with lopinavir	• Permeability of lopinavir-loaded PLGA NPs was increased by 3.04 fold • Also, bioavailability was enhanced by 13.9 folds after oral delivery of lopinavir loaded PLGA NPs	Joshi et al. (2016)
HIV	Acquired immune deficiency syndrome	Efavirenz-loaded lactoferrin NPs (lacto-EFV-nano)	• Lacto-EFV-nano was found to be at least 2 fold less toxic than pure EFV • Anti-HIV-1 activity of lacto-EFV-nano was more than 2 fold higher in comparison to free EFV	Kumar et al. (2017)
HSV	Serious and contagious genital infections	Polymeric NPs of acyclovir drug loaded *in situ* gelling system	• Improved bio-distribution and bioavailability using nanoparticle system	Ramyadevi et al. (2016)
HTLV-1	Acute and favorable chronic Adult T Cell Leukemia (ATL)	Interferon-α-2b loaded chitosan NPs	• Enhanced oral absorption of interferon-α-2b and conserved antiviral activity	Cánepa et al. (2017)
HIV	Acquired immune deficiency syndrome	Darunavir-loaded lipid NPs	• Greatly improved the drug bioavailability • NPs formulation gave increased binding with HIV host cells	Desai and Thakkar (2018)
H1N1 influenza virus	Influenza	Zanamivir loaded seleniumNPs	• Regulated the caspase-3 protein to inhibit the MDCK cells apoptosis	Lin et al. (2017a)
HSV	Serious infections	Acyclovir-loaded solid lipid NPs	• Compared to commercial suspension, acyclovir-loaded SLNs had 4-fold higher plasma concentration	Hassan et al. (2020)

(Continued)

TABLE 10.1 (Continued)
Nanomaterials as Drug Delivery Agents in Antiviral Therapeutics

Virus	Viral Disease	Drug Loaded/Encapsulated Nanomaterial	Outcome	Reference
HSV	Oral infection	Chloroquine-loaded poly(lactic acid) NPs	• Chloroquine-loaded NPs had improved antiviral activity in comparison to chloroquine	Lima et al. (2018)
EV71	Hand-foot-mouth disease (HFMD)	Oseltamivir loaded selenium NPs	• Antiviral NPs entered cells efficiently and EV71 proliferation was inhibited • Antiviral NPs reduced the reactive oxygen species generation in the cells to inhibit the activity of EV71	Zhong et al. (2019)
ZIKV	ZIKV infection	Ivermectin loaded modified PLGA NPs	• Increased ivermectin concentration in plasma • Reduced the expression of the NS1 protein of ZIKV	Surnar et al. (2019)
SARS-CoV-2	COVID-19 infection	IVM-loaded PLGA-b-PEG-MAL NPs	• Inhibition of viral spike protein and down regulation of ACE2 expression • Inhibited nuclear transport activities	Surnar et al. (2020)
HIV	Acquired immune deficiency syndrome	Ritonavir-loaded solid lipid NPs	• Efficient encapsulation of ritonavir, maintained antiviral activity and controlled drug release	Javan et al. (2017)
HSV	Orolabial herpes and genital herpes	Acyclovir-loaded glycosaminoglycan mimetic-functionalizedmesoporous Silica NPs	• Presence of sulfonate group was essential for antiviral activity, and incorporating a benzene group within the ligand activity was enhanced. • Inhibited viral entry and DNA replication	Lee et al. (2018)
HSV	Orolabial herpes and genital herpes	Acyclovir-loaded biopolymeric NPs embedded into mucoadhesive films	• Studies showed the potential of nanoformulation for enhanced bioavailability of the drug	Nair et al. (2017)
H1N1 influenza virus	Influenza	Zanamivir loaded Ag NPs	• Neuraminidase activity of virus was inhibited • Also resisted virus induced cell apoptosis	Lin et al. (2017b)
HIV	Acquired immune deficiency syndrome	Reverse transcriptase inhibitors conjugated graphene quantum dots	• Improved biocompatibility and therapeutic efficacy • RT activity was inhibited and inhibited virus binding to cells	Iannazzo et al. (2018)

was enhanced on using nanoparticle system. Further, studies on vaginal administration of nanoparticle system in a rat model showed that therapeutic drug level was maintained (0.6 ± 0.2 μL) in plasma for 24 hours. On using a nanoparticle system, the mean residence time of the drug was improved significantly, and drug bioavailability was enhanced by 2 folds in comparison to pure drug. Also, 2–3 fold higher tissue distribution was observed in animals treated with a nanoparticles system in comparison to that of pure drug. This study shows sustained drug release with improved biodistribution and bioavailability using nanoparticle system that can be applied for therapy against herpes simplex virus (HSV) infection (Ramyadevi et al. 2016). Joshi et al. (2016) designed an orally administrable drug-delivery system to transport lopinavir drug at HIV reservoir site (lymph) using poly(lactide-co-glycolide) (PLGA) NPs loaded with lopinavir. These NPs were prepared by using the nanoprecipitation method. After oral delivery of lopinavir-loaded PLGA NPs, penetration and absorption in Caco-2 cells and intestine was shown through confocal studies. Studies show that the permeability of lopinavir-loaded PLGA NPs was increased by 3.04 fold in Caco-2 cells. Also, bioavailability in rats was enhanced by 13.9 folds after oral delivery of lopinavir-loaded PLGA NPs (Joshi et al. 2016). Kumar et al. (2017) formulated an oral drug-delivery system in which sol-oil protocol was utilised for the synthesis of efavirenz-loaded lactoferrin nanoparticles (lacto-EFV-nano). Antiviral activity studies revealed that the anti-HIV-1 activity of lacto-EFV-nano was more than 2 fold higher in comparison to free EFV. Further, lacto-EFV-nano was found to be at least 2 fold less toxic than pure EFV to Jurkat T cell, peripheral blood mononuclear cells and B16-F10 cell lines as per cell viability studies. Studies revealed lacto-EFV-nano had improved in vivo pharmacokinetic profile along with improved oral bioavailability. Toxicity related to organs was not shown by lacto-EFV-nano. This study depicts the use of lacto-EFV-nano as a promising oral drug-delivery system for the treatment of HIV-1 viral infection (Kumar et al. 2017). Joshy et al. (2018) synthesised zidovudine (AZT) drug-encapsulated amide functionalized alignate NPs (AZT-GAAD NPs). During the synthesis of NPs, PF-68 as a stabiliser was used. By dialysis method, the loading efficiency of NPs was found to be 29.5 ± 3.2%. The in vitro studies showed that AZT was released slowly in a sustained manner from the NPs in PBS (pH-7.4). Also, studies on cell viability and intercellular uptakes revealed that AZT-GAAD NPs internalisation efficiency was significantly higher in glioma cell lines. Further studies also revealed the biocompatibility of AZT-GAAD NPs. This study depicts that zidovudine drug-loaded alginate conjugate NPs is a promising drug-delivery system for HIV/AIDS treatment (Joshy et al. 2018).

Lin et al. (2017a) explored the antiviral activity of selenium NPs modified with zanamivir against the H1N1 influenza virus. Studies showed that when using selenium NPs modified with zanamivir H1N1 virus, proliferation was restricted and higher viability was attained for MDCK cells. During H1N1 virus infection, induced activation and expression of caspase-3 were depressed on treatment with selenium NPs modified with zanamivir. Also, phosphorylation of JNK and p38 were decreased by selenium NPs modified with zanamivir. This showed the potential of antiviral drug-loaded NPs as promising pharmaceutical for the treatment of infection caused by the H1N1 influenza virus (Lin et al. 2017a). Lin et al. (2017b) designed zanamivir-modified silver NPs as a nanosystem against the H1N1 influenza virus. The study revealed that the neuraminidase activity of the virus was inhibited by the nanosystem. Further, MDCK cells apoptosis induced by the virus was resisted by zanamivir modified silver NPs. Virus-induced reactive oxygen species (ROS) accumulation was efficiently reduced by zanamivir modified silver NPs and also nanosystem reduced the activation of signaling pathways (p53 and p38). This study shows zanamivir-loaded silver NPs as a potential pharmaceutical for the treatment of H1N1 influenza virus infection (Lin et al. 2017b). Zhong et al. (2019) synthesised selenium nanoparticles (NPs) and loaded oseltamivir onto the NPs surface to get antiviral NPs. These antiviral nanoparticles had good stability and showed effective release of drug in PBS and serum. This antiviral NPs entered human astrocyte U251 cells efficiently, and EV71 proliferation was inhibited, which could protect cell apoptosis. The antiviral NPs reduced the reactive oxygen species generation in the cells to inhibit the activity of EV71. This study showed that oseltamivir antiviral activity was enhanced by selenium NPs (Zhong et al. 2019).

Cánepa et al. (2017) prepared chitosan NPs (CT-NPs) using ionotropic gelation method, and interferon-α-2b drug was loaded on these chitosan NPs (IFN-CT NPs). Percentage encapsulation efficiency of the drug in IFN-CT NPs was approximately 100%. In vitro studies showed that IFN-CT-NPs had unchanged antiviral activity against human lymphotropic-T virus type 1 as compared to the commercial drug. After 1 hour of oral administration of IFN-CT NPs (0.3 MIU), IFNα drug was found to be present in detectable levels in the plasma of CF1 mice. But IFNα was not detected in the plasma when the commercial formulation was used (Cánepa et al. 2017). Lima et al. (2018) synthesised chloroquine-loaded poly(lactic acid) NPs using the emulsification-solvent evaporation-modified method. The purpose of these NPs syntheses was to have improved antiviral activity of chloroquine against herpes simplex virus type 1 (HSV-1). The efficiency of spherical NPs for the encapsulation of chloroquine was 64.1%. Bare NPs were found biocompatible to Vero E6 cells in the in vitro study. Conversely, chloroquine-loaded NPs showed different levels of cytotoxicity. In plaque-reduction assay, chloroquine-loaded NPs had improved antiviral activity in comparison to chloroquine. Therefore, this study reveals that NPs can solve the limitations associated with the drugs for better treatment against viral infections (Lima et al. 2018). Hassan et al. (2020) produced solid lipid NPs (SLNs) as a carrier for acyclovir drug that is used for treatment against infection caused by herpes simplex virus. Studies revealed that oral bioavailability of drug was superior in the case of SLNs loaded with acyclovir drug in comparison to the commercial suspension of the drug. Also, when compared to commercial suspension, acyclovir loaded SLNs had 4-fold higher plasma concentration. Therefore, the method used for acyclovir encapsulation in this study showed a pathway for enhancing the bioavailability of the drug, as well as the use of SLNs for oral drug delivery (Hassan et al. 2020) (Table 10.1).

10.3.2 Nanomaterials as Therapeutic Agents against Viruses

Recently, nanomaterials have been widely applied in antiviral research, and a lot of nanomaterials have been developed as therapeutic agents against viruses. Some of the recently developed materials have been discussed under this section, and along with a few others, they have been summarised in Table 10.2.

TABLE 10.2
Nanomaterials as Therapeutic Agents against Viral Infections

Virus	Viral Disease	Nanomaterial	Mechanism/Outcome	Reference
PRRSV	Porcine reproductive and respiratory syndrome	GO modified with silver NPs	• Inhibits virus entry into cells • Production of IFN-stimulating genes and interferon-α that prevents the virus proliferation	Du et al. (2018b)
PRRSV	Porcine reproductive and respiratory syndrome	Te/BSA nanostars modified with mercaptoethane sulfonate	• Mainly inhibited the internalisation process of the virus	Zhou et al. (2020)
HSV-1	Mild skin and mucosal lesions, cancers and fatal neurologic diseases	Derivatives of triazine-functionalized graphene using fatty amines and polyglycerol sulfate	• Antiviral activity of graphene derivatives is due to secondary hydrophobic interactions of alkyl chains • Graphene derivatives with alkyl chain length of 6 and 9 carbon atoms showed antiviral activity with low toxicity	Donskyi et al. (2019)
H1N1 influenza virus	Respiratory infection	Bare zinc oxide NPs and PEGylated zinc oxide NPs	• PEGylated zinc oxide NPs had high antiviral activity and lower cytotoxicity than bare zinc oxide NPs • Antiviral activity of NPs was only after virus enter into the host cells	Ghaffari et al. (2019)
Influenza A virus	Respiratory infection	Porous gold NPs	• Blocking of virus entry by attaching of porous gold nanoparticles with the surface protein of influenza viruses	Kim et al. (2020)
CHIKV	Chikungunya fever	Ag NPs using *Tinospora cordifolia*, *Phyllanthus niruri* and *Andrographis paniculata* extracts	• *Andrographis paniculata* Ag NPs and *Tinospora cordifolia* Ag NPs displayed excellent antiviral activity	Sharma et al. (2019)

(Continued)

TABLE 10.2 (Continued)
Nanomaterials as Therapeutic Agents against Viral Infections

Virus	Viral Disease	Nanomaterial	Mechanism/Outcome	Reference
Human coronavirus HCoV-229E	Respiratory infection	Carbon quantum dots	• Inhibition of the virus entry by interaction of virus entry receptor with the functional groups of carbon quantum dots • Also interference with viral replication step	Łoczechin et al. (2019)
PEDV	Serious respiratory infection	Glutathione-capped Ag_2S nanoclusters	• Viral budding and viral negative-strand RNA synthesis were inhibited • Generation of interferon-stimulating genes and expression of proinflammation cytokines	Du et al. (2018a)
RSV	Respiratory infection	Curcumin loaded on β-cyclodextrin functionalized GO	• Direct inactivation of the virus, inhibiting the attachment of RSV with the host cells and interference with the replication of virus	Yang et al. (2017)
Influenza A virus	Respiratory infection	Chitosan/ siRNA NPs	• Prevented the virus growth by inhibiting replication of influenza virus • Prevented the infection in mice	Jamali et al. (2018)
RSV	Respiratory infection	NPs of copper, silver and gold; and their conjugate with anti-RSV polyclonal antibody	• Antibody-conjugated copper NPs had better RSV detection efficiency	Valdez et al. (2016)
RSV	Respiratory infection	Modified silver NPs by using curcumin (cAgNPs)	• cAgNPs can directly inactivate the RSV to prevent the RSV infection	Yang et al. (2016)
EBOV	Hemorrhagic fever	Graphene	• Strong association of graphene at various interfaces of protein and graphene disrupts the CTD-CTD interface that is vital in forming the matrix of the Ebola virus	GC et al. (2017)

(Continued)

TABLE 10.2 (Continued)
Nanomaterials as Therapeutic Agents against Viral Infections

Virus	Viral Disease	Nanomaterial	Mechanism/Outcome	Reference
H1N1 influenza A virus	Serious respiratory illness	Iron oxide NPs	• Change in viral RNA transcripts	Kumar et al. (2019)
HSV-1	Corneal blindness, orolabial and facial lesions, peripheral nervous system disorders and encephalitis	CuO NPs	• 100 μg/mL was highest concentration found non toxic and results at this concentration was 2.8 log10 TCID50 reduction in virus titer	Tavakoli and Hashemzadeh (2020)
Bovine herpesvirus-1 (BoHV-1)	Infectious bovine rhinotracheitis/ infectious bovine balanopothitis	Ag NPs	• 24 μg/mL concentration of Ag NPs is recommended to protect the cellular culture against viral replications. • Administration of 24 μg/mL concentration of Ag NPs prior to viral infection is capable of inhibiting BoHV-1	El-Mohamady et al. (2018)
HSV-1, HAV-10 virus and Coxsackie B4 virus	-	Ag NPs using *Lampranthus coccineus* and *Malephora lutea* extracts	• NPs were capable of inhibiting the virus infection. • Most probably NPs binded virus envelop proteins and prevented viral penetration	Haggag et al. (2019)
PEDV		Cationic carbon dots from curcumin		Ting et al. (2018)

(Continued)

TABLE 10.2 (Continued)
Nanomaterials as Therapeutic Agents against Viral Infections

Virus	Viral Disease	Nanomaterial	Mechanism/Outcome	Reference
	Serious respiratory infection		• Inhibits proliferation of virus, inhibits viral entry and prevent budding and synthesis of negative-strand RNA • Suppress the ROS accumulation • Production of pro-inflammatory cytokines and interferon-stimulating genes of Vero cells	
HSV-1	Mild skin and mucosal lesions, cancers and fatal neurologic diseases	Functional carbon nanodots	• Inhibits virus entry due to nanodots interaction with virus	Barras et al. (2016)
HIV	HIV infection	Boronic acid-attributed carbon dots	• Enhanced antiviral activity as carbon dots prevented the binding of virus with the target cell	Fahmi et al. (2016)
Enveloped feline coronavirus (FCoV) and without envelope infectious bursal disease virus (IBDV)	-	Ag NPs anchored on GO	• For both virus with envelope and without envelope Ag NPs anchored on GO exhibited antiviral activity • Whereas GO had antiviral activity against enveloped viruses only	Chen et al. (2016)

Gc et al. (2017) investigated interactions of graphene with protein VP40 of Ebola virus matrix by utilising molecular dynamics simulations and graphene pelleting assay. There exist several oligomeric and conformational forms of VP40 protein and, for disrupting the life cycle of the Ebola virus, it is a potential target. The result shows a strong association of graphene at various interfaces of protein, and graphene disrupts the CTD-CTD interface that is vital in forming the matrix of the Ebola virus. This study shows the potential of graphene nanomaterials because their solutions may be used as disinfectants to prevent the disease spread by the Ebola virus (Gc et al. 2017). Yang et al. (2017) synthesised GO composite with excellent antiviral activity against respiratory syncytial virus (RSV). The composite material used had curcumin loaded on β-cyclodextrin functionalized GO. This composite material also had great biocompatibility with host cells. Studies showed that GO composite inhibited RSV infection by three possible mechanisms: direct inactivation of the virus, inhibiting the attachment of RSV with the host cells, and interference with the replication of virus. Against RSV, composite material had prophylactic and therapeutic effects (Yang et al. 2017). Du et al. (2018b) used GO modified with silver NPs against porcine reproductive and respiratory syndrome virus (PRRSV). Results showed that nanocomposite suppressed the viral infection with an inhibition efficiency of 59.2%. Further, production of IFN-stimulating genes and interferon-α was enhanced by the nanocomposite that can directly prevent the virus proliferation (Du et al. 2018b). Donskyi et al. (2019) prepared numerous derivatives of triazine-functionalized graphene using fatty amines and polyglycerol sulfate and studies antiviral activity against herpes simplex virus type-1 (HSV-1) through interactions. Graphene derivatives bind to the virus through electrostatic interactions between the virus and polyglycerol sulfate. Conversely, antiviral activity of graphene derivatives is due to secondary hydrophobic interactions of alkyl chains. Among various graphene derivatives, the derivative having an alkyl chain with 12 carbon atoms displayed the highest activity, but for the Vero cell line, it was toxic. Further, graphene derivatives with alkyl chain length of 6 and 9 carbon atoms showed antiviral activity with low toxicity. Therefore, this study demonstrated that stepwise and controlled functionalization of graphene could be used for the development of antiviral agents against HSV-1 (Donskyi et al. 2019).

Łoczechin et al. (2019) prepared different carbon quantum dots and explored their antiviral activity against human coronavirus HCoV-229E. Ethylenediamine/citric acid were used as carbon precursors and were hydrothermally carbonized for the synthesis of first generation of carbon quantum dots. This first generation of carbon quantum dots was further modified with boronic acid ligands, and nanostructures obtained had concentration dependent antiviral activity. As estimated, EC_{50} for these nanostructures were 52 ± 8 μg mL^{-1}. Whereas, second-generation carbon quantum dots as antiviral nanomaterial were prepared from 4-aminophenylboronic acid without postmodification. These second-generation carbon quantum dots had even lowered EC_{50} value of 5.2 ± 0.7 μg mL^{-1}. Inhibition of the virus entry was the mechanism for the antiviral activity of these carbon quantum dots, which is due to the interaction of a virus-entry receptor with the functional groups of carbon quantum dots. The results also showed that carbon quantum dots had interference with the viral replication step (Łoczechin et al. 2019).

Yang et al. (2016) synthesised modified silver nanoparticles by using curcumin (cAgNPs), which were utilised for antiviral activity against respiratory syncytial virus (RSV) infection. The tissue culture infectious dose assay studies revealed that these cAgNPs had high antiviral activity against RSV infection. Viral titers were decreased by about 2 order of magnitudes with no toxic effects. It was found that cAgNPs can directly inactivate the RSV to prevent the RSV infection (Yang et al. 2016). Valdez et al. (2016) demonstrated a simple and economical method for the detection of respiratory syncytial virus (RSV) using localised surface plasmon resonance (LSPR) spectroscopy. In this method, bare metal NPs of copper, silver, and gold and their conjugate with anti-RSV polyclonal antibody were utilised. Studies revealed that the antibody-conjugated copper NPs had better RSV detection efficiency. Whereas in the presence of nonspecific entities, it was found that antibody-conjugated copper and silver NPs were more specific toward RSV. And quantitative results of the studies revealed that antibody-conjugated copper NPs were better in comparison to the antibody-conjugated silver and gold NPs. Advantage of this study could be used for the development of a detection device for rapid detection of RSV (Valdez et al. 2016). Du et al. (2018a) synthesised biocompatible glutathione-capped Ag_2S nanoclusters with high antiviral activity against porcine epidemic diarrhea virus (PEDV). Results showed infection of PEDV was reduced by 3 order of magnitude by glutathione-capped Ag_2S nanoclusters when used at a noncytotoxic concentration after 12 hours of infection. Viral budding and viral negative-strand RNA synthesis were inhibited on treatment with this nanocluster material. Further, generation of interferon-stimulating genes and expression of proinflammation cytokines were activated on treatment with nanoclusters, which might inhibit the infection caused by PEDV. As PEDV was used as a model of Coronavirus, this study revealed the capability of glutathione-capped Ag_2S nanoclusters to be used as a therapeutic drug against Coronavirus (Du et al. 2018a). Sharma et al. (2019) synthesised silver nanoparticles (Ag NPs) using plant extracts of *Tinospora cordifolia*, *Phyllanthus niruri*, and *Andrographis paniculata* and used them to study the antiviral activity against chikungunya virus (CHIKV). Studies on cytotoxicity in Vero cells showed *Andrographis paniculata* Ag NPs to be the most cytotoxic, *Phyllanthus niruri* Ag NPs to be less cytotoxic, and *Tinospora cordifolia* Ag NPs to be least cytotoxic among them. Testing antiviral activity of different Ag NPs against CHIKV on Vero cells, *Andrographis paniculata* Ag NPs and *Tinospora cordifolia* Ag NPs displayed excellent antiviral activity. This study shows a pathway to use biosynthesized NPs as an alternative treatment method against virus infections for which there are no specific antiviral or vaccines (Sharma et al. 2019).

Kim et al. (2020) utilised the surfactant-free emulsion method for the synthesis of porous gold nanoparticles and applied them against influenza viruses (H1N1, H3N2, and H9N2). The study revealed remarkable antiviral activity of porous gold nanoparticles as cell viability of infected MDCK cells increased after treatment. The study further suggested a mechanism for the viral inhibition to be the blocking of virus entry by attaching of porous gold nanoparticles with the surface protein of influenza viruses and making conformational changes to prevent viral membrane fusion (Kim et al. 2020). Ghaffari et al. (2019) applied bare zinc oxide NPs and PEGylated zinc oxide NPs against influenza virus (H1N1) and estimated the antiviral activity of NPs. The study demonstrated antiviral activity of PEGylated zinc oxide NPs was high in comparison to that of bare zinc oxide NPs. Also, PEGylated zinc oxide NPs had lower cytotoxicity than bare zinc oxide NPs. Study findings also demonstrated that antiviral activity shown by NPs was only after virus enter into the host cells (Ghaffari et al. 2019). Zhou et al. (2020) produced nanostar morphology of bovine serum albumin coated tellurium NPs modified with mercaptoethane sulfonate (Te/BSA nanostars) and applied against porcine reproductive and respiratory syndrome virus (PRRSV). The study demonstrated high antiviral activity of Te/BSA nanostars by inhibiting the internalisation process of the virus. In comparison to the glutathione-modified Te/BSA nanospheres, mercaptoethane sulfonate (MES) modified Te/BSA nanostars had a higher activity that was ascribed to the novel star shape and MES modification. This study shows that Te/BSA nanostars have potential for treatment of viral infections in the future (Zhou et al. 2020).

Jamali et al. (2018) produced nanoparticles complex of siRNA and chitosan, which was utilised against infection caused by influenza virus. This nanoparticle complex was used for both in vitro and in vivo studies. Results from in vitro studies demonstrated efficient uptake of nanoparticle complex by Vero cells, which prevented the virus growth by inhibiting replication of influenza virus. Further in vivo studies, in which nanoparticle complex was delivered to BALB/c mice through nasal method, showed antiviral effect and prevented the influenza virus infection in mice. This study revealed the potential of using chitosan material as a gene carrier for the prevention of viral infections (Jamali et al. 2018) (Table 10.2).

10.4 CONCLUSION

The chapter shows the development of nanomaterials has proven advantageous for the prevention, diagnosis, and treatment of infections caused by viruses over the regular methods. Features that make nanomaterial beneficial in therapeutic applications are easy synthesis, economical, large surface area, and easy surface modification. Nanomaterials act as antiviral agents through several mechanisms. The use of nanomaterials as delivery agents for antiviral drugs has shown superiority in overcoming the limitations associated with the treatment of viral infection by drugs such as low solubility of drugs, low bioavailability of drugs, drug resistance, and frequent drug uptake. Thus, the combination of antiviral drugs with nanomaterials resulted in better therapeutic effects and can provide better and improved antiviral treatment. The thing of major concern is to make nanomaterials more biocompatible and nontoxic. It can be said that progressive development of nanomaterials will provide effective and efficient therapeutics against viral infections in near future.

REFERENCES

Abd Elkodous, M., El-Sayyad, G.S., Abdelrahman, I.Y., et al. 2019. Therapeutic and diagnostic potential of nanomaterials for enhanced biomedical applications. *Colloids and Surfaces B: Biointerfaces* 180:411–428.

Barras, A., Pagneux, Q., Sane, F., et al. 2016. High efficiency of functional carbon nanodots as entry inhibitors of herpes simplex virus type 1. *ACS Applied Materials and Interfaces* 8(14):9004–9013.

Bouvier, N.M. and Palese, P. 2008. The biology of influenza viruses. *Vaccine* 26(S4): D49–D53.

Bule, M., Khan, F. and Niaz, K. 2019. Antivirals: Past, present and future. In *Recent Advances in Animal Virology*, eds. Y. Malik, R. Singh and M. Yadav, 425–446. Springer: Singapore.

Cánepa, C., Imperiale, J.C., Berini, C.A., Lewicki, M., Sosnik, A. and Biglione, M.M. 2017. Development of a drug delivery system based on chitosan nanoparticles for oral administration of interferon-α. *Biomacromolecules* 18(10):3302–3309.

Chakravarty, M. and Vora, A. 2021. Nanotechnology-based antiviral therapeutics. *Drug Delivery and Translational Research* 11(3):748–787.

Chen, Y.N., Hsueh, Y.H., Te Hsieh, C., Tzou, D.Y. and Chang, P.L. 2016. Antiviral activity of graphene–silver nanocomposites against non-enveloped and enveloped viruses. *International Journal of Environmental Research and Public Health* 13(4): 430.

Cheng, S. and Brooks III, C.L. 2013. Viral capsid proteins are segregated in structural fold space. *PLoS Computational Biology* 9(2):e1002905.

Chinchar, V.G. 1999. Replication of viruses. In , eds. *Encyclopedia of Virology* Granoff A., and R.G. Webster, 1471–1478. Amsterdam, Netherlands: Elsevier. 10.1006/rwvi.1999.0245.

Desai, J. and Thakkar, H. 2018. Darunavir-loaded lipid nanoparticles for targeting to HIV reservoirs. *AAPS PharmSciTech* 19(2):648–660.

Donskyi, I.S., Azab, W., Cuellar-Camacho, J.L., et al. 2019. Functionalized nanographene sheets with high antiviral activity through synergistic electrostatic and hydrophobic interactions. *Nanoscale* 11(34):15804–15809.

Du, T., Liang, J., Dong, N., et al. 2018a. Glutathione-capped Ag_2S nanoclusters inhibit coronavirus proliferation through blockage of viral RNA synthesis and budding. *ACS Applied Materials and Interfaces* 10(5):4369 4378.

Du, T., Lu, J., Liu, L., et al. 2018b. Antiviral activity of graphene oxide-silver nanocomposites by preventing viral entry and activation of the antiviral innate immune response. *ACS Applied Bio Materials* 1(5):1286–1293.

El-Mohamady, R.S., Ghattas, T.A., Zawrah, M.F. and Abd El-Hafeiz, Y.G.M. 2018. Inhibitory effect of silver nanoparticles on bovine herpesvirus-1. *International Journal of Veterinary Science and Medicine* 6(2):296–300.

Fahmi, M.Z., Sukmayani, W., Khairunisa, S.Q., et al. 2016. Design of boronic acid-attributed carbon dots on inhibits HIV-1 entry. *RSC Advances* 6(95):92996–93002.

Farman, A., Lal Badshah, S., Khan, K., Ahmad, N. and Naeem, A. 2020. Ebola, the negative stranded RNA virus. In *Some RNA Viruses*, eds. Y. Shah, and E. Abuelzein. IntechOpen, London. doi:10.5772/intechopen.91776.

Faye, O., Freire, C.C.M., Iamarino, A., et al. 2014. Molecular evolution of Zika virus during its emergence in the 20th century. *PLoS Neglected Tropical Diseases* 8(1): e2636.

Gc, J.B., Pokhrel, R., Bhattarai, N., et al. 2017. Graphene-VP40 interactions and potential disruption of the Ebola virus matrix filaments. *Biochemical and Biophysical Research Communications* 493(1):176–181.

Gerber, J.G. 2000. Using pharmacokinetics to optimize antiretroviral drug-drug interactions in the treatment of human immunodeficiency virus infection. *Clinical Infectious Diseases* 30(S2):S123–S129.

Ghaffari, H., Tavakoli, A., Moradi, A., et al. 2019. Inhibition of H1N1 influenza virus infection by zinc oxide nanoparticles: Another emerging application of nanomedicine. *Journal of Biomedical Science* 26(1):70.

Haggag, E.G., Elshamy, A.M., Rabeh, M.A., et al. 2019. Antiviral potential of green synthesized silver nanoparticles of lampranthus coccineus and malephora lutea. *International Journal of Nanomedicine* 14:6217–6229.

Hassan, H., Bello, R.O., Adam, S.K., et al. 2020. Acyclovir-loaded solid lipid nanoparticles: Optimization, characterization and evaluation of its pharmacokinetic profile. *Nanomaterials* 10(9):1785.

Iannazzo, D., Pistone, A., Ferro, S., et al. 2018. Graphene quantum dots based systems as HIV inhibitors. *Bioconjugate Chemistry* 29(9):3084–3093.

Jamali, A., Mottaghitalab, F., Abdoli, A., et al. 2018. Inhibiting influenza virus replication and inducing protection against lethal influenza virus challenge through chitosan nanoparticles loaded by siRNA. *Drug Delivery and Translational Research* 8(1): 12–20.

Javan, F., Vatanara, A., Azadmanesh, K., Nabi-Meibodi, M. and shakouri, M. 2017. Encapsulation of ritonavir in solid lipid nanoparticles: In-vitro anti-HIV-1 activity using lentiviral particles. *Journal of Pharmacy and Pharmacology* 69(8): 1002–1009.

Joshi, G., Kumar, A. and Sawant, K., 2016. Bioavailability enhancement, Caco-2 cells uptake and intestinal transport of orally administered lopinavir-loaded PLGA nanoparticles. *Drug Delivery* 23(9):3492–3504.

Joshy, K.S., Susan, M.A., Snigdha, S., Nandakumar, K., Laly, A.P. and Sabu, T. 2018. Encapsulation of zidovudine in PF-68 coated alginate conjugate nanoparticles for anti-HIV drug delivery. *International Journal of Biological Macromolecules* 107(Part A):929–937.

Kim, J., Yeom, M., Lee, T., et al. 2020. Porous gold nanoparticles for attenuating infectivity of influenza A virus. *Journal of Nanobiotechnology* 18(1):1–11.

Kumar, P., Lakshmi, Y.S. and Kondapi, A.K. 2017. An oral formulation of efavirenz-loaded lactoferrin nanoparticles with improved biodistribution and pharmacokinetic profile. *HIV Medicine* 18(7):452–462.

Kumar, R., Nayak, M., Sahoo, G.C., et al. 2019. Iron oxide nanoparticles based antiviral activity of H1N1 influenza A virus. *Journal of Infection and Chemotherapy* 25(5): 325–329.

Lee, E.C., Nguyen, C.T.H., Strounina, E., Davis-Poynter, N. and Ross, B.P. 2018. Structure-activity relationships of GAG mimetic-functionalized mesoporous silica nanoparticles and evaluation of acyclovir-loaded antiviral nanoparticles with dual mechanisms of action. *ACS Omega* 3(2):1689–1699.

Li, Y., Xiao, Y., Chen, Y. and Huang, K. 2021. Nano-based approaches in the development of antiviral agents and vaccines. *Life Sciences* 265:118761.

Lima, T.L.C., Feitosa, R. de C., Dos Santos-Silva, E., et al. 2018. Improving encapsulation of hydrophilic chloroquine diphosphate into biodegradable nanoparticles: A promising approach against herpes virus simplex-1 infection. *Pharmaceutics* 10(4):255.

Lin, Z., Li, Y., Guo, M., et al. 2017a. Inhibition of H1N1 influenza virus by selenium nanoparticles loaded with zanamivir through p38 and JNK signaling pathways. *RSC Advances* 7(56):35290–35296.

Lin, Z., Li, Y., Guo, M., et al. 2017b. The inhibition of H1N1 influenza virus-induced apoptosis by silver nanoparticles functionalized with zanamivir. *RSC Advances* 7(2): 742–750.

Łoczechin, A., Séron, K., Barras, A., et al. 2019. Functional carbon quantum dots as medical countermeasures to human coronavirus. *ACS Applied Materials and Interfaces* 11(46): 42964–42974.

Lodish, H., Berk, A., Zipursky, S.L., Matsudaira, P., Baltimore, D. and Darnell, J. 2000. *Molecular Cell Biology*, New York: W. H. Freeman and Company.

Machhi, J., Herskovitz, J., Senan, A.M., et al. 2020. The natural history, pathobiology, and clinical manifestations of SARS-CoV-2 infections. *Journal of Neuroimmune Pharmacology* 15(3):359–386.

Mehendale, R., Joshi, M. and Patravale, V. 2013. Nanomedicines for treatment of viral diseases. *Critical Reviews in Therapeutic Drug Carrier Systems* 30(1):1–49.

Nair, A.B., Al-Ghannam, A.A., Al-Dhubiab, B.E. and Hasan, A.A. 2017. Mucoadhesive film embedded with acyclovir loaded biopolymeric nanoparticles: In vitro studies. *Journal of Young Pharmacists* 9(1):100–105.

Oshikoya, K.A., Oreagba, I.A., Ogunleye, O.O., Lawal, S. and Senbanjo, I.O. 2013. Clinically significant interactions between antiretroviral and co-prescribed drugs for HIV-infected children: Profiling and comparison of two drug databases. *Therapeutics and Clinical Risk Management* 9(1):215–221.

Ramyadevi, D., Rajan, K.S., Vedhahari, B.N., Ruckmani, K. and Subramanian, N. 2016. Heterogeneous polymer composite nanoparticles loaded in situ gel for controlled release intra-vaginal therapy of genital herpes. *Colloids and Surfaces B: Biointerfaces* 146:260–270.

Sharma, V., Kaushik, S., Pandit, P., Dhull, D., Yadav, J.P. and Kaushik, S. 2019. Green synthesis of silver nanoparticles from medicinal plants and evaluation of their antiviral potential against chikungunya virus. *Applied Microbiology and Biotechnology* 103(2):881–891.

Sharma, V.K., Jinadatha, C. and Lichtfouse, E. 2020. Environmental chemistry is most relevant to study coronavirus pandemics. *Environmental Chemistry Letters* 18(4):993–996.

Singh, L., Kruger, H.G., Maguire, G.E.M., Govender, T. and Parboosing, R. 2017. The role of nanotechnology in the treatment of viral infections. *Therapeutic Advances in Infectious Disease* 4(4):105–131.

Singh, R. and Lillard, J.W. 2009. Nanoparticle-based targeted drug delivery. *Experimental and Molecular Pathology* 86(3):215–223.

Smith, J.A. and Daniel, R. 2006. Following the path of the virus: The exploitation of host DNA repair mechanisms by retroviruses. *ACS Chemical Biology* 1(4):217–226.

Strasfeld, L. and Chou, S. 2010. Antiviral drug resistance: Mechanisms and clinical implications. *Infectious Disease Clinics of North America* 24(3):809–833.

Surnar, B., Kamran, M.Z., Shah, A.S., et al. 2019. Orally administrable therapeutic synthetic nanoparticle for Zika virus. *ACS Nano* 13(10):11034–11048.

Surnar, B., Kamran, M.Z., Shah, A.S. and Dhar, S. 2020. Clinically approved antiviral drug in an orally administrable nanoparticle for COVID-19. *ACS Pharmacology and Translational Science* 3(6):1371–1380.

Tavakoli, A. and Hashemzadeh, M.S. 2020. Inhibition of herpes simplex virus type 1 by copper oxide nanoparticles. *Journal of Virological Methods* 275:113688.

Ting, D., Dong, N., Fang, L., et al. 2018. Multisite inhibitors for enteric coronavirus: Antiviral cationic carbon dots based on curcumin. *ACS Applied Nano Materials* 1(10):5451–5459.

Valdez, J., Bawage, S., Gomez, I. and Singh, S.R. 2016. Facile and rapid detection of respiratory syncytial virus using metallic nanoparticles. *Journal of Nanobiotechnology* 14(1):13.

Verdonck, K., González, E., Van Dooren, S., Vandamme, A.M., Vanham, G. and Gotuzzo, E. 2007. Human T-lymphotropic virus 1: Recent knowledge about an ancient infection. *Lancet Infectious Diseases* 7(4):266–281.

Yang, X.X., Li, C.M. and Huang, C.Z. 2016. Curcumin modified silver nanoparticles for highly efficient inhibition of respiratory syncytial virus infection. *Nanoscale* 8(5):3040–3048.

Yang, X.X., Li, C.M., Li, Y.F., Wang, J. and Huang, C.Z. 2017. Synergistic antiviral effect of curcumin functionalized graphene oxide against respiratory syncytial virus infection. *Nanoscale* 9(41):16086–16092.

Zhong, J., Xia, Y., Hua, L., et al. 2019. Functionalized selenium nanoparticles enhance the anti-EV71 activity of oseltamivir in human astrocytoma cell model. *Artificial Cells, Nanomedicine and Biotechnology* 47(1):3485–3491.

Zhou, Y., Jiang, X., Tong, T., et al. 2020. High antiviral activity of mercaptoethane sulfonate functionalized Te/BSA nanostars against arterivirus and coronavirus. *RSC Advances* 10(24):14161–14169.

11 *In Vivo* Study of Anti-Influenza Effect of Silver Nanoparticles in a Mouse Model

Ludmila Puchkova, Mohammad Al Farroukh, Ekaterina Ilyechova, and Irina Kiseleva

CONTENTS

11.1 INTRODUCTION

The various aspects of human activities have been negatively affected by viral respiratory infections caused by the influenza A virus (IAV) during the last 100 years. IAVs are a cause of recurrent seasonal epidemics, as well as catastrophic pandemic outbreaks, which have affected hundreds of millions of people and taken a great toll on economies (Kieny et al. 2006). Vaccination is the primary protection against influenza infections, and it is especially important in pandemic cases (Shie and Fang 2019). However, the vaccine cannot be used in all cases; for example, it can't be used when there are medical contraindications for vaccination or ongoing infection. In these cases, the use of antiviral chemical agents becomes the main strategy. These agents tend to suppress the function of important viral proteins, e.g. neuraminidase or M2 protein (van der Vries et al. 2013). However, the viral genome is highly variable, so the virus quickly becomes drug-resistant and evades the immune response caused by vaccination. This significantly limits the effectiveness of infection control; therefore, the search for new means of prevention and effective treatment of acute respiratory viral infections is always relevant.

DOI: 10.1201/9781003136644-14

The search for metabolic links specifically altered by the virus and their antiviral modification is an additional promising modern approach to suppress viral reproduction. It has been shown that the metabolism of an infected cell is globally reprogrammed following by a viral strategy. Recently, these studies have led to the creation of the concept of "virocell" (Forterre 2016). According to this concept, targeting these systems has an indirect effect on viral replication, thus potentially protecting the body from influenza infections, regardless of the virus and any acquired resistance to currently used antiviral agents. Growing evidence demonstrates that IAV replication depends on cellular copper balance, and, possibly, organism-wide (Turpin et al. 2005; Kuss-Duerkop et al. 2017; Dai et al. 2018; Ren et al. 2019). Moreover, the new area of studying and developing agents that affect cell-signaling pathways employed by IAV for its replication also implicates copper-dependent pathways (Ludwig et al. 2003; Ramos et al. 2016; Wang et al. 2016; Schloer et al. 2019). It would be very important to identify the parts of the copper homeostasis system that may be targeted to counteract the influenza infection.

11.2 BIOLOGICAL ACTIVITY OF SILVER NANOPARTICLES

Among modern functional materials, silver nanoparticles (AgNP) are the most used for biological and medical purposes (Natsuki et al. 2015; Zhang et al. 2016). This is due to their bactericidal properties. AgNP effectively suppresses the growth of various pathogens like gram-positive and gram-negative bacteria, fungi, bacteriophages, and viruses in a dose- and time-dependent manner. At the same time, the sensitivity of bacteria to AgNP is significantly lower than the threshold of sensitivity to them in mammals (Munger et al. 2014; Liao et al. 2019; Paladini and Pollini 2019). As a result, AgNP are widely used in nanomedicine, textile, and food industries (Siddiqi et al. 2018; Zivic et al. 2018).

The mechanism responsible for the antibacterial action of AgNP remains unclear. The most experimentally supported is the assumption that AgNP are adsorbed on the surface of the bacterial cell, cause the formation of pores, and penetrate through the porins into the cell; from there, the surface layer of AgNP undergoes chemical modification, resulting in the formation of silver ions Ag(+) and free radicals. Corrosion products induce membrane destruction, reactive oxygen species (ROS) generation, cytoplasmic leakage, DNA and RNA damage, protein denaturation, cellular respiration, and apoptosis-like process (Morones et al. 2005; Orlov et al. 2016; Pareek et al. 2018).

The threshold of sensitivity of fungi to AgNP is lower than that of bacteria (Sankova et al. 2015) but high enough to be used as effective drugs against pathogenic fungi, which in the human body, as a rule, are part of biofilms characterised by high resistance to antibiotics (Kim et al. 2007; Sardi et al. 2013). The bactericidal effect of AgNP on yeast is manifested by decreased transcription, reduced endocytosis, and led to a misbalance of ATP production (Tambosi et al. 2018).

The virucidal activity of AgNP has been demonstrated in numerous studies, the results of which are summarised in modern reviews (Galdiero et al. 2011; Gurunathan et al. 2020). It has been shown that RNA and DNA containing bacteriophages are not very sensitive to AgNP treatment. Thus, long prior exposure of MS2 to AgNP did not inactivate MS2, even at their high concentrations. Moreover, the virulence of the MS2 phage increased by several orders of magnitude (You et al. 2011). Phage φ80 after treatment with AgNP for a day at a concentration of nanoparticles that completely prevents the growth of *E. coli* retains 90% virulence (Sankova et al. 2015).

In contrast, eukaryotic viruses of the families *Orthomyxoviridae* and *Pneumoviridae*, containing a membrane envelope, are highly sensitive to AgNP. Treatment of viruses with AgNP reduces their infectivity in vitro and in vivo (Xiang et al. 2011; Fatima et al. 2016; Park et al. 2018; Morris et al. 2019). This is since, upon direct contact of the IAV with AgNP, the structure of hemagglutinin and neuraminidase is disrupted by 80% and 20%, respectively. Dramatic disturbances in hemagglutinin prevent the virus from entering cells in vitro and in vivo. Another group of studies shows that AgNP-treated mice have increased resistance to influenza infection (Xiang et al. 2013; Kiseleva et al. 2020a). It is difficult to assume that intraperitoneally injected AgNP can enter the upper respiratory tract and directly contact the IAV in the upper respiratory tract. Therefore, it can be thought that the resistance of AgNP-treated mice to influenza infection is a consequence of the effect of AgNP on the mice. The mechanism of the antiviral activity of AgNP remains unexplored. There is also little information about how AgNP affects mammals. The issue under discussion is closely related to the problem of toxicity of AgNP to mammals and humans.

In experiments aimed at revealing the toxicity of AgNP, high concentrations of nanoparticles were used. They showed that, regardless of the route of entry (intraperitoneal, oral, intranasal), AgNP cause the accumulation of silver in various organs, affect the function of the lungs and brain, cause hyperplasia of the bile ducts, have a genotoxic effect, and increase the concentration of ROS (Loeschner et al. 2011; Sharma 2019; Hadrup et al. 2020; Báez et al. 2021). The toxic effect of AgNP is found at concentrations that are many times higher than the toxic dose for prokaryotes. However, AgNP, even at low concentrations, can affect the mammalian health because they interfere with copper metabolism. This is because, in the body, AgNP dissociate and form Ag(+), which are isoelectronic to copper in the Cu(+) oxidation state, and so can replace Cu(I) in copper-transport proteins, cuproenzymes, and regulatory proteins that require copper for their activity (Puchkova et al. 2019).

The central idea of this chapter is the assumption that the antiviral effect of AgNP is the result of their interference with the host copper metabolism and, possibly, disruption of those chains, which the IAV uses to produce offspring. This thought is supported by the following data showing the relationship between copper metabolism and the reproduction of the IAV (Figure 11.1):

i. It has been shown that holo-ceruloplasmin (holo-Cp), the main extracellular cuproenzyme that binds 95% of serum copper in humans (Bernevic et al. 2018), is an extracellular factor influencing the early and late stages of IAV reproduction (Tomas et al. 1983; Tomas and Topârceanu 1986).
ii. At the early stages of infection, the IAV causes oxidative stress and, in parallel, suppresses the activity of the main participant in the antioxidant system of the host cell, Cu(II)/Zn(II)-superoxide dismutase (SOD1) (Pyo et al. 2014; Jung et al. 2018).

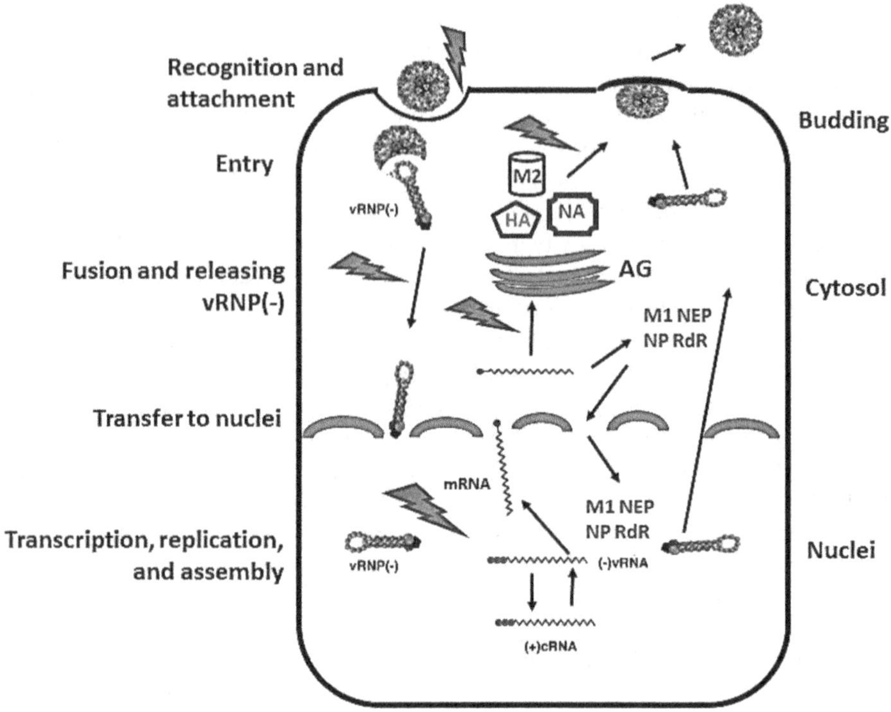

FIGURE 11.1 Schematic representation of the important steps of the IAV life cycle. *At the stage of entry* into the cell, viral HA recognises the terminal α-2,6- or α-2,3-linked N-acetylneuraminic acid residues in glycan of the sialo-containing receptor on the plasma membrane, and attaches to them, facilitating viral entry to the host cell *via* endocytosis. Once inside the acidic endosomal compartment and after the structural rearrangement of HA, the lipid bilayers of the virus and endosome fuse, a pore is formed, through which the viral genome packed in vRNPs is released into the cytosol. Later they are transferred to the nucleus. *Transcription and replication* of the IAV genome occur at the genomic (-)vRNA in the nucleus. v-mRNAs are transported into the cytosol and translated on the membrane-bound or free polyribosomes. Viral proteins that are needed in virus assembly (M1, NP, NEP, RNA-dependent RNA-polymerase subunits) return to the nucleus to be packaged as vRNPs. Later they are exported to the cytosol and delivered to the plasma membrane. Envelope viral proteins (HA, NA, and M2) are synthesised on the rough ER and enter the cellular secretory pathway. Where, HA and NA co-translationally acquire the N-carbohydrate block, which completes their maturation in Golgi. After that, they are transported *via* post-Golgi transport machinery to the PM and meet the vRNPs. Where, budding and release of progeny viruses occurs. Protein M1 assists in the formation of mature virions, using fragments of the lipid bilayer and some proteins of the host cell. The viral NA cleaves sialic acid from cell-surface glycoproteins and glycolipids, and virions are released. "Lightning" marks the points where the reproduction cycle of the influenza virus intersects with the pathways of cellular copper metabolism (Adapted from Puchkova et al. 2021).

iii. Normal prion protein, the conformational changes of which are responsible for the development of prion brain diseases, protects the cells of the lung epithelium from viral infection. Possibly, the phenomenon is based on the ability of a prion protein localised on the plasma membrane of pulmonary epithelial cells to increase the level of intracellular copper required for SOD1, which contributes to a decrease in the concentration of ROS induced by the virus (Sakaguchi and Chida 2018). IAV decreases levels of the copper sensing transcription factor Sp1, responsible for the expression of the CTR1 gene, which encodes a universal copper importer (Ghoshal et al. 2001; Ebrahimie et al. 2015; Moheimani et al. 2018).
iv. Expression products of CTR1 and ATP7A genes are necessary for IAV reproduction in A549 cells (Rupp et al. 2017).
v. Ions of Zn, Mg, and Cu ions are required for the formation of the virion (Chaturvedi and Shrivastava 2005).
vi. IAV RNA-transcriptase activity is inhibited by copper chelators in vitro and in vivo (Oxford and Perrin 1974).
vii. IAV, like the SARS-CoV-2 virus, causes loss of smell (Doty et al. 2014; Moein et al. 2020) and the co-activators of the odorant receptors are copper ions (Haag et al. 2020). In addition, the IAV uses cell-signaling cascades for its reproduction, including copper-dependent signaling pathways (Kardos et al. 2018; Wang et al. 2018).

11.3 THE INTERVENTION OF SILVER NANOPARTICLES IN MAMMALIAN COPPER METABOLISM

Copper is an essential micronutrient; in mammals, it has two biological functions. First, it is a catalytic and structural cofactor of such vitally important enzymes as cytochrome-c-oxidase (CCO), cytosolic and extracellular Cu(II)/Zn(II)-superoxide dismutase's (SOD1 and SOD3 respectively), dopamine-beta-hydroxylase, tyrosinase, monoaminoxidase, peptidylglycine alpha hydroxylating monooxygenase, multi-copper blue (ferr)oxidase, lysyl oxidase, and other (Puchkova et al. 2018). Second, copper is a component of several signaling pathways and a co-activator of key transcription factors (Kudrin 2000; Mufti et al. 2007; Turski et al. 2012; Gaier et al. 2013; Wu et al. 2019).

Paradoxically, the indispensability of copper is accompanied by high toxicity as free copper ions catalyse the Fenton type reactions result in the formation of ROS and cause oxidative stress, which promotes the development of neurodegenerative, oncological, metabolic, and cardiovascular diseases (Pereira et al. 2016; Krzywoszyńska et al. 2020). So, mammalian organisms possess precisely controlled systems, which safely transport Cu from the gastrointestinal tract to sites of cuproenzyme formation (Pierson et al. 2019).

Absorbed dietary copper is transported via the portal vein to the hepatocyte cell membrane (Wapnir 1998). Hepatocytes import copper through mostly CTR1 (high-affinity copper transporter 1) (Figure 11.2). Inside hepatocytes, copper is distributed across several compartments (Chen et al. 2020). From CTR1, copper is transported directly to copper chaperones, which terminally deliver it to apo-SOD1 in cytosol

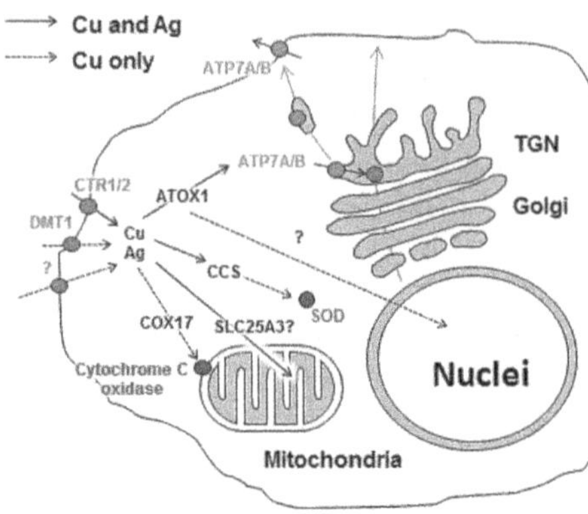

FIGURE 11.2 Scheme of copper and silver distribution in a mammalian cell. Copper is taken up via CTR1, DMT1, or putative transporter. After being imported into the cell the copper is transferred to chaperones ATOX1, CCS, and COX17, which ferry it to ATP7A/B in the Golgi, to SOD1 in the cytosol and to CCO in the mitochondria respectively. In the Golgi ATP7A/B load Cu on newly synthesised cuproenzymes, which traffic along the biosynthetic pathway. A significant increase in intracellular Cu induces the export of ATP7A/B toward post-Golgi compartments and plasma membrane where ATP7A/B drives the excretion of excessive Cu from the cell. Silver uses similar copper-transporting routes. However, several copper-transporting pathways cannot be invaded by silver (Puchkova et al. 2019).

and to form CCO in mitochondria. A fraction of copper is also delivered into the lumen of the Golgi complex by ATP7A/B (ATPase copper transporting alfa polypeptide also known as Menkes disease protein and ATPase copper transporting beta polypeptide also known as Wilson disease protein, respectively), where it metallates apo-Cp. Holo-Cp is secreted to the bloodstream and comprises the major extracellular copper pool.

Besides, copper is trapped by metallothionein-Cu(I), glutathione (Cu(I)/Cu(II)), and COMMD1-Cu(II) system (Bhattacharjee et al. 2017). In this system, copper can change its oxidation state. The system may share copper with Cu-chaperones for cuproenzyme metalation, provide copper for copper-required signaling factors, transcription factors (HIF1α; NF-κB, Sp1, and other), and replenish the mitochondrial pool of non-catalytic copper. The excess of absorbed copper is excreted from the liver into the bile by the ATP7B protein.

There is enough information to distinguish the links between the copper balance supported system, in which Ag(+) nanoparticles are involved (Figure 11.2). So, in the same way with copper ions, the absorbed Ag(+) is transported through the bloodstream to hepatocytes using alpha-2-macroglobulin (Skomorokhova et al. 2020). Ag(+) is captured by the extracellular CTR1 copper-binding sites and transported via its cuprophilic pore to the cytosolic copper-binding domain (Sankova et al. 2017). Copper chaperon superoxide dismutase (CCS) and ATOX1 (copper chaperon ATP7A/B) but not copper chaperon for CCO (COX17), accept Ag(+) from the cytosolic domain of CTR1 (Puchkova et al. 2019). However, CCS does not transfer Ag (+) to the SOD1 active center, thence AgNP treatment in mice does not cause a loss of SOD1 and CCO enzymatic activity (Zatulovskiy et al. 2012; Ilyechova et al. 2014;

Orlov et al. 2016). ATOX1 delivers Ag(+) to ATP7A/B, which, in turn, transfers Ag(+) to the Golgi apparatus lumen, where they are incorporated into the active centers of Cp during its maturation and folding (Orlov et al. 2016; Skomorokhova et al. 2020). The Cp molecule can bind up to four silver atoms. This leads to unfolding and loss of enzymatic activity (Ilyechova et al. 2014). At the same time, AgNP do not affect the expression of the Cp gene at the level of transcription, translation, secretion, and T½ Ag-Cp life span in the bloodstream (Orlov et al. 2016). Ag(+) are formed from AgNP are excreted mainly through bile, which is partially associated with Cp (Skomorokhova et al. 2020).

Biocomputer and experimental analysis of the active centers of cuproenzymes showed that Ag(+) can replace copper in the active centers of the blue multicopper oxidases: Cp, glycosylphosphatidylinositol-anchored Cp, hefestin, cyclopene, laccase (Puchkova et al. 2019; Jimenez-Arroyo et al. 2021). It seems that AgNP affect the activity of only those cuproenzymes that are formed in the Golgi apparatus and their coordination sphere contains cysteine residues. Therefore, the manifestation of the low AgNP concentration can be considered an oxidase activity decrease in the blood serum. At the same time, a decrease in the enzymatic activity of Cp is observed, even after the single injection, and the oxidase activity is restored within hours after the injection of copper ions (Ilyechova et al. 2011). Cp, in addition to ferroxidase activity, is involved in the transport of copper to cells of nonhepatocytic lineages (Ramos et al. 2016). Because in mice treated with AgNP, atomic silver is found in the cells of internal organs and the brain, it can be assumed that the Cu/Ag transport function of Cp does not change. These data suggest that a low concentration of AgNP reduces the level of oxidase Cp only.

11.4 INFLUENCE OF THE MODULATED COPPER STATUS ON VIRUS INFECTION: PROOF OF CONCEPT

In this chapter, the concept of the indirect influence of copper metabolism on the pathogenesis of viral infection is considered. Proof of this concept is the in vivo realisation of nanosilver antiviral effect on mice infected with a representative of the RNA viruses having negative-sense, single-stranded genomes, or influenza A virus. Why exactly AgNP were used as an impact agent? They reduce the enzymatic activity level of Cp, the main extracellular cuproenzyme.

The effect of the treatment with AgNP on the IAV infection was evaluated by two scientific groups (Xiang et al. 2013; Kiseleva et al. 2020a). Methodological approaches used in these studies (influenza viruses, route, and schedule of administration of AgNP and their concentration) differed. Xiang et al. (2013) administered AgNP to mice intranasally at concentrations of 5 mg/kg or 20 mg/kg for a total of three inoculations post intranasal infection with the mouse-adapted subtype A(H3N2) virus. Kiseleva et al. (2020a) treated mice with AgNP intraperitoneally at a dose of 0.2 mg of atomic silver per 1 kg body mass daily according to three schemes: scheme 1 – mice were pretreated four days before infection with highly pathogenic for mice IAV of subtype A(H1N1)pdm09 (Kiseleva et al. 2020b), on the day of infection with IAV (day 0), and during 13 days post-infection; scheme 2 – mice were pretreated four days before infection with A(H1N1)pdm09 IAV and on day 0; scheme 3 – mice were treated on day 0 and during 13 days post-infection. The most effective scheme was found to be scheme 1.

Interestingly, the results obtained by both groups were comparable, and the effectiveness of AgNP against influenza infection was confirmed. In addition, unlike the work of our group (Kiseleva et al. 2020a), Xiang et al. (2013) did not link the possible mechanism of anti-influenza action of AgNP to copper metabolism.

In the study of Kiseleva et al. (2020a), the AgNP concentration for the mice treatment was selected so that it was able to significantly decrease copper status indexes while remaining safe for health. It was shown that in mice, daily intraperitoneal injections of AgNP for nine days did not cause pathological changes in the morphology of the brain, lungs, heart, and internal organs, except testicular tissues. The equivalent silver dose was well below the toxicity threshold for mammalian cells (Małaczewska 2014). In control mice, after three days of infection, the oxidase activity increased approximately two-fold.

These results correspond with the data showing that during influenza infection, the holo-Cp level in the blood increases together with other acute-phase proteins and remains high throughout the infection (van Lenten et al. 2001; Gevkaliuk et al. 2019). This finding demonstrates that the increase in Cp gene expression occurs at the transcriptional level. However, the mechanism of the anti-influenza action of AgNP has not been exactly established. It is known that during influenza infection, the level of pro-inflammatory cytokines, including interleukins 1 and 6, increases (Schmitz et al. 2005; Lauder et al. 2013; Dutta et al. 2019). Both upregulate Cp gene activity at the transcription level (Gitlin et al. 1992; Di Bella et al. 2017). Treatment of mice with AgNP for four and seven days leads to a decrease in the holo-Cp level by factors of ~2 and ~6, respectively.

As mentioned above, the AgNP-mediated decrease in Cp oxidase activity occurred because Ag(+) displaced copper ions in the Cp active centers during its folding in the Golgi complex lumen, and oxidase activity decreased. However, the concentration of Cp-mRNA, its polypeptide level in the Golgi complex, in addition to the secretion rate and half-life of Ag-Cp, did not change (Orlov et al. 2016). Thus, AgNP do not affect any of the levels of Cp gene expression. Therefore, AgNP and the IAV may influence copper status in two different directions: (i) AgNP inhibit the formation of Cp active centers and decrease the holo-Cp level, (ii) the IAV induces the Cp gene at the level of transcription and promotes the increase in the holo-Cp level.

The mice treated with AgNP by the most effective scheme 1 demonstrated statistically significant lower mortality; in mice inoculated intranasally with 1 LD_{50} and 10 LD_{50} of A(H1N1)pdm09 IAV, the protection index reached 60–70% at the end of the experiment (Figure 11.3, left panel), and mean lifespan was prolonged. In addition, the treatment of the mice with AgNP resulted in the normalisation of the body weight dynamics. Interestingly, despite the amelioration of the infection, AgNP treatment did not influence IAV replication (Figure 11.3, right panel). In mice treated by AgNP, the progression of infection occurred against the background of low oxidase activity. On day 3 post-infection, the oxidase activity in treated with AgNP mice was several times lower than that in untreated animals. In the infected Ag-mice, the oxidase activity was higher than in the uninfected mice.

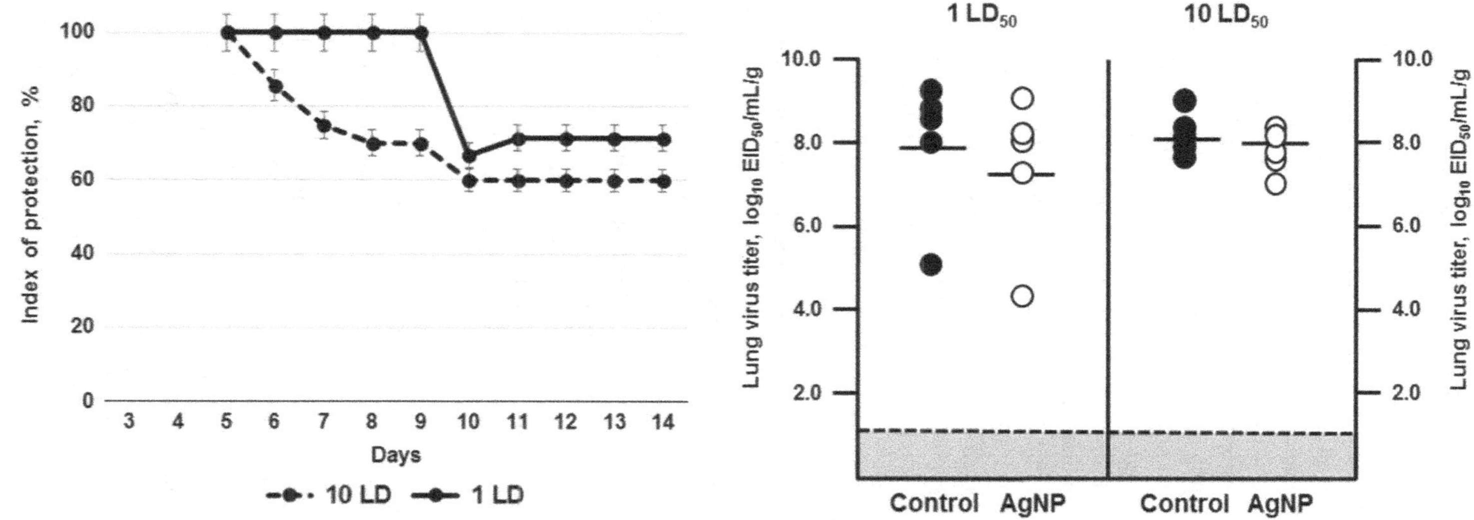

FIGURE 11.3 Protection of mice from lethal primary viral pneumonia by AgNP. Mice were inoculated intranasally with 1 LD_{50} and 10 LD_{50} of A(H1N1)pdm09 influenza virus and were monitored daily for 14 days. Left panel – the protection index (ratio of mortality in the control group over mortality in the experimental group). Right panel – influenza virus replication in the lungs of mice on day 3 post-infection as measured by titration in developing chicken eggs (Adapted from Kiseleva et al. 2020a, b).

The above data suggest that the effectiveness of AgNP protection correlates with lower elevation of oxidase activity induced by the IAV. This raises the key question of how AgNP inhibit the virus-mediated elevation of oxidase activity. As of today, little is known to answer this question. However, several indirect pieces of evidence exist: influenza infection causes an increase in the level of neutrophils in the lungs (Nakajima et al. 2013), the apoptosis of which leads to an increase in the concentration of pro-inflammatory interleukins (El Kebir and Filep 2010; Dutta et al. 2019); and holo-Cp induces apoptosis of neutrophils in lungs (Nakajima et al. 2013).

These data allow suggesting that, in Ag-mice, the holo-Cp level decreases while Cp protein concentration does not change (Orlov et al. 2016), and the ratio of holo-Cp/apo-Cp is strongly biased toward apo-Cp, which cannot induce apoptosis of neutrophils (Golenkina et al. 2018). The high level of Ag-Cp may prevent both neutrophil apoptosis and induction of the activity of its gene. This thinking is supported by data showing that AgNP-treated IAV-infected mice have decreased levels of the pro-inflammatory cytokines (i.e. IL-1 alfa, IL-6, TNF-alfa) and pro-inflammatory chemokines (i.e. CCL2, CL3, CCL5) (Morris et al. 2019). Moreover, anti-inflammatory M2 macrophage protected mice from the lethal effect of IAV, possibly by preventing the development of a cytokine storm (Chida et al. 2020).

11.5 CONCLUSIONS

All antivirals can be divided into two main target-dependent categories: directly acting antivirals (DAAs) and indirectly acting antivirals (IAAs). DAAs act on the different viral proteins and/or stages of the viral life cycle. IAAs act as symptomatic and etiotropic drugs (for instance, vitamins, fever reducers, pain relievers, etc.) or through a change of the host cell metabolism that was modified under the influence of a viral infection. When using DAAs, the correlation between the effectiveness of inhibition of the IAV replication in vivo models and the level of reduction of lethality is pronounced (Stephen et al. 1975; Belser et al. 2015).

In contrast, when the mortality does not depend on the level of IAV replication in vivo, the IAAs are successfully used. In studies devoted to the effect of nanoparticles on the outcome of influenza infection, there is no direct evidence of their effect on the virus replication (Mehrbod et al. 2008; Xiang et al. 2011; Ghaffari et al. 2019). Possibly, the inhibition of the replication of the IAV in the lower respiratory tract of lab animals treated with AgNP is not the main mechanism of their anti-influenza action.

All licensed anti-influenza compounds, such as rimantadine, oseltamivir, zanamivir, peramivir, and baloxavir, belong to the group of DAAs (Toots and Plemper 2020). However, acquired resistance of viruses to DAA drugs is a substantial challenge in fields such as clinical practice and virology. From this perspective, screening of IAAs that are independent of the genetic variability of the target virus is of importance. The inhibitory effects of AgNP on the IAV in vitro and in vivo experiments (Mehrbod et al. 2008; Xiang et al. 2011; Xiang et al. 2013; Kiseleva et al. 2020a) allows concluding that AgNP as an IAA can be a viable and perspective tool for the development of a formulation that efficiently combats any influenza infection.

The positive properties of the AgNP as IAAs agent should be emphasised. The reducing effect of AgNP on the Cp oxidase activity develops quickly and can also be quickly canceled. In terms of development time, this effect is comparable to the influenza disease progression.

ACKNOWLEDGMENTS

The authors gratefully acknowledge Ekaterina Skomorokhova, Andrei Rekstin, Ekaterina Bazhenova, Daria Magazenkova, and Iurii Orlov for their expert technical support to fulfill the biochemical and virological part of this research and helpful discussions. This work was supported by the Russian Science Foundation [grant number 20-74-10087].

REFERENCES

Báez, D.F., Gallardo-Toledo, E., Oyarzún, M.P., Araya, E. and Kogan, M.J. 2021. The influence of size and chemical composition of silver and gold nanoparticles on *in vivo* toxicity with potential applications to central nervous system diseases. *International Journal of Nanomedicine* 16:2187–2201. 10.2147/ijn.S260375.

Belser, J.A., Maines, T.R., Creager, H.M., Katz, J.M. and Tumpey, T.M. 2015. Oseltamivir inhibits influenza virus replication and transmission following ocular-only aerosol inoculation of ferrets. *Virology* 484:305–312. 10.1016/j.virol.2015.06.020.

Bernevic, B., El-Khatib, A.H., Jakubowski, N. and Weller, M.G. 2018. Online immunocapture ICP-MS for the determination of the metalloprotein ceruloplasmin in human serum. *BMC Research Notes* 11(1):213. 10.1186/s13104-018-3324-7.

Bhattacharjee, A., Chakraborty, K. and Shukla, A. 2017. Cellular copper homeostasis: Current concepts on its interplay with glutathione homeostasis and its implication in physiology and human diseases. *Metallomics* 9(10):1376–1388. 10.1039/c7mt00066a.

Chaturvedi, U.C. and Shrivastava, R. 2005. Interaction of viral proteins with metal ions: Role in maintaining the structure and functions of viruses. *FEMS Immunology and Medical Microbiology* 43(2):105–114. 10.1016/j.femsim.2004.11.004.

Chen, J., Jiang, Y., Shi, H., Peng, Y., Fan, X. and Li, C. 2020. The molecular mechanisms of copper metabolism and its roles in human diseases. *Pflügers Archiv: European Journal of Physiology* 472(10):1415–1429. 10.1007/s00424-020-02412-2.

Chida, J., Hara, H., Uchiyama, K., et al. 2020. Prion protein signaling induces M2 macrophage polarization and protects from lethal influenza infection in mice. *PLoS Pathogens* 16(8):e1008823. 10.1371/journal.ppat.1008823.

Dai, J.P., Wang, Q.W., Su, Y., et al. 2018. Oxymatrine inhibits influenza A virus replication and inflammation via TLR4, p38 MAPK and NF-κB Pathways. *International Journal of Molecular Sciences* 19(4). 10.3390/ijms19040965.

Di Bella, L.M., Alampi, R., Biundo, F., Toscano, G. and Felice, M.R. 2017. Copper chelation and interleukin-6 proinflammatory cytokine effects on expression of different proteins involved in iron metabolism in HepG2 cell line. *BMC Biochemistry* 18(1):1. 10.1186/s12858-017-0076-2.

Doty, R.L., Berman, A.H., Izhar, M., et al. 2014. Influenza vaccinations and chemosensory function. *American Journal of Rhinology and Allergy* 28(1):50–53. 10.2500/ajra.2014.28.3987.

Dutta, M., Dutta, P., Medhi, S., Borkakoty, B. andBiswas, D.2019. Immune response during influenza virus infection among the population of Assam, Northeast India. *Indian Journal of Medical Microbiology* 37(4):549–556. 10.4103/ijmm.IJMM_19_211.

Ebrahimie, E., Nurollah, Z., Ebrahimi, M., Hemmatzadeh, F. and Ignjatovic, J. 2015. Unique ability of pandemic influenza to downregulate the genes involved in neuronal disorders. *Molecular Biology Reports* 42(9):1377–1390. 10.1007/s11033-015-3916-4.

El Kebir, D. and Filep, J.G. 2010. Role of neutrophil apoptosis in the resolution of inflammation. *The Scientific World Journal* 10:1731–1748. 10.1100/tsw.2010.169.

Fatima, M., Zaidi, N.U., Amraiz, D. and Afzal, F. 2016. *In vitro* antiviral activity of *Cinnamomum cassia* and its nanoparticles against H7N3 influenza A virus. *Journal of Microbiology and Biotechnology* 26(1):151–159. 10.4014/jmb.1508.08024.

Forterre, P. 2016. To be or not to be alive: How recent discoveries challenge the traditional definitions of viruses and life. *Studies in History and Philosophy of Biological and Biomedical Sciences* 59:100–108. 10.1016/j.shpsc.2016.02.013.

Gaier, E.D., Eipper, B.A. and Mains, R.E. 2013. Copper signaling in the mammalian nervous system: Synaptic effects. *Journal of Neuroscience Research* 91(1):2–19. 10.1002/jnr.23143.

Galdiero, S., Falanga, A., Vitiello, M., Cantisani, M., Marra, V. and Galdiero, M. 2011. Silver nanoparticles as potential antiviral agents. *Molecules* 16(10):8894–8918. 10.3390/molecules16108894.

Gevkaliuk, N.O., Sydliaruk, N.I., Posolenyk, L.Y., Vydoinyk, O.Y. and Kuchyrka, L.I. 2019. The state of oxidative homeostasis in children with influenza stomatitis. *Wiadomości Lekarskie* 72(3):405–408.

Ghaffari, H., Tavakoli, A., Moradi, A., et al. 2019. Inhibition of H1N1 influenza virus infection by zinc oxide nanoparticles: Another emerging application of nanomedicine. *Journal of Biomedical Science* 26(1):70. 10.1186/s12929-019-0563-4.

Ghoshal, K., Majumder, S., Zhu, Q., et al. 2001. Influenza virus infection induces metallothionein gene expression in the mouse liver and lung by overlapping but distinct molecular mechanisms. *Molecular and Cellular Biology* 21(24):8301–8317. 10.1128/mcb.21.24.8301-8317.2001.

Gitlin, J.D., Schroeder, J.J., Lee-Ambrose, L.M. and Cousins, R.J. 1992. Mechanisms of caeruloplasmin biosynthesis in normal and copper-deficient rats. *Biochemical Journal* 282(Pt 3):835–839. 10.1042/bj2820835.

Golenkina, E.A., Viryasova, G.M., Galkina, S.I., Gaponova, T.V., Sud'ina, G.F. and Sokolov, A.V. 2018. Fine regulation of neutrophil oxidative status and apoptosis by ceruloplasmin and its derivatives. *Cells* 7(1):8. 10.3390/cells7010008.

Gurunathan, S., Qasim, M., Choi, Y., et al. 2020. Antiviral potential of nanoparticles - can nanoparticles fight against coronaviruses? *Nanomaterials (Basel)* 10(9):1645. 10.3390/nano10091645.

Haag, F., Ahmed, L., Reiss, K., Block, E., Batista, V.S. and Krautwurst, D. 2020. Copper-mediated thiol potentiation and mutagenesis-guided modeling suggest a highly conserved copper-binding motif in human OR2M3. *Cellular and Molecular Life Sciences* 77(11):2157–2179. 10.1007/s00018-019-03279-y.

Hadrup, N., Sharma, A.K., Loeschner, K. and Jacobsen, N.R. 2020. Pulmonary toxicity of silver vapours, nanoparticles and fine dusts: A review. *Regulatory Toxicology and Pharmacology* 115:104690. 10.1016/j.yrtph.2020.104690.

Ilyechova, E., Skvortsov, A., Zatulovsky, E., et al. 2011. Experimental switching of copper status in laboratory rodents. *Journal of Trace Elements in Medicine and Biology* 25(1):27–35. 10.1016/j.jtemb.2010.08.002.

Ilyechova, E.Y., Saveliev, A.N., Skvortsov, A.N., et al. 2014. The effects of silver ions on copper metabolism in rats. *Metallomics* 6(10):1970–1987. 10.1039/c4mt00107a.

Jimenez-Arroyo, N., Cardona-Felix, C.S., Rojas-Trejo, S.P. and Rudino-Pinera, E. 2021. *Structural Function of the Methionine-rich beta-hairpin in the Laccase for T thermophilus*. Protein Data Bank. https://www.rcsb.org/structure/5AFA.

Jung, K.I., Pyo, C.W. and Choi, S.Y. 2018. Influenza A virus-induced autophagy contributes to enhancement of virus infectivity by SOD1 downregulation in alveolar epithelial cells. *Biochemical and Biophysical Research Communications* 498(4):960–966. 10.1016/j.bbrc.2018.03.089.

Kardos, J., Heja, L., Simon, A., Jablonkai, I., Kovacs, R. and Jemnitz, K. 2018. Copper signalling: Causes and consequences. *Cell Communication and Signaling* 16(1):71. 10.1186/s12964-018-0277-3.

Kieny, M.P., Costa, A., Hombach, J., et al. 2006. A global pandemic influenza vaccine action plan. *Vaccine* 24(40-41):6367–6370. 10.1016/j.vaccine.2006.07.021.

Kim, J.S., Kuk, E., Yu, K.N., et al. 2007. Antimicrobial effects of silver nanoparticles. *Nanomedicine* 3(1):95–101. 10.1016/j.nano.2006.12.001.

Kiseleva, I.V., Farroukh, M.A., Skomorokhova, E.A., et al. 2020a. Anti-influenza effect of nanosilver in a mouse model. *Vaccines (Basel)* 8(4):679. 10.3390/vaccines8040679.

Kiseleva, I., Rekstin, A. and Al Farroukh, M. 2020b. Non-mouse-adapted H1N1pdm09 virus as a model for influenza research.*Viruses* 12(6):590. 10.3390/v12060590.

Krzywoszyńska, K., Witkowska, D., Swiatek-Kozlowska, J., Szebesczyk, A. and Kozłowski, H. 2020. General aspects of metal ions as signaling agents in health and disease. *Biomolecules* 10(10):1417. 10.3390/biom10101417.

Kudrin, A.V. 2000. Trace elements in regulation of NF-kappaB activity. *Journal of Trace Elements in Medicine and Biology* 14(3):129–142. 10.1016/s0946-672x(00)80001-2.

Kuss-Duerkop, S.K., Wang, J., Mena, I., et al. 2017. Influenza virus differentially activates mTORC1 and mTORC2 signaling to maximize late stage replication. *PLoS Pathogens* 13(9):e1006635. 10.1371/journal.ppat.1006635.

Lauder, S.N., Jones, E., Smart, K., et al. 2013. Interleukin-6 limits influenza-induced inflammation and protects against fatal lung pathology. *European Journal of Immunology* 43(10):2613–2625. 10.1002/eji.201243018.

Liao, C., Li, Y. and Tjong, S.C. 2019. Bactericidal and cytotoxic properties of silver nanoparticles. *International Journal of Molecular Sciences* 20(2). 10.3390/ijms20020449.

Loeschner, K., Hadrup, N., Qvortrup, K., et al. 2011. Distribution of silver in rats following 28 days of repeated oral exposure to silver nanoparticles or silver acetate. *Particle and Fibre Toxicology* 8:18. 10.1186/1743-8977-8-18.

Ludwig, S., Planz, O., Pleschka, S. and Wolff, T. 2003. Influenza-virus-induced signaling cascades: Targets for antiviral therapy? *Trends in Molecular Medicine* 9(2):46–52. 10.1016/s1471-4914(02)00010-2.

Małaczewska, J. 2014. The *in vitro* effect of commercially available noble metal nanocolloids on the splenocyte proliferative response and cytokine production in mice. *Polish Journal of Veterinary Sciences* 17(1):37–45. 10.2478/pjvs-2014-0005.

Mehrbod, P., Motamed N., Tabatabaian, M., et al. 2008. *In vitro* antiviral effect of "nanosilver" on influenza virus. *DARU Journal of Pharmaceutical Sciences* 17(2): 88–93.

Moein, S.T., Hashemian, S.M., Mansourafshar, B., Khorram-Tousi, A., Tabarsi, P. and Doty, R.L. 2020. Smell dysfunction: A biomarker for COVID-19. *International Forum of Allergy and Rhinology* 10(8):944–950. 10.1002/alr.22587.

Moheimani, F., Koops, J., Williams, T., et al. 2018. Influenza A virus infection dysregulates the expression of microRNA-22 and its targets; CD147 and HDAC4, in epithelium of asthmatics. *Respiratory Research* 19(1):145. 10.1186/s12931-018-0851-7.

Morones, J.R., Elechiguerra, J.L., Camacho, A., et al. 2005. The bactericidal effect of silver nanoparticles. *Nanotechnology* 16(10):2346–2353. 10.1088/0957-4484/16/10/059.

Morris, D., Ansar, M., Speshock, J., et al. 2019. Antiviral and immunomodulatory activity of silver nanoparticles in experimental RSV infection. *Viruses* 11(8):732. 10.3390/v11080732.

Mufti, A.R., Burstein, E. and Duckett, C.S. 2007. XIAP: Cell death regulation meets copper homeostasis. *Archives of Biochemistry and Biophysics* 463(2):168–174. 10.1016/j.abb.2007.01.033.

Munger, M.A., Radwanski, P., Hadlock, G.C., et al. 2014. *In vivo* human time-exposure study of orally dosed commercial silver nanoparticles. *Nanomedicine* 10(1):1–9. 10.1016/j.nano.2013.06.010.

Nakajima, N., Van Tin, N., Sato, Y., et al. 2013. Pathological study of archival lung tissues from five fatal cases of avian H5N1 influenza in Vietnam. *Modern Pathology* 26(3):357–369. 10.1038/modpathol.2012.193.

Natsuki, J., Natsuki, T. and Hashimoto, Y. 2015. A review of silver nanoparticles: Synthesis methods, properties and applications. *International Journal of Materials Science and Applications* 4:325. 10.11648/j.ijmsa.20150405.17.

Orlov, I.A., Sankova, T.P., Babich, P.S., et al. 2016. New silver nanoparticles induce apoptosis-like process in *E. coli* and interfere with mammalian copper metabolism. *International Journal of Nanomedicine* 11:6561–6574. 10.2147/ijn.S117745.

Oxford, J.S. and Perrin, D.D. 1974. Inhibition of the particle-associated RNA-dependent RNA polymerase activity of influenza viruses by chelating agents. *Journal of General Virology* 23(1):59–71. 10.1099/0022-1317-23-1-59.

Paladini, F. and Pollini, M. 2019. Antimicrobial silver nanoparticles for wound healing application: Progress and future trends. *Materials (Basel)* 12(16):2540. 10.3390/ma12162540.

Pareek, V., Gupta, R. and Panwar, J. 2018. Do physico-chemical properties of silver nanoparticles decide their interaction with biological media and bactericidal action? A review. *Materials Science and Engineering C* 90:739–749. 10.1016/j.msec.2018.04.093.

Park, S., Ko, Y.S., Lee, S.J., Lee, C., Woo, K. and Ko, G. 2018. Inactivation of influenza A virus via exposure to silver nanoparticle-decorated silica hybrid composites. *Environmental Science and Pollution Research* 25(27):27021–27030. 10.1007/s11356-018-2620-z.

Pereira, T.C., Campos, M.M. and Bogo, M.R. 2016. Copper toxicology, oxidative stress and inflammation using zebrafish as experimental model. *Journal of Applied Toxicology* 36(7):876–885. 10.1002/jat.3303.

Pierson, H., Yang, H. and Lutsenko, S. 2019. Copper transport and disease: What can we learn from organoids? *Annual Review of Nutrition* 39:75–94. 10.1146/annurev-nutr-082018-124242.

Puchkova, L.V., Babich, P.S., Zatulovskaia, Y.A., Ilyechova, E.Y. and Di Sole, F. 2018. Copper metabolism of newborns is adapted to milk ceruloplasmin as a nutritive source of copper: Overview of the current data. *Nutrients* 10(11):1591. 10.3390/nu10111591.

Puchkova, L.V., Broggini, M., Polishchuk, E.V., Ilyechova, E.Y. and Polishchuk, R.S. 2019. Silver ions as a tool for understanding different aspects of copper metabolism. *Nutrients* 11(6):1364. 10.3390/nu11061364.

Puchkova, L.V., Kiseleva, I.V., Polishchuk, E.V., Broggini, M. and Ilyechova, E.Y. 2021. The cross-roads between host copper metabolism and influenza infection. *International Journal of Molecular Science* 22(11):5498. 10.3390/ijms22115498.

Pyo, C.W., Shin, N., Jung, K.I., Choi, J.H. and Choi, S.Y. 2014. Alteration of copper-zinc superoxide dismutase 1 expression by influenza A virus is correlated with virus replication. *Biochemical and Biophysical Research Communications* 450(1):711–716. 10.1016/j.bbrc.2014.06.037.

Ramos, D., Mar, D., Ishida, M., et al. 2016. Mechanism of copper uptake from blood plasma ceruloplasmin by mammalian cells. *PLoS One* 11(3):e0149516. 10.1371/journal.pone.0149516.

Ren, L., Zhang, W., Han, P., et al. 2019. Influenza A virus (H1N1) triggers a hypoxic response by stabilizing hypoxia-inducible factor-1α via inhibition of proteasome. *Virology* 530:51–58. 10.1016/j.virol.2019.02.010.

Rupp, J.C., Locatelli, M., Grieser, A., et al. 2017. Host cell copper transporters CTR1 and ATP7A are important for influenza A virus replication. *Virology Journal* 14 (1):11. 10.1186/s12985-016-0671-7.

Sakaguchi, S. and Chida, J. 2018. Roles of prion protein in virus infections. *DNA and Cell Biology* 37(10):808–811. 10.1089/dna.2018.4402.

Sankova, T.P., Orlov, I.A., Saveliev, A.N., et al. 2017. The extracellular domain of human high affinity copper transporter (hNdCTR1), synthesized by *E. coli* cells, chelates silver and copper ions *in vivo*. *Biomolecules* 7(4):78. 10.3390/biom7040078.

Sankova, T.P., Sosnin, I.M., Karpenko, M.N., et al. 2015. On the biological activity of silver nanoparticles. *Materials Physics and Mechanics* 24:289–296.

Sardi, J.C.O., Scorzoni, L., Bernardi, T., Fusco-Almeida, A.M. and Mendes Giannini, M.J.S. 2013. Candida species: Current epidemiology, pathogenicity, biofilm formation, natural antifungal products and new therapeutic options. *Journal of Medical Microbiology* 62:10–24. 10.1099/jmm.0.045054-0.

Schloer, S., Hübel, N., Masemann, D., et al. 2019. The annexin A1/FPR2 signaling axis expands alveolar macrophages, limits viral replication, and attenuates pathogenesis in the murine influenza A virus infection model. *The FASEB Journal* 33(11): 12188–12199. 10.1096/fj.201901265R.

Schmitz, N., Kurrer, M., Bachmann, M.F. and Kopf, M. 2005. Interleukin-1 is responsible for acute lung immunopathology but increases survival of respiratory influenza virus infection. *Journal of Virology* 79(10):6441–6448. 10.1128/jvi.79.10.6441-6448.2005 .

Sharma, M. 2019. Transdermal and intravenous nano drug delivery systems: Present and future. In *Applications of Targeted Nano Drugs and Delivery Systems*, eds. S.S. Mohapatra, S. Ranjan, N. Dasgupta, R.K. Mishra, and S. Thomas. 499–550. Netherlands: Elsevier. 10.1016/B978-0-12-814029-1.00018-1.

Shie, J.J. and Fang, J.M. 2019. Development of effective anti-influenza drugs: Congeners and conjugates - a review. *Journal of Biomedical Science* 26(1):84. 10.1186/s12929-019-0567-0.

Siddiqi, K.S., Husen, A. and Rao, R.A.K. 2018. A review on biosynthesis of silver nanoparticles and their biocidal properties. *Journal of Nanobiotechnology* 16(1):14. 10.1186/s12951-018-0334-5.

Skomorokhova, E.A., Sankova, T.P., Orlov, I.A., et al. 2020. Size-dependent bioactivity of silver nanoparticles: Antibacterial properties, influence on copper status in mice, and whole-body turnover. *Nanotechnology, Science and Applications* 13:137–157. 10.2147/nsa.S287658.

Stephen, E.L., Dominik, J.W., Moe, J.B., Spertzel, R.O. and Walker, J.S. 1975. Treatment of influenza infection of mice by using rimantadine hydrochlorides by the aerosol and intraperitoneal routes. *Antimicrobial Agents and Chemotherapy* 8(2):154–158. 10.1128/aac.8.2.154.

Tambosi, R., Liotenberg, S., Bourbon, M.L., et al. 2018. Silver and copper acute effects on membrane proteins and impact on photosynthetic and respiratory complexes in bacteria. *mBio* 9(6):e01535-18. 10.1128/mBio.01535-18.

Tomas, E., Samuel, I., Popescu, A. and Cajal, N. 1983. Comparative study of some characteristics of influenza virus A/PR8/34 (H1N1) cultivated on chorioallantoic membrane fragments in the presence of ceruloplasmin or of parainfluenza type I (Sendai) virus. *Virologie* 34(4):295–301.

Tomas, E. and Topârceanu, F. 1986. Considerations about the possible function of ceruloplasmin in influenza and parainfluenza virus infections. *Virologie* 37(4):279–287.

Toots, M. and Plemper, R.K. 2020. Next-generation direct-acting influenza therapeutics. *Translational Research* 220:33–42. 10.1016/j.trsl.2020.01.005.

Turpin, E., Luke, K., Jones, J., Tumpey, T., Konan, K. and Schultz-Cherry, S. 2005. Influenza virus infection increases p53 activity: Role of p53 in cell death and viral replication. *Journal of Virology* 79(14):8802–8811. 10.1128/jvi.79.14.8802-8811.2005.

Turski, M.L., Brady, D.C., Kim, H.J., et al. 2012. A novel role for copper in Ras/mitogen-activated protein kinase signaling. *Molecular and Cellular Biology* 32(7):1284–1295. 10.1128/mcb.05722-11.

van der Vries, E., Schutten, M., Fraaij, P., Boucher, C. and Osterhaus, A. 2013. Influenza virus resistance to antiviral therapy. *Advances in Pharmacology* 67:217–246. 10.1016/b978-0-12-405880-4.00006-8.

van Lenten, B.J., Wagner, A.C., Nayak, D.P., Hama, S., Navab, M. and Fogelman, A.M. 2001. High-density lipoprotein loses its anti-inflammatory properties during acute influenza A infection. *Circulation* 103(18):2283–2288. 10.1161/01.cir.103.18.2283.

Wang, K., Langevin, S., O'Hern, C.S., et al. 2016. Anomaly detection in host signaling pathways for the early prognosis of acute infection. *PLoS One* 11(8):e0160919. 10.1371/journal.pone.0160919.

Wang, L., Jiang, H., Shen, S.M., Wen, C.X., Xing, Z. and Shi, Y. 2018. Inhibition of autophagy and chemokine induction by sphingosine 1-phosphate receptor 1 through NF-κB signaling in human pulmonary endothelial cells infected with influenza A viruses. *PLoS One* 13(10):e0205344. 10.1371/journal.pone.0205344.

Wapnir, R.A. 1998. Copper absorption and bioavailability. *American Journal of Clinical Nutrition* 67(5S):1054S–1060S. 10.1093/ajcn/67.5.1054S.

Wu, Z., Zhang, W. and Kang, Y.J. 2019. Copper affects the binding of HIF-1α to the critical motifs of its target genes. *Metallomics* 11(2):429–438. 10.1039/c8mt00280k.

Xiang, D.X., Chen, Q., Pang, L. and Zheng, C.L. 2011. Inhibitory effects of silver nanoparticles on H1N1 influenza A virus *in vitro*. *Journal of Virological Methods* 178(1–2): 137–142. 10.1016/j.jviromet.2011.09.003.

Xiang, D., Zheng, Y., Duan, W., et al. 2013. Inhibition of A/Human/Hubei/3/2005 (H3N2) influenza virus infection by silver nanoparticles *in vitro* and *in vivo*. *International Journal of Nanomedicine* 8:4103–4113. 10.2147/ijn.S53622.

You, J., Zhang, Y. and Hu, Z. 2011. Bacteria and bacteriophage inactivation by silver and zinc oxide nanoparticles. *Colloids and Surfaces B: Biointerfaces* 85(2):161–167. 10.1016/j.colsurfb.2011.02.023.

Zatulovskiy, E.A., Skvortsov, A.N., Rusconi, P., et al. 2012. Serum depletion of holo-ceruloplasmin induced by silver ions *in vivo* reduces uptake of cisplatin. *Journal of Inorganic Biochemistry* 116:88–96. 10.1016/j.jinorgbio.2012.07.003.

Zhang, X.F., Liu, Z.G., Shen, W. and Gurunathan, S. 2016. Silver nanoparticles: Synthesis, characterization, properties, applications, and therapeutic approaches. *International Journal of Molecular Sciences* 17(9). 10.3390/ijms17091534.

Zivic, F., Grujović, N., Mitrovic, S., Ahad, I. and Brabazon, D. 2018. Characteristics and applications of silver nanoparticles. In *Commercialization of Nanotechnologies – A Case Study Approach*, eds. D. Brabazon, E. Pellicer, F. Zivic, and J. Sort, 227–273. Switzerland: Springer International Publishing. 10.1007/978-3-319-56979-6_10.

12 Antiviral Drugs as Tools for Nanomedicine

Kavita Pal and Mayuri Iyer

CONTENTS

12.1 INTRODUCTION

Cancer is an abnormal condition in which a few cells behave against the normal physiological rules and divide and grow uncontrollably. It remains a leading cause of death worldwide. Infectious agents, such as viruses, bacteria, and parasites, behave as etiological agents in ~20% of human cancers, out of which nearly 12% of cancers worldwide are caused by oncovirus. The conventional treatments individually are not effective, and therefore, combinatorial therapies were introduced to control cancer; however the side effects still remain associated with them due to their nonspecificity. Use of a nano-carrier system proves to be effective in cancer treatment due to its drug-entrapment efficiency, target specificity, and long circulation half-life. Various nano drug-delivery platforms and current research are thoroughly described further. This chapter emphasises the advancement in use of nano-delivery platforms for toxic antiviral drugs against viruses which are known to cause cancer directly or indirectly.

12.1.1 Cancer Epidemiology: Global and Indian Scenario

Cancer is a multifactorial, precisely tuned disorder, which requires the active participation of the cells, dysregulated levels of various regulatory proteins, and

DOI: 10.1201/9781003136644-15

micro-environment (https://www.ncbi.nlm.nih.gov/books/NBK54025/; Pavlova and Thompson 2016). It remains a leading cause of death worldwide, accounting for approximately 9.5 million deaths in 2018, as reported by the International Agency for Research on Cancer (IARC). Similarly, there is an increase, with 12.7 million new cases per year. By 2040, the number of new cancer cases per year is expected to rise to 29.5 million. World Health Organisation estimates cancer to be the first or second prominent cause of death in age > 70 years in 112 countries and 3rd or 4th in 23 countries out of a total of 183 countries. The rising cases of cancer and related deaths partially reflect a significant decrease in mortality due to strokes and heart diseases. In general, cancer incidence and mortality is growing rapidly all over the world, considering the growth of population and age, and associated socioeconomic development (Sung et al. 2021). Figure 12.1 shows the worldwide distribution of Cases and Deaths for 36 Cancers in 185 Countries in both sexes and each individual sex, as well for the Top 10 Most Common Cancers in 2020. Cancer is the principal cause of death equally in developed and underdeveloped countries, but it is more prevalent

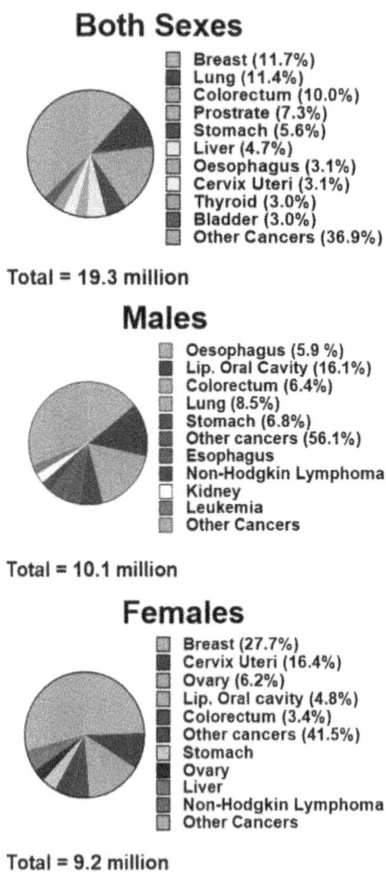

FIGURE 12.1 Worldwide distribution of Cases and Deaths for 36 Cancers in 185 Countries for the Top 10 Most Common Cancers in 2020 for (A) Both Sexes, (B) Men, and (C) Women. For each sex, the area of the pie chart reflects the proportion of the total number of cases or deaths; nonmelanoma skin cancers (excluding basal cell carcinoma for incidence) are included in the "other" category. Source: Redrawn from GLOBOCAN 2020.

in middle-income countries, probably due to prevailing poor socioeconomic conditions (https://www.ncbi.nlm.nih.gov/books/NBK54028). The geographic differences in the prevalence of cancer can be explained by many factors, such as early diagnosis, age, occurrence of risk factors, accessibility and reliability on screening tests, and availability of effective treatment plan (https://www.cancer.gov/about-cancer/understanding/statistics).

The National Cancer Registry Programme Report, 2020, released by Indian Council of Medical Research (ICMR) estimates 13.9 lakh cases of cancer in India in 2020 and predicted the rise to 15.7 lakh by 2025. The estimates are based on the data retrieved from all over India during 2012 and 2016, from 28 Population-Based Cancer Registries (PBCRs), which measure the cancer incidence for a defined population, and 58 Hospital-Based Cancer Registries (HBCRs), which provided data on the clinical presentation, diagnosis, and cancer care. The prediction highlights increase in 27.1% of India's cancer burden in 2020 related to tobacco, followed by gastrointestinal cancer (19.7%) and cancer of the cervix uteri (5.4%). Figure 12.2 shows the distribution of cases and deaths for the most common cancers in 2018 in India for (A) Both Sexes, (B) Men, and (C) Women. For each sex, the area of the pie chart reflects the proportion of the total number of cases or deaths; nonmelanoma skin cancers (excluding basal cell carcinoma for incidence) are included in the "other" category (http://cancerindia.org.in/globocan-2018-india-factsheet/).

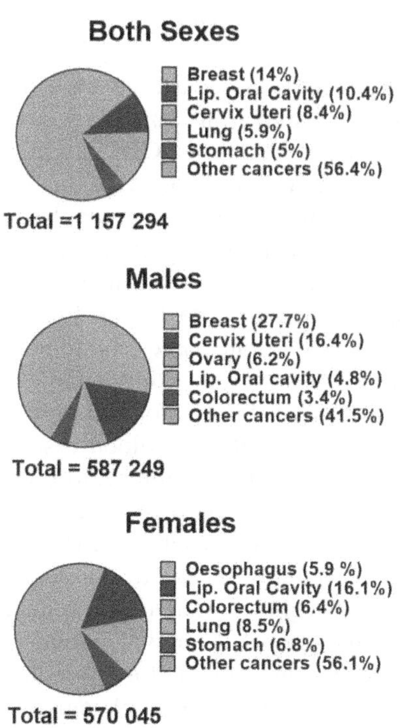

FIGURE 12.2 Distribution of Cases and Deaths for the Most Common Cancers in 2018 in India for (A) Both Sexes, (B) Men, and (C) Women. For each sex, the area of the pie chart reflects the proportion of the total number of cases or deaths; nonmelanoma skin cancers (excluding basal cell carcinoma for incidence) are included in the "other" category. Source: Redrawn from GLOBOCAN 2018: India Factsheet).

12.1.2 Understanding the Cancer

Cancer is an abnormal condition in which a few cells behave against the normal physiological rules and divide and grow uncontrollably. The cancerous cells bypass normal cell cycle various signals. They develop a high degree of self-sufficiency that makes them able to proliferate and transform (Fouad and Aanei 2017), leading to a fatal condition, wherein cells spread to distant tissues/organs called metastasis.

In normal cells, during mitosis, cells divide and grow in an interdependent manner, depending on the availability of various stimulants or growth factors and ambient conditions. As the supply is on or off, normal cells switch to proliferation or stop dividing; however, such switch mechanisms are disrupted in transformed cells, due to which they become independent of any signals (Hanahan and Weinberg 2000, 2011). Such transformed cells also lose contact inhibition that is seen in normal cells, where cells stop dividing after attaining a specific threshold. On the contrary, cancer cells do not have any contact inhibition, leading to an uncontrolled mass of cells (Pavel et al. 2018). The life of a normal cell is limited and well-programmed; it dies when function is over, by apoptosis, and is replaced by a new cell. This is due to the limited efficiency of DNA replication, as, at each time, there is a shortening of telomeres. However, cancer cells show the very high activity of telomerase, which keeps restoring the worn-out ends of the telomere, thus granting the cell to proliferate unlimitedly (Trybek et al. 2020). Witsch et al. (2011) has described the stepwise process of cancer development and the role of various growth factors in each step.

Figure 12.3 depicts the various steps of cancer development (Witsch et al. 2011). In brief, cancer is instigated by a somatic mutation conferring considerable survival and growth advantages to the initiated cell (1). GFs like EGF and IGF1 help the resulting expansion of clones having mutations (2), this results in intra-luminal lesions (3), such as carcinoma at the site or neoplasia, which are surrounded by the basal membrane. (4) Next step is migration and penetration by cancer cells into neighbouring tissues called invasion. Loss of epithelial polarity, motility attainment, changing phenotype to mesenchymal-like, and secretion of enzymes like proteases. The various oncogenes and tumour suppressors, with the involvement of different growth factors, control the critical phase of tumour development. The extravasation of transformed cells - cancer cells from the site and intravasation into the lymphatic and blood vessels results in metastasis (5) to distant organs. Extra- and intravasation require the assistance of macrophages, platelets, and endothelial cells. At this stage of resulting micro-metastases, (6) the cells usually are sensitive to chemotherapy and radiotherapy. However, constant accumulation of mutation leads to the acquisition of further mutations and makes the cancer cells proficient to produce growth factors. This leads to autocrine loops that propel the development of resistant clones (7). Angiogenesis (8) is an indispensable factor for the establishment of secondary tumours. In the last phase, large metastases (9) move to a distinct set of target organs leading to metastases.

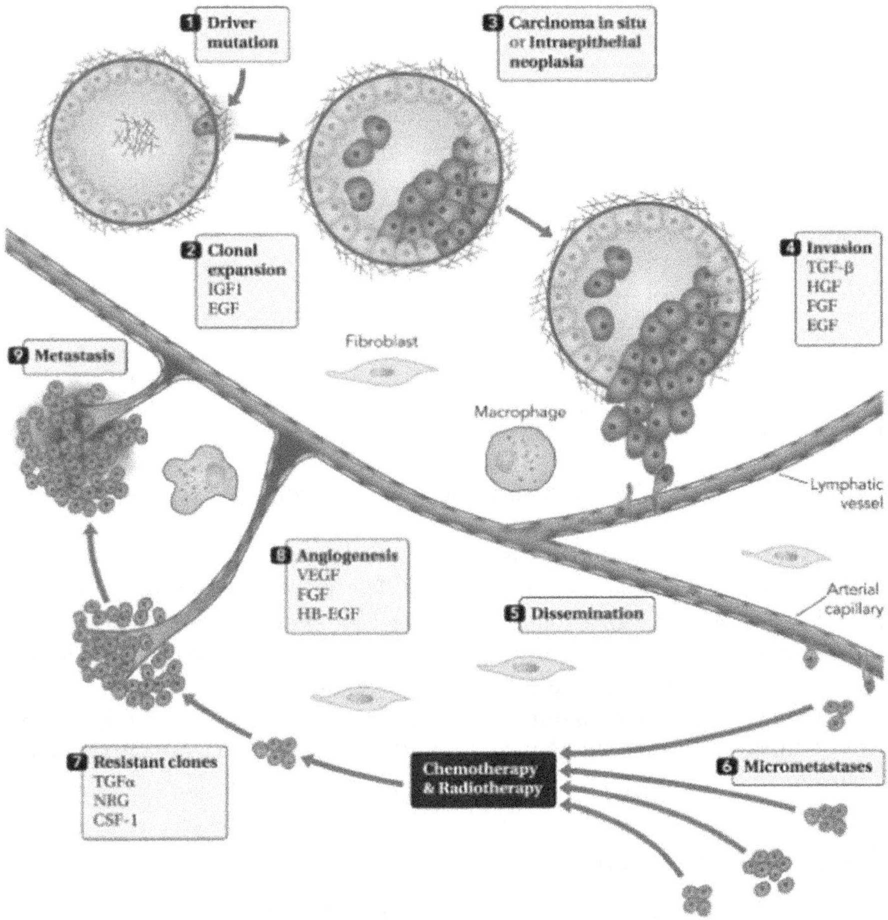

FIGURE 12.3 The process of cancer development and role of different growth factors (GFs) reprinted from Witsch et al. (2011) (Abbreviations: CSF-1, colony stimulating factor 1; EGF, epidermal growth factor; FGF, fibroblasts growth factor; HB-EGF, heparin-binding EGF; NRG, neuregulin; TGF, transforming growth factor; VEGF, vascular endothelial growth factor).

12.1.3 Causes of Cancer

Genetic changes are the cause of cancer especially DNA damage and genomic instability (Ferguson et al. 2015; Basu 2018). A few are due to genetic inheritance However, mostly the environment, i.e. anything outside the human body that interacts with it, lifestyle, or behavioural exposures lead to cancer. In general, cancer is not contagious in humans, though it can be caused by viruses and cancer bacteria. Avoiding exposure to such factors can contribute to a 75–80% reduction in cancer deaths. Environmental factors include exposure to different chemical and physical agents; for example, tobacco (accounts ~30% of cancer deaths), pollutants, diet and obesity (accounts ~35% of cancer deaths),

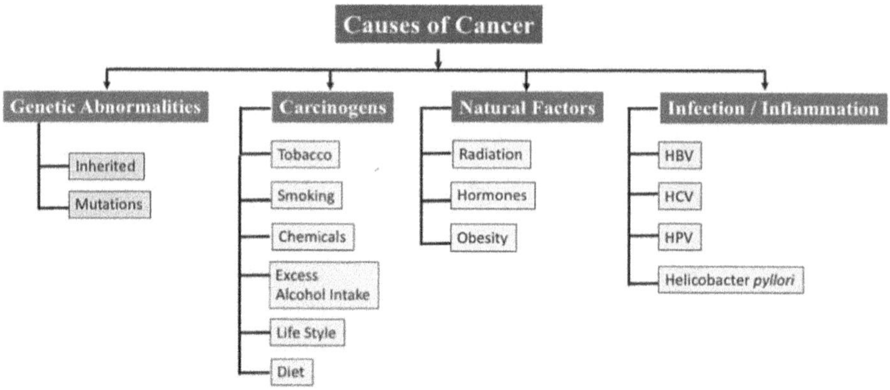

FIGURE 12.4 Causative factors of cancer.

infections (accounts ~20% of cancer deaths) and radiation (accounts ~10% of cancer deaths) (Merlo et al. 2012; Parsa 2012).

However, these may not be solely responsible, but they at least play a role in altering the gene functions of the cells. The DNA damage-and-repair mechanism can take care of such DNA damage; however, many such changes, which remain uncorrected, accumulate and lead to a transformation of a normal cell. One of the important risk factors is ageing, in which many molecular and cellular changes may manifest as cancer (Maslov and Vijg 2009; Campisi 2013; Aunan et al. 2017). Figure 12.4 illustrates various causative factors of cancer.

12.1.4 Types of Cancer

Cancer is classified according to the kind of fluid or tissue from which it originates, or according to the tissue/organ in the body where it first originated or developed. In addition, a few cancers are of mixed types. The following are the five broad categories of cancer:

1. *Carcinoma*: It is cancer of the epithelial tissue that lines surfaces of organ/glands or body structures. They account for 80–90% of all cancer cases, e.g. melanoma, basal cell carcinoma, squamous cell skin carcinoma, merkel cell carcinoma.
2. *Sarcoma*: A sarcoma is a malignant tumour growing from connective tissues, such as cartilage, fat, muscle, tendons, and bones. The most common sarcoma, a tumour on the bone, usually occurs in young adults, e.g. osteosarcoma (bone), chondrosarcoma (cartilage), Erwig's sarcoma, soft tissue sarcoma.
3. *Lymphoma*: These refer to cancer of nodes or glands of the lymphatic system. The lyphatic systems are responsible for producing WBCs and clean body fluids, e.g. Hodgkin's lymphoma, non-Hodgkin's lymphoma and cutaneous lymphoma.

4. *Leukemia*: It is also known as blood cancer. It is a cancer of the bone marrow, which produces normal red and white blood cells and platelets. White blood cells are known as the body's fighter cells, which resist infection. Red cells contain a special protein called haemoglobin, which carries oxygen from the lungs to the rest of the body and then returns carbon dioxide from the body to the lungs, where it gets exhaled. Platelets support the body. Platelets also called as thrombocytes play a vital role in normal blood clotting, e.g. acute lymphocytic leukemia, acute myeloid leukemia, agnogenic myeloid leukemia, chronic lymphocytic leukemia, chronic myeloid leukemia, essential thrombocythemia (ET), hairy cell leukemia, myelodysplastic syndromes (MDS).
5. *Myeloma*: It grows in the plasma cells of bone marrow. Sometimes, it is plasmacytoma, i.e. the myeloma cells collect in one bone and form a single tumour. In other cases, it is multiple myeloma – where the cells collect in many bones, forming many bone tumours.

12.1.5 Conventional and Emerging Cancer Therapies

Use of anti-cancer drugs – chemotherapeutic agents began way back in 1940s, prior to which the only modality for treatment was surgical removal of the cancerous tissue. Initial drugs were the nitrogen mustards – a powerful alkylating agent, sulfur mustard gas, and later, a second class of cancer chemotherapeutics – antimetabolites (Gilman and Philips 1946; Gilman 1963). With the advancement of these initial drugs, there was a development of other antimetabolites that target DNA replication and other biological cellular targets that are necessary for cell division. Chemotherapy drugs mainly produce reactive oxygen species, which lead to genotoxicity and thus destruction of tumour cell (Teppo et al. 2017). However, chemotherapy cannot discriminate between ordinary cells and tumour cells, which results in side effects such as fatigue, nausea, hair loss, and vomiting, or even death in extreme cases (Abbas and Rehman 2018).

In 1960, radiation therapy was introduced to control the local disease. In radiation therapy, there is high energy radiation that stops cell division and blocks its ability to grow by damaging the genetic material (Gianfaldoni et al. 2017). However, then, it was realised that these treatments individually were not effective and therefore combinatorial therapies were introduced to control cancer. Radiation therapy done before surgery shrinks the tumour, and if performed after, it destructs the tumour cells that are left behind, resulting in a reduction in the cancer relapse (Delaney et al. 2005; Baskar et al. 2012). The work of Hanahan and Weinberg opened new avenues for understanding and knowledge about some hidden secrets of this aggressive disease. They reported ten traits that make a normal cell become a cancerous one, namely sustaining proliferative signalling, evading growth suppressors, avoiding immune destruction, enabling replicative immortality, tumour-promoting inflammation, activating invasion and metastasis, angiogenesis, resisting cell death, deregulating cellular energetics, and genome instability and mutation (Hanahan and Weinberg 2000, 2011). This led to development of new approaches for cancer treatment.

With the progress and inventions in treatment modalities, different drugs, biological molecules, and immune-mediated therapies are introduced for treatment. But, until today, a well accepted therapeutic modality is not achieved that can reduce the mortality rate. Then came the era of gene therapy in which there is direct insertion of exogenous genes into the benign tumours (Kaufmann et al. 2013); hormonal therapy – widely used in malignancy (Fairchild et al. 2015; Himabindhu 2020), wherein, hormones restrict the tumour progression; stem cell therapy – in which stem cells are modified in vitro by introducing specifically customised genes having antitumor effects (Chu et al. 2020); dendritic cell-based immunotherapy – manipulates/trains the immune system to destroy tumours (Weinstock et al. 2017).

12.2 VIRUSES, CANCER AND TREATMENT

12.2.1 Mechanism of Viral Cancer

Infectious agents such as viruses, bacteria, and parasites behave as etiological agents in ~20% of human cancers, out of which nearly 12% of cancers worldwide are caused by oncovirus, mostly seen in the developing world (de Martel et al. 2017) (Figure 12.5). Viruses are very small particles nearly in size range of 20–30 nm, obligate parasite with either ds- or ss-DNA or RNA enclosed within a protein shell called capsid. Understanding and managing such cancers is facing daunting challenges due to unavailability of suitable animal models for pre-clinical studies to evaluate the potential immuno-prophylactic therapies. Another difficulty

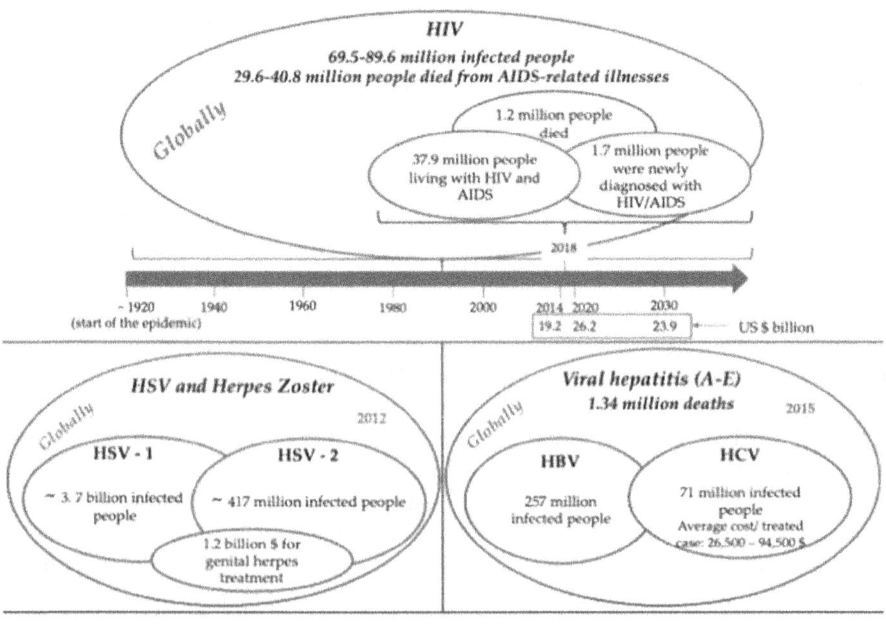

FIGURE 12.5 Global impact of viral diseases. Adopted from Cojocaru et al. (2020)

lies with the contrasting nature of viruses, their distinct types, and their interaction with host cell.

During the 20th Century, transmittable agents that can cause tumors in chickens and rabbits were identified as rous sarcoma virus (RSV). RSV was the first RNA tumour virus, and shope papillomavirus was the first DNA tumor virus studied for its oncogenic efficiency. In recent years, focus has been given to viruses that can cause human cancers. The role of viruses in human cancer led to identification and characterisation of various oncogenes and tumour-suppressor genes. With the help of these studies, a few generalised principles were deduced as:

1. In the virally associated cancers, long latent periods follow initial infection, indicating that the viruses are not sufficient to cause cancer on their own. During this long latent period, these viruses must be able to survive in the host.
2. The virus-induced cancers do not constitute a natural part of the viral life cycle. However, virally associated cancer cells are like dead-end streets, which are not of any advantage to the virus.
3. The viral properties responsible for the induction of tumors often play critical roles in the life cycle of the virus.

A few common traits have been identified with viral-induced oncogenesis as: (1) Oncoviruses are necessary to cause cancer, but by themselves cannot lead to cancer development; (2) Persistent infection that occurs many years after acute infection is required for virus-induced cancer; (3) Immune system acts as a double-edged sword, meaning it can play a deleterious or a protective role, i.e. in some cases, it may have immunosuppression, or in others, it may have chronic inflammation. Weinberg and Hanahan have described ten different Hallmarks of Cancer that help in understanding their roles in the carcinogenic process (Hanahan and Weinberg 2000, 2011), wherein, each hallmark represents oncogenic alterations underlying the alterations happening in the cell leading to tumorigenic properties (Figure 12.6).

Viruses on the basis of the type of their genome are classified into DNA viruses and RNA viruses. In DNA viruses, viral DNA enters into the host cell nucleus, gets transcribed into mRNA using host cell RNA polymerase, and translates into various virus-specific proteins. The proteins so formed contain some enzymes, which aid in synthesising more viral DNA as well as proteins. Once the virion particles are assembled, these are released by budding or after cell lysis. A retrovirus is a type of virus that has RNA genome; it inserts RNA into the DNA of a host cell, leading to genomic alterations. Within the cytoplasm, the virus uses its reverse transcriptase enzyme and produces DNA from its RNA genome; that is why named as *retro* (backwards). This new DNA gets incorporated into the host genome by an integrase enzyme. The host cell cannot distinguish it and treats the viral DNA as part of its own genome, resulting in transcription and translation of the integrated viral genes along with its own genes, producing the proteins required to assemble new copies of the virus. Two genes, namely viral genes (*v-onc*) and cellular genes (*c-onc*), have been identified, wherein *c-onc* genes regulate

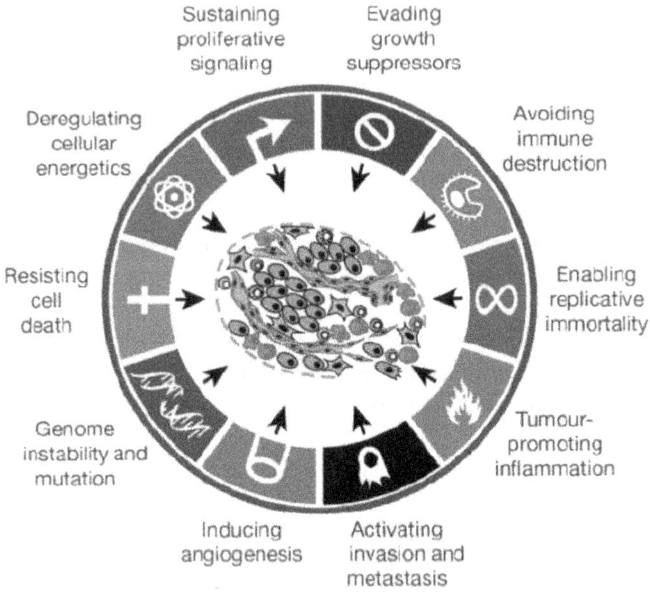

FIGURE 12.6 Hallmarks of cancer by Hanahan and Weinberg (Reprinted from Hanahan and Weinberg 2011).

cellular growth (by encoding for growth factors, cellular receptors, signalling and regulatory proteins, DNA binding proteins) and differentiation, but like their viral counterparts, these are mutated. In case of HTLVs, these viruses have a trans-activating gene (*tax gene),* which is essential for replication and transactivates several cellular genes leading to transformation. Such multigenic events occurring result in the development of HTLV-induced leukemia.

12.2.2 Human Cancer-Causing Viruses

1. Hepatitis B (HBV) and hepatitis C viruses (HBC), which are associated with hepatocellular carcinoma (HCC). HBV is a dsDNA virus of the hepadnaviridae family. It is estimated that approximately 350 million people worldwide are infected with HBV worldwide. Hepatitis B can lead to liver diseases, ranging from the acute hepatitis to chronic hepatitis, cirrhosis, and hepatocellular carcinoma (HCC) (Sarkar et al. 2015). HCV is a +ve stranded RNA virus belonging to the *Flaviviridae* family. Around 180 million people are infected with hepatitis C virus worldwide (Goto et al. 2020). Liver cancer is the 6th most common cancer worldwide; it is the 2nd most common cause of cancer deaths. It is not realised that chronic hepatitis B is the primary global risk factor for developing liver cancer. Hepatitis B viruses cause hepatitis, which translates to "inflammation of liver." The virus attacks the liver and weakens its detoxifying performance and ability to maintain the level of sugar in the blood. Chronic (long-term) infection with hepatitis B or hepatitis C viruses can lead to liver cancer. Worldwide, hepatitis B is much more

widespread than hepatitis C, making it a priority when it comes to the prevention of liver cancer. Approximately 292 million people around the world are living with hepatitis B infection (Chang 2011).

2. Epstein–Barr virus (EBV), which are associated with B cell lymphomas, nasopharyngeal and gastric carcinomas. The Epstein–Barr virus (EBV), also called human herpesvirus 4 (HHV-4), is the virus of herpes family, and it is one of the most common viruses in humans. It is best known as the cause of infectious mononucleosis – glandular fever. The Epstein–Barr virus (EBV) infection is linked to the development of a variety of human malignancies, including Burkitt's lymphoma, Hodgkin's disease, nasopharyngeal carcinoma, some T cell lymphomas, post-transplant lymphoproliferative disease, and certain cancers of the stomach and smooth muscle and central nervous system lymphomas associated with HIV (Baumforth et al. 1999; Kanda et al. 2019; Wang et al. 2021; Vockerodt et al. 2015).
3. Human papilloma viruses (HPVs), which are associated with cervical carcinoma, other anogenital cancers and a subset of head and neck cancers. The human papillomavirus (HPV) is a viral infection that is passed between people through skin-to-skin contact. There are over 100 varieties of HPV, more than 40 of which are passed through sexual contact and can affect the genitals, mouth, or throat. HPV can cause cervical and other cancers, including that of the vulva, vagina, penis, or anus. It can also cause cancer in the back of the throat, including the base of the tongue and tonsils; that is called oropharyngeal cancer (de Martel et al. 2017; Timbang et al. 2019). Cancer often takes years, even decades, to develop after a person gets HPV. The types of HPV that can cause genital warts are not the same as the types of HPV that can cause cancers.
4. Human T cell lymphotropic virus I (HTLV-1), which are associated with adult T cell lymphomas. Human T-cell leukemia virus type 1 (HTLV-1) is the first human retrovirus to be discovered; its infection is usually not given much attention in the health-care system. These are a group of human retroviruses that are known to cause a type of cancer called adult T-cell leukemia/lymphoma and a demyelinating disease called HTLV-1 associated myelopathy (Biswas et al. 2010). Approximately 20 million people worldwide are infected with HTLV-1 (de Thé and Kazanji 1996), 90% of whom are infected individuals who remain asymptomatic carriers during their lives.
5. The human herpesvirus-8 (HHV-8), also called Kaposi sarcoma herpesvirus (KSHV), belongs to the family of DNA viruses, herpeseviridae. It causes Kaposi sarcoma (a vascular malignancy) and B cell lymphoproliferative diseases, such as primary effusion lymphoma (PEL) and multicentric Castleman disease (MCD) (Weiss et al. 1998).

HIV (human immunodeficiency virus) is a virus that attacks the immune system. If HIV remains untreated, it can lead to AIDS (acquired immunodeficiency syndrome). There is currently no effective cure for this syndrome; once infected, it is for life. HIV-infected people are at higher risk of developing cancer, although it is globally diminished in the era of highly active antiretroviral treatment (HAART) (Shmakova et al. 2020) (Figure 12.7).

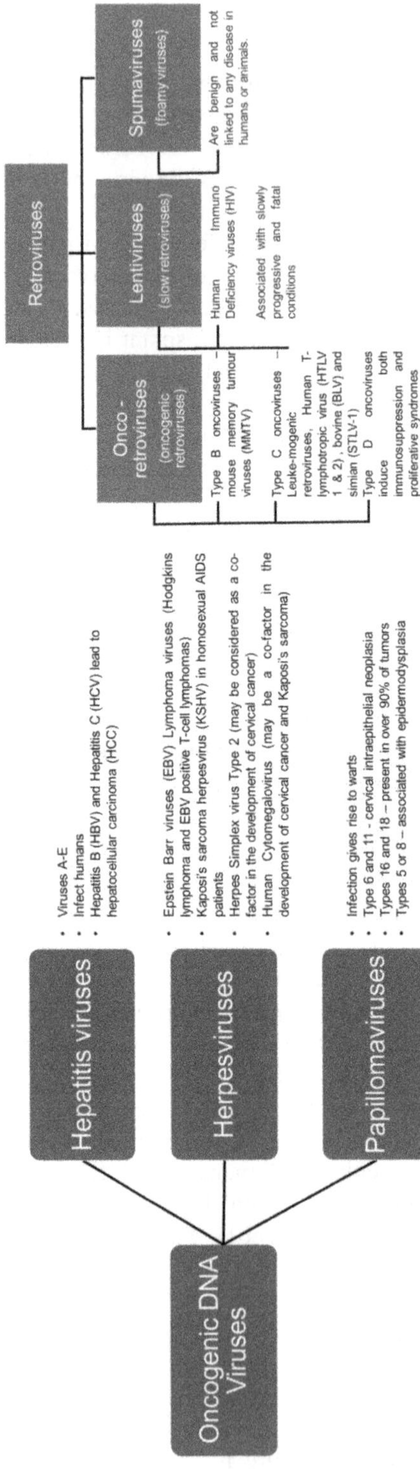

FIGURE 12.7 Pathogenic DNA and RNA viruses.

12.2.3 Current Viral Vaccines and Drugs

Currently, preventative vaccines are being developed using weakened or harmless virus forms to impart and invoke the immune system, which can then recognise and fight with the potential threats. These therapeutic vaccines compel the immune system to attack cancer cells. Four vaccines have been designed to treat or prevent cancer:

- Sipuleucel-T: The first therapeutic cancer vaccine that received the U.S. Food and Drug Administration (FDA) approval for treatment of some forms of prostrate cancer. It uses a patient's re-engineered cells, which are then injected back into the body to help activate the immune system.
- Bacille Calmette-Guerin (BCG): It is widely used as a preventative vaccine for tuberculosis, but also used as a therapeutic vaccine to treat very early stages of bladder cancer.
- Hepatitis B vaccine (HBV): HBV became the first FDA-approved vaccine to prevent cancer. Children receive the HBV vaccine soon after birth to prevent liver cancer, as recommended by the U.S. Centers for Disease Control and Prevention.
- Human papillomavirus (HPV) vaccine: These are the preventive vaccines designed for protection against infections from HPV strains responsible for many cancers.

Many researches and clinical trials are studying vaccines for various types of cancers: bladder cancer, brain cancer, breast cancer, cervical cancer, colorectal cancer, kidney cancer, leukemia, lung cancer, melanoma, myeloma, pancreatic cancer and prostrate cancer (Thomas and Prendergast 2016; Lopes et al. 2019; Perez and Palma 2019; Zhang et al. 2020). Treatment of viral diseases include drugs, namely, cidofovir, adefovir dipivoxil, tenofovir, penciclovir, famciclovir, valganciclovir, valaciclovir, ganciclovir, acyclovir, entecavir, tenofovir, lamivudine, adefovir and telbivudine, all of which can help fight the virus but at the same time damage the liver and kidney. The exact frequency of nephrotoxicity induced by these antiviral drugs is difficult to determine. Antiviral drugs cause renal failure through several mechanisms. Direct renal tubular toxicity with unique effects on epithelial cells of the kidney have been reported with use of antiviral drugs (Izzedine et al. 2005). Additionally, crystal deposition in the kidney may promote the development of renal failure. Drug-induced kidney injury is a major side effect in clinical practice, frequently leading to acute renal failure (ARF). It accounts for more than 2% to 15% of cases of ARF in patients admitted to the hospital or in the intensive care unit, respectively (Leowattana 2019).

12.3 NANOTECHNOLOGY IN CANCER TREATMENT

Significant progress has been made in the field of cancer therapies, but the mortality rate remains high. This strongly demands further research. As mentioned earlier, current cancer therapies, such as surgery on many occasions, are unable to completely remove all cancer cells in the human body; chemotherapy and radiotherapy show severe toxic side effects on normal cells. Recently, there has been significant progress in the application of nanotechnology in diagnosis and therapeutics (Niemeyer and Mirkin 2004; McNeil 2011; Baetke et al. 2015; El-Sayed and Kamel 2020). The field wherein nanomaterials are used in diagnostic or therapeutic applications is known as nanomedicine (Brewer et al. 2007; Kim et al. 2010). Different kinds of nanoparticles, such as polymeric (Amreddy et al. 2017; Afsharzadeh et al. 2018), carbon nanotubes (Li and Al-Jamal 2021; Riley and Narayan 2021), liposomes (Lawrie et al. 2013; Edwards et al. 2015), metallic (Ma et al. 2019; Peng and Liang 2019; Kim et al. 2021), ceramic (Huang et al. 2011; Thomas et al. 2015), semiconductor (Zhu et al. 2017) etc. Many of them have proved their potential inefficient diagnostic and/or therapeutic tools for cancer. The need for the development of novel drug-delivery systems resulted in several innovative delivery systems using the nanotechnology approach. In the present chapter, emphasis is given to nano-delivery approaches for antiviral agents in cancer therapy.

The inherent problems with antiviral drugs/agents are low efficacy, less solubility, low bioavailability, short half-life, and systemic toxic side-effects. Taking into consideration these factors, novel drug-delivery approaches are explored by improving the design, formulation, and delivery systems for successful and efficacious therapy. Figure 12.8 depicts such drug-delivery systems studied for antiviral and antiretroviral therapy, and Table 12.1 enlists recent studies reported on nano drug-delivery systems for cancer therapeutics. Nanoparticles, due to their unique physical, chemical, and biological characteristics, can successfully accumulate at the site of tumours through enhanced permeability and retention (EPR), thereby decreasing the dose and side effects of toxic drugs and increasing treatment efficiency.

Nanoparticles have different shapes and morphologies, surface charge, biomimetic properties (Gagliardi 2017), increased target specificity, and controlled drug release (Muthu and Singh 2009), which make them ideal drug-delivery agents. Such properties impart efficient delivery of antiviral drugs through impermeable barriers (Petros and DeSimone 2010; Mahajan et al. 2012). Their large surface area to volume ratio encourages a large amount of drug incorporation (McNeil 2011) with enhanced stability and bioavailability (Goldberg et al. 2007). Surface conjugation with target specific markers can result in reduced toxicity to normal healthy cells. Due to these properties, nano drug-delivery systems show promising virucidal activity against variety of viruses viz. HIV, HSV, HBV, etc. Such targeted, sustained, and efficient delivery through the engineered nano-delivery systems result in reduction in drug resistance and adverse side effects.

However, several limitations can cause the nano-drug antiviral delivery systems to fail. These issues therefore need to be addressed. They include avoiding degradation in stomach acid pH following oral administration or minimal absorption due to difficulty in getting into the mucus barrier, interaction with immune cells and getting opsonised, adverse immunological responses may lead to renal toxicity, and the last but not the least, their designing and manufacturing cost. Considering the

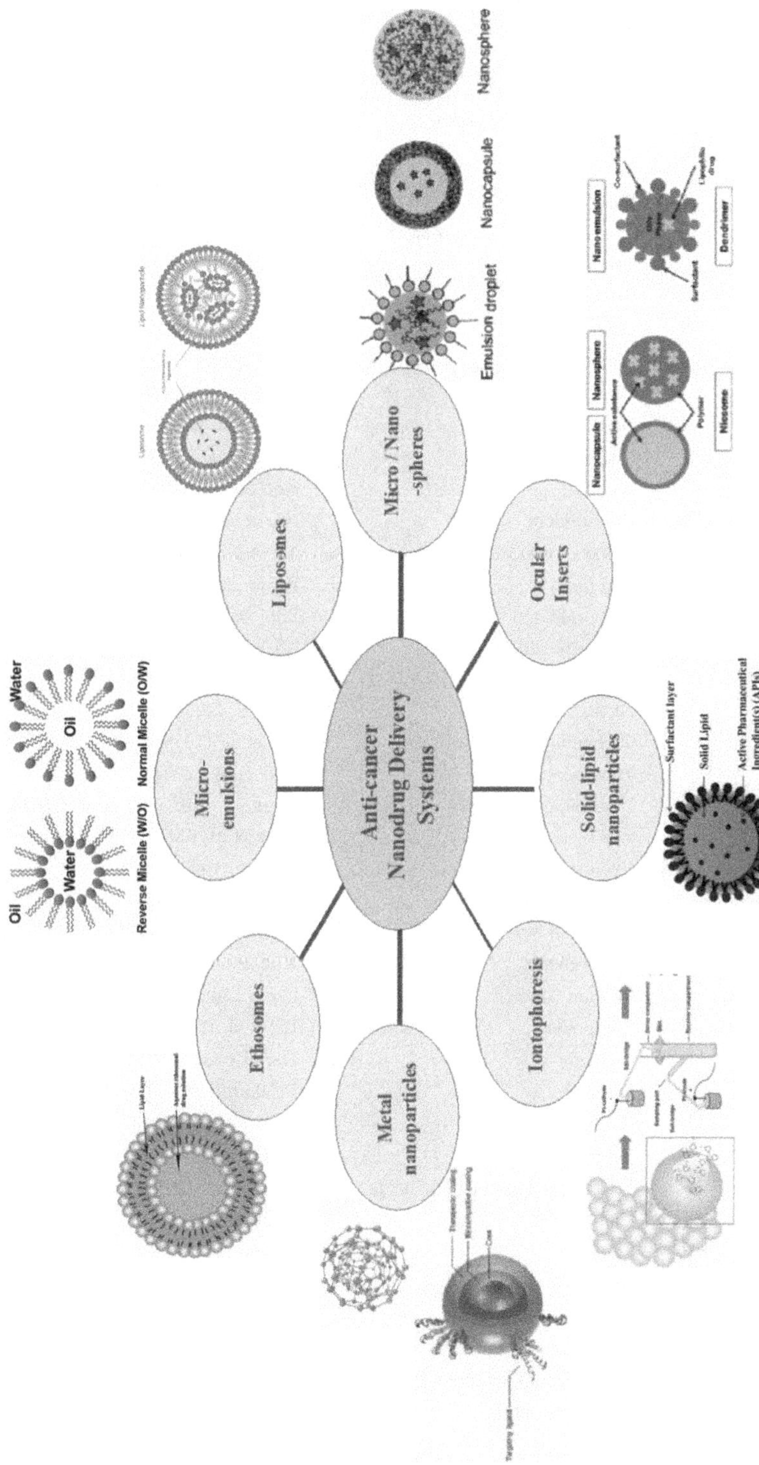

FIGURE 12.8 Various forms of anti-cancer nano drug-delivery systems.

TABLE 12.1
Nano-Drug Delivery Systems for Cancer Therapeutics

Type of Nano-Drug Carrier Systems	Drug Loaded	References
Sub-micron particles	Paclitaxel	Chavanpatil et al. (2007)
	Rhodamine and doxorubicin	Constantinides et al. (2000)
	Cisplatin	Bai and Liu (2014)
	Doxorubicin	Choukrani et al. (2020)
Micro-emulsions	Elemene	Zhai et al. (2018)
	Methotrexate	Flores et al. (2016)
	Celecoxib	Mojeiko et al. (2019)
	Gemcitabine	Tsai et al. (2010)
Liposomes	Curcumin	Feng et al. (2017)
	ATB	Kou et al. (2020)
	Doxorubicin	Niu et al. (2010)
	Pclitaxel and siRNA	Yu et al. (2019)
	Curcumin	Peram et al. (2019)
	Curcumin and sulphoraphane	Soni et al. (2020)
	Berberine chloride and evodiamine	Lin et al. (2020)
Ethosomes	Oxaliplatin	Li et al. (2020)
	Curcumine and doxorubicin	Ma et al. (2019)
	Tocotrienol	Nair et al. (2021)
Micro and nanospheres	Doxorubicin	Lv et al. (2013); Li et al. (2016); Tan et al. (2016)
	Cyclosporine	Melero et al. (2017)
	CR-lycosin-I	Zhang et al. (2020)
	Rutin and benzamide	Deepika et al. (2019)
	Doxorubucin	Kumar et al. (2020)
	Methotrexate	Costa Lima et al. (2017)
	Campothecin	He et al. (2019)
	Cisplatin	Darini et al. (2019)
Ocular inserts	Methotrexate	Palakurthi et al. (2011)
	Carboplatin	Shome et al. (2014)
	Voriconazole	Qin et al. (2021)
	Short interfering nucleic acids	Jiang et al. (2019)
Solid Lipid nanoparticles	Gadolinium	Wang et al. (2020)
	Tamoxifen	Guney Eskiler et al. (2018)
	Nutlin	Grillone et al. (2019)
	Irinotecan	Rajpoot and Jain (2019)
	Letrozole	Yassemi et al. (2020)
	si-RNA and paclitaxel	Buyukkoroglu et al. (2019)
	Andrographolide	Li et al. (2020)

(Continued)

TABLE 12.1 (Continued)
Nano-Drug Delivery Systems for Cancer Therapeutics

Type of Nano-Drug Carrier Systems	Drug Loaded	References
	Curcumin	Wang et al. (2018)
	Etoposide	Athawale et al. (2014)
Iontophoresis	5-FU	Petrilli et al. (2018)
	antigen peptide	Toyoda et al. (2015)
	5-FU and leucovorin	Gratieri and Kalia (2014)
	miproxifen phosphate	Komuro et al. (2013)
	CpG-oligodeoxynucleotide	Kigasawa et al. (2011)
	Doxorubicin	Taveira et al. (2014)
Metal/ceramic nanoparticles	Nucleic acid	Ferreira et al. (2020)
	Docetaxel	Thambiraj et al. (2019)
	Curcumin	Mahalunkar et al. (2019)
	Doxorubicin	Gonzalez-Rodriguez et al. (2019); Mukherjee et al. (2020)
	siRNA	Shaat et al. (2016); Rahme et al. (2019)
	Cisplatin	Zhao et al. (2018)
	Paclitaxel	Manivasagan et al. (2016); Feng et al. (2019)

above limitations, a multidisciplinary team needs to focus on the efficient designing and functioning parameters to make an 'ideal' nanocarrier system. Recently, personalised nanomedicine is also become possible via a synergistic approach.

Nanomaterials have tunable size, which should be large enough to avoid getting washed off the body and small enough to avoid deposition in the reticulo-endothelial system (RES) (Alyautdin et al. 2014; Fonseca-Santos et al. 2015). The ideal size range is 10–200 nm, particles >10 nm in diameter to avoid first-pass elimination through the kidneys and <200 nm to avoid sequestration by sinusoids in the spleen and fenestra of the liver (Hoshyar et al. 2016). The size of NPs, therefore, determines their circulation half-life and tumour accumulation. In addition to size, surface characteristics also play a critical role in determining the long half-life in circulation and their fate. Surface modification with charged moiety/functionalization increases nanoparticles' stability, determines its interaction, prolongs its circulation time, and increases passive accumulation in tumours (Breus et al. 2009; Muro et al. 2010). For example, PEGylation – imparting a hydrophilic surface helps to escape opsonisation and thus macrophage invasion (Harris et al. 2001; Vonarbourg et al. 2006). These nanocarriers can further be targeted by conjugating with target receptor molecules/proteins; this process is known as functionalization.

However, the EPR effect depends on several factors, such as the surface characteristics of NPs, immunogenicity, tumour characteristics, which thus may result in many challenges in the optimisation of passive targeting (Rosenblum et al. 2018; Attia et al. 2019). Therefore, a full understanding of the degree of these factors in the EPR effect would be worthwhile (Scheetz et al. 2019).

Passive targeting could facilitate the efficient localisation of NPs in the tumour interstitium; however, it cannot further promote cellular uptake by cancer cells, which can be achieved by actively targeting NPs functionalized with ligands, such as antibodies, peptides, nucleic acid aptamers, and small molecules, to receptors or other surface membrane proteins overexpressed on target cells (Sanita et al. 2020). This strategy exploits the differences between malignant cells and normal cells. Specific interactions between the ligands on the surface of NPs and receptors expressed on the tumour cells facilitate NP internalisation by triggering receptor-mediated endocytosis. Tumour-targeted nanomaterials are being employed for early tumour diagnosis and therapy, which could reduce or eliminate the delivery of potentially toxic agents to healthy tissues, resulting in increased diagnostic or therapeutic effectiveness and decreased potential side effects (Phillips et al. 2010). Furthermore, active targeting has shown great potential in overcoming a variety of obstacles such as bypassing the blood-brain barrier (Veiseh et al. 2009; Rabiei et al. 2020) and multi-drug resistance (MDR) in tumours (Kim et al. 2015; Yao et al. 2020).

12.4 CURRENT OVERVIEW OF ANTIVIRAL THERAPY IN CANCER

Liposomes are nanosized concentric bilayered vesicles, with aqueous volume inside enclosed by a membranous phospholipid bilayer; these structures are widely used for drug delivery (Duzguneş et al. 2005). A comparative intracellular uptake study of zalcitabine, a reverse transcriptase inhibitor encapsulated in liposomes, demonstrated that a liposome-carrying drug penetrated mouse macrophage cell line more quickly and easily compared to the free conventional drug. This was due to the altered anionic charge engineered on the liposomal nanocarriers (Makabi-Panzu et al. 1998). Bronshtein et al. (2011) studied the cell-derived liposomes for targeting HIV-infected cells on HIV and HIV-infected cells. These cells remain sources for obtaining these cell-derived liposomes, which demonstrate greater target specificity to HIV infected cells. Cell-derived liposomes reduced 60% viability of gp120-expressing cells compared to that of control cells (Bronshtein et al. 2011). Other interesting nanoengineered liposomal carriers include elastic liposomes, magnetic liposomes, galactosylated liposomes, pegylated liposomes, and mannosylated liposomes. The efficacy of elastic liposomes of zidovudine (azidothymidine) (AZT) for transdermal administration was studied by Jain et al. (2006). Better permeability through rat skin of the nanoformulation was demonstrated to that of the free drug, which was indicated by a higher transdermal flux of elastic liposomes carrying AZT. Also, the plasma concentration of liposomal AZT was 12-folds higher than the control after 24 hours (Jain et al. 2006). Clayton et al. (2009) engineered liposomes with PEG on the membrane surface as drug vehicles for HIV-1 protein inhibitor (PI) delivery. These sterically stable pegylated liposomes comprised of ligands obtained from the Fab' fragment of HIV-gp120-directed monoclonal antibody F105, which enabled target specificity to infected cells. The cellular specificity incorporated on the liposome enables selective uptake by HIV-1 infected cells, thereby increasing the cytoplasmic concentration of PI and intracellular localisation of the drug. Hence, the team observed that the immunoliposomes carrying antiviral drugs showed greater and longer antiviral activity than that of free drug or non-cell-specific drug formulations (Clayton et al. 2009). In another study, azidothymidine 5'-triphosphate (AZTTP) was reported to be incorporated in a magnetic liposomes outer encompassment showing ~54.5% of drug entrapment efficiency (Saiyed et al. 2010). They applied an external magnetic field to study the ability of nanoformulation to transmigrate directly or via monocyte-mediated transport across an in vitro blood-brain barrier (BBB) model. Table 12.2 shows recent advancements using various nano-carrier systems for antiviral drug delivery.

TABLE 12.2
Recent Advancements in Nano-Carrier Systems for Antiviral Delivery

Drug Name	Virus	Nano Carrier System	Reference
Efavirenz	HIV	Polymeric micelles, mixed polymeric micelles, dendrimer, Chitosan-g-HPβCD NPs	Chiappetta et al. (2010); Chiappetta et al. (2011); Chiappetta et al. (2013); Calderón et al. (2013)
Nevirapine	HIV	Liposomes, polymeric NPs grafted with transferrin, SLNs	Ramana et al. (2010); Kuo et al. (2011); Kuo and Chung (2011)
Lamivudine	HIVHIV	Polymeric micelles, polymeric NPs, SLNs, LNCs (lipid NP)	Kuo and Chen (2006); Li et al. (2010); Sankar et al. (2012); Cavalcanti et al. (2017)
Stavudine	HIV	Liposomes, SLNs	Garg et al. (2006); Shegokar and Singh (2011)
Zidovudine	HIV	Liposomes, polymeric NPs, niosomes, alginate NPs, microemulsion	Gopinath et al. (2001); Jin et al. (2005); Jain et al. (2008); Singh et al. (2010); Mainardes and Gremiao, (2012); Carvalho et al. (2016); Joshy et al. (2018)
Lopinavir	HIV	SLNs, niosomes, ethosomes	Alex et al. (2011); Patel et al. (2012)
Saquinavir	HIV	Polymeric NPs grafted with γ–PGA	Kuo and Yu (2011a)
Acyclovir	HSV, HBV	SLNs, Chitosan NPs, gel nanoemulsion, microemulsion, liposome, recombinant HDL	Alsarra et al. (2008); Feng et al. (2008a); Peira et al. (2009); Mahboobian et al. (2020); Parthiban et al. (2020)

(Continued)

TABLE 12.2 (Continued)
Recent Advancements in Nano-Carrier Systems for Antiviral Delivery

Drug Name	Virus	Nano Carrier System	Reference
Indinavir and lactoferrin	HIV	Nanoemulsion	Karami et al. (2019)
Ritonavir	HIV	SLNs	Javan et al. (2017)
Adefovir dipivoxil	Hepatitis B virus	SLNs	Xing et al. (2008)
Saquinavir	HIV	Chitosan NPs	Ramana et al. (2008)
Dolutegravir sodium	HIV	β-cyclodextrin-based NPs	Belgamwar et al. (2019)
Enfuvirtide and protoporphyrin IX	HIV-1	Liposome	Figueira et al. (2020)
Tenofovir	HIV	Thiolated chitosan (TCS) core/shell nanofibers	Meng et al. (2016)
siRNA	HCV	Cationic liposomes	Kim et al. (2009)
Nosiheptide	HBV	Recombinant HDL	Feng et al. (2008b)
Interferon		Magnetic microspheres	Zhou et al. (2008)
Saquinavir	HIV	Nps	Mahajan et al. (2010)
Indinavir	HIV	Immunoliposomes	Gagné et al. (2002)

On the other hand, solid lipid nanoparticles (SLN) also show several benefits, for example, protecting chemically labile substances against degradation (Owens and Peppas 2006). Zhang et al. (2008) prepared adefovir dipivoxil loaded saquinavir (SQV), which possessed an inhibitory action in vitro. Atazanavir, an HIV protease inhibitor (PI), has low brain permeability; to enhance brain uptake, the SLNs incorporating atazanavir were prepared (Chattopadhyay et al. 2008). The designed NPs (avg size ~167 nm) were evaluated on a well-characterised human brain microvessel endothelial cell line (hCMEC/D3) that represented the BBB. Kuo and Chen (2009) developed cationic solid lipid nanoparticles (CSLNs) carrying SQV.

Polymeric micelles are nanospheres less than 100 nm diameter that were aimed to improve aqueous solubility, intestinal permeability, and disease site targeting of several drug molecules. Different polymeric microspheres formulations are reported as drug-delivery systems for antivirals. Saquinavir-loaded poly (ethylene oxide)-modified poly (epsilon-caprolactone) (PEO-PCL) polymeric nanoparticulate system was developed by Shah and Amiji (2006). This novel formulation aimed for intracellular delivery of SQV at a significantly higher level through the medium of nanoparticles formulation in THP-1 human monocyte/macrophage (Mo/Mac). ARV drug release from poly (DL-lactide-co-glycolide) NPs in BALB/c mice was evaluated by Destache et al. (2010). They administered free ritonavir, lopinavir, and efavirenz (EFV) intraperitoneally and compared with the same drugs' nanoparticles formulation. Serum-free ARV drug concentrations peaked 4 h post-injection and were eliminated by 72 hours, whereas poly (DL-lactide-co-glycolide) NPs, animals had detectable ritonavir, lopinavir, and EFV concentrations in all tissues for 28 days. ARV nanoparticles treatment of monocyte-derived macrophages demonstrated sustained inhibition of HIV-1 replication (Destache et al. 2010). Tenofovir-loaded chitosan-based nanoparticles showed mucoadhesive properties due to the cationic charge on the chitosan molecule (Meng et al. 2011). Meng et al used the Box-Behnken design to evaluate the effect of formulation variables on the NPs size and drug encapsulation efficiency (EE%). This chitosan-based NP formulation was found to be non-cytotoxic to the vaginal epithelial cell line for 48 hours. The study demonstrated that the percentage of mucoadhesion increased from 6% to 12% as the diameter reduced from 900 nm to 188 nm, respectively (Meng et al. 2011). The study by Neves et al. (2010) demonstrated that polyethylene oxide-coated nanoparticles reduced permeability across monolayers/tissues, while modification of nanosystems with cetyltrimethylammonium bromide (CTAB) enhanced transport. Lamivudine stearate (LAS) was studied by Li et al. (2010). The team also studied the antiviral activity of LAS encompassed in stearic acid-γ-chitosan oligosaccharide (CSO-SA) polymeric micelles. LAS was synthesised with an ester linkage between lamivudine (LA) and stearic acid. CSO-SA/LAS presented low cytotoxicity and high cellular uptake percentage of LAS against HBV-transfected tumour cells (HepG2.2.15) (Li et al. 2010). Chiappetta et al. (2011) studied oral pharmacokinetics of the anti-HIV efavirenz (EFV) encapsulated within polymeric micelles in adult healthy volunteers. The micelles demonstrated consistent results and significantly higher absorption rates and the pharmacokinetic (PK) parameters increased 3-fold higher.

Another approach for antiviral and antiretroviral agents in the form of transdermal drug delivery includes the Iontophoresis technique, which provides the penetration of hydrophilic and charged molecules across the skin. Drug-delivery systems synthesised via the Iontophoresis technique can be delivered through the ocular, translingual, buccal, or nasal route. Ethosomes are made from phospholipids, alcohol (ethanol and isopropyl alcohol), were reported to possess remarkably higher transdermal flux of antiviral and antiretroviral drugs. Ethosomes can impact the ultrastructure of the stratum corneum and thus improve skin permeation of antiviral agents applied topically (DeJalon et al. 2001; Verma and Pathak 2010).

Dendrimers are also reported as effective target-specific nano delivery systems, especially in the elimination of HIV-AIDS. Dutta et al. (2008) evaluated target specificity and anti-HIV activity of efavirenz (EFV)-loaded, tuftsin-conjugated fifth-generation poly (propylene imine) dendrimers (TuPPI) in vitro. The entrapment efficiency of PPI (37.43 ± 0.3%) was slightly lower than TuPPI (49.31 ± 0.33%). TuPPI showed slow sustained EFV release in vitro at 144 hours compared to PPI, which excreted the drug in 24 hours itself. TuPPI possessed insignificant cytotoxicity as compared to PPI. The cellular uptake of TuPPI was 34.5 times higher than that of the free drug in HIV-infected macrophages and showed higher specificity, too (Dutta et al. 2008). de Las Cuevas et al. (2012) reported the development of carbosilane dendrimers as delivery vehicles for gene therapy against HIV and hepatocarcinoma. The water-stable cationic carbosilane dendrimers formed nanoconjugates with nucleic acids stably overtime under a wide range of pH values suggesting a strong interaction between the designed dendrimer and nucleic acid. These systems were nontoxic in both peripheral blood mononucleated cell (PBMC) and HepG2 cell lines under the experimental conditions. A high degree of 70–90% nanoconjugates transfection was observed, depending on the generation (de Las Cuevas et al. 2012) (Table 12.2).

Niosomes are microscopic lamellar structures made from non-ionic surfactant of the alkyl or dialkyl polyglycerol ether class and cholesterol with subsequent hydration in aqueous media. Ruckmani et al. (2010) studied the tissue distribution, pharmacokinetics, and stability studies of AZT delivered by niosomes and found the distribution of AZT in lungs, kidney, heart, liver, and spleen of mice after intravenous bolus injection. Zidan et al. (2011) studied a novel pediatric anti-HIV tenofovir niosomes with a high-pressure homogeniser and performed its characterisation of various parameters, such as vesicular sizing parameters, electrical properties, drug entrapment data, and drug-release characteristics. They found that the release of the drug was more affected by the compositional factors rather than the physical factors (Zidan et al. 2011).

Recently, transdermal drug delivery has proved to be a convenient route of administration for a variety of clinical therapeutics, including antiviral drugs (Benson 2005; Bala et al. 2014). The stratum corneum is the outermost layer of skin that prevents absorption/penetration of drugs through the skin. However, transdermal drug delivery or topical delivery avoids direct effects on the stomach and intestine, by bypassing the first-pass metabolism (Ramadon et al. 2021). Colloidal systems, microparticulate systems, and nano-delivery systems are commonly used in transdermal drug-delivery systems. Microemulsions, thermodynamically stable dispersions of oil and water stabilised by a co-surfactant, enhance the bioavailability of weakly soluble antiviral drugs, such as acyclovir, penciclovir, and ritonavir. Microemulsions help in maintaining these drugs in molecular dispersion in the GI tract, thereby extending the absorption window available in the GI lumen (Peira et al. 2009; Chudasaina et al. 2011).

12.5 CONCLUSION

There are many concerns and challenges, such as biocompatibility, pharmacokinetics, and in vivo targeting efficacy, that need to be addressed while using antiviral nanoparticles in cancer therapy into clinical practices. Since these are a new chemical entity, a proper and thorough investigation on nanomaterial toxicity and potential health risks needs to be performed. In the present chapter, an overview of cancer, causes and types, and its conventional therapies and drawbacks are discussed. Later, the role of nanotechnology in cancer treatment is discussed, with special emphasis given to cancer due to viral infection, its treatment, and applications of nano drug delivery for antiviral delivery in cancer. Of all the nanomaterials discussed, liposomes prove to be a promising candidate because they are clinically approved and currently possess the greatest number of clinical trials; some formulations are already available in the marketplace, while many others are yet to receive approval. A variety of nanomaterials, which are not biodegradable, may retain for long within the body. Therefore, a biodegradable and biocompatible antiviral drug-delivery system remains the choice that has a much higher clinical value. It is therefore strongly believed that such antiviral nanocarriers will get through the strict evaluation parameters for successful human application with the help of well-designed experimental approach and clinical trials.

REFERENCES

Abbas, Z. and Rehman, S. 2018. An overview of cancer treatment modalities. In *Neoplasm*, ed. Hafiz Naveed Shahzad, 139–157. London: IntechOpen.

Afsharzadeh, M., Hashemi, M., Mokhtarzadeh, A., Abnous, K. and Ramezani, M. 2018. Recent advances in co-delivery systems based on polymeric nanoparticle for cancer treatment. *Artificial Cells, Nanomedicine and Biotechnology* 6(6):1095–1110.

Alex, A., Paul, W., Chacko, A.J. and Sharma, C.P. 2011. Enhanced delivery of lopinavir to the CNS using compritol-based solid lipid nanoparticles. *Therapeutic Delivery* 2(1):25–35.

Alsarra, I.A., Hamed, A.Y. and Alanazi, F.K. 2008. Acyclovir liposomes for intranasal systemic delivery: Development and pharmacokinetics evaluation. *Drug Delivery* 15(5):313–321.

Alyautdin, R., Khalin, I., Nafeeza, M.I., Haron, M.H. and Kuznetsov, D. 2014. Nanoscale drug delivery systems and the blood-brain barrier. *The International Journal of Nanomedicine* 9:795–811.

Amreddy, N., Babu, A., Muralidharan, R., Munshi, A. and Ramesh, R. 2017. Polymeric nanoparticle-mediated gene delivery for lung cancer treatment. *Topics in Current Chemistry (Cham)* 375(2):35–58.

Athawale, R.B., Jain, D.S., Singh, K.K. and Gude, R.P. 2014. Etoposide loaded solid lipid nanoparticles for curtailing B16F10 melanoma colonization in lung. *Biomedical Pharmacotherapy* 68(2):231–240.

Attia, M.F., Anton, N., Wallyn, J., Omran, Z. and Vandamme, T.F. 2019. An overview of active and passive targeting strategies to improve the nanocarriers efficiency to tumour sites. *Journal of Pharmacy and Pharmacology* 71(8):1185–1198.

Aunan, J.R., Cho, W.C. and Søreide, K. 2017. The biology of aging and cancer: A brief overview of shared and divergent molecular hallmarks. *Aging and Disease* 8(5): 628–642.

Baetke, S.C., Lammers, T. and Kiessling, F. 2015. Applications of nanoparticles for diagnosis and therapy of cancer. *British Journal of Radiology* 88(1054): 20150207–20150219.

Bai, M.Y. and Liu, S.Z. 2014. A simple and general method for preparing antibody-PEG-PLGA sub-micron particles using electrospray technique: An in vitro study of targeted delivery of cisplatin to ovarian cancer cells. *Colloids and Surfaces B: Biointerfaces* 117:346–353.

Bala, P., Jathar, S., Kale, S. and Pal, K. 2014. Transdermal drug delivery system (TDDS) - A multifaceted approach for drug delivery. *Journal of Pharmacy Research* 8(12): 1805–1835.

Baskar, R., Lee, K.A., Yeo, R. and Yeoh, K.W. 2012. Cancer and radiation therapy: Current advances and future directions. *International Journal of Research in Medical Sciences* 9(3):193–199.

Basu, A.K. 2018. DNA damage, mutagenesis and cancer. *International Journal of Molecular Science* 19(4):970–983.

Baumforth, K.R., Young, L.S., Flavell, K.J., Constandinou, C. and Murray, P.G. 1999. The Epstein-Barr virus and its association with human cancers. *Molecular Pathology* 52(6):307–322.

Belgamwar, A.V., Khan, S.A. and Yeole P.G. 2019. Intranasal dolutegravir sodium loaded nanoparticles of hydroxypropyl-beta-cyclodextrin for brain delivery in Neuro-AIDS. *Journal of Drug Delivery Science and Technology* 52:1008–1020.

Benson, H.A. 2005. Transdermal drug delivery: Penetration enhancement techniques. *Current Drug Delivery* 2:23–33.

Biswas, H.H., Kaidarova, Z., Garratty, G., et al. 2010. HTLV outcomes study: Increased all-cause and cancer mortality in HTLV-II infection. *Journal of Acquired Immune Deficiency Syndromes* 54(3):290–296.

Breus, V.V., Heyes, C.D., Tron, K. and Nienhaus, G.U. 2009. Zwitterionic biocompatible quantum dots for wide pH stability and weak nonspecific binding to cells. *ACS Nano* 3:2573–2580.

Brewer, M., Zhang, T., Dong, W., Rutherford, M. and Tian, Z.R. 2007. Future approaches of nanomedicine in clinical science. *Medical Clinics of North America* 91(5): 963–1016.

Bronshtein, T., Toledano, N., Danino, D., Pollack, S. and Machluf, M. 2011. Cell derived liposomes expressing CCR5 as a new targeted drug-delivery system for HIV infected cells. *The Journal of Controlled Release* 151:139–148.

Buyukkoroglu, G., Şenel, B. and Yenilmez, E. 2019. Vaginal suppositories with siRNA and Paclitaxel-incorporated solid lipid nanoparticles for cervical cancer: Preparation and in vitro evaluation. *Methods in Molecular Biology* 1974:303–328.

Calderón, L., Harris, R., Cordoba-Diaz, M., et al. 2013. Nano and microparticulate chitosan-based systems for antiviral topical delivery. *European Journal of Pharmaceutical Sciences* 48(1-2):216–222.

Campisi, J. 2013. Aging, cellular senescence, and cancer. *Annual Review of Physiology* 75:685–705.

Carvalho, A.L., Silva, J.A., Lira, A.A., et al. 2016. Evaluation of microemulsion and lamellar liquid crystalline systems for transdermal zidovudine delivery. *Journal of Pharmaceutical Sciences* 5:2188–2193.

Cavalcanti, S.M.T., Nunes, C., Lima, S.A.C., Soares-Sobrinho, J.L. and Reis, S. 2017. Multiple Lipid Nanoparticles (MLN), a new generation of lipid nanoparticles for drug delivery systems: Lamivudine-MLN experimental design. *Pharmaceutical Research*34(6):1204–1216.

Chang, M.H. 2011. Hepatitis B virus and cancer prevention. *Recent Results in Cancer Research* 188:75–84.

Chattopadhyay, N., Zastre, J., Wong, H.L. and Bendayan, R. 2008. Solid lipid nanoparticles enhance the delivery of the HIV protease inhibitor, atazanavir, by a human brain endothelial cell line. *Pharmaceutical Research* 25:2262.

Chattopadhyay, N., Zastre, J., Wong, H.L., Wu, X.Y. and Bendayan, R. 2008. Solid lipid nanoparticles enhance the delivery of the HIV protease inhibitor, atazanavir, by a human brain endothelial cell line. *Pharmaceutical Research* 25:2262–2271.

Chavanpatil, M.D., Khdair, A. and Panyam, J. 2007. Surfactant-polymer nanoparticles: A novel platform for sustained and enhanced cellular delivery of water-soluble molecules. *Pharmaceutical Research* 24(4):803–810.

Chiappetta, D.A., Facorro, G., de Celis, E.R. and Sosnik, A. 2011. Synergistic encapsulation of the anti-HIV agent efavirenz within mixed poloxamine/poloxamer polymeric micelles. *Nanomedicine* 7(5):624–637.

Chiappetta, D.A., Hocht, C., Opezzo, J.A. and Sosnik, A. 2013. Intranasal administration of antiretroviral-loaded micelles for anatomical targeting to the brain in HIV. *Nanomedicine (London)* 8(2):223–237.

Chiappetta, D.A., Hocht, C., Taira, C. and Sosnik, A. 2010. Efavirenz-loaded polymeric micelles for pediatric anti-HIV pharmacotherapy with significantly higher oral bioavailability. *Nanomedicine (London)* 5(1):11–23.

Chiappetta, D.A., Hocht, C., Taira, C. and Sosnik, A. 2011. Oral pharmacokinetics of the anti-HIV efavirenz encapsulated within polymeric micelles. *Biomaterials* 32: 2379–2387.

Choukrani, G., Maharjan, B., Park, C.H., Kim, C.S. and Kurup Sasikala, A.R. 2020. Biocompatible superparamagnetic sub-micron vaterite particles for thermo-chemotherapy: From controlled design to in vitro anticancer synergism. *Materials Science and Engineering* 106:110226–110234.

Chu, D.T., Nguyen, T.T., Tien, N.L.B., et al. 2020. Recent progress of stem cell therapy in cancer treatment: Molecular mechanisms and potential applications. *Cells* 9 (3): 563–582.

Chudasaina, A., Patel, V.K., Nivsarkar, M., Vasu, K. and Shishoo, C. 2011. Investigation of microemulsion system for transdermal delivery of itraconazole. *Journal of Advanced Pharmaceutical Technology & Research* 2:30–38.

Clayton, R., Ohagen, A. and Nicol, F. 2009. Sustained and specific in-vitro inhibition of HIV-1 replication by a protease inhibitor encapsulated in gp120- targeted liposomes. *Antiviral Research* 84:142–149.

Constantinides, P.P., Lambert, K.J., Tustian, A.K., et al. 2000. Formulation development and antitumor activity of a filter-sterilizable emulsion of paclitaxel. *Pharmaceutical Research* 17(2):175–182.

Contreras, L., Villarroel, I., Torres, C. and Rozas, R. 2021. Doxorubicin encapsulation in carbon nanotubes having Haeckelite or Stone-Wales defects as drug carriers: A molecular dynamics approach. *Molecules* 26(6):1586–1604.

Costa Lima, S.A., Gaspar, A., Reis, S. and Durães L. 2017. Multifunctional nanospheres for co-delivery of methotrexate and mild hyperthermia to colon cancer cells. *Materials Science and Engineering: C Materials for Biological Applications* 75:1420–1426.

Cojocaru. F.D., Botezat. D., Gardikiotis. I. et.al. 2020. Nanomaterials Designed for Antiviral Drug Delivery Transport across Biological Barriers. *Pharmaceutics* 12:171–205.

Damasco, J.A., Ravi, S., Perez, J.D., Hagaman, D.E. and Melancon, M.P. 2020. Understanding nanoparticle toxicity to direct a safe-by-design approach in cancer nanomedicine. *Nanomaterials (Basel)* 10(11):2186–2227.

Darini, A., Eslaminejad, T., Nematollahi Mahani, S.N. and Ansari, M. 2019. Magnetogel nanospheres composed of cisplatin-loaded alginate/B-cyclodextrin as controlled release drug delivery. *Advanced Pharmaceutical Bulletin* 9(4):571–577.

Deavall, D.G., Martin, E.A., Horner, J.M. and Roberts, R. 2012. Drug-induced oxidative stress and toxicity. *Journal Toxicology* 2012:1–13.

Deepika, M.S., Thangam, R., Sheena, T.S., Vimala, R.T.V. 2019. Dual drug loaded PLGA nanospheres for synergistic efficacy in breast cancer therapy. *Materials Science and Engineering: C Materials for Biological Applications* 103:109716–109734.

De Jalon, E.G., Blanco-Prieto, M.J., Ygartua, P. and Santoyo, S. 2001. Topical application of acyclovir-loaded microparticles: Quantification of the drug in porcine skin layers. *The Journal of Controlled Release* 75:191–197.

de Las Cuevas, N., Garcia-Gallego, S. and Rasines, B. 2012. In-vitro studies of water-stable cationic carbosilane dendrimers as delivery vehicles for gene therapy against HIV and hepatocarcinoma. *Current Medicinal Chemistry* 19:5052–5061.

de Martel, C., Plummer, M., Vignat, J. and Franceschi, S. 2017. Worldwide burden of cancer attributable to HPV by site, country and HPV type. *International Journal of Cancer* 41(4):664–670.

de Thé, G. and Kazanji, M. 1996. An HTLV-I/II vaccine: From animal models to clinical trials?. *Journal of Acquired Immune Deficiency Syndromes* 13:191–198.

Dehghan, M.H.G. 2008. Advances in iontophoresis for drug delivery. *Journal of Health Research* 1:115–127.

Delaney, G., Jacob, S., Featherstone, C. and Barton, M. 2005. The role of radiotherapy in cancer treatment: Estimating optimal utilization from a review of evidence-based clinical guidelines. *Cancer* 104(6):1129–1137.

Destache, C.J., Belgum, T., Goede, M., Shibata, A. and Belshan, M.A. 2010. Antiretroviral release from poly(DL-lactide-co-glycolide) nanoparticles in mice. *Journal of Antimicrobial Chemotherapy* 65:2183–2187.

Dutta, T., Garg, M. and Jain, N.K. 2008. Targeting of efavirenz loaded tuftsin conjugated poly(propyleneimine) dendrimers to HIV infected mac- rophages in vitro. *European Journal of Pharmaceutical Sciences* 34:181–189.

Duzguneş, N., Simões, S., Slepushkin, V., et al. 2005. Delivery of antiviral agents in liposomes. *Methods Enzymol* 391:351–373.

Edwards, S.J., Barton, S., Thurgar, E. and Trevor, N. 2015. Topotecan, pegylated liposomal doxorubicin hydrochloride, paclitaxel, trabectedin and gemcitabine for advanced recurrent or refractory ovarian cancer: A systematic review and economic evaluation. *International Journal of Technology Assessment in Health Care* 19(7):1–480.

El-Sayed, A. and Kamel, M. 2020. Advances in nanomedical applications: Diagnostic, therapeutic, immunization, and vaccine production. *Environmental Science and Pollution Research International* 27(16) :19200–19213.

Fairchild, A., Tirumani, S.H., Rosenthal, M.H., et al. 2015. Hormonal therapy in oncology: A primer for the radiologist. *American Journal of Roentgenology* 204(6):W620–W630.

Feng, M., Cai, Q., Huang, H. and Zhou, P. 2008a. Liver targeting and anti-HBV activity of reconstituted HDL-acyclovir palmitate complex. *European Journal of Pharmaceutics and Biopharmaceutics* 68(3):688–693.

Feng, M., Cai, Q., Shi, X., Huang, H., Zhou, P. and Guo, X. 2008b. Recombinant high-density lipoprotein complex as a targeting system of nosiheptide to liver cells. *Journal of Drug Target* 16(6):502–508.

Feng, T., Wei, Y., Lee, R.J. and Zhao, L. 2017. Liposomal curcumin and its application in cancer. *The International Journal of Nanomedicine* 12:6027–6044.

Feng, Z.Q., Yan, K., Li, J., Xu, X., Yuan, T., Wang, T. and Zheng, J. 2019. Magnetic Janus particles as a multifunctional drug delivery system for paclitaxel in efficient cancer treatment. *Materials Science and Engineering: C Materials for Biological Applications* 104:109716–109734.

Ferguson L.R., Chen, H., Collins A.R., et al. 2015. Genomic instability in human cancer: Molecular insights and opportunities for therapeutic attack and prevention through diet and nutrition. *Seminars in Cancer Biology* 35:S5–S24.

Ferreira, D., Fontinha, D., Martins, C., Pires, D., Fernandes, A.R. and Baptista, P.V. 2010. Gold nanoparticles for vectorization of nucleic acids for cancer therapeutics. *Molecules* 25(15):3489–3507.

Figueira, T.N., Domingues, M.M., Illien, F., et al. 2020. Enfuvirtide-Protoporphyrin IX dual-loaded liposomes: In vitro evidence of synergy against HIV-1 entry into cells. *ACS Infectious Diseases* 6(2):224–236.

Flores, S.E., Rial-Hermida, M.I., Ramirez, J.C., Pazos, A., Concheiro, A., Alvarez-Lorenzo, C. and Peralta, R.D. 2016. Microemulsions for colorectal cancer treatments: General considerations and formulation of methotrexate. *Mini-Reviews in Medicinal Chemistry* 16(6):498–508.

Fonseca-Santos, B., Gremião, M.P.D. and Chorilli, M. 2015. Nanotechnology-based drug delivery systems for the treatment of Alzheimer's disease. *International Journal of Nanomedicine* 10:4981–5003.

Fouad, Y.A. and Aanei, C. 2017. Revisiting the hallmarks of cancer. *American Journal of Cancer Research* 7(5):1016–1036.

Gagliardi, M. 2017. Biomimetic and bioinspired nanoparticles for targeted drug delivery. *Therapeutic Delivery* 8:289–299.

Gagné, J.F., Désormeaux, A., Perron, S., Tremblay, M.J. and Bergeron, M.G. 2002.

Targeted delivery of indinavir to HIV-1 primary reservoirs with immunoliposomes. *Biochemistry Biophysics Acta* 1558(2):198–210.

Garg, M., Asthana, A., Agashe, H.B., Agrawal, G.P. and Jain, N.K. 2006. Stavudine-loaded mannosylated liposomes: In-vitro anti-HIV-I activity, tissue distribution and pharmacokinetics. *Journal of Pharmacy and Pharmacology* 58(5):605–616.

Gianfaldoni, S., Gianfaldoni, R., Wollina, U., Lotti, J., Tchernev, G. and Lotti, T. 2017. An overview on radiotherapy: From its history to its current applications in dermatology. *Open Access Macedonian Journal of Medical Sciences* 5(4):521–525.

Gilman, A. 1963. The initial clinical trial of nitrogen mustard. *American Journal of Surgery* 105:574–578.

Gilman, A. and Philips, F.J. 1946. The biological actions of therapeutic applications of the B-chloroethyl amines and sulfides. *Science* 103:409–436.

Goldberg, M., Langer, R. and Jia, X. 2007. Nanostructured materials for applications in drug delivery and tissue engineering. *Journal of Biomaterials Science, Polymer Edition* 18:241–268.

Gonzalez-Rodriguez, R., Campbell, E. and Naumov, A. 2019. Multifunctional graphene oxide/iron oxide nanoparticles for magnetic targeted drug delivery dual magnetic resonance/fluorescence imaging and cancer sensing. *PLoS One* 14(6):e0217072–e0217090.

Gopinath, D., Ravi, D., Karwa, R., Rao, B.R., Shashank, A. and Rambhau, D. 2001. Pharmacokinetics of zidovudine following intravenous bolus administration of a novel niosome preparation devoid of cholesterol. *Arzneimittelforschung* 51(11):924 -930.

Goto, K., Roca Suarez, A.A., Wrensch, F., Baumert, T.F. and Lupberger, J. 2020. Hepatitis C virus and hepatocellular carcinoma: When the host loses its grip. *International Journal of Molecular Sciences* 21(9):3057–3075.

Gratieri, T. and Kalia, Y.N. 2014. Targeted local simultaneous iontophoresis of chemotherapeutics for topical therapy of head and neck cancers. *International Journal of Pharmacy* 460(1-2):24–27.

Grillone, A., Battaglini, M., Moscato, S., et al. 2019. Nutlin-loaded magnetic solid lipid nanoparticles for targeted glioblastoma treatment. *Nanomedicine (Lond).* 14(6): 727–752.

Guney Eskiler, G., Cecener, G., Dikmen, G., Egeli, U. and Tunca, B. 2018. Solid lipid nanoparticles: Reversal of tamoxifen resistance in breast cancer. *European Journal of Pharmaceutical Sciences* 120:73–88.

Hanahan, D. and Weinberg, R.A. 2000. The hallmarks of cancer. *Cell* 100:57–70.

Hanahan, D. and Weinberg, R.A. 2011. Hallmarks of cancer: The next generation. *Cell* 144: 646–674.

Harris, J.M., Martin, N.E. and Modi, M. 2001. Pegylation: A novel process for modifying pharmacokinetics. *Clinical Pharmacokinetics* 40(7):539–551.

He, G., Ma, Y., Zhou, H. et al. 2019. Mesoporous NiS_2 nanospheres as a hydrophobic anticancer drug delivery vehicle for synergistic photothermal-chemotherapy. *Journal of Materials Chemistry B* 7(1):143–149.

Himabindhu, G. 2020. Hormonal therapy. *Endocrinology Metabolism Syndrome* 9:315–316.

Hoshyar, N., Gray, S., Han, H. and Bao, G. 2016. The effect of nanoparticle size on in vivo pharmacokinetics and cellular interaction. *Nanomedicine* 11:673–692.

Huang, H.C., Barua, S., Sharma, G., Dey, S.K. and Rege, K. 2011. Inorganic nanoparticles for cancer imaging and therapy. *The Journal of Controlled Release* 155(3):344–357.

Izzedine, H., Launay-Vacher, V. and Deray, G. 2005. Antiviral drug-induced nephrotoxicity. *American Journal of Kidney Diseases* 45(5):804–817.

Jain, S., Tiwary, A.K. and Jain N.K. 2006. Sustained and targeted delivery of an anti-HIV agent using elastic liposomal formulation: Mechanism of action. *Current Drug Delivery* 3:157–166.

Jain, S., Tiwary, A.K. and Jain, N.K. 2008. PEGylated elastic liposomal formulation for lymphatic targeting of zidovudine. *Current Drug Delivery* 5(4):275–281.

Javan, F., Vatanara, A., Azadmanesh, K., Nabi-Meibodi, M. and Shakouri, M. 2017. Encapsulation of ritonavir in solid lipid nanoparticles: In-vitro anti-HIV-1 activity using lentiviral particles. *Journal of Pharmacy and Pharmacology* 69(8): 1002–1009.

Jiang, K., Hu, Y., Gao, X., et al. 2019. Octopus-like flexible vector for noninvasive intraocular delivery of short interfering nucleic acids. *Nano Letter* 19(9): 6410–6417.

Jin, S.X., Bi, D.Z., Wang, J., Wang, Y.Z., Hu, H.G. and Deng, Y.H. 2005. Pharmacokinetics and tissue distribution of zidovudine in rats following intravenous administration of zidovudine myristate loaded liposomes. *Pharmazie* 60(11):840–843.

Joshy, K.S., Susan, M.A., Snigdha, S., Nandakumar, K., Laly, A.P. and Sabu, T. 2018. Encapsulation of zidovudine in PF-68 coated alginate conjugate nanoparticles for anti-HIV drug delivery. *International Journal of Biological Macromolecules* 107(Pt A): 929–937.

Kanda, T., Yajima, M., Ikuta, K. 2019. Epstein-Barr virus strain variation and cancer. *Cancer Science* 110(4):1132–1139.

Karami, Z., Saghatchi Zanjani, M.R., Rezaee, S., Rostamizadeh, K., Hamidi, M. 2019. Neuropharmacokinetic evaluation of lactoferrin-treated indinavir-loaded nanoemulsions: Remarkable brain delivery enhancement. *Drug Development and Industrial Pharmacy* 45(5):736–744.

Kaufmann, K. B., Büning, H., Galy, A., Schambach, A. and Grez, M. 2013. Gene therapy on the move. *EMBO Molecular Medicine* 5(11):1642–1661.

Kaur, B. and Newton, M.J. 2016. Impact of Pluronic F-68 vs tween 80 on fabrication and evaluation of acyclovir SLNs for skin delivery. *Recent Patents on Drug Delivery and Formulation* 10(3):207–221.

Kigasawa, K., Kajimoto, K., Nakamura, T., Hama, S., Kanamura, K., Harashima, H. and Kogure, K. 2011. Noninvasive and efficient transdermal delivery of CpG-oligodeoxynucleotide for cancer immunotherapy. *The Journal of Controlled Release* 150(3):256–265.

Kim, B.Y., Rutka, J.T. and Chan, W.C. 2010. Nanomedicine. *The New England Journal of Medicine*. 363(25):2434–2443.

Kim, D., Amatya, R., Hwang, S., Lee, S., Min, K.A. and Shin, M.C. 2021. BSA-Silver nanoparticles: A potential multimodal therapeutics for conventional and photothermal treatment of skin cancer. *Pharmaceutics* 13(4):575–589.

Kim, S.I., Shin, D., Lee, H., Ahn, B.Y., Yoon, Y. and Kim, M. 2009. Targeted delivery of siRNA against hepatitis C virus by apolipoprotein A-I-bound cationic liposomes. *Journal of Hepatology* 50(3):479–488.

Kim, S.S., Harford, J.B., Pirollo, K.F. and Chang, E.H. 2015. Effective treatment of glioblastoma requires crossing the blood-brain barrier and targeting tumors including cancer stem cells: The promise of nanomedicine. *Biochemical and Biophysical Research Communications* 468(3):485–489.

Komuro, M., Suzuki, K., Kanebako, M., et al. 2013. Novel iontophoretic administration method for local therapy of breast cancer. *The Journal of Controlled Release* 168(3): 298–306.

Kou, L., Huang, H., Lin, X., et al. 2020. Endocytosis of $ATB^{0,+}$(SLC6A14)-targeted liposomes for drug delivery and its therapeutic application for pancreatic cancer. *Expert Opinion on Drug Delivery* 17(3):395–405.

Kumar, K., Kumar Shyamlal, B.R., Verma, R., Kondaiah, P. and Chaudhary, S. 2020. Reduction-triggered doxorubicin delivery by self-assembled nanospheres of lipoylated caffeine. *Chemical Medical Chemistry* 15(9):733–737.

Kuo, Y.C. and Chen, H.H. 2006. Effect of nanoparticulate polybutylcyanoacrylate and methylmethacrylate-sulfopropylmethacrylate on the permeability of zidovudine and lamivudine across the in vitro blood-brain barrier. *International Journal of Pharmaceutics* 327(1-2):160–169.

Kuo, Y.C. and Chen, H.H. 2009. Entrapment and release of saquinavir using novel cationic solid lipid nanoparticles. *International Journal of Pharmaceutics* 365:206–213.

Kuo, Y.C. and Chung, J.F. 2011. Physicochemical properties of nevirapine-loaded solid lipid nanoparticles and nanostructured lipid carriers. *Colloids and Surfaces B: Biointerfaces* 83(2):299–306.

Kuo, Y.C. and Hsin-Wei Y. 2011. Expression of ornithine decarboxylase during the transport of saquinavir across the blood–brain barrier using composite polymeric nanocarriers under an electromagnetic field. *Colloids and Surfaces B: Biointerfaces* 88(2):627–634.

Kuo, Y.C. and Kuo, C.Y. 2008. Electromagnetic interference in the permeability of saquinavir across the blood-brain barrier using nanoparticulate carriers. *International Journal of Pharmaceutics* 351:271–281.

Kuo, Y.C., Lin, P.I. and Wang, C.C. 2011. Targeting nevirapine delivery across human brain microvascular endothelial cells using transferrin-grafted poly(lactide-co-glycolide) nanoparticles. *Nanomedicine (Lond)* 6(6):1011–1026.

Lawrie, T.A., Bryant, A., Cameron, A., Gray, E. and Morrison, J. 2013. Pegylated liposomal doxorubicin for relapsed epithelial ovarian cancer. *Cochrane Database of Systematic Reviews* 2013(7):1–124.

Leowattana, W. 2019. Antiviral drugs and Acute Kidney Injury (AKI). *Infect Disord Drug Targets* 19(4):375–385.

Li, D. and Al-Jamal, K.T. 2021. siRNA design and delivery based on carbon nanotubes. *Methods in Molecular Biology* 2282:181–193.

Li, J., Zhou, M., Liu, F., Xiong, C., et al. 2016. Hepatocellular carcinoma: Intra-arterial delivery of doxorubicin-loaded hollow gold nanospheres for photothermal ablation-chemoembolization therapy in rats. *Radiology* 281(2):427–435.

Li, Q., Du, Y.Z., Yuan, H., Zhang, X.G., Miao, J., Cui, F.D. and Hu, F.Q. 2010. Synthesis of lamivudine stearate and antiviral activity of stearic acid-g-chitosan oligosaccharide polymeric micelles delivery system. *European Journal of Pharmaceutical Sciences* 41:498–507.

Li, Q., Du, Y.Z., Yuan, H., Zhang, X.G., Miao, J., Cui, F.D. and Hu, F.Q. 2010. Synthesis of lamivudine stearate and antiviral activity of stearic acid-γ-chitosan oligosaccharide polymeric micelles delivery system. *European Journal of Pharmaceutical Sciences* 41(3-4):498–507

Li, Y., Xu, P., He, D., Xu, B., Tu, J. and Shen, Y. 2020. Long-circulating thermosensitive liposomes for the targeted drug delivery of oxaliplatin. *The International Journal of Nanomedicine* 15:6721–6734.

Lin, H., Lin, L., Choi, Y. and Michniak-Kohn B. 2020. Development and in-vitro evaluation of co-loaded berberine chloride and evodiamine ethosomes for treatment of melanoma. *International Journal of Pharmaceutics* 581:119278–119285.

Lopes, A., Vandermeulen, G. and Préat, V. 2019. Cancer DNA vaccines: Current preclinical and clinical developments and future perspectives. *Journal of Experimental & Clinical Cancer Research* 38(1):146–170.

Lv, R., Gai, S., Dai, Y., Niu, N., He, F. and Yang, P. 2013. Highly uniform hollow GdF3 spheres: Controllable synthesis, tuned luminescence, and drug-release properties. *ACS Applied Materials & Interfaces* 5(21):10806–10818.

Ma, B., Wang, S., Liu, F., Zhang, S., et al. 2019. Self-assembled copper-amino acid nanoparticles for in situ glutathione "and" H_2O_2 sequentially triggered chemodynamic therapy. *Journal of the American Chemical Society* 141(2):849–857.

Ma, L., Wang, X., Wu, J., et al. 2019. Polyethylenimine and sodium cholate-modified ethosomes complex as multidrug carriers for the treatment of melanoma through transdermal delivery. *Nanomedicine (Lond)* 18:2395–2408.

Maeda, E., Akahane, M., Kiryu, S., et al. 2009. Spectrum of Epstein-Barr virus-related diseases: A pictorial review. *The Japanese Journal of Radiology* 27:4–19.

Mahajan, S.D., Aalinkeel, R., Law, W.C., et al. 2012. Anti-HIV-1 nanotherapeutics: Promises and challenges for the future. *The International Journal of Nanomedicine* 7: 5301–5314.

Mahajan, S.D., Roy, I. and Xu, G. 2010. Enhancing the delivery of anti retroviral drug "Saquinavir" across the blood brain barrier using nanoparticles. *Current HIV Research* 8(5):396–404.

Mahalunkar, S., Yadav, A.S., Gorain, M., et al. 2019. Functional design of pH-responsive folate-targeted polymer-coated gold nanoparticles for drug delivery and in vivo therapy in breast cancer. *The International Journal of Nanomedicine* 14:8285–8302.

Mahboobian, M.M., Mohammadi, M. and Mansouri, Z. 2020. Development of thermo-sensitive in situ gel nanoemulsions for ocular delivery of acyclovir. *Journal of Drug Delivery Science and Technology* 55:101400–101436.

Mainardes, R.M. and Gremiao, M.P. 2012. Nanoencapsulation and characterization of zidovudine on poly(L-lactide) and poly(L-lactide)-poly(ethylene glycol)-blend nanoparticles. *Journal of Nanoscience and Nanotechnology* 12(11):8513–8521.

Makabi-Panzu, B., Gourde, P., Desormeaux, A. and Bergeron, M.G. 1998. Intracellular and serum stability of liposomal 2',3'-dideoxycytidine: Effect of lipid composition. *Cell and Molecular Biology* 44:277–284.

Manivasagan, P., Bharathiraja, S., Bui, N.Q., Lim, I.G. and Oh, J. 2016. Paclitaxel-loaded chitosan oligosaccharide-stabilized gold nanoparticles as novel agents for drug delivery and photoacoustic imaging of cancer cells. *International Journal of Pharmaceutics* 511(1):367–379.

Maslov, A.Y. and Vijg, J. 2009. Genome instability, cancer and aging. *Biochemical Biophysics Acta* 1790(10):963–969.

McNeil, S.E. 2011. Unique benefits of nanotechnology to drug delivery and diagnostics. *Methods in Molecular Biology* 697:3–8.

Melero, A., Draheim, C., Hansen, S., et al. 2017. Targeted delivery of Cyclosporine A by polymeric nanocarriers improves the therapy of inflammatory bowel disease in a relevant mouse model. *European Journal of Pharmaceutics and Biopharmaceutics.* 119:361–371.

Meng, J., Sturgis, T.F., Youan, B.B. 2011. Engineering tenofovir loaded chitosan nanoparticles to maximize microbicide mucoadhesion. *European Journal of Pharmaceutical Sciences* 44:57–67.

Meng, J., Agrahari, V., Ezoulin, M.J., et al. 2016. Tenofovir containing thiolated chitosan core/shell nanofibers: In vitro and in vivo evaluations. *Molecular Pharmaceutics* 13(12):4129–4140.

Milla, P., Dosio, F. and Cattel, L. 2012.PEGylation of proteins and liposomes: A powerful and flexible strategy to improve the drug delivery. *Current Drug Metabolism* 13(1):105–109.

Merlo, D.F., Filiberti, R., Kobernus, M., et al. 2012. Cancer risk and the complexity of the interactions between environmental and host factors: HENVINET interactive diagrams as simple tools for exploring and understanding the scientific evidence. *Environ Health* 11(S9):1–12.

Mojeiko, G., de Brito, M., Salata, G.C. and Lopes, L.B. 2019. Combination of microneedles and microemulsions to increase celecoxib topical delivery for potential application in chemoprevention of breast cancer. *International Journal of Pharmaceutics* 560: 365–376.

Mukherjee, S., Kotcherlakota, R., Haque, S., Bhattacharya, D., Kumar, J.M., Chakravarty, S. and Patra, C.R. 2020. Improved delivery of doxorubicin using rationally designed PEGylated platinum nanoparticles for the treatment of melanoma. *Materials Science and Engineering: C Materials for Biological Applications*. 108:110375–110416.

Muro, E., Pons, T., Lequeux, N., et al. 2010. Small and stable sulfobetaine zwitterionic quantum dots for functional live-cell imaging. *Journal of the American Chemical Society* 132:4556–4557.

Muthu, M.S. and Singh, S. 2009. Targeted nanomedicines: Effective treatment modalities for cancer, AIDS and brain disorders. *Nanomedicine* 4:105–118.

Nair, R.S., Billa, N., Leong, C.O. and Morris, A.P. 2021. An evaluation of tocotrienol ethosomes for transdermal delivery using Strat-M® membrane and excised human skin. *Pharmaceutical Development and Technology* 26(2):243–251.

Neves, J.D., Amiji, M.M., Bahia, M.F. and Sarmento, B. 2010. Nanotechnology-based systems for the treatment and prevention of HIV/AIDS. *Advanced Drug Delivery Reviews* 62:458–477.

Niemeyer, C.M. and Mirkin, C.A. 2004. *Nanobiotechnology: Concepts, Applications and Perspectives*, Vol. 1. Weinheim: John Wiley & Sons.

Niu, G., Cogburn, B. and Hughes, J. 2010. Preparation and characterization of doxorubicin liposomes. *Methods in Molecular Biology* 624:211–219.

Niu, X.Q., Zhang, D.P., Bian, Q., et al. 2019. Mechanism investigation of ethosomes transdermal permeation. *International Journal of Pharmaceutics* 1:102019–100040.

Owens, D.E. and Peppas, N.A. 2006. Opsonization, biodistribution, and pharmacokinetics of polymeric nanoparticles. *International Journal of Pharmaceutics* 307:93–102.

Palakurthi, N.K., Correa, Z.M., Augsburger, J.J. and Banerjee, R.K. 2011. Toxicity of a biodegradable microneedle implant loaded with methotrexate as a sustained release device in normal rabbit eye: A pilot study. *Journal of Ocular Pharmacology and Therapeutics* 27(2):151–156.

Parsa, N. 2012. Environmental factors inducing human cancers. *Iran Journal of Public Health* 41(11):1–9.

Parthiban, R., Shanmugapriya, I., Sathishkumar, S., Prakash, R. 2020. Design and evaluation of acyclovir-loaded solid lipid nanoparticles for sustained release. *Drug Invention Today* 14(1):108–111.

Patel, K.K., Kumar, P. and Thakkar, H.P. 2012. Formulation of niosomal gel for enhanced transdermal lopinavir delivery and its comparative evaluation with ethosomal gel. *AAPS Pharmacy SciTech* 13(4):1502–1510.

Pavel, M., Renna, M., Park, S.J., et al. 2018. Contact inhibition controls cell survival and proliferation via YAP/TAZ-autophagy axis. *Nature Communications* 9:2961–2979.

Pavlova, N.N. and Thompson, C.B. 2016. The emerging hallmarks of cancer metabolism. *Cell Metabolism* 23(1):27–47.

Peira, E., Chirio, D., Carlotti, M.E., Spagnolo, R. and Trotta, M. 2009. Formulation studies of microemulsions for topical applications of acyclovir. *Journal of Drug Delivery Science and Technology* 19:191–196.

Peng, J. and Liang, X. 2019. Progress in research on gold nanoparticles in cancer management. *Medicine (Baltimore)* 98(18):e15311–e15318.

Peram, M.R., Jalalpure, S., Kumbar, V., et al. 2019. Factorial design based curcumin ethosomal nanocarriers for the skin cancer delivery: In vitro evaluation. *Journal of Liposome Research* 29(3):291–311.

Perez, C.R. and De Palma, M. 2019. Engineering dendritic cell vaccines to improve cancer immunotherapy. *Nature Communication* 10(1):5408–5418.

Petrilli, R., Eloy, J.O., Saggioro, F.P., et al. 2018. Skin cancer treatment effectiveness is improved by iontophoresis of EGFR-targeted liposomes containing 5-FU compared with subcutaneous injection. *The Journal of Controlled Release* 283:151–162.

Petros, R.A. and DeSimone, J.M. 2010. Strategies in the design of nanoparticles for therapeutic applications. *Journal of Liposome Research* 9:615–627.

Phillips, M.A., Gran, M.L., Peppas, N.A. 2010. Targeted nanodelivery of drugs and diagnostics. *Nano Today* 5:143–159.

Qin, T., Dai, Z., Xu, X., et al. 2021. Nanosuspension as an efficient carrier for improved ocular permeation of voriconazole. *Current Pharmaceutical Biotechnology* 22(2): 245–253.

Rabiei, M., Kashanian, S., Samavati, S.S., Jamasb, S. and McInnes, S.J.P. 2020. Active targeting towards and inside the brain based on nanoparticles: A review. *Current Pharmaceutical Biotechnology* 21(5):374–383.

Rahme, K., Guo, J. and Holmes, J.D. 2019. Bioconjugated gold nanoparticles enhance siRNA delivery in prostate cancer cells. *Methods in Molecular Biology* 1974: 291–301.

Rajpoot, K. and Jain, S.K. 2019. Irinotecan hydrochloride trihydrate loaded folic acid-tailored solid lipid nanoparticles for targeting colorectal cancer: Development, characterization, and in vitro cytotoxicity study using HT-29 cells. *Journal of Microencapsulation* 36(7):659–676.

Ramadon, D., McCrudden, M.T.C., Courtenay, A.J., et al. 2021. Enhancement strategies for transdermal drug delivery systems: Current trends and applications. *Drug Delivery and Translational Research Current Pharmaceutical Biotechnology* 20:1–34.

Ramana, L.N., Sethuraman, S., Ranga, U., Krishnan, U.M. 2010. Development of a liposomal nanodelivery system for nevirapine. *Journal of Biomedical Science* 17(1): 57–66.

Ramana, L.N., Sharma, S., Sethuraman, S., Ranga, U. and Krishnan, U.M. 2014. Evaluation of chitosan nanoformulations as potent anti-HIV therapeutic systems. *Biochim Biophys Acta* 1840(1):476–484.

Riley, P.R. and Narayan, R.J. 2021. Recent advances in carbon nanomaterials for biomedical applications: A review. *Current Opinion in Biomedical Engineering* 17: 100262–100272.

Roberts, S.A. and Gordenin, D.A. 2014. Hypermutation in human cancer genomes: Footprints and mechanisms. *Nature Reviews Cancer* 14(12):786–800.

Rosenblum, D., Joshi, N., Tao, W., Karp, J.M. and Peer, D. 2018. Progress and challenges towards targeted delivery of cancer therapeutics. *Nature Communication* 9(1): 1410–1422.

Ruckmani K., Sankar V. and Sivakumar M. 2010. Tissue distribution, pharmacokinetics and stability studies of zidovudine delivered by niosomes and proniosomes. *Journal of Biomedical Nanotechnology* 6:43–51.

Saiyed, Z.M., Gandhi, N.H. and Nair, M.P. 2010. Magnetic nanoformulation of azidothymidine 5'-triphosphate for targeted delivery across the blood-brain barrier. *The International Journal of Nanomedicine* 5:157–166.

Sanita, G., Carrese, B. and Lamberti, A. 2020. Nanoparticle surface functionalization: How to improve biocompatibility and cellular internalization. *Frontiers in Molecular Biosciences* 7:587012–587032.

Sankar, V., Nareshkumar, P.N., Ajitkumar, G.N., Penmetsa, S.D. and Hariharan, S. 2012. Comparative studies of lamivudine-zidovudine nanoparticles for the selective uptake by macrophages. *Current Drug Delivery* 9(5):506–514.

Sarkar, N., Pal, A., Das, D., et al. 2015. Virological characteristics of acute hepatitis B in Eastern India: Critical differences with chronic infection. *PLoS One* 10(11): e0141741–e0141756.

Scheetz, L., Park, K.S., Li, Q., Lowenstein, P.R., Castro, M.G., Schwendeman, A. and Moon, J.J. 2019. Engineering patient-specific cancer immunotherapies. *Nature Biomedical Engineering* 3(10):768–782.

Shaat, H., Mostafa, A., Moustafa, M., Gamal-Eldeen, A., Emam, A., El-Hussieny, E. and Elhefnawi, M. 2016. Modified gold nanoparticles for intracellular delivery of anti-liver cancer siRNA. *International Journal of Pharmaceutics* 504(1-2):125-133.

Shah, L.K. and Amiji, M.M. 2006. Intracellular delivery of saquinavir biodegradable polymeric nanoparticles for HIV/AIDS. *Pharmaceutical Research* 23:2638–2645.

Shegokar, R. and K.K. Singh. 2011. Stavudine entrapped lipid nanoparticles for targeting lymphatic HIV reservoirs. *International Journal of Pharmaceutical Sciences* 66(4): 264–271.

Shmakova, A., Germini, D. and Vassetzky, Y. 2020. HIV-1, HAART and cancer: A complex relationship. *International Journal of Cancer*. 146(10):2666–2679.

Shome, D., Kalita, D., Jain, V., Sarin, R., Maru, G.B., Bellare, J.R. 2014. Carboplatin loaded polymethylmethacrylate nano-particles in an adjunctive role in retinoblastoma: An animal trial. *Indian Journal of Ophthalmology* 62(5):585–589.

Singh, S., Dobhal, A.K., Jain, A., Pandit, J.K. and Chakraborty, S. 2010. Formulation and evaluation of solid lipid nanoparticles of a water soluble drug: Zidovudine. *Chemical Pharmacology Bullet (Tokyo)* 58(5):650–655.

Sloan, F.A. and Gelband, H. 2007. Institute of medicine (US) committee on cancer control in low- and middle-income countries. In *Cancer Control Opportunities in Low- and Middle-Income Countries*. eds. Frank A. Sloan and Hellen Gelband, National Academics Press (US).

Soni, K., Mujtaba, A., Akhter, M.H., Zafar, A. and Kohli, K. 2020. Optimisation of ethosomal nanogel for topical nano-CUR and sulphoraphane delivery in effective skin cancer therapy. *Journal of Micro-encapsulation* 37(2):91–108.

Sung, H., Ferlay, J., Siegel, R.L., Laversanne, M., Soerjomataram, I., Jemal. A. and Bray, F. 2021. Global cancer statistics 2020: GLOBOCAN estimates of incidence and mortality worldwide for 36 cancers in 185 countries. *A Cancer Journal for Clinicians* 71(3): 209–249.

Tan, L., Liu, T., Fu, C., Wang, S., Fu, S., Ren, J. and Meng, X. 2016. Hollow ZrO_2/PPy nanoplatform for improved drug delivery and real-time CT monitoring in synergistic photothermal-chemo cancer therapy. *Journal of Materials Chemistry* 4(5):859–866.

Taveira, S.F., De Santana, D.C., Araújo, L.M., Marquele-Oliveira, F., Nomizo, A., Lopez, R.F. 2014. Effect of iontophoresis on topical delivery of doxorubicin-loaded solid lipid nanoparticles. *Journal of Biomedical Nanotechnology* 10(7):1382–1390.

Teppo, H.R., Soini, Y. and Karihtala, P. 2017. Reactive oxygen species-mediated mechanisms of action of targeted cancer therapy. *Oxidative Medicine and Cellular Longevity* 2017:1485283–1485294.

Thambiraj, S., Shruthi, S., Vijayalakshmi, R. and Ravi Shankaran, D. 2019. Evaluation of cytotoxic activity of docetaxel loaded gold nanoparticles for lung cancer drug delivery. *Cancer Treatment and Research Communications* 21:100157–100165.

Thomas, S. and Prendergast, G.C. 2016. Cancer vaccines: A brief overview. *Methods in Molecular Biology* 1403:755–761.

Thomas, S.C., Harshita, P.K. and Talegaonkar, S. 2015. Ceramic nanoparticles: Fabrication methods and applications in drug delivery. *Current Pharmaceutical Design* 21(42): 6165–6188.

Timbang, M.R., Sim, M.W., Bewley, A.F., Farwell, D.G., Mantravadi, A., Moore, M.G. 2019. HPV-related oropharyngeal cancer: A review on burden of the disease and opportunities for prevention and early detection. *Human Vaccines & Immunotherapeutics* 15(7-8): 1920–1928.

Toyoda, M., Hama, S., Ikeda, Y., Nagasaki, Y. and Kogure, K. 2015. Anti-cancer vaccination by transdermal delivery of antigen peptide-loaded nanogels via iontophoresis. *International Journal of Pharmaceutics* 483(1-2):110–114.

Trybek, T., Kowalik, A., Góźdź, S. and Kowalska, A. 2020. Telomeres and telomerase in oncogenesis. *Oncology Letters* 20:1015–1027.

Tsai, Y.H., Hsieh, Y.H., Huang, Y.B., Chang, J.S., Huang, C.T., Wu, P.C. 2010. Microemulsions for intravesical delivery of gemcitabine. *Chemical and Pharmaceutical Bulletin (Tokyo)* 58(11):1461–1465.

Veiseh O., Sun C., Fang C., et al. 2009 Specific targeting of brain tumors with an optical/magnetic resonance imaging nanoprobe across the blood-brain barrier. *Cancer Research* 69:6200–6207.

Verma, P. and Pathak, K. 2010. Therapeutic and cosmeceutical potential of ethosomes: An overview. *Journal of Advanced Pharmaceutical Technology & Research* 1: 274–282.

Vockerodt, M., Yap, L.F., Shannon-Lowe, C., Curley, H., Wei, W., Vrzalikova, K. and Murray, P.G. 2015. The Epstein-Barr virus and the pathogenesis of lymphoma. *Journal of Pathology* 235(2):312–322.

Vonarbourg, A., Passirani, C., Saulnier, P. and Benoit, J.P. 2006. Parameters influencing the stealthiness of colloidal drug delivery systems. *Biomaterials* 27:4356–4373.

Wang, H., Ding, W., Peng, L., et al. 2020. Gadolinium-loaded solid lipid nanoparticles for colorectal tumor in MR colonography. *Journal of Biomedical Nanotechnology* 16(5): 594–602.

Wang, H., Zhou, L., Yang, Y. and Luo, B. 2021. Screening and identification of key genes in EBV-associated gastric carcinoma based on bioinformatics analysis. *Pathology - Research and Practice*. 222:153439–153446.

Wang, W., Chen, T., Xu, H., et al. 2018. Curcumin-loaded solid lipid nanoparticles enhanced anticancer efficiency in breast cancer. *Molecules* 23(7):1578–1591.

Weinstock, M., Rosenblatt, J. and Avigan, D. 2017. Dendritic cell therapies for hematologic malignancies. *Molecular Therapy Methods & Clinical Development* 5: 66–75.

Weiss, R.A., Whitby, D., Talbot, S., Kellam, P. and Boshoff, C. 1998. Human herpesvirus type 8 and Kaposi's sarcoma. *Journal of the National Cancer Institute* 23: 51–54.

Yao, Y., Zhou, Y., Liu, L., et al. 2020. Nanoparticle-based drug delivery in cancer therapy and its role in overcoming drug resistance. *Frontiers in Molecular Biosciences* 7: 193–207.

Yassemi, A., Kashanian, S. and Zhaleh, H. 2020. Folic acid receptor-targeted solid lipid nanoparticles to enhance cytotoxicity of letrozole through induction of caspase-3 dependent-apoptosis for breast cancer treatment. *Pharmaceutical Development and Technology* 25(4):397–407.

Yu, S., Bi, X., Yang, L., et al. 2019. Co-delivery of Paclitaxel and PLK1-targeted siRNA using aptamer-functionalized cationic liposome for synergistic anti-breast cancer effects in vivo. *Journal of Biomedical Nanotechnology* 15(6):1135–1148.

Zhai, B., Zeng, Y., Zeng, Z., et al. 2018. Drug delivery systems for elemene, its main active ingredient β-elemene, and its derivatives in cancer therapy. *The International Journal of Nanomedicine* 13:6279–6296.

Zhang, Q., Zhang, P., Jian, S., et al. 2020. Drug-bearing peptide-based nanospheres for the inhibition of metastasis and growth of cancer. *Molecular Pharmaceutics* 17(9): 3165–3176.

Zhang, X.G., Miao, J., Li, M.W., Jiang, S.P., Hu, F.Q. and Du, Y.Z. 2008. Solid lipid nanoparticles loading adefovir dipivoxil for antiviral therapy. *Journal of Zhejiang University: Science B* 9:506–510.

Zhang, Y., Ma, J.A., Zhang, H.X., Jiang, Y.N. and Luo, W.H. 2020. Cancer vaccines: Targeting KRAS-driven cancers. *Expert Review of Vaccines* 19(2):163–173.

Zhao, X., Pan, J., Li, W., Yang, W., Qin, L. and Pan, Y. 2018. Gold nanoparticles enhance cisplatin delivery and potentiate chemotherapy by decompressing colorectal cancer vessels. *The International Journal of Nanomedicine* 13:6207–6221.

Zhou, S., Sun, J., Sun, L., et al. 2008. Preparation and characterization of interferon-loaded magnetic biodegradable microspheres. *Journal of Biomedical Materials Research - Part B Applied Biomaterials* 87(1):189–196.

Zhu, L., Zhou, Z., Mao, H. and Yang, L. 2017. Magnetic nanoparticles for precision oncology: Theranostic magnetic iron oxide nanoparticles for image-guided and targeted cancer therapy. *Nanomedicine (London)* 12(1):73–87.

Zidan, A.S., Rahman, Z. and Khan, M.A. 2011. Product and process under-standing of a novel pediatric anti-HIV tenofovir niosomes with a high-pressure homogenizer. *European Journal of Pharmaceutical Sciences* 44:93–102.

13 Emerging Nanotechnology-Enabled Approaches to Mitigate COVID-19 Pandemic

Maria Shoukat, Samiullah Khan, Arshad Islam, Maleeha Azam, and Malik Badshah

CONTENTS

DOI: 10.1201/9781003136644-16

13.1 INTRODUCTION

The coronavirus disease of 2019 (COVID-19), caused by the Severe Acute Respiratory Syndrome Coronavirus 2 (SARS-CoV-2), has posed a serious health threat to the globe. The total number of cases around the globe had reached 178,587,177 by June 19, 2020, with mortality cases of around 3,866,643 in numbers inspiring a sprint to find new approaches to prevent, diagnose, and treat this viral infection (Worldometer 2021). This viral infection of COVID-19 is a pulmonary disease with symptoms of fever, pneumonia, sore throat, respiratory illness, conjunctivitis, and shortness of breath, or dyspnea. Since the preliminary epidemic starting in December 2019 in Wuhan (China), the virus has spread globally with an exponential growth rate. Keeping in view, the pandemic situation, the World Health Organisation (WHO) revealed it as "Public Health Emergency of International Concern (PHEIC)" on January 30, 2020. Furthermore, the outbreak was declared as "Pandemic" later, on March 11, 2020 (Boopathi et al. 2020).

During the last two decades, Coronavirus outbreaks severely affected the world with viral pneumonia, such as severe acute respiratory syndrome (SARS) and Middle East respiratory syndrome (MERS). The outbreak of SARS first occurred at China in 2002, resulting in hundreds of deaths worldwide with an 11% mortality rate. MERS first originated in Saudi Arabia and afterward transmitted to other countries with a 37% fatality rate. The causative agent of SARS, MERS, and COVID-19 belong to the β-lineage of Coronaviruses.

The virus causing COVID-19 belongs to the subfamily *Orthocoronavirinae* in family *Coronaviridae* and order *Nidovirales*. Subfamily *Orthocoronavirinae* includes 4 genera named as alpha, beta, gamma, and delta Coronavirus (α-, β-, γ- and δ-CoV). The genome of Coronavirus is enveloped, positive sense, single-stranded RNA with the largest known genome in RNA viruses ranging from 26–32 kb. Both α- and β-CoV infect mammals, while γ- and δ-CoV infect birds (Qin et al. 2020).

The global scientific community is in a continuous race to develop fast and effective testing procedures, drugs, and vaccines to prevent and control this viral infection. A vast number of promising diagnostic and treatment methods, as well as vaccines, are under investigation.

In this continuous race of R&D, nanotechnology has played an important role in all aspects to combat the viral infections caused by viruses structurally similar to SARS-CoV-2 (Campos et al. 2020). This chapter combines the studies previously reported and recent research conducted on COVID-19 to highlight the nanoparticles based strategies to combat COVID-19.

13.2 DIAGNOSIS OF SARS COV-2: CURRENTLY USED METHODS AND LIMITATIONS

Diagnosis and identification of SARS-CoV-2 is carried out by reverse transcription polymerase chain reaction (RT-PCR), which is the most widely accepted method worldwide followed by computed tomography (CT) scan and next-generation sequencing (NGS) (Lukas et al. 2020; Mahapatra and Chandra 2020Mahapatra 2020; World Health Organisation 2020a). However, the accuracy and specificity of RT-PCR for identification of SARS-CoV-2 is not universal due to the rapid genetic mutation, diversity of the virus, kit sensitivity, technical expertise, and extended time needed to get results. Moreover, the sample collection and its transport is considered as vital hallmarks for accurate results. Developing countries are facing the problem more profoundly due to the lack of medical facilities and services (Ali et al. 2020). In addition, the CT scan-based diagnostic method cannot distinguish other viral phenomenon from SARS-CoV-2, and it has a low specificity (Ding et al. 2020; Hao and Li 2020). Also, serological assays cannot identify the antibodies in serum before specific duration of infection onset. Therefore, there is a crucial requirement for the development of efficient, cost-effective, simple, and point-of-care (PoC) devices, without the need for technical specialists, to provide fast and reliable diagnosis for COVID-19 (Gupta et al. 2020).

13.3 USE OF NONTECHNOLOGY FOR DETECTION AND DIAGNOSIS OF SARS-COV-2

SARS-CoV-2 is 60–140 nm (Udugama et al. 2020) having intracellular life cycle; therefore, it's difficult to isolate and detect by ordinary methods and optical microscopes (Draz and Shafiee 2018; Chen and Liang 2020). The detection and diagnosis are vital tools to control COVID-19 through rapid employment of control actions. Nucleic acids, antibodies, and proteins (serology) are the core repertoire of techniques used for the diagnosis of viral infections. Protein-based tests are broadly accepted, but the reliability of method may get compromised due to cross-reactivity of the structurally similar antibodies, which can lead to false positive results. Secondly, mutation during the propagation period is also an important limitation in viral detection and diagnosis. Nanotechnology can be used to enhance the proficiency of these tests for identification of SARS-CoV-2 as a cost-effective tool.

The U.S. Food and Drug Administration (FDA) authorised 22 in vitro diagnostic methods under emergency use authorisation (EUA) for the identification of COVID-19 infection; some of them are based on nanotechnology (Gupta et al. 2020). In addition, several diagnostic techniques are under trial phases for early detection of viruses (Alteri et al. 2020; Calabrese et al. 2020; Carter et al. 2020; Gupta et al. 2020). A variety of nanomaterials are already being used for diagnosing viral infections (Draz and Shafiee 2018). NPs can improve the reaction rate due to their high surface-to-volume ratio, the decrease in detection time, and their increased efficiency (Layqah and Eissa 2019; Talebian et al. 2020).

13.3.1 Nanotechnology-Based Point of Care (PoC) Testing for COVID-19 Detection

PoC testing is a testing or diagnosis of a disease at patient's home or current location. The biosensors are the most critical components employed in PoC testing to perform a biochemical assay for the detection of pathogens (Choi 2020). PoC testing needs minimal storage space and technical expertise (Yang et al. 2020).

Lateral Flow Assay (LFA), a PoC method, is under development in which nanoparticles are conjugated to increase the sensitivity and reliability of the device (Rudolf et al. 2020). Studies have shown very promising results with 100% specificity and 82% overall sensitivity for immunoglobulin IgM and IgG (Xiang et al. 2020). A gold nanoparticle-lateral flow (AuNP-LF) assay was developed for the fast identification of the antibody IgM produced against SARS-CoV-2. The assay was based on antihuman IgM–AuNP conjugate as a detector. The specificity and sensitivity of the AuNP-LF assay were found to be 93.3% and 100%, respectively (Huang et al. 2020).

The FDA has recently authorised many other antigen-based assays, such as BD Veritor System followed by Sofia 2 SARS Antigen FIA test for PoC detection of SARS-CoV2 (FDA Newsroom 2020).

13.3.2 Nanotechnology Based Biosensors for Detection of COVID-19

Biosensors are often used to analyse the analytes within body fluids and liquid solutions by using a physical transducer and a biorecognition element. When an analyte reacts with a biological material, an indicator or signal is generated that is transformed into a quantifiable and measurable entity by the physical transducer. A biorecognition element includes proteins, e.g. enzymes or antibodies, biological and organic receptors, nucleic acids, e.g. DNA or RNA, whole cells, and tissues, while physical transducers are particularly electrochemical, optical, and piezoelectric systems (Asif et al. 2020).

Nanomaterials are usually used as transductors due to their biocompatibility with biological fluids and stability in various media. Besides, nanomaterials are beneficial for the bioconjugation of molecules due to their surface chemistry, strong amplification effect on signals, and high surface energy. Because of these properties, metallic NPs including silver and gold NPs, carbon-based NPs (graphene and nanotubes), nanogels, microgels, and photonic crystals are commonly used biosensors. Carbon-based NPs are previously being used in microfluidic-based kits for viral and bacterial detection due to their electrochemistry and piezoelectric nature (Castillo-Henríquez et al. 2020).

13.3.3 Microfluidic Devices

Microfluidic devices are another PoC testing-based approach (Syedmoradi et al. 2017). Microfluidic devices are tiny chips containing small channels and reaction cavities, such as poly-dimethyl sulfoxide, paper, or glass. They require only a small amount of sample and detection time (Gupta et al. 2020). These microfluidic devices or chips work on the principal of electrokinetic and capillary action to segregate liquid samples (Bhalla et al. 2020). In a recent study, the microfluidic devices were modified by employing the smartphone application, resulting in much faster diagnosis of various sexually transmitted viral infections with sensitivity and specificity of 100%, and 87%, respectively (Zhang and Liu 2016). In another study, a highly throughput microfluidic device was developed that was able to access the level of IgG and IgM antibodies against four different antigenic proteins of SARS-CoV-2 with 95% sensitivity and 91% specificity. It was further demonstrated that both of the parameters were enhanced up to 100% when the human sera was collected from the 3rd week of symptom onset. Further research is needed to develop new microfluidic devices to diagnose COVID-19.

13.3.4 Colorimetric Assays

A specific "naked-eye," easy and fast colorimetric test for the identification of SARS-CoV-2 is a crucial and immediate need. Several colorimetric assays have been developed. To understand the optical properties of AuNPs, a study was conducted in which specific antisense oligonucleotides (ASOs) were reorganised by thiol for its conjugation with N phosphoprotein of SARS-CoV-2. After RNA extraction and isolation, detection was performed within 10 minutes. The presence of target RNA sequence in sample was detected by aggregate formation due to capping of AuNP by thiol-modified ASOs, resulting in a change in surface plasmon resonance. The addition of RNase breaks down the RNA–DNA hybrid, resulting in the formation of a detectable precipitate (Mauriz 2020; Moitra et al. 2020).

13.3.5 Magnetic Nanoparticle-Based Separation

Magnetic nanoparticles modified with carboxyl polymer (pcMNPs) have been established to make the SARS-CoV-2 nucleic acid extraction from clinical sample simpler and easier (Zhao et al. 2020). This pcMNP–RNA complex has also been employed in RT-PCR protocol to increase the sensitivity by 10-fold. Furthermore, it is significantly important for the development of PoC devices (Qin et al. 2020; Zhao et al. 2020). Apart from the above discussion, Table 13.1 provides a summary of current research on nano-based diagnostic approaches for COVID-19.

13.3.6 Detection Based on Biomolecules

Biorecognition elements or biomolecules are employed in biosensors to detect the target, depending upon type of assay used. These biomolecules can be enzymes, antibodies, nucleic acids, or aptamers (Nadzirah et al. 2020). The role of nanoparticles to employ techniques and commonly used biomolecules for diagnosis and detection is summarised below.

13.3.6.1 Nucleic Acid and Antibody Based Biosensors

The biological sensors have been employed to modify the gene-editing technique by using CRISPR-Chip. The chip is combined with a graphene-based field effect transistor (FET) having the ability to identify up to 1.7 fM quantity of nucleic acid without amplification within 15 minutes (Samson et al. 2020).

The CRISPR–Cas12-based lateral flow assay technique is recently established for diagnosis of COVID-19 infection. It is a good substitution for real-time RT-PCR technique, and it is easy to implement in short time duration of

TABLE 13.1
Nanotechnology Based Diagnostic Approaches for SARS-CoV-2

Sensor Type	Samples	Analytical Data for SARS-CoV-2
Polymeric NP containing coated streptavidin dye	Oropharynx swab	LOD: 12 copies 100% sensitivity and specificity 1 hour for diagnosis
Lanthanide coated polysterene NP	Serum derived from blood	92% specificity with ≈10 minutes detection time
Poly (amino ester) containing carboxyl groups incorporated with magnetic nanoparticles (pcMNP)	Pseudovirus samples	LOD: 10 copies with linear range: 10–105 copies ≈20 minutes detection time
Gold nanoparticles	Venous blood	87% sensitivity and 91% specificity ≈15 minutes duration for test
Gold nanoislands	Synthetic oligonucleotides	LOD: 0.22 pmol/L
Zinc ferrite nanoparticles	Oropharyngeal swabs through RT-PCR	Reaction time: ≈15 minutes
Nanopore target sequencing	Oropharyngeal swabs	LOD: 10 copies with 100% specificity; ability to detect other respiratory viruses simultaneously; process is little bit lengthy, taking 6–10 hours
Nanoplasmonic resonance sensor designed to detect spike protein	Pseudovirus diluted in bovine serum albumin (BSA), 15 minutes.	Linear range: 103–106 virus particle/mLLOD: 750 vp/mL with ≈15 minutes reaction time

Sources: Chen et al. (2020); Huang et al. (2020); Li et al. (2020); Qiu et al. (2020); Somvanshi et al. (2020); Wang et al. (2020); Zhao et al. (2020); Zhu et al. (2020).

40 minutes (Broughton et al. 2020). The FET-based biosensors use monoclonal antibodies against the SARS-CoV-2 spike protein with coating of the graphene sheets of the FET. This FET biosensor device can detect SARS-CoV-2 spike protein 100 fg/mL concentration in the viral transport medium and 1 fg/mL concentration in phosphate-buffered saline (PBS) (Seo et al. 2020).

Recently a biosensor-based device was developed for the identification of COVID-19 infection, which was fabricated with a fluorine-doped tin oxide (FTO) electrode in combination with gold nanoparticles (AuNPs) and antibodies against spike proteins. The device shows high specificity to identify spike antigen, even in minute quantities ranging from 1 fM to 1 μM, and provides the results within a minute (Mahari et al. 2020).

Surface plasmon resonance (SPR) and localised surface plasmon resonance (LSPR) based viral biosensors are used for the identification of SARS-CoV-2. These are newly reported biosensors that detect nucleocapsid antibodies in undiluted human serum in place of an oropharynx swab. This SPR sensor is functionalized with SARS-CoV-2 nucleocapsid recombinant protein and coated with a peptide monolayer for diagnosis of COVID-19 within 15 minutes of reaction (Djaileb et al. 2020). In addition, a dual-functional, plasmonic biosensor, plasmonic photothermal (PPT), and LSPR has also been developed for the detection of SARS-CoV-2 with high specificity and detection limit up to the concentration of 0.22 pM (Qiu et al. 2020). The 2D gold nanoislands (AuNIs) functionalized with complementary DNA (cDNA) receptors. The LSPR sensing technique and PPT effect provide a promising solution to diagnose COVID-19 by nucleic acid hybridisation (Samson et al. 2020).

13.3.6.2 Aptamer-Based Biosensors

Aptamer-based biosensors are commercially used for virus diagnosis due to their ability to detect multiple targets with high specificity and sensitivity. These consist of ssDNA or ssRNA and interact conformationally with viral proteins, genes, or any other markers robustly (Zhou et al. 2014). Aptamer-based biosensors have significant dominance over antibodies due to easy synthesis through exponential enrichment methods and stability at a broad range of temperatures (Liu et al. 2020).

A recent study has reported promising results for diagnosis of COVID-19, in which ssDNA aptamer was developed and modified by adding biotin at its 5′ terminal that specifically interacts with N-protein immobilised on an ELISA plate. The reported affinity was very high, with KD of 4.93 ± 0.30 nM. The avidin labeled with horseradish peroxidase was introduced for color indication, and tetramethyl benzidine was used for measuring the absorbance (Cho et al. 2011; Gupta et al. 2020).

13.4 NANOMATERIAL-BASED VACCINE DEVELOPMENT AND IMMUNOMODULATION

Recently, nanoparticles-based approaches are proving to be highly promising in immunotherapeutic development due to their physiochemical properties. They play the role of adjuvant and a carrier for the antigen. Meanwhile, nanoparticles prevent premature degradation of antigen by increasing its exposure and uptake by antigen-presenting cells (APCs) and help in its sustained release, stability, and targeted delivery (Mehrabi et al. 2020). Furthermore, nanoparticles induce cellular and humoral immune response by interacting with immune machineries. Small nanoparticles ranging from 20 and 200 nm in size induce T cell response due to preferential internalisation by endocytosis into APCs, while large particles ranging from 0.5–5 μm are best in inducing humoral immune response due to internalisation by phagocytosis. Nano-based systems can be developed for targeting specific subcellular locations to enhance the potential of nano-based vaccines. Currently, many nanotechnology-based vaccines are available commercially, such as epaxal R, cervarix TM, gardasil TM, and inflexal RV. A description of vaccine candidates based on NPs are summarised in Table 13.2.

TABLE 13.2
Summary of Important Nano-Based Vaccines for COVID-19 Infection

Vaccine and Developer	Composition of Vaccine
Moderna coronavirus vaccine developed by the cooperation of Moderna (United States) and National Institutes of Health	Spike protein encoding mRNA vaccine encapsulated in lipid NPs
Vaccine introduced by Novavax, Inc. (United States)	S protein in combination with adjuvant matrix-M, virus-like NPs
Ad5-nCoV developed by CanSino Biologics, Inc. (China)	S protein NPs in adenovirus type-5 vector
Pfizer-BioNTech COVID-19 Vaccine introduced by Pfizer (Germany), BioNTech and Fosun Pharma	Lipid-based NPs and mRNA based
Ad26.COV2.S introduced in Belgium by Janssen Pharmaceuticals, Inc.	Recombinant vaccine developed by using AdVac® technology
Vaccine developed by LLC and Ufovax (United States)	Virus-like NPs of S protein
Vaccine introduced by Viroclinics Xplore at Netherlands	S protein+NPs
COVID-19 vaccine by IMV, Inc. (Canada)	LNPs formulated with DPX platform, comprising peptides epitopes from S protein
Vaccine by the cooperation of Precision NanoSystems (Canada) and CanSino Biologics (China)	LNPs and mRNA based vaccine
Vaccine introduced by Shanghai Jiao Tong University/ Fudan University /RNACure Biopharma (China)	LNPs and mRNA based
Vaccine developed by LakePharma, Inc. (United States)	S protein and NPs based
Vaccine introduced by St. Petersburg Scientific Research Institute of Vaccines and Serums (Russia)	S protein + LNPs + other epitopes of SARS-CoV-2
Vaccine by University of Tokyo/Daiichi-Sankyo (Japan)	LNPs + mRNA
Vaccine introduced by BIOCAD (Russia)	LNPs + mRNA
Translate Bio/Sanof Pasteur (United States)	LNPs + mRNA

World Health Organisation (2020b).

13.4.1 Virus-Like Particles (VLP)

These are the vaccines in which virus-like particles (VLPs) are used to induce immune response. In addition, these NPs are also employed as vehicles for RNA delivery to innate immune systems. VLPs contain viral proteins and can mimic whole virus without associated genetic material. They are more stable than soluble antigens in diverse environments (Bogani et al. 2020).

MERS-CoV spike (S) protein NPs have developed recently and show promising results in preclinical studies to prevent pathogen entry in cell and stop virus replication in lungs (Nascimento Junior et al. 2020). The spike (S) protein are the effective target for vaccine development due to their primary role in receptor binding, cell entry, and inducing neutralising antibodies. Another study was conducted by the same researchers to develop VLPs (~25 nm diameter) of SARS-CoV S proteins and full-length MERS-CoV. The level of neutralising antibodies against VLPs was significantly increased by 15-fold and 68-fold in combination of alum and Matrix M1 adjuvants, respectively (Coleman et al. 2017; Chen et al. 2020).

This nano-based approach has been examined by Epivax, aiming to produce a cocktail vaccine to provide protection against COVID-19 (Epivax 2020). Another nanotechnology-based vaccine candidate developed by Nanovax Company is presently in phase 3 clinical trial (NCT04368988). It is formulated by using full-length S protein of SARS-CoV-2 in combination with saponin-based Matrix-M™ adjuvant to stimulate the entrance of APCs into the vaccine injected site, thus increasing the immunological response by presenting the antigen in targeted sites, such as local lymph nodes (Magnusson et al. 2018). Currently, 16 more nano-based VLP vaccine candidates against SARS-CoV-2 have been developed and are under preclinical phase (WHO 2020b). Table 13.3 depicts the recently developed and under trial VLPs as vaccine candidates against SARS-CoV-2.

13.4.2 Other Nano-Based Vaccines

Organic NPs: Organic NPs such as liposomes are commonly used nanoparticles due to their promising properties with clinically authorised products for influenza A and

TABLE 13.3
Research and Development on VLPs as Vaccine Candidates against SARS-CoV-2

Developer	Composition of Vaccine
University of Manitoba	VLPs based vaccine targeting dendritic cells
VBI Vaccines Inc.	Enveloped VLPs developed against all type of coronaviruses
Saiba GmbH	VLPs
Navarrabiomed, Onco immunology group	baculovirus vehicles + VLP + lentivirus
OSIVAX	VLPs and Coronavirus
University of Sao Paulo	VLPs peptides/whole virus
Tampere University	VLPs
Shiraz University	Plant derived VLPs
Bezmialem Vakif University	VLPs
Imophoron Ltd and Bristol University	VLPs
Doherty Institute	VLPs
ARTES Biotechnology	Enveloped VLPs
Max Planck Institute	VLPs

Source: WHO (2020b).

B and hepatitis A (Itani et al. 2020). Further research is underway to develop vaccines against SARS-CoV-2.

Polymeric NPs: Examples of polymeric NPs include PEI NPs developed for plasmid DNA nasal vaccination and chitosan NPs for nasal DNA immunisation against SARS-CoV (Zhou et al. 2020). Currently, these NPs are being studied in vaccine development against COVID-19 in combination with other respiratory viruses.

Lipid-based NPs: Lipid nanoparticles, due to their high effectivity at delivering nucleic acids, are already in use for administration of mRNA for MERS-CoV S proteins, indicating strong humoral immune responses (Liu et al. 2020). Both the Pfizer and Moderna vaccines against SARS-CoV-2 are approved by the U.S. government. These RNA-based vaccines are carried by lipid nanoparticles due to their ability to protect RNA from being broken down at high temperatures (FDA 2020; Krause and Gruber 2020).

Biopolymer-based NPs: These NPs are highly effective because of having biodegradability and less toxicity and e.g. protein-based NPs. SARS-CoV S protein NPs in combination with Matrix M1 adjuvant-induced antibody production are in preclinical studies (Gupta et al. 2020).

13.4.3 Nano-Based Vaccine Delivery

Nanoparticles help correct antigenic display to immune cells by conjugation or physical entrapment, thereby enhancing the efficacy of nano-based delivered vaccines as compared to the conventionally delivered vaccines. Therefore, modulating APCs by employing nanoparticles could be very beneficial for offering the key advantage of having a diverse route of administration and addressing the specific targets, such as lymph nodes etc. (Chauhan et al. 2020). Several nano-based vaccine-delivery mechanisms are explained by various researchers, as depicted in Figure 13.1.

13.5 NANO-BASED THERAPEUTICS AND DRUGS DELIVERY FOR COVID-19

Nanoparticle-based therapeutic drugs have a strong ability to reduce and hinder the effects of viral infection in different ways, e.g. by preventing viral entry and blocking host cell receptors, inhibiting viral replication, and inactivating the virus.

13.5.1 Nano-Based Drug Delivery

Drugs delivery via nano-carriers helps to overcome the challenges linked with traditional medicines, such as less availability and probability of in vivo degradation of drug and toxicity. These nano-based delivery systems are sensitised to external environmental stimuli, such as magnetic field and internal environment, including temperature and pH. Most common examples of drug delivery through nano-carrier systems are dendrimers, microspheres, liposomes, micelles etc. In the case of COVID-19, inhalable nanoparticles can be administered to deliver drugs against SARS-CoV-2 at the site of action, i.e. lungs (Chauhan et al. 2020).

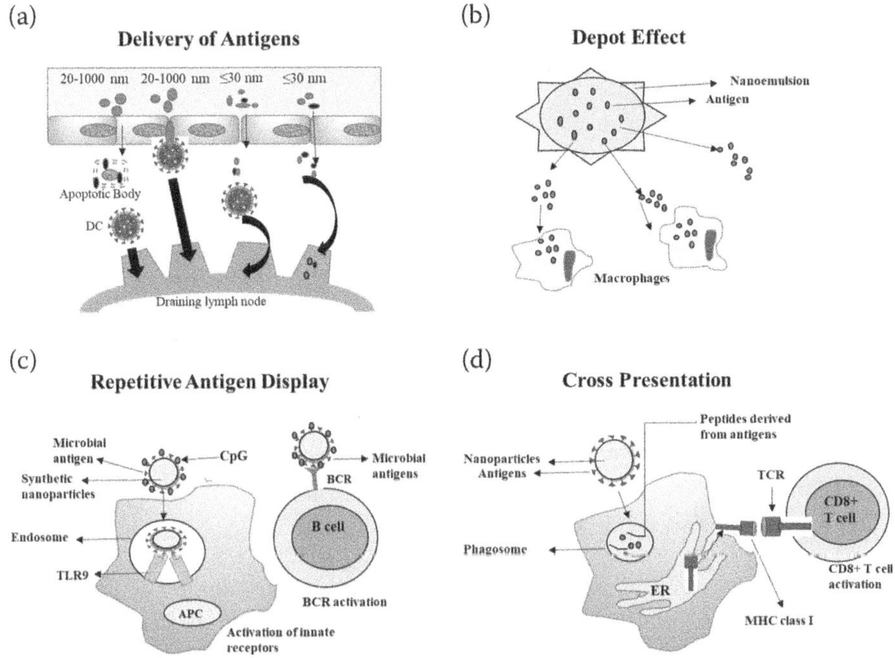

FIGURE 13.1 Nanoparticle-based modulation in normal immune response. (a) NPs-based antigen delivery targeting organ and tissues. (b) Depot effect for continued release of antigen without changing its integrity. (c) Increasing the immune response by repetitive display of antigenic targets to immune cells in the presence of nanoparticle. (d) Modulation of immune response by activation of $CD8^+$ T cells specific for antigen. Adopted and modified from Smith et al. (2013).

(Abbreviations: Antigen-presenting cell (APC); dendritic cell (DC); endoplasmic reticulum (ER); B cell receptor (BCR); T cell receptor (TCR) due to cross penetration of nano-delivered antigens).

13.5.2 Nanoparticles for Blocking Cell Attachment and Viral Entry

Chloroquine is a well studied drug used to block the attachment of SARS-CoV-2 with the host cell, thus preventing its entry. In general, chloroquine inhibits endocytosis of NPs; since SARS-CoV-2 is comparable to NPs in morphology, its binding with ACE2 receptor and endocytosis can be blocked (Medhi et al. 2020), as illustrated in Figure 13.2. Encapsulation with polymeric NPs further increases the efficacy of chloroquine by aiding in its cellular uptake. Polylactic acid (PLA) polymeric nanoparticles are generally used for encapsulating chloroquine (Lima et al. 2018).

Furthermore, previous studies reported that cationic carbon dots (~1.6 nm) produced from curcumin (CCM-CDs) have strong ability to inhibit the entry of a Coronavirus model named as porcine epidemic diarrhea virus (PEDV) (Ting et al. 2020). The principal behind blocking was the electrostatic interactions between the negatively charged PEDV and cationic CDs, thus neutralising the effective charge

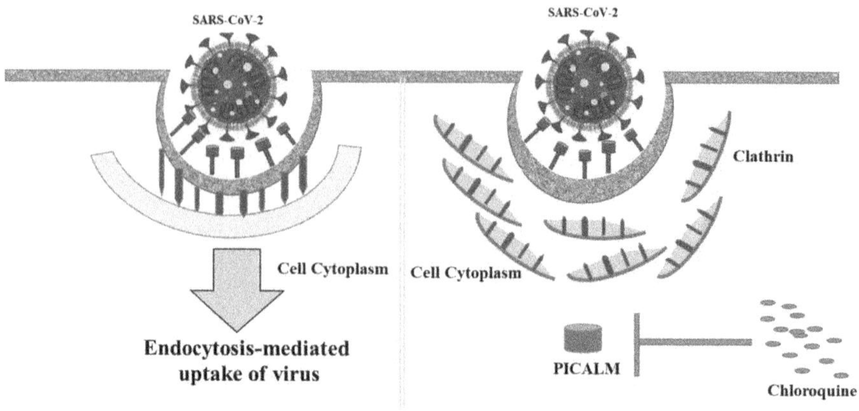

FIGURE 13.2 Mechanism of action of chloroquine-induced suppression of PICALM to inhibit endocytosis-mediated uptake of NPs and virus particles, e.g. SARS-CoV-2. Adopted and modified from Hu et al. (2020).

on the virus entity resulting in virus aggregation. The CDs also inhibit cell apoptosis by suppressing the accumulation of reactive oxygen species (ROS) in the cell. In addition, curcumin plays the role of capping and reducing agents in the formation of curcumin-modified AgNPs (cAgNPs), which are used to inhibit cell entry of respiratory viruses.

Recent studies demonstrated the role of AuNPs to inhibit the cell-entry mechanism of MERS-CoV. The S2 protein of MERS-CoV is much similar to SARS-CoV-2 S2 containing heptad repeat 1 (HR1), heptad repeat 2 (HR2), and a fusion protein (FP). As showed in Figure 13.3, once the FP binds with the cell membrane

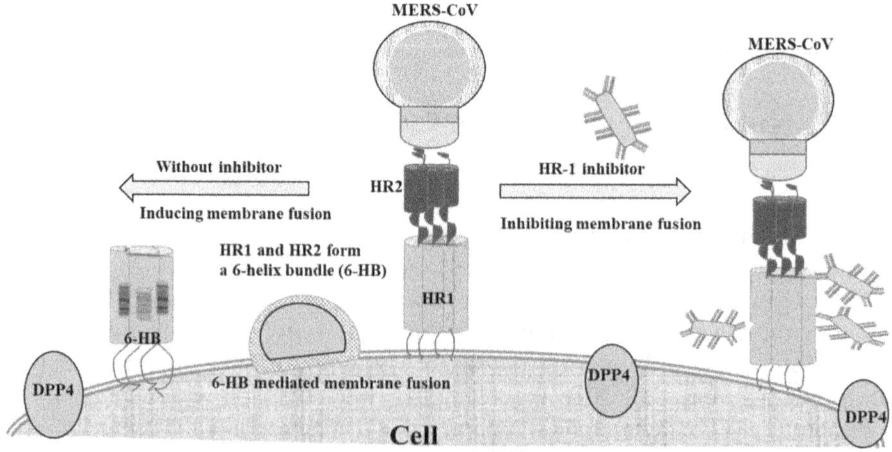

FIGURE 13.3 Inhibition of membrane fusion of MERS-CoV S2-subunit with an HR1 inhibitor. The HR1 inhibitor prevent HR1/HR2 complex (6-HB), thus preventing membrane fusion and infection of MERS-CoV. Adopted and modified from Huang et al. (2019).

of the host cell, HR1 and HR2 combine to form a six-helix bundle by binding together (6-HB). The six-helix bundle pulls together the host cell membrane and viral envelope supporting fusion. A peptide, named pregnancy-induced hypertension (PIH), has been identified with similar conformation as of HR2. This peptide can bind with HR1 to inhibit the development of 6-HB, preventing the cell-fusion process. The efficiency of PIH was further increased by immobilising it on the gold nanorods (PIH–AuNRs), resulting in a 10-fold increase in inhibitory affects and complete blockage of cell fusion. The PIH–AuNRs also depict high biocompatibility (Huang et al. 2019).

Porous silicon NPs (SiNPs) are also well-studied antiviral agents. They block the SARS-CoV entry into the host cell. They are easily biodegradable and are converted into nontoxic silicic acid after dissolving in water. SiNPs are potential antiviral agents against SARS-CoV-2 as their binding with virions is universal for a diverse group of enveloped viruses (Osminkina et al. 2014). Other examples of nanoparticles showing significant results for blocking viral attachment and entry into host cells are mesoporous selenium NPs (SeNPs) (Li et al. 2019).

13.5.3 Nanoparticles as a Blocking Agent for Viral Replication and Proliferation

Therapeutics having a strong ability to inhibit viral proliferation and replication are of significant interest. Glutathione-capped Ag_2S nanoclusters (Ag_2S NCs) are very effective in inhibiting viral proliferation by suppressing RNA synthesis. Furthermore, Ag_2S NCs activate the expression of cytokine and interferon (IFN)-stimulating genes (ISGs). The study was conducted on porcine epidemic diarrhea virus (PEDV), a model for the Coronavirus family due to its high similarity (Du et al. 2016).

Cellulose nanocrystals (CNCs) modified with tyrosine sulfate mimetic ligands showed promising results to inhibit the proliferation of alphaviruses, a genus of RNA viruses having structure similarity with Coronaviruses (Du et al. 2018). These therapeutic discoveries for treatment of viruses similar to Coronaviruses propose the potential applications of nanoparticles for the treatment of SARS-CoV-2.

13.6 NANOPARTICLES BASED DISINFECTANTS WITH VIRICIDAL EFFECTS

COVID-19 is highly contagious due to its diverse routes of transmission. Current research demonstrates that SARS-CoV-2 can easily spread by touching contaminated surfaces and micro-droplets emitted from infected persons. It can survive for 3 hours in air, 24 hours on cardboard, 72 hours on plastics, 4 hours on copper and 48 hours on stainless steel (Van Doremalen et al. 2020). It is recommended to use chemical and physical disinfectants to mitigate contamination. In this context, disinfectants, sanitisers, and PPE are significantly important (WHO 2020a). Many different disinfectant agents are recommended in literature, including soaps/surfactants, alcohols, sodium hypochlorite, and hydrogen peroxide, etc. (Du et al. 2018). According to a recent study on perspectives of COVID-19, drawbacks of currently used disinfectants are highlighted to search for more effectual disinfectants and sanitisers (Lei et al. 2020).

13.6.1 Nanomaterials for Surface Decontamination

Gold nanoparticles (AuNPs) covered with 3-mercaptoethylsulfonate (MES) inhibit the viral infection when used in high concentration consistent to EC90, but infectivity can be entirely recovered upon dilution. This procedure is known as reversible viral inhibition, just as with the example of heparin, a commonly used virustatic element targeting virus–cell interactions. Some metallic nanoparticles, especially silver nanoparticles, are broad-spectrum antiviral agents. However, the mechanism of action of these metallic NPs is still largely unknown (Pereira et al. 2020).

The Nanotech Surface Company formulated a disinfectant by combining titanium dioxide and silver nanoparticles. This formulation was used for sterilisation of surfaces and buildings in Milan during the current pandemic. Another company, the nanoSeptic, developed a formulation with composition of crystal nanoparticles for self-cleaning of surfaces. These nanoparticles disinfect the surfaces by activating the oxidation reactions that were further potentialized by light (Statnano 2020a, b).

The viral decoy molecule sialic acid also has strong antiviral activity to cause irreversible physical damage by selectively targeting envelop of RNA viruses. Grafting onto the nanodiscs further enhances the antiviral activity of sialic acid due to the availability of multivalent interacting sites. Furthermore, these functionalized nanodiscs induce self-disruption of virus (Bhattacharya et al. 2010).

Ferromagnetic (Fe_3O_4) NPs are also well studied and highly effective antiviral agents, often known as nanozymes (IONzymes) due to their strong catalytic activity and unique enzyme-like properties. They are 200 nm in diameter with research evidence to target the viral envelops of 12 different subtypes (H1–H12) of animal influenza viruses (IAVs). These NPs destroy the integrity of surface proteins by inducing lipid peroxidation, thus leading to virus inactivation. These IONzyme nanomaterials exhibit a highly potential approach against SARS-CoV-2 due to their high compatibility, safety, and easy synthesis (Ding et al. 2020).

In addition to the above-mentioned nanoparticles, several other studies also suggest the promising results of NPs as viricidal agents for SARS-CoV-2 and other viruses.

13.7 FUTURE PERSPECTIVES

Based on the emergency research efforts carried out in the last few months to combat COVID-19, it is concluded that the maturity of antiviral research has always depended on consistent efforts, long-term study data, and population dynamics.

Recent studies on nanoparticles as antimicrobial agents can provide better insights for viricidal agents and surface disinfectants with enhanced effectiveness of PPEs to prevent the infection of COVID-19. Furthermore, the role of nanotechnology for diagnosis and treatment is also promising in all aspects, including diagnostic tools, drugs development, drug delivery, vaccine development, vaccine delivery, and immunoglobulin development. Nano-based drugs are beneficial over conventional drugs due to less chance of antiviral resistance. However, more research is needed to ensure the safety of NPs because most of the studies performed so far are based on in vitro studies. Therefore, in vivo studies need time to evaluate the toxicokinetic behavior for long-term exposure of drugs and therapeutics.

REFERENCES

Abbas, A., Abdelsamea, M.M. and Gaber, M.M. 2021. Classification of COVID-19 in chest X-ray images using DeTraC deep convolutional neural network. *Applied Intelligence* 51(2):54–64.

Ai, T., Yang, Z., Hou, H., et al. 2020. Correlation of chest CT and RT-PCR testing for coronavirus disease 2019 (COVID-19) in China: A report of 1014 cases. *Radiology* 296(2):32–40.

Alteri, C., Cento, V., Antonello, M., et al. 2020. Detection and quantification of SARS-CoV-2 by droplet digital PCR in real-time PCR negative nasopharyngeal swabs from suspected COVID-19 patients. *PLoS One* 15(9):0236311.

Asif, M., Ajmal, M., Ashraf, G., et al. 2020. The role of biosensors in COVID-19 outbreak. *Current Opinion in Electrochemistry* 23:174–184.

Bhalla, N., Pan, Y., Yang, Z. and Payam, A.F. 2020. Opportunities and challenges for biosensors and nanoscale analytical tools for pandemics: COVID-19. *ACS Nano* 14(7):7783–7807.

Bhattacharya, P., Grimme, S., Ganesh, B., et al. 2010. Nanodisc-incorporated hemagglutinin provides protective immunity against influenza virus infection. *Journal of Virology* 84(1):361–371.

Bogani, G., Raspagliesi, F., Ditto, A. and de la Fuente, J. 2020. The adoption of viral capsid derived virus-like particles (VLPs) for disease prevention and treatments. *Vaccines* 432.

Boopathi, S., Poma, A.B. and Kolandaivel, P. 2020. Novel 2019 coronavirus structure, mechanism of action, antiviral drug promises and rule out against its treatment. *Journal of Biomolecular Structure and Dynamics* 2020 1–10.

Broughton, J.P., Deng, X., Yu, G., et al. 2020. CRISPR–Cas12-based detection of SARS-CoV-2. *Nature Biotechnology* 38(7):870–874.

Calabrese, F., Pezzuto, F., Fortarezza, F., et al. 2020. Pulmonary pathology and COVID-19: Lessons from autopsy. The experience of European Pulmonary Pathologists. *European Journal of Pathology* 477:359–372.

Campos, E.V., Pereira, A.E., De Oliveira, J.L., et al. 2020. How can nanotechnology help to combat COVID-19? Opportunities and urgent need. *Journal of Nanobiotechnology* 18(1):1–23.

Carter, L.J., Garner, L.V., Smoot, J.W., et al. 2020. Assay techniques and test development for COVID-19 diagnosis. *ACS Central Science* 6(5):591–605.

Castillo-Henríquez, L., Brenes-Acuña, M., Castro-Rojas, A., Cordero-Salmerón, R., Lopretti, M. and Vega-Baudrit, J.R. 2020. Biosensors for the detection of bacterial and viral clinical pathogens and COVID-19 diagnosis. *Nanotechnology* 20(23):6926.

Chauhan, G., Madou, M.J., Kalra, S., Chopra, V., Ghosh, D. and Martinez-Chapa, S.O. 2020. Nanotechnology for COVID-19: Therapeutics and vaccine research. *ACS Nano* *14*(7):7760–7782.

Chen, L. and Liang, J. 2020. An overview of functional nanoparticles as novel emerging antiviral therapeutic agents. *Materials Science and Engineering* 112:110924.

Chen, Z., Zhang, Z., Zhai, X., et al. 2020. Rapid and sensitive detection of anti-SARS-CoV-2 IgG, using lanthanide-doped nanoparticles-based lateral flow immunoassay. *Analytical Chemistry* 92(10):7226–7231.

Cho, S.J., Woo, H.M., Kim, K.S., Oh, J.W. and Jeong, Y.J. 2011. Novel system for detecting SARS coronavirus nucleocapsid protein using an ssDNA aptamer. *Journal of Bioscience and Bioengineering* 112(6):535–540.

Choi, J.R. 2020. Development of point-of-care biosensors for COVID-19. *Frontiers in Chemistry* 8:517.

Coleman, C.M., Venkataraman, T., Liu, Y.V., et al. 2017. MERS-CoV spike nanoparticles protect mice from MERS-CoV infection. *Vaccine* 35(12):1586–1589.

Ding, X., Yin, K., Li, Z., et al. 2020. Ultrasensitive and visual detection of SARS-CoV-2 using all-in-one dual CRISPR-Cas12a assay. *Nature Communications* 11(1):1–10.

Djaileb, A., Charron, B., Jodaylami, M.H., et al. 2020. A rapid and quantitative serum test for SARS-CoV-2 antibodies with portable surface plasmon resonance sensing. *ChemRvix* 10:26434.

Draz, M.S. and Shafiee, H. 2018. Applications of gold nanoparticles in virus detection. *Theranostics* 8(7):1985.

Du, T., Liang, J., Dong, N., et al. 2016. Carbon dots as inhibitors of virus by activation of type I interferon response. *Carbon* 110:278–285.

Du, T., Liang, J., Dong, N., et al. 2018. Glutathione-capped Ag_2S nanoclusters inhibit coronavirus proliferation through blockage of viral RNA synthesis and budding. *ACS Applied Materials and Interfaces* 10(5):4369–4378.

Epivax. 2020. EPV-CoV19: HCW Vaccine - EpiVax, Inc. - *Informatics and Immunology*. https://epivax.com/pipeline/epv-cov19. Accessed 28 January 2021.

Food and Drug Administration U.S. (FDA). 2020. *COVID-19 Vaccines*, FDA. https://www.fda.gov/emergency-preparedness-and-response/coronavirus-disease-2019-covid-19/covid-19-vaccines. Accessed 29 January 2021.

Food and Drug Administratio U.S. FDA Newsroom. 2020. *Coronavirus (COVID-19) Update: FDA Issued Emergency Use Authorization for Point of Care Antigen Test*, FDA. https://www.fda.gov/news-events/press-announcements/coronavirus-covid-19-update-fda-issued-emergency-use-authorization-point-care-antigen-test. Accessed 27 January 2021.

Franck, C.O., Fanslau, L., Bistrovic Popov, A., Tyagi, P. and Fruk, L. 2021. Biopolymer-based carriers for DNA vaccine design. *Angewandte Chemie* 133(24): 13333–13351.

Gupta, R., Sagar, P., Priyadarshi, N., et al. 2020. Nanotechnology-based approaches for the detection of SARS-CoV-2. *Frontiers in Nanotechnology* 2:589832.

Hao, W. and Li, M., 2020. Clinical diagnostic value of CT imaging in COVID-19 with multiple negative RT-PCR testing. *Travel Medicine and Infectious Disease* 34:101627.

Hu, T.Y., Frieman, M. and Wolfram, J. 2020. Insights from nanomedicine into chloroquine efficacy against COVID-19. *Nature Nanotechnology* 15(4):247–249.

Huang, C., Wen, T., Shi, F.J., Zeng, X.Y. and Jiao, Y.J. 2020. Rapid detection of IgM antibodies against the SARS-CoV-2 virus via colloidal gold nanoparticle-based lateral-flow assay. *ACS Omega* 5(21):12550–12556.

Huang, L., Ding, L., Zhou, J., et al. 2021. One-step rapid quantification of SARS-CoV-2 virus particles via low-cost nanoplasmonic sensors in generic microplate reader and point-of-care device. *Biosensors and Bioelectronics* 171:112685.

Huang, X., Li, M., Xu, Y., et al. 2019. Novel gold nanorod-based HR1 peptide inhibitor for middle East respiratory syndrome coronavirus. *ACS Applied Materials and Interfaces* 11(22):19799–19807.

Itani, R., Tobaiqy, M. and Al Faraj, A. 2020. Optimizing use of theranostic nanoparticles as a life-saving strategy for treating COVID-19 patients. *Theranostics* 10(13):5932.

Krause, P.R. and Gruber, M.F. 2020. Emergency use authorization of Covid vaccines - safety and efficacy follow-up considerations. *New England Journal of Medicine* 383(19):107.

Layqah, L.A. and Eissa, S. 2019. An electrochemical immunosensor for the corona virus associated with the Middle East respiratory syndrome using an array of gold nanoparticle-modified carbon electrodes. *Microchimica Acta* 186(4):1–10.

Lei, H., Ye, F., Liu, X., et al. 2020. SARS-CoV-2 environmental contamination associated with persistently infected COVID-19 patients. *Influenza and Other Respiratory Viruses* 14(6):688- 699.

Li, Y., Lin, Z., Gong, G., et al. 2019. Inhibition of H1N1 influenza virus-induced apoptosis by selenium nanoparticles functionalized with arbidol through ROS-mediated signaling pathways. *Journal of Materials Chemistry* 7(27):4252–4262.

Li, Z., Yi, Y., Luo, X., et al. 2020. Development and clinical application of a rapid IgM-IgG combined antibody test for SARS-CoV-2 infection diagnosis. *Journal of Medical Virology* 92(9):1518–1524.

Lima, T.L.C., Feitosa, R.D.C., Santos-Silva, D., et al. 2018. Improving encapsulation of hydrophilic chloroquine diphosphate into biodegradable nanoparticles: A promising approach against herpes virus simplex-1 infection. *Pharmaceutics* 10(4):255.

Liu, Q., Zhang, W., Chen, S., et al. 2020. SELEX tool: A novel and convenient gel-based diffusion method for monitoring of aptamer-target binding. *Journal of Biological Engineering* 14(1):1–13.

Liu, Z., Xu, W., Xia, S., et al. 2020. RBD-Fc-based COVID-19 vaccine candidate induces highly potent SARS-CoV-2 neutralizing antibody response. *Signal Transduction and Targeted Therapy* 5(1):1–10.

Lukas, H., Xu, C., Yu, Y. and Gao, W. 2020. Emerging telemedicine tools for remote COVID-19 diagnosis, monitoring, and management. *ACS Nano* 14(12):16180–16193.

Magnusson, S.E., Altenburg, A.F., Bengtsson, K.L., et al. 2018. Matrix-M™ adjuvant enhances immunogenicity of both protein-and modified vaccinia virus Ankara-based influenza vaccines in mice. *Immunologic Research* 66(2):224–233.

Mahapatra, S. and Chandra, P. 2020. Clinically practiced and commercially viable nanobio engineered analytical methods for COVID-19 diagnosis. *Biosensors and Bioelectronics* 165:112361.

Mahari, S., Roberts, A., Shahdeo, D. and Gandhi, S. 2020. eCovSens-ultrasensitive novel in-house built printed circuit board based electrochemical device for rapid detection of nCovid-19 antigen, a spike protein domain 1 of SARS-CoV-2. *BioRxiv*. 10.1101/2020.04.24.059204.

Mauriz, E. 2020. Recent progress in plasmonic biosensing schemes for virus detection. *Sensors* 20(17):4745.

Medhi, R., Srinoi, P., Ngo, N., Tran, H.V. and Lee, T.R. 2020. Nanoparticle-based strategies to combat COVID-19. *ACS Applied Nano Materials* 3(9):8557–8580.

Mehrabi, M., Dounighi, N.M., Mohammadi, M. and Masoudi, A. 2020. Nanoparticles and vaccine development. *Pharmaceutical Nanotechnology* 8(1):6–21.

Moitra, P., Alafeef, M., Dighe, K., Frieman, M.B. and Pan, D. 2020. Selective naked-eye detection of SARS-CoV-2 mediated by N gene targeted antisense oligonucleotide capped plasmonic nanoparticles. *ACS Nano* 14(6):7617–7627.

Nadzirah, S., Gopinath, S.C., Parmin, N.A., et al. 2020. State of the-art on functional titanium dioxide-integrated nano-hybrids in electrical biosensors. *Critical Reviews in Analytical Chemistry*, XX1–12.

Nascimento Junior, J.A.C., Santos, A.M., Quintans-Júnior, L.J., Walker, C.I.B., Borges, L.P. and Serafini, M.R. 2020. SARS, MERS and SARS-CoV-2 (COVID-19) treatment: A patent review. *Expert Opinion on Therapeutic Patents* 30(8):567–579.

Osminkina, L.A., Timoshenko, V.Y., Shilovsky, I.P., et al. 2014. Porous silicon nanoparticles as scavengers of hazardous viruses. *Journal of Nanoparticle Research* 16(6):1–10.

Pereira, A.E., De Oliveira, J.L., Carvalho, L.B., Guilger-Casagrande, M., De Lima, R. and Fraceto, L.F. 2020. How can nanotechnology help to combat COVID-19? Opportunities and urgent need. *Journal of Nanobiotechnology* 18:125.

Qin, Z., Peng, R., Baravik, I.K. and Liu, X. 2020. Fighting COVID-19: Integrated micro-and nanosystems for viral infection diagnostics. *Matter* 3(3):628–651.

Qiu, G., Gai, Z., Tao, Y., Schmitt, J., Kullak-Ublick, G.A. and Wang, J. 2020. Dual-functional plasmonic photothermal biosensors for highly accurate severe acute respiratory syndrome coronavirus 2 detection. *ACS Nano* 14(5):5268–5277.

Rodriguez-Moncayo, R., Cedillo-Alcantar, D.F., Guevara-Pantoja, P.E., et al. 2020. A high-throughput multiplexed microfluidic device for COVID-19 serology assays. *Lab on a Chip* 21(1):93–104.

Rudolf, F., Kaltenbach, H.M., Linnik, J., et al. 2020. Clinical characterisation of eleven lateral flow assays for detection of COVID-19 antibodies in a population. *MedRxiv. The Preprint Server for Health Sciences.* 10.1101/2020.08.18.20177204.

Samson, R., Navale, G.R. and Dharne, M.S. 2020. Biosensors: Frontiers in rapid detection of COVID-19. *3Biotech* 10(9):1–9.

Seo, G., Lee, G., Kim, M.J., et al. 2020. Rapid detection of COVID-19 causative virus (SARS-CoV-2) in human nasopharyngeal swab specimens using field-effect transistor-based biosensor. *ACS Nano* 14(4):5135–5142.

Smith D.M., Simon J.K. and Baker J.R. Jr. 2013. Applications of nanotechnology for immunology. *Nature Reviews Immunology* 13(8):592–605.

Somvanshi, S.B., Kharat, P.B., Saraf, T.D., Somwanshi, S.B., Shejul, S.B. and Jadhav, K.M. 2020. Multifunctional nano-magnetic particles assisted viral RNA-extraction protocol for potential detection of COVID-19. *Materials Research Innovations* 25(3):169–174.

Statnano 2020a. *Coronavirus: Nanotech surface sanitizes Milan with nanomaterials remaining self-sterilized for years, STATNANO.* https://statnano.com/news/67531/Coronavirus-Nanotech-Surface-Sanitizes-Milan-with-Nanomaterials-Remaining-Self-sterilized-for-Years. Accessed 28 January 2021.

Statnano 2020b. *Mineral nanocrystal-based coating activated by light kills coronavirus, STATNANO.* https://statnano.com/news/67583/Mineral-Nanocrystal-based-Coating-Activated-by-Light-Kills-Coronavirus. Accessed 28 January 2021.

Syedmoradi, L., Daneshpour, M., Alvandipour, M., Gomez, F.A., Hajghassem, H. and Omidfar, K. 2017. Point of care testing: The impact of nanotechnology. *Biosensors and Bioelectronics* 87:373–387.

Talebian, S., Wallace, G.G., Schroeder, A., Stellacci, F. and Conde, J. 2020. Nanotechnology-based disinfectants and sensors for SARS-CoV-2. *Nature Nanotechnology* 15(8): 618–621.

Ting, D., Dong, N., Fang, L., et al. 2020. Erratum: Multisite inhibitors for enteric coronavirus: Antiviral cationic carbon dots based on curcumin. *ACS Applied Nano Materials* 1(10):5451–5459.

Udugama, B., Kadhiresan, P., Kozlowski, H.N., et al. 2020. Diagnosing COVID-19: The disease and tools for detection. *ACS Nano* 14(4):3822–3835.

Van Doremalen, N., Bushmaker, T., Morris, D.H., et al. 2020. Aerosol and surface stability of SARS-CoV-2 as compared with SARS-CoV-1. *New England Journal of Medicine* 382(16):1564–1567.

Wang, M., Fu, A., Hu, B., et al. 2020. Nanopore targeted sequencing for the accurate and comprehensive detection of SARS-CoV-2 and other respiratory viruses. *Small* 16(32): 2002169.

World Health Organization (WHO) 2020a. *Laboratory testing for coronavirus disease (COVID-19) in suspected human cases: Interim guidance.* 19 March 2020. (No. WHO/COVID-19/laboratory/2020.5).

World Health Organization (WHO) 2020b. *Draft landscape and tracker of COVID-19 candidate vaccines.* https://www.who.int/publications/m/item/draft-landscape-of-covid-19-candidate-vaccines. Accessed 26 January 2021.

Worldometer. 2021. *COVID-19 coronavirus Pandemic-COVID live update: 178,678,871 cases and 3,868,661 deaths from the coronavirus.* https://www.worldometers.info/coronavirus/. Accessed 19 June 2021.

Xiang, J., Yan, M., Li, H., et al. 2020. Evaluation of enzyme-linked immunoassay and colloidal gold-immunochromatographic assay kit for detection of novel coronavirus (SARS-Cov-2) causing an outbreak of pneumonia (COVID-19). *MedRxiv*, doi: 10.1101/2020.02.27.20028787.

Yang, T., Wang, Y.C., Shen, C.F. and Cheng, C.M. 2020. Point-of-care RNA-based diagnostic device for COVID-19. *Diagnostics* 10:165.

Zhang, D. and Liu, Q. 2016. Biosensors and bioelectronics on smartphone for portable biochemical detection. *Biosensors and Bioelectronics* 75:273–284.

Zhao, W., Zhong, Z., Xie, X., Yu, Q. and Liu, J. 2020. Relation between chest CT findings and clinical conditions of coronavirus disease (COVID-19) pneumonia: A multicenter study. *American Journal of Roentgenology* 214(5):1072–1077.

Zhao, Z., Cui, H., Song, W., Ru, X., Zhou, W. and Yu, X. 2020. A simple magnetic nanoparticles-based viral RNA extraction method for efficient detection of SARS-CoV-2. *BioRxiv*. doi: 10.1101/2020.02.22.961268.

Zhou, W., Huang, P.J.J., Ding, J. and Liu, J. 2014. Aptamer-based biosensors for biomedical diagnostics. *Analyst* 139(11):2627–2640.

Zhou, X., Jiang, X., Qu, M., et al. 2020. Engineering antiviral vaccines. *ACS Nano* 14(10):12370–12389.

Zhu, X., Wang, X., Han, L., et al. 2020. Reverse transcription loop-mediated isothermal amplification combined with nanoparticles-based biosensor for diagnosis of COVID-19. *MedRxiv*. doi:10.1021/acsnano.0c06109.

14 Applications of Antiviral Nanoparticles in Cancer Therapy

Anusha Konatala, Sai Brahma Penugonda, Fain Parackel, and Sudhakar Pola

CONTENTS

DOI: 10.1201/9781003136644-17

14.1 INTRODUCTION

Nanoparticles (NPs) are submicron-sized inorganic/organic particles in the range of nanometers (10^{-7} to 10^{-9} m). They can be produced through top-down or bottom-up chemical, physical, and biological manufacturing approaches (Zhang et al. 2016; Chugh et al. 2018). Inorganic nanoparticles include Au, Ag, Zinc Oxide, and Silica (SiO_2). Polymer-based NPs include chitosan, gelatin, micelles, cyclodextrin, liposomes, and dendrimers, etc. Then there are nucleic acid-based NPs like RNA tetramers, siRNA, plasmid DNA (pDNA), etc. Nanoparticles have a multitude of applications ranging from antiviral therapy, antimicrobial therapy, drug delivery, bio-sensing, imaging, bioremediation, and clean energy, among several other physical applications. Nanoparticles can be multifunctional, especially in cancer therapy and diagnostics, i.e. used to image tumours, detect disease-specific biomarkers, deliver drug molecules to the target, and, finally, monitor disease progression. The first polymeric nanoparticle-drug conjugate was created back in the 1950s (Jatzkewitz 1955). Since then, throughout the decades, several nanoparticles have been studied and identified for cancer applications. The first nanoparticle that was approved by the Food and Drug Administration (FDA) for clinical use is doxil (liposomal doxorubicin) in 1995 (Barenholz 2012). Since then, numerous nanoparticles have been approved and are in use for various stages in cancer treatment.

There are several types of nanoparticles, and they can be classified on the basis of properties such as size, structure, chemical nature/composition, and mechanism of action. A certain subset of NPs are particularly effective and have a wide range of applications in antiviral technology. Types of nanoparticles that have been proven to have antiviral activity include quantum dots, gold and silver nanoparticles, nanoclusters, carbon dots, graphene oxide, silicon materials, and dendrimers. In addition to their antiviral nature, these NPs are also used in many cancer applications. Starting all the way from diagnostics (imaging and biomarkers), to applications in therapy (gene, chemo, nucleic acid, radio, hormone and immune) and drug delivery, antiviral nanoparticles (AVNPs) have shown potential in additional applications such as post-therapy monitoring and surveillance. Some AVNPs are effective in mediating both therapy and diagnostics and hence are called theranostics. A specific subset of viruses that cause cancers in their hosts, such as EBV, HIV, and HPV, can also be targeted using the multimodal approach of AVNPs.

Due to the incredibly complex nature of cancer, there are numerous challenges in developing therapeutic molecules or even detecting/diagnosing cancer at an early stage to prevent severe illness. Some of the most common challenges include the tumour microenvironment's inhibitory effects on therapeutic molecules, poor scalability of the manufacture of nanoparticles, and a lack of success in translational research. Other challenges include drug resistance, poor bioavailability, high off-target effects, and cytotoxicity.

Nanoparticles hold the potential to overcome the current challenges in cancer therapy. Their physical and chemical nature can not only be customised but also be manipulated when introduced into biological systems. NPs can also be used to deliver multiple therapeutic molecules with different pharmacokinetic and physiochemical properties. Therapeutic molecules can be delivered via burst release or pulsatile release. Liposomal and polymeric NPs have received significant attention over the past couple of decades due to their application as biocompatible nanoscale drug carriers. A few of the challenges with NPs in cancer therapy include toxicity due to tissue accumulation of nonbiodegradable NPs, phagocytosis by mononuclear cells (Dadwal et al. 2018), et cetera. The addition of polyethylene glycol (PEG) or PEGylation avoids the phagocytosis by mononuclear cells.

This chapter focuses on the utility of nanoparticles with antiviral properties in cancer-related applications. Here we will outline the definition and describe nanoparticles, mention the antiviral nature of nanoparticles, give a brief description of their history, and discuss current applications. We then move into challenges in cancer therapy and outline what are the major obstacles that have to be overcome. We next outline the potential utility of antiviral nanoparticles to overcome those challenges. We also discuss the challenges in utilising AVNPs in cancer-related applications. The next section will delve deeper into AVNPs and their characteristics that make them ideal candidates for applications in cancer therapy. Next, we briefly go over some of the applications of AVNPs in various aspects of cancer treatment (diagnostics, therapy, vaccines, and theranostics).

14.2 ANTIVIRAL NANOPARTICLES

In gene therapy, viral-vector systems have the most site-specific binding as they target the protein from the capsids or the glycoprotein from the viral envelope to attach to the host cells (Chen and Liang 2020). The most commonly used adenovirus vectors in gene therapy are highly immunogenic. The host antibodies and immune cells affect vectors carrying the therapeutic gene (Kaspar 2009). Hence, during gene therapy, a high number of viral vectors can be produced, posing a risk to post-mitotic and dividing cells (Kaspar 2009). Besides, there is an unavailability of specific viral-vector systems for various vaccines and therapeutics. There are occurrences of resistant viral strains and unfavorable reactions associated with antiviral therapies (Milovanovic et al. 2017). One-third of the mortality of infectious disease is posed by viral infections, causing significant global health challenges (Gurunathan et al. 2020). Moreover, conventional antiviral drugs have limited solubility and shorter half-life due to size hindrance in reaching specific sites. A major challenge is the nonacceptance of the drug by host cells. There have been studies carried out to increase the bioavailability and pharmacokinetics. There are FDA-approved viral drugs; however, unfortunately, they are known to cause side effects and are not suitable for children (Milovanovic et al. 2017).

To overcome these challenges, nanoparticles have been studied for a long time in medicine for the advantage of their size. In addition, nanoscale materials have unique chemical and physical properties (Milovanovic et al. 2017; Patra et al. 2018). One of the benefits of nanoparticles over conventional antiviral therapies is the absence of viral components. The other notable advantages of nanomaterials compared to traditional methods are higher bioavailability, better in vivo instability, increased absorption in the body, target-specific delivery, and, most significantly, their controlled drug release (Gurunathan et al. 2020). However, the efficacy of antiviral activity of nanoparticles varies depending on their size, shape, and type of material used in therapy.

Most NPs exhibit antiviral activity through the following two mechanisms (Parayath et al. 2020): (a) binding to the viral genome and blocking penetration into the host cell, and (b) binding with the replication factors and inhibiting viral replication.

The antiviral activity of nanoparticles can be measured through different assays. To name a few, techniques like flow cytometry, western blot assay, and real-time PCR are frequently used. Among the imaging techniques applicable, transmission electron microscopy gives clear images of nanoparticles and their binding sites. Some of them are viral quantification assays measuring the virus, while some measure cell death. The plaque assay determines the functionality of nanoparticles by calculating the plaque-forming units per milliliter (PFU value of the virus) (Rigotto et al. 2011; Chen and Liang 2020).

Nanotechnology-based methods are used in a variety of medical tests, for example, home pregnancy tests kits. An advantage of nanoparticle-mediated treatment is its higher surface-to-volume ratio, increasing capacity of nanoparticles to carry more antibodies, small molecules, peptides, aptamers, and targeted delivery (Swain et al. 2016).

14.2.1 Types of Antiviral NPs

One method of classification of nanomaterials is on the type of base used in nanomaterial formulation. At the nanoscale, the physical and chemical properties differ from the properties shown by larger sizes. The methods of preparation of nanoparticles can be divided into three categories: physical, chemical, and biological. In the physical method, ultraviolet (UV) radiation, microwave irradiation, sonochemical, thermal decomposition (thermolytic), and photochemical approaches are most commonly used. The chemical synthesis involves the sol-gel method, microemulsion, hydrothermal synthesis, and polyol synthesis, which results in better size, structure, and composition (Dhand et al. 2015; Khandel et al. 2018; Odularu 2018b). However, these methods are generally toxic in nature and expensive to scale up (Augustine and Hasan 2020). Finally, in biological methods, nanoparticles are synthesised in bacteria, virus, fungi, algae, and also plants. These methods are easy to scale up, eco-friendly, and also biocompatible.

14.2.1.1 Inorganic/Metal NPs

Silver nanoparticles (AgNPs) are extensively studied for their medical property in a broad range due to their unique physico-chemical property. Silver nanoparticles have shown antimicrobial, antiviral, anti-inflammatory, and anti-cancer properties (Gurunathan et al. 2020).

AgNPs have shown to decrease the pro-apoptotic factor (Bax protein) in the virus that triggers mitochondrial-mediated apoptosis. Some research suggested that silver nanoparticles form bonds with sulfhydryl groups on glycoproteins, which denatures the protein structure, thus changing the virus binding site of the protein. Researchers have suggested two different modes of AgNP action: by inhibiting replication of virus through binding with DNA and RNA of the virus, or AgNPs blocking the recognition site by binding to the outer coat of the virus (Gurunathan et al. 2020).

It is thought that silver nanoparticles produced by green biosynthesis using plant extracts or microorganisms may be more suited for clinical applications rather than AgNPs produced through physical or chemical methods. Additionally, green biosynthesis of AgNPs is a simple, safe, cost-effective, and eco-friendly approach (Lara et al. 2010).

Iron oxide and gold are among other extensively studied inorganic nanoparticles due to their optical and magnetic properties. These nanoparticles are extensively studied in antitumuor therapy and as drug carriers. Iron nanoparticles of size 10–15 nm range showed 8-fold reduction in viral infection against H1N1 virus (Kumar et al. 2019).

14.2.1.2 Organic NPs

Organic nanoparticles are widely used for drug-delivery and therapeutic purposes in humans due to their biocompatibility, biodegradability, and easy surface modification. The most common organic nanoparticles are polymeric nanoparticles, which are solids ranging from 10 to 1000 nm in diameter. Their small size can facilitate capillary penetration and uptake by cells resulting in increased concentrations at target sites. Drugs are incorporated into nanoparticles by dissolution, entrapment, adsorption, and attachment, or by encapsulation, and the nanoparticles provide sustained release of the drugs for longer periods.

Various synthetic polymers, including poly(lactic-co-glycolic) acid (PLGA), polyanhydrides, and dextrans, as well as natural polymers, including elastin-like polypeptides, have been used for preparing nanoparticles. The PEG-cloaked drugs have higher bioavailability compared to naked drugs.

Coating of nanoparticles with organic polymers is often done to reduce the clearance by immune cells. PEG is approved by the FDA for the stealth process. Other polymers like dextran, di-block copolymers, and albumin are also used as stealth polymers. The benefit of this process is that it decreases its uptake by immune cells because the polymers are highly hydrophobic and thus can have increased blood circulation. Organic nanoparticles have shown great promise through several studies conducted on their characteristics and mode of action. Some drawbacks are limited sustained release, potential for aggregation, limited storage stability, and challenges of loading a broad variety of drugs (Mitragotri and Stayton 2014).

14.3 CANCER VACCINES

Cancer immunotherapy has exploded in the past few decades. However, standalone cancer therapy or combination therapy have shown minimal efficacy. Unlike preventative vaccines, cancer vaccines affect cancer-causing cells by helping immune cells recognise tumour associated antigens present on cancer cells. Nanomaterials emulate cellular components in their biophysical and biochemical properties (Yang et al. 2016).

NP technology has provided fresh insights into the development of vaccines over traditional vaccines and increased stability. Nanotechnology provides a wonderful opportunity to develop vaccines with a broad range of composition, sizes, shapes, and surface properties. (Shen et al. 2018).

Vaccines can be classified into two categories, depending on either carriers or the basis of immune reaction, immuno-stimulatory reaction, and delivery modes. The trend in vaccine development is moving toward minimalist composition, which consists of a subunit and its protein bases, which results in reduced immunogenic reaction toward the foreign materials introduced into host cells. Numerous nanomaterials have been approved for human use, as nanomaterials have targeted delivery, controlled drug delivery, less immunogenicity, higher solubility, and more stability compared to conventional vaccines (Mamo and Poland 2012).

14.4 APPLICATIONS OF ANTIVIRAL NANOPARTICLES IN CANCER

NPs with antiviral properties play a remarkable role in cancer treatment and diagnosis. Various NPs with these properties include silver, gold, carbon dots, quantum dots, and dendrimers. They have applications in several aspects of cancer treatment. Some NPs, like quantum dots (QD), have high antiviral effect with increase in their size (Chen and Liang 2020). Silver NPs are well known for their antiviral property and low toxicity (Chen and Liang 2020). Gold NPs (AuNPs) have the reliable capability of entering tumour cells with negligible systemic toxicity and the added advantage of their customisable size. (Fan et al. 2020). Carbon dots have the capability of regulating or enhancing the properties of cancer therapeutics, thereby showing great potential in cancer theranostics with further investigation and application in the life sciences (Jia et al. 2020). Quantum dots also have significant potential for application as antiviral agents, but they are currently only utilised in detection and cell labelling. Current studies aim to understand the nature of QD in cancer therapy (Wang et al. 2012; Jia et al. 2020). Graphene oxide is purported to be more reliable as an antiviral agent, but it needs further evidence through preclinical and clinical studies to establish anticancer efficacy (Jia et al. 2020). These engineered NPs provide improved therapeutic effects over the traditional methods of treating cancer. Hence, these NPs with antiviral effects are widely considered to be the future of cancer medicine (Chen et al. 2020). These antiviral-natured NPs play an important role in cancer theranostics. The following section briefly describes the applications of AVNPs in the context of cancer.

14.4.1 Chemotherapy

The process of cancer treatment using chemical drugs is known as chemotherapy. One of the major drawbacks of chemotherapy is the impact on noncancerous cells. Tumour cells can also adapt to the overexpression of drug-efflux pumps. Targeting the efficiency of chemotherapeutics can be enhanced by exaggerating the loading capacity of drugs to the tumour cells. To curb toxic side effects (Kumar et al. 2013), Liang et al. formed a scheme to carry platinum (IV) compounds to prostate tumour cells. Gold NPs that are glutathione-modified targeting the CRGDK peptide sequence at the destination were employed. These AuNPs have shown increased nuclear factor Kappa-B (NF-κB) proteins and initiate apoptosis (Kumar et al. 2014).

Yook et al. created AuNPs with doxorubicin (DOX) along with anti-PD-L1 for studying their effects on colorectal tumours in photo-thermal and chemotherapies. The reports showed that the highly stable AuNPs expressed PD-L1 in CT-26 cells, and the therapeutic efficacy of DOX was also studied. In this combination therapy, AuNPs were irradiated by near infrared light and in vitro proliferation of CT-26 cells was observed to be inhibited due to the cell cycle breakage and high levels of programmed cell death (Emami et al. 2019).

Ding et al. developed lipoic acid (LA)-modified polyethylene glycol (mPEG) as a DOX linker besides the traditional standalone PEG. The amino and carbonyl groups of DOX were modified with mPEG and LA respectively. This approach resulted in the release of DOX-mPEG from the AuNPs in the acidic lysosomes. Through this approach, DOX was delivered inside the target cells through a sustained release process (Cui et al. 2017).

Ehrlich et al. produced AuNPs bound with the drug heat shock protein 90 inhibitor, 17-AAG, and Cullin-5 (Cul-5) DNA-encoding sequence, this was found to increase the receptiveness of 17 AAG in Cul-5 deficient AU565 cells (Fan et al. 2020). To treat the breast tumour cells, Arunakaran et al. developed AuNPs (3nm) with drug quercetin in MDA-MB-231 and MCF-7 breast cancer cell lines. The results depicted that free-drug quercetin is less potent compared to the AuNP-quercetin conjugate in battling breast cancer. Studies conducted by Coulter et al. explain that the AuNPs conjugated with the drug maytansine analog DM1 (MTC-100038) to treat murine hepatocellular cancer. This conjugated chemotherapy in hepatocellular tumours enhanced systemic tolerability (Hale et al. 2019).

14.4.2 Radiation Therapy

An application of the ionizing radiations like X-rays and gamma rays on the tumor area is considered as a primary antiproliferative treatment method (Kwatra et al. 2013). The mechanism involved in killing cancer cells is through the generation of reactive oxygen species (ROS), formed due to the aerobic metabolism of the oxygen molecules. These ROS in turn affect the exposed cells (Racca and Cauda 2021). Challenges like off-target effects are traditionally overcome by increasing radiation to promote cell death in hypoxic cells.

NP-mediated radiation increases the exposure of the cells to the radiation and intensifies the toxicity (Racca and Cauda 2021). NPs assist in generating photoelectrons, which can enhance the oxidative stress. Chen et al. observed that the selenium NPs enhance the radiation toxic effects on breast cancer cells. There is a consequent enhancement in the ROS production (Racca and Cauda 2021). Zhao et al. studied Fe_3O_4/Ag/C225 composed of the epidermal growth-factor receptor with the combination of x-ray treatment to increase the sensitivity of human nasopharyngeal carcinoma cell lines (CNEs) to the irradiation (Pizzino et al. 2017).

14.4.3 Gene Therapy

The replacement or modification of the defective genes with the healthy version of the gene is called gene therapy. Cancer as a disease also has a genetic aspect to it. It is driven by the overexpression or inhibition of certain genes (for example, the inhibition of tumour-suppressing genes). Earlier, there were concerns about exploring gene therapy for cancer treatment among the scientific community. In recent years, gene therapy has been among the most promising treatments for challenging diseases globally. A threat in traditional cancer treatment is multidrug resistance (MDR). This resistance is caused by the repeated influx of proteins on the cell membrane, and, as a consequence, the therapeutic efficacy of the drug is depleted (Quader and Kataoka 2017). Most commonly in patients with MDR, unexpected results due to traditional therapies are reported. These hurdles can be mitigated with the use of gene therapy in combination with the drugs. Li et al. studied cancer treatment for non-small-cell lung cancer (NSCLC), caused mostly due to a mutated P53 gene by the advantage of the combination therapy of p53 gene and bortezomib (BTZ). This combined treatment was masterfully used with mesoporous hollow silica nanospheres (Quader and Kataoka 2017). The reports explained the refurbishment and revitalisation of the p53 signalling pathway. Xu et al. studied the pigment epithelium derived factor (PEDF) in combination with paclitaxel (PTX) to inhibit cancer development in a colon cancer C26 subcutaneous model using PEG-Poly (lactic-co-glycolic acid) (PLGA) NP-based co-delivery system. The reports identified aggregate PEDF expression, high apoptosis, depleted angiogenesis, and, importantly, G2/M cell-cycle phase breakdown (Xu et al. 2016).

The artificial oligonucleotides are injected into the compact tumours and sarcomas and were able to reduce nontarget effects in cancer therapy (Dias and Stein 2002; Rosi et al. 2006; Zuhorn et al. 2007). A profound antitumour effect was found in vivo from a research groups' study in a mouse model by administration of multistage siRNA-di-oleylphosphatidylcholine. This experiment reported the silencing of the gene-inducing ovarian cancer for three weeks. Besides, there was no considerable toxicity or off-target effects reported (Miele et al. 2012). The ultimate goal of this combination therapy is to overcome MDR and protect healthy cells (Parvanian et al. 2017).

14.4.4 Targeted Therapy

The treatment of cancer can be targeted to prevent damage to healthy cells since this will increase the treatment quality and life expectancy of the affected individuals. Approaches toward targeted cancer therapy hold a lot of promise with respect to the outcomes and side effects of treatment. Cancer therapy and delivery of diagnostic molecules can be targeted either actively (through specific cell surface antigens/markers) or passively (through the enhanced permeability and retention or EPR effect). Cancer diagnosis is usually determined by the moderate accumulation of the imaging probes (Li et al. 2021). Among the different NPs involved in actively targeting cancer diagnosis, dendrimers are a part the frontrunners for being molecules of interest due to their high density and several terminal groups (Li et al. 2021). A few of the advantages with using dendrimers are as follows:

- Different surface groups provide different resolutions of images
- Longer retention of dendrimers in lymph nodes due to the phosphate groups at the terminals provides more specific imaging (Nishimoto et al. 2020)
- The carboxy-terminal groups are noted to have a disposition to the osteolytic bone lesions and bone tumours (Yan et al. 2019)

Another strategy for the targeted therapy is passive targeting through NPs. The NPs mostly depend on the EPR effect in solid tumours (Fang et al. 2020) for passive targeting. The endothelial cells in the immature tumour neovascular structures and lymphatic reflux nonavailability make the NPs take advantage of the EPR effect (Li et al. 2021).

14.4.5 Immunotherapy

To overcome tumour tolerance, vaccination can be used to enhance antitumour responses. Immunotherapy could be a potential approach in targeting the specific antigens of the tumours and can be sensitised to the host immune system (Jonuleit et al. 2016). Dendritic cells are well known for triggering naive T-cells and the adaptive immune responses (Schuler et al. 2003). This generally results in the reduction of cancer progression.

14.5 AVNPs IN CANCER DIAGNOSIS

The major challenge in effectively treating cancer is the ability to not diagnose it at an early stage. Cancers are initially asymptomatic or mildly symptomatic, which can be mistaken for other ailments. This leads to diagnosing the cancer at a later stage, thereby decreasing the survival rate of the individual. Two of the primary measures for effectively treating cancers are not ignoring persistent symptoms and not getting diagnosed at the onset of even mild symptoms. The effectiveness of the cancer treatment relies on the early diagnosis of the cancer. This ensures that the cancer doesn't become metastatic and can be treated at the initial site. Therefore, cancer diagnosis at an early stage is essential.

Among the different nanoparticles involved in the diagnosis, this section outlines dendrimers and their importance in cancer diagnosis. Essentially, dendrimers are branched molecules with more surface units that can actively interact with more cancer receptors/markers available on the cancer cells to effectively identify and quantify the cancer in an affected individual.

Dendrimer is derived from the Greek word "dendron," which stands for *tree*. They are branched molecules with recurring units of 5–10 nm (Kumar and Sharma 2018). Dendrimers are composed of three architectural units consisting of the core part, multifunctional units, and terminal or end groups (Kumar and Sharma 2018). The scope of dendrimer applications has increased significantly in the past few years. A few of the applications of dendrimers in cancer diagnosis are discussed below:

Dendrimers are mostly spherical nanomolecules with a branched architecture. The size-to-volume ratio, and their heat-stable capability, makes them a unique material in diagnosing tumours (Rai et al. 2017). A commercially available dendrimer is poly-propyleneimine (PPI) (Chakravarty and Vora 2020). It has antimicrobial activity and has been proven to be highly effective against cancers (Chakravarty and Vora 2020). Both neutral and negative charges are found to be more biologically compatible, although positive-charge dendrimers can exhibit some noxious effect (Kumar and Sharma 2018).

Dendrimers target receptors to diagnose cancers at an early stage but seek out different cancer markers, like cell-free tumour DNA (ct DNA), circulating tumour cells (CTCs), as well as tumour-derived exosomes (Bu et al. 2020). Dendrimers remarkably detect the initial cancer stage and details about the cancer prognosis, invasiveness, and also characterisation of the tumours (Li et al. 2020). These are available in different subclasses based on their structure and property. The types vary from dendrons to dendritic polymers, which have a few to several sub-branches (Li et al. 2021). The characteristics of the dendrimers, such as low viscosity, high water solubility, and diverse functional groups, make them exceptional particles for cancer diagnosis (Li et al. 2021).

14.5.1 Preparation of Dendrimers

Dendrimers are prepared in a multistage process. The functional groups of the dendrimers can be modified, based on the type of cancer target cells and their biomarkers, to increase the therapeutic index. The synthesis of the dendrimers by traditional ways is through the convergent and divergent preparation methods. The major concern with these methods of dendrimer manufacture is low yield and defective terminals (Li et al. 2021). Other popular preparation methods used in dendrimer construction are phase inversion temperature, solvent evaporation, and polymerisation-solvent displacement (Chakravarty and Vora 2020). The preparation methods of dendrimers are outlined in Figure 14.1.

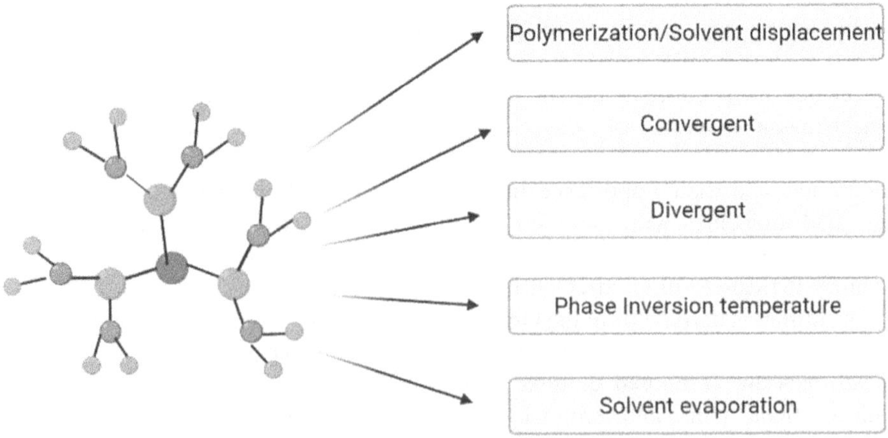

FIGURE 14.1 Preparation methods of dendrimers.

14.5.2 Properties of Dendrimers

The unique characteristics of the dendrimers make them a good fit for cancer theranostics, a few of which were outlined earlier. These properties affect the pharmacokinetic properties of dendrimers. The utility of dendrimers resides in the following capabilities (Kumar and Sharma 2018).

- Drugs can be loaded onto the functional groups, increasing the loading capability
- Ability to recognise various receptors of the cancer cells for precise targeting
- Provision of modifying the terminal groups for target-specific delivery
- High number of branched units enable good entrapment of the targeted site
- Weight of dendrimers can be controlled based on the target-specific cells
- Dendrimers can be as small as 1–5 nm in size, which enables the possibility of travelling through initial cancer-stage cell, thereby detecting cancers at an early stage

Each functional group attached to the branched terminal units of the dendrimers impart different activities. The functional groups and their activities are illustrated in Table 14.1.

14.5.3 Dendrimers in MRI

Molecular resonance imaging (MRI) is a noninvasive diagnostic technology that provides 3D images. The technology involved in MRI is to excite and detect the gadolinium (Gd) particles and the direction of the excited protons found in the water. The sensor of the MRI captures proton energy and projects it as an anatomical image (MRI). In general, Gd is injected intravenously before the scan. When exposed to the magnetic field, excitation of the protons happens, and they move in the direction of the strong magnetic field. When the radio waves are passed as a counter field, the protons rotate in either 90° or 180° to the direction of the existing magnetic field. Once the radio frequency is paused, the protons tend to go back to their original direction. At this point, the release of an energy occurs, which is

TABLE 14.1
The Functional Groups and Their Activities Are Illustrated in a Tabular Format

Functional group	Activity
Carboxylated and Sulphated group	Preventive activity
Ammonium	Self assemble
Phosphoric acid	Anti-HIV activity
Thiolated	Sustained release
Tryptophan	Anti HIV-I (more potent)
Tyrosine	Anti EV71

captured by the MRI sensors as a three-dimensional anatomical image. The quicker the change in the rotation on the onset of the radio frequency, the brighter the image of the anatomical images.

Contrast agents like Gd are used to ensure accuracy of the MRI report (Nikzamir et al. 2021). However, there are some challenges associated with using Gd, such as quick excretion, low specificity in the tissues, as well as low contrast ability. In order to mitigate these hurdles, contrast agents are added to a single core scaffold (Li et al. 2021; Nikzamir et al. 2021).

Dendrimers are emerging as novel NPs in cancer theranostics. They are used as the polymers with a huge scaffold for carrying the paramagnetic ions. These particles help to increase the cell stability within the cell, increase concentration at the target site, and enable long-term monitoring of cancer progression (Li et al. 2021). MRI scans with and without dendrimers are detailed in Figure 14.2.

A characteristic feature in cancer is hypoxia in solid tumours. Hypoxia is related to a set of events like the metastasis of the tumour, radiation, and drug resistance to multiple drugs (Qin et al. 2018; Kumari et al. 2020). A case study by researchers used dendrons coated with iron oxide NPs and targeted metronidazole-based ligands to get the anatomical images of hypoxia-activated MRI. In vitro, the adipose tissue derived stem cells (ASCs) and MCF-7 cells, are tested for normoxia and hypoxia by the variation in the relaxivity percentage detected. For the MCF-7 cells, it is found to be 170%, and for the ASCs cells, it is 100% (Filippi et al. 2019). Another study group found that the dendritic MRI contrast agents tend to have higher blood circulation time than the regular Gd contrast agents.

Besides the applicability of the dendritic contrasting agents, they are an exceptional agent for MRI to generate high-quality anatomical images in 3D. Dendrimers are also well known for their biosafety and biocompatibility, making excellent candidates in cancer imaging.

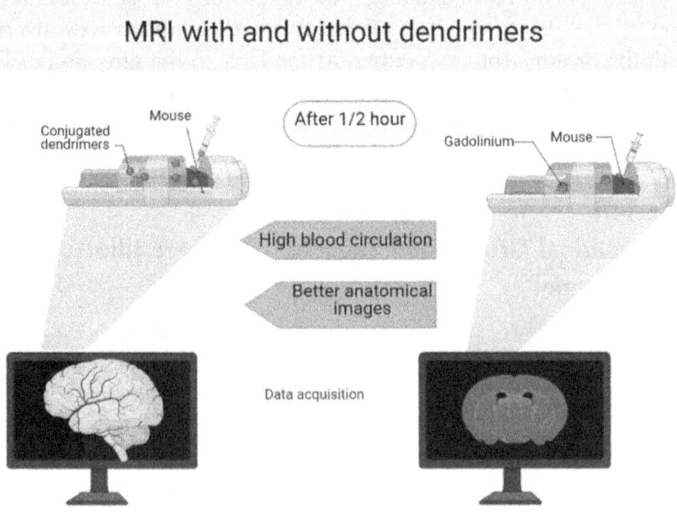

FIGURE 14.2 MRI with and without dendrimers.

14.6 AVNPs IN VACCINES AND THERANOSTICS

Nanoparticles (NPs) have a scope for application in delivery systems for cancer vaccines as they potentiate the co-delivery of tumor-associated antigens and proteins to the specific cells to generate a stronger immune response. Moreover, several sophisticated NP systems are used for optimisation of vaccines to improve efficiency. Synthetic vaccines using nanomaterials have been explored for a long time because of properties like biocompatibility, biodegradability, and improved bioavailability for a long-lasting immune response (Sulczewski et al. 2018). Unlike viral vaccines, nanoparticle vaccines show improved safety and efficacy. These nanoformulations are not only similar to the viruses' size and shape but are also consistent with their ability to deliver antigen and activate immunological response (Shen et al. 2018; Sulczewski et al. 2018).

An effective delivery system should concentrate the antigen, protect it from degradation, increase its uptake and processing in dendritic cells (DCs), and induce the production of cytokines that create a robust immune response (Shen et al. 2018). Liposomal, polymeric nanoparticles, gold nanoparticles, mesoporous nanoparticles, and carbon nanotubes are most commonly used in vaccination. In liposomes, the composition of lipids and fatty acids decides the liposome stability and performance. These carriers can incorporate hydrophilic molecules in the inner core, and hydrophobic molecules can be embedded within bilayers. Theranostics is a portmanteau derived from therapeutics and diagnostics. Nanoparticle technology has initiated a smart theranostics platform that can diagnose, monitor response, and initiate primary as well as secondary treatment, as required (Sneider et al. 2017).

14.7 AVNP MEDIATED CANCER THERAPY

The various types of NPs used in cancer therapy include: liposomes, polymeric NPs, polymeric micelles, dendrimers, microspheres, carbon NPs, and carbon-based nano systems (nano diamond, carbon nanotubes, graphene oxide), and metallic NPs. As mentioned previously, there are a multitude of antiviral nanoparticles used in cancer treatment applications. Gold nanoparticles were used in a nanophotolysis approach to selectively target and kill breast cancer cells (Ashiq 2018). Gold nanoparticles have also been used in a nanotheranostic approach to both diagnose and selectively deliver drugs in a targeted manner to treat melanoma (Bagheri et al. 2018). Silver NPs have been extensively studied for their anticancer and antiviral properties. AgNP-related studies conducted in-vitro showed promise in drug delivery, imaging, and therapy in several cancers, including breast cancer, leukemia, hepatocellular carcinoma, and gliomas, to name a few (Zhang et al. 2016). A preclinical study in mice showed that targeted delivery of miRNA through RNA nanotechnology effectively controlled the growth of prostate cancer (Binzel et al. 2016). The following example describes the use of nanoparticles in cancers caused by viruses (oncoviruses), specifically nanoparticles used to fight hepatocellular carcinoma.

14.7.1 Hepatocellular Carcinoma and AVNP-Mediated Treatment

Liver cancer is the third leading cause of cancer deaths worldwide (American Cancer Society). The most common form of primary liver cancer in adults is hepatocellular carcinoma (HCC), or hepatoma. It accounts for approximately 75% of all liver cancers. Hepatitis B virus is the leading cause of HCC, whereas hepatitis C virus can also cause non-Hodgkin's lymphoma. The risk of developing HCC increases with liver cirrhosis that can be caused by HBV/HCV infection and/or alcoholism. It must be noted that up to 25% of patients diagnosed have no history of cirrhosis or risk factors for HCC. Asia and Africa see a higher endemic prevalence of HBV and HCV, thereby increasing the incidence of HCC through an increased risk of chronic liver disease. HBV is an enveloped DNA virus of the hepadnaviridae family. The various treatment options available for HCC include liver resection/transplant surgery, radiofrequency/cryo-ablation, targeted chemotherapy (chemo-/radio-embolization), radiation, immunotherapy, etc., depending on the size and location, liver function, and overall health (Mayo Clinic).

14.7.2 Current Therapy Approaches to Treat HCC

The first priority to treat HCC is to surgically resect the cancer for patients who are suitable candidates. For patients who cannot undergo respective surgery, performing a transplant might be the best option to perform curative therapy. The United Network of Organ Sharing (UNOS) has set out some specific criteria that are used to evaluate transplant eligibility for nonmetastatic (localised) yet unresectable tumours. Ablation, radiation, and embolization are prescribed to patients that do not meet the UNOS criteria as the next priority. Finally, for patients who are not suitable for any of the above treatment strategies and have an unresectable tumour, a management-type approach is taken through systemic treatment. Some of the common systemic treatment options include sorafenib (Colagrande et al. 2016), lenvatinib, bevacizumab (in combination with atezolizumab). Some other commonly used chemotherapeutics include regorafenib, nivolumab, pembrolizumab, cabozantinib, and ramucirumab. Some of the chemotherapeutics that have shown marginal activity in small clinical trials include gemcitabine, capecitabine, and doxorubicin.

14.7.3 AVNPs-Mediated Therapy in HCCs

NP systems such as PEG, di-thiodipropionate, hyaluronic acid, polycaprolactone, chitosan, O-carboxymethyl chitosan, hematoporphyrin, apotransferrin, and glycyrrhizic acid have been used in the delivery of doxorubicin (DOX), docetaxel, paclitaxel, curcumin, fluorescein isothiocyanate (FITC), gemcitabine, and sorafenib in the treatment of HCC. Several PEGylated polymeric NPs have been used in the delivery of siRNAs to treat HCC (Varshosaz and Farzan 2015). AVNPs use either active or passive targeting to effectively invade the tumour cells while carrying the payload of chemotherapeutic molecules. In active targeting, there are several cell surface receptors used, transferrin, epidermal growth factor receptor (EGFR), low-density lipopolysaccharide, being just a few examples. Gold NPs were shown to have cyto- and genotoxicity effects through the 3-[4,5-dimethylthiazol-2-yl]-2,5-diphenydl tetrazolium bromide (MTT) and alkaline comet assays (also known as single-cell gel electrophoresis) on healthy hepatic cell lines (HepG2) and peripheral blood mononuclear cells (PBMCs) (Paino et al. 2012).

14.7.3.1 Doxorubicin

In vitro and in vivo study in rats conducted by (Pranatharthiharan et al. 2017), HCC cells were targeted using the asialoglycoprotein (ASGPR) receptor to deliver doxorubicin-loaded NPs that targeted various ligands such as pullulan (Pul) and arabinogalactan (AGn). Pul-Dox NPs showed superior efficacy, lower toxicity, and better intrahepatic distribution when compared to non-NP-mediated therapy. (Xia et al. 2019) studied the delivery of Dox using galactose-modified selenium NPs in vitro through the clathrin-mediated endocytosis pathway, which showed significant apoptotic activity in HepG2 cells. This actively targeted, pH-controlled delivery system proved to have better antitumour efficacy than the passive targeting through selenium-mediated Dox delivery in vivo while also ensuring that no damage occurred to several vital organs. An interesting phase 3 randomised controlled trial conducted by researchers showed that in patients with advanced HCC (where sorafenib treatment failed), NP-assisted delivery of Dox didn't improve patient outcomes. This clinical trial shows that while NP-assisted therapies show a lot of promise in overcoming the challenges of traditional chemotherapy, they are by no means a cure for treating cancer.

14.7.3.2 Sorafenib

Sorafenib is the leading chemotherapeutic drug used clinically to treat HCC. However, it has its drawbacks, including poor bioavailability, cytotoxicity, and off-target effects. All of these drawbacks can theoretically be overcome using AVNPs. In an attempt to improve delivery efficiency, (Thapa et al. 2015) have used liquid crystalline NPs. They were also able to use these liquid crystalline NPs for controlled release of the sorafenib in an acidic tumour microenvironment. However, the NP-assisted sorafenib was found to be more cytotoxic to HepG2 cells than the free drug. Another study by (Cao et al. 2015) involved leveraging the hydrophobic interactions among sorafenib, curcumin, and a PEG derivative of Vitamin E succinate to create a self-assembled sorafenib-curcumin nanoparticle (SCN), which enhanced in vitro cytotoxicity and exhibited anti-angiogenesis activity. An in vivo mouse study conducted by (Wang et al. 2019) showed significant antitumor efficacy through the co-administration of sorafenib and NPs of poly(L-glutamic acid)-graft-methoxy poly (ethylene glycol)/combretastatin A4 sodium salt (CA4-NPs). They compared the efficacy of monotherapy of each of the therapeutic molecules in subcutaneous and orthotopic H22 hepatic tumour models. Finally, in a study conducted by Depalo et al. (2017), sorafenib was delivered using a magnetically targeted iron-oxide-based nano platform. PEGylated phospholipid micelles loaded with superparamagnetic iron oxide NPs (SPIONs) were delivered in vitro into HepG2 cells, and their effects were studied. The NP-assisted therapy was able to enhance sorafenib's antitumor activity through magnetic targeting.

14.7.3.3 RNA-Based

Huang et al. (2018) studied the delivery of siRNA through a galactose derivative-modified lipid/calcium/phosphate (LCP) NPs to treat HCC. The LCP-NPs were conjugated with eight types of galactosidase derivatives to study their antitumour efficacy. VEGF expression was successfully downregulated through the LCP-NP-mediated delivery of the associated siRNA.

14.7.3.4 Polymeric NPs

In a fascinating study conducted by Ou et al. (2017), low density lipoprotein NPs were conjugated with docosahexaenoic acid (LDL-DHA), and their anticancer activity was measured. They described a novel anti-apoptotic iron-mediated cell-death pathway called ferroptosis, caused by the LDL-DHA AVNPs through which rat and human HCC cell lines were killed. Three features that stood out in this cell-death pathway were the lipid peroxidation, depletion of glutathione, and the inactivation of the lipid antioxidant glutathione peroxidase (GPX4). In an extension of their in vitro study, the group also studied the in vivo effect of LDL-DHA on tumour xenografts in mice and found that the intratumoral injections inhibited the growth of HCC long-term and caused ferroptic cell death. In a study to understand the therapeutic potential of, and to create an efficient delivery system for, umbelliferone β-D-galactopyranoside (UFG), Wistar rats with diethyl nitrosamine (DEN)-induced HCC were investigated in vivo, and HuH-7 and Hep-G2 were investigated in vitro. The group experimented with a PLGA NP-based delivery system prepared and loaded with UFG through sonication. DEN-induced reactive oxygen species generation, mitochondrial dysfunction, and pro-inflammatory cytokine alteration were used to characterise the anticancer potential of UFG-PLGA-NPs. In a similar study by a group of researchers, rutin-loaded PLGA NPs were assessed for their anticancer activity in in vitro, in vivo, and biochemical studies. They found that the RT-PLGA-NPs administered orally reduced the incidence of hepatic nodules and exhibited significant anticancer activity.

Natural polymers like chitosan and pectin were also used to effectively mediate cancer therapy in several studies. (Salem et al. 2018) found that chitosan-coated gold NPs used to deliver 5-fluorouracil (5-Fu) showed significant anticancer activity in HepG2 cell lines. The NP-mediated treatment methodology reduced the concentration of 5-Fu required to inhibit growth by 50% when compared to free 5-FU. They also found that in the presence of laser radiation, the concentration required for the same inhibition was even lower. Another chitosan-based NPs (CNPs) were used to induce necrosis in HepG2 cells, which had DEN-induced HCC (Elkeiy et al. 2020). They posited that the CNPs derived from *Artemia salina* induced necrosis, not apoptosis, and induced the antioxidant enzyme activity in vivo and through the stimulation of ROS in vitro. A pectin-based drug-delivery system was evaluated by (Yu et al. 2014) to deliver 5-FU in rats and mice and to study cytotoxicity in HepG2 and A549 cell lines. They found that the NP-mediated delivery system induced prolonged-circulation and asialoglycoprotein (ASGPR) receptor-mediated targeting to lower off-target effects and toxicity of 5-FU. A few polymer-based NP-mediated treatment methodologies also can be applied to image/diagnose HCC.

Liposomal NPs targeting CD44 antibodies were used to image and treat HCC by (Wang et al. 2012). The group delivered Dox as a therapeutic and a triple-fusion (TF) gene containing thymidine kinase, renilla luciferase, and red fluorescent protein to monitor optical bioluminescence. They specifically studied the effect of NP-mediated therapy in targeting tumorigenic cancer stem cells (CSCs), rather than the more sought-after, non-tumorigenic progeny. They demonstrate the potential advantages of using this novel approach to target CSCs as a way to prevent relapse and effectively eliminate cancer.

14.8 CONCLUSIONS

As described throughout this chapter, nanoparticles that exhibit antiviral properties have shown significant potential in the entire process of cancer management, starting from imaging, diagnosis, drug delivery, treatment, monitoring, and even prevention. Several challenges with traditional approaches to cancer management, like poor bioavailability, off-target effects, high cytotoxicity, and many more, have been overcome on some scale through the application of a NP-mediated approach. However, there is still a long way ahead for NPs to be fully integrated into cancer management since the translatability of these findings still must be proven, and challenges surrounding scalability of manufacture must be overcome. AVNPs have the potential to transform cancer management through a highly personalised approach; however, it is important to remember that they are no silver bullet for all the issues plaguing cancer management. Extensive research, clinical studies at scale, and an efficient and reliable method of manufacture are just some of the factors that will play a key role in enabling AVNP-mediated cancer management.

ACKNOWLEDGMENTS

The corresponding author Dr. Pola Sudhakar would like to express his deepest thanks to authorities of Andhra University Visakhapatnam, India for extending constant support.

REFERENCES

Ashiq, M.G.B. 2018. Breast cancer treatment by nanophotolysis approach. *Results in Physics* 9:982–986. 10.1016/j.rinp.2018.03.058

Augustine, R. and Hasan, A. 2020. Multimodal applications of phytonanoparticles. *Phytonanotechnology* 195–219. 10.1016/b978-0-12-822348-2.00011-5

Bagheri, S., Yasemi, M., Safaie-Qamsari, E., et al. 2018. Using gold nanoparticles in diagnosis and treatment of melanoma cancer. *Artificial Cells, Nanomedicine and Biotechnology* 46(S1):462–471. 10.1080/21691401.2018.1430585

Barenholz, Y. 2012. Doxil® - The first FDA-approved nano-drug: Lessons learned. *Journal of Controlled Release* 160(2):117–134. 10.1016/j.jconrel.2012.03.020

Binzel, D.W., Shu, Y., Li, H., et al. 2016. Specific delivery of MiRNA for high efficient inhibition of prostate cancer by RNA nanotechnology. *Molecular Therapy* 24(7): 1267–1277. 10.1038/mt.2016.85

Bu, J., Nair, A., Kubiatowicz, L.J., et al. 2020. Surface engineering for efficient capture of circulating tumor cells in renal cell carcinoma: From nanoscale analysis to clinical application. *Biosensors and Bioelectronics* 162:112250. 10.1016/j.bios.2020.112250

Cao, H., Wang, Y., He, X., et al. 2015. Codelivery of sorafenib and curcumin by directed self-assembled nanoparticles enhances therapeutic effect on hepatocellular carcinoma. *Molecular Pharmaceutics* 12(3):922–931. 10.1021/mp500755j

Chakravarty, M. and Vora, A. 2020. Nanotechnology-based antiviral therapeutics. *Drug Delivery and Translational Research*. 10.1007/s13346-020-00818-0

Chen, L. and Liang, J. 2020. An overview of functional nanoparticles as novel emerging antiviral therapeutic agents. *Materials Science and Engineering C* 112:110924. 10.1016/j.msec.2020.110924

Chen, Y., Ma, J., Xu, M. and Liu, S. 2020. Antiviral nanoagents: More attention and effort needed? *Nano Today* 35:100976. 10.1016/j.nantod.2020.100976

Chugh, H., Sood, D., Chandra, I., Tomar, V., Dhawan, G. and Chandra, R. 2018. Role of gold and silver nanoparticles in cancer nano-medicine. *Artificial cells, Nanomedicine and Biotechnology* 46(S1):1210–1220. 10.1080/21691401.2018.1449118

Colagrande, S., Inghilesi, A.L., Aburas, S., Taliani, G.G., Nardi, C. and Marra, F. 2016. Challenges of advanced hepatocellular carcinoma. *World Journal of Gastroenterology* 22(34):7645–7659. 10.3748/wjg.v22.i34.7645

Cui, T., Liang, J.J., Chen, H., et al. 2017. Performance of doxorubicin-conjugated gold nanoparticles: Regulation of drug location. *ACS Applied Materials and Interfaces* 9(10):8569–8580. 10.1021/acsami.6b16669

Dadwal, A., Baldi, A. and Kumar Narang, R. 2018. Nanoparticles as carriers for drug delivery in cancer. *Artificial Cells, Nanomedicine and Biotechnology* 46(S2):295–305. 10.1080/21691401.2018.1457039

Depalo, N., Iacobazzi, R.M., Valente, G., et al. 2017. Sorafenib delivery nanoplatform based on superparamagnetic iron oxide nanoparticles magnetically targets hepatocellular carcinoma. *Nano Research* 10(7):2431–2448. 10.1007/s12274-017-1444-3

Dhand, C., Dwivedi, N., Loh, X.J., et al. 2015. Methods and strategies for the synthesis of diverse nanoparticles and their applications: A comprehensive overview. *RSC Advances* 5(127):105003–105037. 10.1039/c5ra19388e

Dias, N. and Stein, C.A. 2002. Antisense oligonucleotides: Basic concepts and mechanisms. *Molecular Cancer Therapeutics* 1(5):347–355.

Elkeiy, M.M., Khamis, A.A., El-Gamal, M.M., et al. 2020. Chitosan nanoparticles from *Artemia salina* inhibit progression of hepatocellular carcinoma in vitro and in vivo. *Environmental Science and Pollution Research* 27(16):19016–19028. 10.1007/s11356-018-3339-6

Emami, F., Banstola, A., Vatanara, A., et al. 2019. Doxorubicin and anti-PD-L1 antibody conjugated Gold nanoparticles for colorectal cancer photochemotherapy. *Molecular Pharmaceutics* 16(3):1184–1199. 10.1021/acs.molpharmaceut.8b01157

Fan, M., Han, Y., Gao, S., et al. 2020. Ultrasmall gold nanoparticles in cancer diagnosis and therapy. *Theranostics* 10(11):494–4957. 10.7150/thno.42471

Fang, J., Islam, W. and Maeda, H. 2020. Exploiting the dynamics of the EPR effect and strategies to improve the therapeutic effects of nanomedicines by using EPR effect enhancers. *Advanced Drug Delivery Reviews* 157(2020):142–160. 10.1016/j.addr.2020.06.005

Filippi, M., Nguyen, D.V., Garello, F., et al. 2019. Metronidazole-functionalized iron oxide nanoparticles for molecular detection of hypoxic tissues. *Nanoscale* 11(46): 22559–22574. 10.1039/c9nr08436c

Gurunathan, S., Qasim, M., Choi, Y., et al. 2020. Antiviral potential of nanoparticles: Can nanoparticles fight against *coronaviruses*? *Nanomaterials* 10(9):1–29. 10.3390/nano10091645

Hale, S.J.M., Perrins, R.D., Garcia, C.E., et al. 2019. DM1 loaded ultrasmall Gold nanoparticles display significant efficacy and improved tolerability in murine models of hepatocellular carcinoma. *Bioconjugate Chemistry* 30(3):703–713. 10.1021/acs.bioconjchem.8b00873

Huang, K.W., Lai, Y.T., Chern, G.J., et al. 2018. Galactose derivative-modified nanoparticles for efficient siRNA delivery to hepatocellular carcinoma. *Biomacromolecules* 19(6): 2330–2339. 10.1021/acs.biomac.8b00358

Jatzkewitz, H. 1955. An ein kolloidales Blutplasma-Ersatzmittel (Polyvinylpyrrolidon) gebundenes Peptamin (Glycyl-L-leucyl-mezcalin) als neuartige Depotform für biologisch aktive primäre Amine (Mezcalin). *Zeitschrift Für Naturforschung B* 10(1):27–31. 10.1515/znb-1955-0106

Jia, Q., Zhao, Z., Liang, K., et al. 2020. Recent advances and prospects of carbon dots in cancer nanotheranostics. *Materials Chemistry Frontiers* 4(2):449–471. 10.1039/c9qm00667b

Jonuleit, H., Bopp, T. and Becker, C. 2016. Treg cells as potential cellular targets for functionalized nanoparticles in cancer therapy. *Nanomedicine* 11(20):2699–2709. 10.2217/nnm-2016-0197

Kaspar, B. 2009. Gene therapy: Direct viral delivery. In *Encyclopedia of Neuroscience*, ed. L.R. Squire, 633–639. London: Academic Press.

Khandel, P., Yadaw, R.K., Soni, D.K., Kanwar, L. and Shahi, S.K. 2018. Biogenesis of metal nanoparticles and their pharmacological applications: Present status and application prospects. *Journal of Nanostructure in Chemistry* 8(3):217–254. 10.1007/s40097-018-0267-4

Kumar, A., Huo, S., Zhang, X., et al. 2014. Neuropilin-1-targeted gold nanoparticles enhance therapeutic efficacy of platinum(IV) drug for prostate cancer treatment. *ACS Nano* 8(5):4205–4220. 10.1021/nn500152u

Kumar, A., Zhang, X. and Liang, X.J. 2013. Gold nanoparticles: Emerging paradigm for targeted drug delivery system. *Biotechnology Advances* 31(5):593–606. 10.1016/j.biotechadv.2012.10.002

Kumar, R., Nayak, M., Sahoo, G.C., et al. 2019. Iron oxide nanoparticles based antiviral activity of H1N1 *influenza* A virus. *Journal of Infection and Chemotherapy* 25(5): 325–329. 10.1016/j.jiac.2018.12.006

Kumar, D. and Sharma, P.K. 2018. Nanoparticulate system for cancer therapy: An updated review. *Int J Nanomater Nanotechnol Nanomed* 4(2):22–34. 10.17352/2455-3492.000027

Kumari, R., Sunil, D. and Ningthoujam, R.S. 2020. Hypoxia-responsive nanoparticle based drug delivery systems in cancer therapy: An up-to-date review. *Journal of Controlled Release* 319:135–156. 10.1016/j.jconrel.2019.12.041

Kwatra, D., Venugopal, A. and Anant, S. 2013. Nanoparticles in radiation therapy: A summary of various approaches to enhance radiosensitization in cancer. *Translational Cancer Research* 2(4):330–342. 10.3978/j.issn.2218-676X.2013.08.06

Lara, H.H., Ayala-Nuñez, N.V., Ixtepan-Turrent, L. and Rodriguez-Padilla, C. 2010. Mode of antiviral action of silver nanoparticles against HIV-1. *Journal of Nanobiotechnology* 8:1–10. 10.1186/1477-3155-8-1

Li, C., Wang, J., Lu, X., et al. 2020. Hydrogen peroxide-response nanoprobe for CD44-targeted circulating tumor cell detection and H_2O_2 analysis. *Biomaterials* 255:120071. 10.1016/j.biomaterials.2020.120071

Li, H., Sun, J., Zhu, H., Wu, H., Zhang, H., Gu, Z. and Luo, K. 2021. Recent advances in development of dendriticpolymer-based nanomedicines for cancer diagnosis. *Wiley Interdisciplinary Reviews: Nanomedicine and Nanobiotechnology* 13(2): 1–28. 10.1002/wnan.1670

Mamo, T. and Poland, G.A. 2012. Nanovaccinology: The next generation of vaccines meets 21st Century materials science and engineering. *Vaccine* 30(47): 6609–6611. 10.1016/j.vaccine.2012.08.023

Miele, E., Spinelli, G.P., Miele, E., et al. 2012. Nanoparticle-based delivery of small interfering RNA: Challenges for cancer therapy. *International Journal of Nanomedicine* 7:3637–3657. 10.2147/IJN.S23696

Milovanovic, M., Arsenijevic, A., Milovanovic, J., Kanjevac, T. and Arsenijevic, N. 2017. Nanoparticles in antiviral therapy. In *Antimicrobial Nanoarchitectonics* ed. A.M. Grumezescu , 383–410. Amsterdam, Netherlands: Elsevier.

Mitragotri, S. and Stayton, P. 2014. Organic nanoparticles for drug delivery and imaging. *MRS Bulletin* 39(3):219–223. 10.1557/mrs.2014.11

Nelson, O., Patrick, O. and Ndidi, N.2009. Nanotechnology and drug delivery part 2: Nanostructures for drug delivery. *Tropical Journal of Pharmaceutical Research* 8(3). 10.4314/tjpr.v8i3.44547

Nikzamir, M., Hanifehpour, Y., Akbarzadeh, A. and Panahi, Y. 2021. Applications of dendrimers in nanomedicine and drug delivery: A review. *Journal of Inorganic and Organometallic Polymers and Materials*, 0123456789. 10.1007/s10904-021-01925-2

Nishimoto, Y., Nagashima, S., Nakajima, K., et al. 2020. Carboxyl-, sulfonyl-, and phosphate-terminal dendrimers as a nanoplatform with lymph node targeting. *International Journal of Pharmaceutics* 576:119021. 10.1016/j.ijpharm.2020.119021

Odularu, A.T. 2018b. Metal nanoparticles: Thermal decomposition, biomedicinal applications to cancer treatment, and future perspectives. *Bioinorganic Chemistry and Applications* 2018:1–6. 10.1155/2018/9354708

Ou, W., Mulik, R.S., Anwar, A., McDonald, J.G., He, X. and Corbin, I.R. 2017. Low-density lipoprotein docosahexaenoic acid nanoparticles induce ferroptotic cell death in hepatocellular carcinoma. *Free Radical Biology and Medicine* 112(58):597–607. 10.1016/j.freeradbiomed.2017.09.002

Paino, I.M.M., Marangoni, V.S., de Oliveira, R. de C.S., Antunes, L.M.G. and Zucolotto, V. 2012. Cyto and genotoxicity of gold nanoparticles in human hepatocellular carcinoma and peripheral blood mononuclear cells. *Toxicology Letters* 215(2):119–125. 10.1016/j.toxlet.2012.09.025

Parayath, N.N., Stephan, S.B., Koehne, A.L., Nelson, P.S. and Stephan, M.T. 2020. In vitro-transcribed antigen receptor mRNA nanocarriers for transient expression in circulating T cells in vivo. *Nature Communications* 11(1). 10.1038/s41467-020-19486-2

Parvanian, S., Mostafavi, S.M. and Aghashiri, M. 2017. Multifunctional nanoparticle developments in cancer diagnosis and treatment. *Sensing and Bio-Sensing Research* 13:81–87. 10.1016/j.sbsr.2016.08.002

Patra, J.K., Das, G., Fraceto, L.F., et al. 2018. Nano based drug delivery systems: Recent developments and future prospects. *Journal of Nanobiotechnology* 16(1):1–33. 10.1186/s12951-018-0392-8

Pizzino, G., Irrera, N., Cucinotta, M., et al. 2017. Oxidative stress: Harms and benefits for human health. *Oxidative Medicine and Cellular Longevity* 2017. 10.1155/2017/8416763

Pranatharthiharan, S., Patel, M.D., Malshe, V.C., et al. 2017. Asialoglycoprotein receptor targeted delivery of doxorubicin nanoparticles for hepatocellular carcinoma. *Drug Delivery* 24(1):20–29. 10.1080/10717544.2016.1225856

Qin, Y., Liu, H., Juan Li, M., et al. 2018. Salidroside improves the hypoxic tumor microenvironment and reverses the drug resistance of platinum drugs via HIF-1α signaling pathway. *EBioMedicine* 38:25–36. 10.1016/j.ebiom.2018.10.069

Quader, S. and Kataoka, K. 2017. Nanomaterial-enabled cancer therapy. *Molecular Therapy* 25(7):1501–1513. 10.1016/j.ymthe.2017.04.026

Racca, L. and Cauda, V. 2021. Remotely activated nanoparticles for anticancer therapy. *Nano-Micro Letters* 13(1). 10.1007/s40820-020-00537-8

Rai, A., Jenifer, J. and Upputuri, R.T.P. 2017. Nanoparticles in therapeutic applications and role of albumin and casein nanoparticles in cancer therapy. *Asian Biomedicine* 11(1):3–20. 10.5372/1905-7415.1101.534

Rigotto, C., Hanley, K., Rochelle, P.A., De Leon, R., Barardi, C.R.M. and Yates, M.V. 2011. Survival of *adenovirus* types 2 and 41 in surface and ground waters measured by a plaque assay. *Environmental Science and Technology* 45(9):4145–4150. 10.1021/es103922r

Rosi, N.L., Giljohann, D.A., Thaxton, C.S., Lytton-Jean, A.K.R., Han, M.S. and Mirkin, C.A. 2006. Oligonucleotide-modified gold nanoparticles for infracellular gene regulation. *Science* 312(5776):1027–1030. 10.1126/science.1125559

Salem, D.S., Sliem, M.A., El-Sesy, M., Shouman, S.A. and Badr, Y. 2018. Improved chemo-photothermal therapy of hepatocellular carcinoma using chitosan-coated gold nanoparticles. *Journal of Photochemistry and Photobiology B: Biology* 182:92–99. 10.1016/j.jphotobiol.2018.03.024

Schuler, G., Schuler-Thurner, B. and Steinman, R.M. 2003. The use of dendritic cells in cancer immunotherapy. *Current Opinion in Immunology* 15(2):138–147. 10.1016/S0952-7915(03)00015-3

Shen, Y., Hao, T., Ou, S., Hu, C. and Chen, L. 2018. Applications and perspectives of nanomaterials in novel vaccine development. *MedChemComm* 9(2):226–238. 10.1039/c7md00158d

Sneider, A., Vandyke, D., Paliwal, S. and Rai, P. 2017. Remotely triggered nano-theranostics for cancer applications. *Nanotheranostics* 1(1):1–22. 10.7150/ntno.17109

Sulczewski, F.B., Liszbinski, R.B., Romão, P.R.T. and Rodrigues Junior, L.C. 2018. Nanoparticle vaccines against viral infections. *Archives of Virology* 163(9):2313–2325. 10.1007/s00705-018-3856-0

Swain, S., Sahu, P.K., Beg, S. and Babu, S.M. 2016. Nanoparticles for cancer targeting: Current and future directions. *Current Drug Delivery* 13(8):1290–1302. 10.2174/1567201813666160713121122

Thapa, R.K., Choi, J.Y., Poudel, B.K., et al. 2015. Multilayer-coated liquid crystalline nanoparticles for effective sorafenib delivery to hepatocellular carcinoma. *ACS Applied Materials and Interfaces* 7(36):20360–20368. 10.1021/acsami.5b06203

Varshosaz, J. and Farzan, M. 2015. Nanoparticles for targeted delivery of therapeutics and small interfering RNAs in hepatocellular carcinoma. *World Journal of Gastroenterology* 21(42):12022–12041. 10.3748/wjg.v21.i42.12022

Wang, J., Han, S., Ke, D. and Wang, R. 2012. Semiconductor quantum dots surface modification for potential cancer diagnostic and therapeutic applications. *Journal of Nanomaterials* 2012:129041. 10.1155/2012/129041

Wang, L., Su, W., Liu, Z., et al. 2012. CD44 antibody-targeted liposomal nanoparticles for molecular imaging and therapy of hepatocellular carcinoma. *Biomaterials* 33(20):5107–5114. 10.1016/j.biomaterials.2012.03.067

Wang, Y., Yu, H., Zhang, D., et al. 2019. Co-administration of combretastatin A4 nanoparticles and sorafenib for systemic therapy of hepatocellular carcinoma. *Acta Biomaterialia* 92:229–240. 10.1016/j.actbio.2019.05.028

Xia, Y., Zhong, J., Zhao, M. et al. 2019. Galactose-modified selenium nanoparticles for targeted delivery of doxorubicin to hepatocellular carcinoma. *Drug Delivery* 26(1):1–11. 10.1080/10717544.2018.1556359

Xu, B., Xia, S., Wang, F., et al. 2016. Polymeric nanomedicine for combined gene/chemotherapy elicits enhanced tumor suppression. *Molecular Pharmaceutics* 13(2):663–676. 10.1021/acs.molpharmaceut.5b00922

Yan, Y., Gao, X., Zhang, S., et al. 2019. A carboxyl-terminated dendrimer enables osteolytic lesion targeting and photothermal ablation of malignant bone tumors. *ACS Applied Materials and Interfaces* 11(1):160–168. 10.1021/acsami.8b15827

Yang, C., Ren, X., Ding, D., Wang, L. and Yang, Z. 2016. Enzymatic induction of supramolecular order and bioactivity. *Nanoscale* 8(20):10768–10773. 10.1039/c6nr02330d

Yu, C.Y., Wang, Y.M., Li, N.M., et al. 2014. In vitro and in vivo evaluation of pectin-based nanoparticles for hepatocellular carcinoma drug chemotherapy. *Molecular Pharmaceutics* 11(2):638–644. 10.1021/mp400412c

Zhang, X.F., Liu, Z.G., Shen, W. and Gurunathan, S. 2016. Silver nanoparticles: Synthesis, characterization, properties, applications, and therapeutic approaches. *International Journal of Molecular Sciences* 17(9). 10.3390/ijms17091534

Zuhorn, I.S., Engberts, J.B.F.N. and Hoekstra, D. 2007. Gene delivery by cationic lipid vectors: Overcoming cellular barriers. *European Biophysics Journal* 36(4–5):49–362. 10.1007/s00249-006-0092-4

15 Nano Delivery of Antiviral Plant Bioactives as Cancer Therapeutics

Haripriya Shanmugam, Badma Priya, Manickam Senguttuvan Swetha, and Janani Semalaiyappan

CONTENTS

DOI: 10.1201/9781003136644-18

15.1 INTRODUCTION

Plants have been used as therapeutics since ancient times, and they are the main components of traditional systems of medicines across the globe. With the emergence of new viral diseases in the current scenario, a global concern for bioprospecting new antiviral agents that are less toxic and easy to metabolize within the human system is on the rise. Due to the mutating character of the viral genomes, it is challenging to control their infection and is evolving into a considerable threat to human health. Herbs and medicinal plant bioactives are used against arduous viral infections due to their effectiveness and lack of any negative side effects. The mechanism by which plant bioactives exhibit antiviral activities depends on the type of viruses and their mode of infection at the molecular level. Plant bioactives exhibit antiviral activity by affecting common pathways of the human body in eliminating the virus. There are earlier reports suggesting plant bioactives to be safe, affordable, and, most importantly, with void or minimal side effects, depending on the right dose and frequency of administration with an appropriate delivery system.

In general, most plant bioactive-based drugs are designed for oral delivery because intestinal mucosa has a higher absorption capacity and is consumers' most preferred route of drug administration. Though a lot of benefits are present in oral drug delivery, certain significant challenges limit its effectiveness; these include poor aqueous solubility of the drug molecule, rapid gastric emptying, lower rate of diffusion in the intestinal membranes, enzyme activity in the intestine region, and chemical degradation. Such challenges in the conventional oral drug-delivery system can be overcome by designing nano drug delivery systems (NDDS). NDDS are colloidal systems with an average particle size between 10–1000 nm diameter, designed to improve drug efficacy by increasing its half-life, aqueous solubility of hydrophobic drugs, and releasing the loaded drugs in a specific pattern within the body. Of late, NDDS approved by FDA are on the rise since Professor Peter Speiser's work on targeted nanoparticles-based drug delivery during 1996. As of now, NDDS patents, predominantly on noncommunicable diseases and cancer, get the most attention. In this chapter, plant bioactives possessing antiviral properties with anticancer activities, their mechanism of action, limitations on delivering them through oral administration, need for NDDS, and advanced NDDS suited for cancer therapeutics are discussed.

15.2 ANTIVIRAL PLANT BIOACTIVES WITH ANTICANCER ACTIVITY

Viruses are obligate parasites and are acellular. They require a host for their gene expression, metabolism, and multiplication. They generally attach to the hosts through their spike proteins, fuse their membrane with the host membrane, and make an entry. Some viruses, like human papillomavirus and Kaposi sarcoma-associated herpes virus can cause cancer, while some can also flame up cancer metastasis (Bost et al. 2020; Al-Sarraj et al. 2021). Rota virus, a type of RNA virus, causes severe diarrhoea and vomiting in children, infecting millions of people every year and causing 200,000 deaths approximately (Liu et al. 2021). Viruses such as HIV, Ebola, COVID-19, and influenza show very aggressive infections compared to other viruses. Viral enzymes play a major function in viral infections since they are needed for multiplication as well as mutation of viruses; that makes the drug-development process tedious (Behl et al. 2021). Currently, the COVID-19 or SARS-CoV-2 pandemic is catastrophic to human health with global spread affecting millions of people across the globe. Several serum-based monoclonal and polyclonal antibodies, and recombinant vaccines, have been developed. But the viral variants constantly mutate and challenge each combating vaccine/antibody developed (Chen et al. 2021). Various in silico and in vitro studies are being carried out to find more effective drugs to fight the pandemic. One such study reports that 40 different plant-based bioactives, such as sarsasapogenin, rutin, berberine, ursonic acid, and aranotin, show promising results in inhibiting the Nsp15 viral protein. Nsp15 (nonstructural protein 15) is one of the virulent drug targets for SARS-CoV-2 of the *Coronaviridae* family (Kumar et al. 2021). The extensive history of plant-based traditional medicines corroborating the use of plant bioactives to treat various viral diseases is always a rudimentary start for the search of alternatives to conventional drugs.

Numerous phytochemicals are available from plant sources, and they are classified as flavonoids, alkaloids, terpenes, polyphenols, and many others. They are often found in nature in their conjugated forms. Several studies report the importance of plant bioactives for viral therapeutics. For example, *Ipomopsis aggregate* (scarlet trumpet) extracts can be used to treat parainfluenza viral infections, roots extracts of *Lomatiumdis sectum* (biscuitroot or Indian parsley) to treat herpes simplex virus infections, *Eleutherococcus senticosus* (Siberian ginseng) extracts against human retro viruses (Behl et al. 2021). With the need for effective strategies to combat viral infections and their spread globally, potential plant bioactives with antiviral properties that can be used for cancer therapeutics are presented in the following section.

15.2.1 Polyphenols

Compounds containing more than one phenolic group are called polyphenols and are extensively consumed in the human diet. They are highly ubiquitous in many kinds of fruits, in vegetables, and in dark chocolates. There are several kinds of polyphenols like flavonoids, phenolic acid, polyphenolic amides, to mention a few. Most of the phenolic compounds play a major role in signalling pathways like NF-κB, JAK-STATs, and MAPK pathways.

Curcumin, a major bioactive found in the rhizomes of turmeric, *Curcuma longa L.*, is broadly used in the traditional system of Indian medicine owing to its antibacterial, antiviral, anticancer, and antifungal activities. Curcumin is suggested to be safe even at higher dosages (12 grams per day) and has fewer side effects. Curcumin and its bioconjugates have shown strong antiviral effects on different viruses, such as influenza viruses [PR8, H1N1, and H6N1], HSV-1, HSV-2, coxsackie virus B3 (CVB3), hepatitis B virus (HBV), human papilloma viruses (HPVs), and Japanese encephalitis virus (JEV) (ZorofchianMoghadamtousi et al. 2014). Studies suggest that curcumin majorly exhibits its antiviral effects by interfering with replication of various viruses (Kim and Kim 2015). Curcumin exhibits antitumour characteristics by downregulating NF-kB pathway in prostate cancer, arresting cell cycles in G2 and S phase of breast cancer cells, and colorectal cancer cells, and is involved in shrinking colon cancer cells and in hepatocellular carcinoma (Aggarwal et al. 2003).

Ellagic acid is present in strawberries, raspberries, blackberries, walnuts, and pecans occur in conjugation with glucose moieties. Ellagic acid inhibits the activity of proteases present in the HCV (hepatitis C virus) virus, thereby inhibiting its multiplication. It is studied extensively for its antioxidant and antiproliferative potential. The anti-proliferative nature of ellagic acid has made it a very potent anticancer bioactive; it activates the intrinsic apoptotic pathway and works against several cancers, like carcinoma, cervical, and breast cancers (Reddy et al. 2014).

Caffeine, from another commonly consumed beverage, coffee, has potent antioxidant properties. Extensive studies have been conducted on chronic consumption of caffeine and reported it to have side effects, although moderate consumption has few health-promoting effects. Caffeine has proved to be effective against very few viruses like herpes simplex virus. Caffeine retards the viral proliferation and hence reduces its infection (Shiraki and Rapp 1988). Caffeine also interferes with receptor pathways in few cancer cells, delineating its anticancer nature (Tej et al. 2019).

Tannic acid exhibits anti-oxidant, anticancer, antimicrobial, and anti-mutagenic bioactivities, and these properties are exploited in Chinese traditional medicine. However, a report suggests that tannins from few food sources, such as betel nuts and herbal tea, cause cancer (Chung et al. 1998). However, tannic acids are effective against HIV, HSV, influenza virus, noro viruses, and papilloma viruses. Tannic acid inhibits viral replication by targeting multiple stages of viral infection (Kaczmarek 2020). It is reported to inhibit cancer cell viability by halting the cell cycle and induce apoptotic pathways majorly in small cell lung cancer (Nipin et al. 2020).

Coumarin plays a major role in HIV treatment by interfering with the HIV enzymes, such as protease, integrase, and reverse transcriptase. It also curbs viral protein R activity in HIV, inducing antiviral responses with the help of signalling proteins, interferon. Different conjugates of coumarin inhibit the viruses at different stages (Hassan et al. 2016). It has limitless anticancer properties with very few side effects; they target cancer cells in different ways. They induce apoptosis, arrest telomerase and also different stages of cell cycles, inhibit kinases, and possess high mitotic activity (Thakur et al. 2015).

Emodin, an anthraquinone extracted from aloe vera is used to treat liver injuries, necrosis, inflammation, and kidney diseases and to treat carcinoma. Studies report that aloe-emodin exhibits antiviral properties against enveloped viruses like influenza A virus and herpes simplex virus. It is also shown to obstruct attachment, entry, and replication of the influenza virus (Gansukh et al. 2018). The anticancer properties of aloe-emodin have been found to inhibit cancer cells like hepatoma cells, glioma, and carcinoma cells. It also possesses strong anti-proliferative and antineoplastic reaction on cancer cells due to its apoptotic nature (Chou and Liang 2009).

15.2.2 Flavanoids

Flavanoids, a kind of polyphenols, are found in a vast range of fruits, vegetables, roots, flowers, and stems. Quercetin and kaempferol are the most predominant flavanoids occurring in nature. Three aromatic amino acids, like phenylalanine, tyrosine, and tryptophan, are the source for their synthesis within the plant system. They play a major role as an antiviral bioactive and also in basic plant metabolism by inhibiting the enzymes like xanthine oxidase, cyclooxygenase, and many enzymes that produce reactive oxygen species (Pietta 2000).

Quercetin, a commonly occurring secondary metabolite in almost all parts of various plants, stalls the replication of many perilous viruses. It has been reported that quercetin, along with myricetin, quercetagetin, baicalein present in wild berries, onions, green tea, and apples, have been studied for their effects in various viral cells in vitro as well as in vivo and reported to inhibit the cell cycles of viruses by controlling their replication (Ben-Shabat et al. 2020). Quercetin exhibits anticancer characteristics by hindering the proliferation of various kinds of cancers. For example, quercetin in breast cancer increases the rate of apoptosis, sets off the cell cycle arrest in cervical cancer, reduces cell viability in colon cancer, and interferes with the epigenetic modifications, thereby initiating apoptosis in lung cancer cells (Rauf et al. 2018).

Tea (*Camellia sinensis*) leaves, used for a commonly consumed beverage worldwide, contain tannic acids that play a vital role in controlling viral replication by terminating the cell-cycle processes and instigating apoptosis. The polyphenol, catechin in tea, has been reported to have anti-proliferative effect against the influenza virus; they mainly target its viral membrane (Song et al. 2005). Green tea, in specific, has been studied extensively and reported to have antioxidant with anticancer properties. Also, green tea helps in treating cardiovascular diseases, hypertension, and atherosclerosis (Fujimura et al. 2017). Polyphenols in tea leaves can be a good prospect in antiviral and anticancer therapeutics.

Myricetin, structurally related to kaempferol and quercetin, exhibits a wide range of bioactivities such as antiviral, anticancer, anti-inflammatory, and antioxidant properties. Myricetin is reported to regulate the viral enzymes of HIV, Coronavirus, and Rauscher murine leukaemia virus. It is also a potent anticancer agent since it inhibits the enzymes responsible for various types of cancer, like skin cancer, colorectal cancer, and hepatic cancer (Semwal et al. 2016).

A common flavonoid, rutin, inhibits a broad range of viruses primarily by targeting viral polymerases and other enzymes essential for their replication, as well as the metabolic pathways. Rutin also has antineoplastic effects on various cancer cells that includes hepatic carcinoma, leukaemia, colon cancer, colorectal cancer, and melanoma by scavenging free radicals (Ganeshpurkar and Saluja 2017).

Genistein, an isoflavone commonly present in soy products, is a phytoestrogen that has anti-inflammatory, antimicrobial, and antioxidant properties. It has antiviral effects against a wide range of viruses like herpes virus, human papilloma virus, HIV, swine flu and can act synergistically along with other existing antiviral drugs like acyclovir and reduce their side effects (LeCher et al. 2019). It also hampers tyrosine phosphorylation, which is essential for various cellular processes. In this way, the viability of cancer cells is reduced; thus, genistein also acts as a potent anticancer bioactive (Chae et al. 2019).

Daidzein, primarily extracted from soy, is a naturally occurring phytoestrogen and a potent anti-oxidant. It is an inactive analogue of genistein. Reports suggest that the anti-influenza effects of daidzein are due to its interference with genes that regulate influenza replication (Horio et al. 2020). Daidzein is currently under clinical trials exhibiting anti-tumour effect against breast and ovarian cancer, and significantly increases the levels of apoptosis. Daidzein's anticancer property is targeted toward cell cycle arrest, hampering the signalling pathways and also controlling metastasis of the tumour cells (Hua et al. 2018).

15.2.3 Carotenoids

Carotenoids are plant pigments with more than 750 different members and are highly important to sustain normal human metabolism. They are not produced in animal systems, so humans predominantly depend on plant sources for carotenoids. Carotenoids exhibit anti-oxidant properties, pro-oxidant properties, anti-mutagenic activities, and anti-tumour activities (Krinsky and Johnson 2005; Nisar et al. 2015).

The unsaturated carotenoid, lycopene, is a potent singlet-oxygen that inactivates the free radicals, resulting in prevention of tissue damages. Lycopene induces oxidative stress in hepatitis C virus infections and is also reported to benefit infected patients in treating the viral infection through anti-oxidant therapies (Seren et al. 2008). Lycopene also inhibits several cancers by impeding their proliferation and triggering apoptosis. It induces programmed cell death in androgen-independent prostate cancer, detaining the cell cycle followed by apoptosis in leukaemia, and interfering with the cyclin-dependent kinases in human breast cancer cells (Bhuvaneswari and Nagini 2005).

Astaxanthin has been reported to have antioxidant properties because they readily attack the free radicals, thereby preventing tissue damages. In vitro as well as in vivo studies with astaxanthin have shown it to alleviate the effects of COVID-19 infection by surging cytokine storms. Many preclinical studies on astaxanthin in both in vitro as well as in vivo studies have reported to exhibit great promise in cancer therapeutics. Astaxanthin expresses an apparent anti-proliferative effect on human hepatoma, induces caspase-mediated apoptosis in hepatocellular carcinoma, and induces oxidative-stress induction in leukaemia (Zhang and Wang 2015).

Lutein, another effective antioxidant mostly used to treat ocular degeneration in aged people, is present in almost all green leafy vegetables and decreases the expression of hepatitis B surface antigens. It retards the replication of viral DNA by disturbing the transcription of HBV (hepatitis B virus) Fp promoter and is also said to have very fewer negative effects on the human cells (Pang et al. 2010). Studies on tumour cell lines show that they exhibit cytotoxic effects and induce programmed cell death in non-small-cell lung cancer. Lutein also inhibits proliferation of cervical cancer cells by augmenting reactive oxygen species production and by upregulating the caspase-mediated pathways (Zhang et al. 2018; Gansukh et al. 2019).

15.2.4 Terpenes

Terpenes are produced by latex-producing plants, and they protect the infected plant sites upon damages. Based on their anti-oxidant properties, terpenes are categorized into oxygenated monoterpenes, monoterpene hydrocarbons, oxygenated sesquiterpenes, and sesquiterpene hydrocarbons. They are reported to exhibit bioactivities like neuroprotection, cardio protection, hepatoprotection, and protection against diabetes (Gonzalez-Burgos and Gómez-Serranillos 2012).

Ursolic acid, a pentacyclic triterpene acid present in opulence in most vegetables and fruits, protects cells from oxidative damage and is reported to curb replication of rotavirus in both in vitro as well as in vivo conditions (Tohmé et al. 2019). It has also been reported to inhibit polymerase activities in hepatitis C virus, having significant suppression in its replication (Kong et al. 2013). Mostly present as a free acid, ursolic acid has momentous importance in breast cancer treatment by inhibiting the proteins involved in cell-cycle checkpoints and is currently advanced to clinical trials (Yin et al. 2018). It is also reported to induce cell death and arrest the cancer cell cycle, thus is effectively used in cancer therapies.

Indole derivatives are very common in occurrence (broccoli, cabbage, cauliflower, brussels sprouts, and other cruciferous vegetables), and the indole core is essential for many biosynthetic pathways. Indole-3-carbinol found commonly in Brassicaceae vegetables is an effective antiviral agent against HCV by inhibiting the viral entry and its replication (Zhang et al. 2015). The plant hormone, indole-3-carbinol, also has anti-tumour effects against a wide range of tumours, and it inhibits ligases, thereby combating tumours (Katz et al. 2018).

15.2.5 Alkaloids

Alkaloids are commonly occurring bioactives present in potatoes, tomatos, and fungi. Several preclinical studies have evidence on the antitumour significance of alkaloids, such as vinblastine, vincristine, tryptamine, liriodenine, and berberine (Mondal et al. 2019). Berberine, predominantly used as anti-diabetic bioactive, also has anti-neoplastic, anti-inflammatory, and antiviral effects with very low cytotoxic effects. Like other alkaloids, it interferes with the signalling pathway and disrupts viral replication. It works against SARS-CoV-2, HSV, human cytomegalo viruses, and chikungunya viruses (Warowicka et al. 2020). Berberine is also exploited in the treatment of amnesia and Alzheimer's disease. It has also been studied extensively for its antitumour effects exhibiting high cytotoxicity in tumour cells and is known to amplify the cytotoxicity of other anticancer drugs. It mainly targets the kinases responsible for cell cycle and transcription, eventually inhibiting cell growth (Habtemariam 2020).

15.2.6 Glycosides

Digoxin, a cardiac glycoside extracted from *Digitalis purpurea* (foxglove), is extensively used to treat several heart conditions. It strongly inhibits the expression of viral mRNA in SARS-CoV-2 chikungunya, MERS-CoV, HIV, and other viruses

(Cho et al. 2020). In recent years, various studies showed the anticancer effects of digoxin, as individual administration or a combination with other chemosynthetic drugs. It has also been reported to down-regulate the signalling pathways necessary for tumour proliferation in several breast cancer cases (Wang et al. 2020).

Hesperidin, commonly found in the citrus family, is of greater pharmacological significance as a powerful anti-inflammatory and anti-diabetic bioactive. It is a commonly available drug across the world used for treating vascular dilation (Haggag et al. 2020). In COVID-19 research, hesperidin helps in triggering the host immune system and acts as an antiviral bioactive in other respiratory-related viral infections. Hesperidin, customarily used in Chinese traditional medicine, is also shown to display antitumour effects against a broad spectrum of tumours, especially breast cancer, by downregulating few genes in signalling pathways necessary for cancer cell growth, multiplication, and metastasis (Kongtawelert et al. 2020).

Kaempferol reported to be effective against Coronavirus can also be used as an antiviral agent against various viral infections. Kaempferol glycosides interfere with the 3a protein pathway and have a greater potency compared to kaempferol (Schwarz et al. 2014). This common flavonoid is used as an antitumour agent against breast cancer, as it interferes with the cell cycle mechanism of breast cancer cells like other flavonoids and induces apoptosis by preventing metastasis (Wang et al. 2019).

15.2.7 Thiosulfinates

Allicin present in garlic cloves is primarily responsible for its unique smell and taste. It is a sulphur-containing bioactive that has various pharmacological activities and is primarily used in the traditional system of medicines. It primarily targets the thiol enzymes and exhibits antiviral activity against a wide class of viruses that includes rhinoviruses, HCM, herpes simplex virus, and vaccinia virus (Ankri and Mirelman 1999). Allicin is involved in inducing apoptosis and prevents cancer cell metastasis in a vast range of cancer cells, especially in colorectal and pancreatic cancer (El-Sabre et al. 2020).

15.2.8 Glucoside

Nomilin, a glucoside present in citrus fruit seeds, interferes with many signalling pathways, mostly the MAPK pathway. Cell culture studies report nomilin to exhibit antiviral effects against HIV infection by inhibiting HIV proteases, thereby stalling their replication (Battinelli et al. 2003). Nomilin targets many metabolic pathways and is said to have anti-obesity as well as anti-hyperglycemic activities. It also acts as a promising drug against many kinds of cancer since it inhibits cell proliferation and arrests cell cycle, leading to programmed cell death (Shanmugam et al. 2017).

15.2.9 Other Plant Molecules

Glutathione, a very common anti-oxidant present in most plants and fungi, plays a major role in immune system regulation. It is studied extensively for its antiviral, anti-oxidant, anti-inflammatory, and anti-coagulant properties. It helps in regulating the immune system as it deals with cytokine storms. It helps in surging large number of immune cells that are responsible for inflammation and subsequent immune responses helping in combating the virus. Glutathione, like other anti-oxidants, exhibits an anticancer mechanism against a broad range of cancer, such as breast, colorectal, and cervical cancer (Desideri et al. 2019; Polonikov 2020).

Among all the plant bioactives described above, flavonoids, a group of polyphenols, are extensively studied and exploited for their antiviral potency against various viruses. About 10–15% of human cancers all over the world are caused by viruses such as hepatitis B virus, hepatitis C virus, and papilloma viruses. And there have been incidences where antiviral drugs have been effective in cancer treatments too. For example, when people with chronic hepatitis infections were treated with antiviral drugs, there was a remarkable decrease in the occurrence of carcinomas. Since viral infections are a very crucial cause of human cancers, antiviral plant bioactives along with their anticancer properties need to be studied extensively to fight against both incidences. Understanding phytochemicals along with their bio-conjugates, dosages, and mechanism of action can help researchers develop safe and effective plant-based therapeutics to treat viral infections, as well as cancer therapeutics.

15.3 NANO DRUG-DELIVERY SYSTEMS FOR CANCER THERAPEUTICS

Cancer, the top most death-causing global disease, has chemotherapy as the most preferred and prescribed treatment. Currently, treatment for this deadliest disease includes chemotherapy, surgery, and radiation. The conventional system of cancer treatment with drugs leads to lethal damage of normal healthy cells on long-term usage. Limitations in conventional oral delivery of drugs include reduced selectivity of anticancer drugs, insufficient concentration of drug at tumour cells, limited oral bioavailability, non-specific distribution, resistance to multidrug, and toxicity to normal cells (Chidambaram et al. 2011). Oral drug deliveries are considered the predominantly accepted route of drug administration for its ease in administration and patient compliance. Despite being advantageous, it certainly has limitations like expulsion of the drug via intestinal drug transporter, prevention of drug penetration by P-glycoprotein (P-gp) mucus barrier, and low stability in the gastric region. To address these limitations existing in conventional oral drug delivery, NDDS are being explored with great interest.

Nano drug-delivery system (NDDS) are drug-loaded nanocarriers with at least one dimension in nanometre scale. Nanocarriers of size 10–400 nm are designed for drug localization and cellular uptake by providing uninterrupted and controlled drug release at tumour site (Huda et al. 2020; Edis et al. 2021). These NDDS are designed to increase drug stability and increase the cycle time and rate of uptake by the target cells (Deng et al. 2020). Nano drug-delivery system (NDDS) provides control over transportation and organ/tissue-specific onsite action. It also maximizes the drug concentration between the cells, either through enhanced permeability and retention effect (EPR) or by means of endocytosis. NDDS guarantees that the drug molecule is properly channelled at the desired site and rate at which drug it is to be delivered as per the need for the body (Saraf 2010). This way, NDDS are effective for cancer patients in alleviating drug resistance, precision in targeting tumour cells with desired pharmacokinetics, without or reduced drug side effects, and extended circulation time in vivo (Yao et al. 2020).

Plant bioactives are, in most cases, poorly absorbed (hydrophobic molecules) due to their higher molecular weight leading to a lower passive absorption and also lower lipid solubility (hydrophilic biomolecules), making it difficult for them to pass through the biological membrane. To resolve these drawbacks, NDDS are more explored to realize the therapeutic efficiency of many plant bioactives. Through NDDS, the therapeutic efficacy of both hydrophilic and hydrophobic plant bioactives can be improved by downsizing the plant bioactives and encapsulating them in various nanoformulations (Alexander et al. 2016). A plant bioactive may be conjugated to a nanocarrier or complexed ionically to the molecule of opposite charge or physically loaded inside a matrix (polymer or lipid or protein). On the basis of matrices used for developing NDDS, they can be classified as organic, inorganic, and composite nanomaterials. This chapter focuses only on organic and composite-based NDDS. Advanced NDDS like targeted release, stimuli-responsive release, and controlled release of plant bioactives are discussed in the forthcoming section. The graphical representation of types of advanced NDDS suitable for oral delivery of plant bioactives for cancer treatment are presented in Figure 15.1.

15.3.1 Targeted Nano Drug-Delivery System

The targeted nano drug-delivery system (targeted NDDS) releases the loaded plant bioactives at a chosen tumour site, protecting the normal cells from drug cytotoxicity. It selectively targets the site of action to deliver the drug/plant bioactive, thereby avoiding the nontarget healthy organs, tissues, or cells. A targeted NDDS focus on pharmacokinetic behaviour of the plant bioactives loaded nanocarrier, whereas the conventional targeted drug-delivery system primarily focuses only on delivering the drug at the targeted site. In the targeted NDDS, drug molecules are only intended to be delivered at the tumour cells to minimize their negative side effects, especially to healthy cells. Passive and active targeted systems are two different types of targeted NDDS, which are elaborately explained in the section below. Targeted nano drug-delivery systems for loading plant bioactives are presented in Table 15.1 and Figure 15.2.

15.3.1.1 Passive Drug Targeting (PDT)

Passive drug targeting means site-specific plant bioactives loaded into NDDSs that depend on the physico-chemical or pharmacological factors of the delivery system. It involves pathophysiological factors, physico-chemical factors, anatomical opportunities, and chemical approaches (Vasir et al. 2005). In passive targeting NDDS, the nanocarriers with physical properties like size and zeta potential are considered important for the delivery of drug bioactives to the site of action, which is being targeted (Babazadeh et al. 2020). The drug pharmacokinetics of NDDS ensures longer systemic circulation of the delivered drugs through nanosized

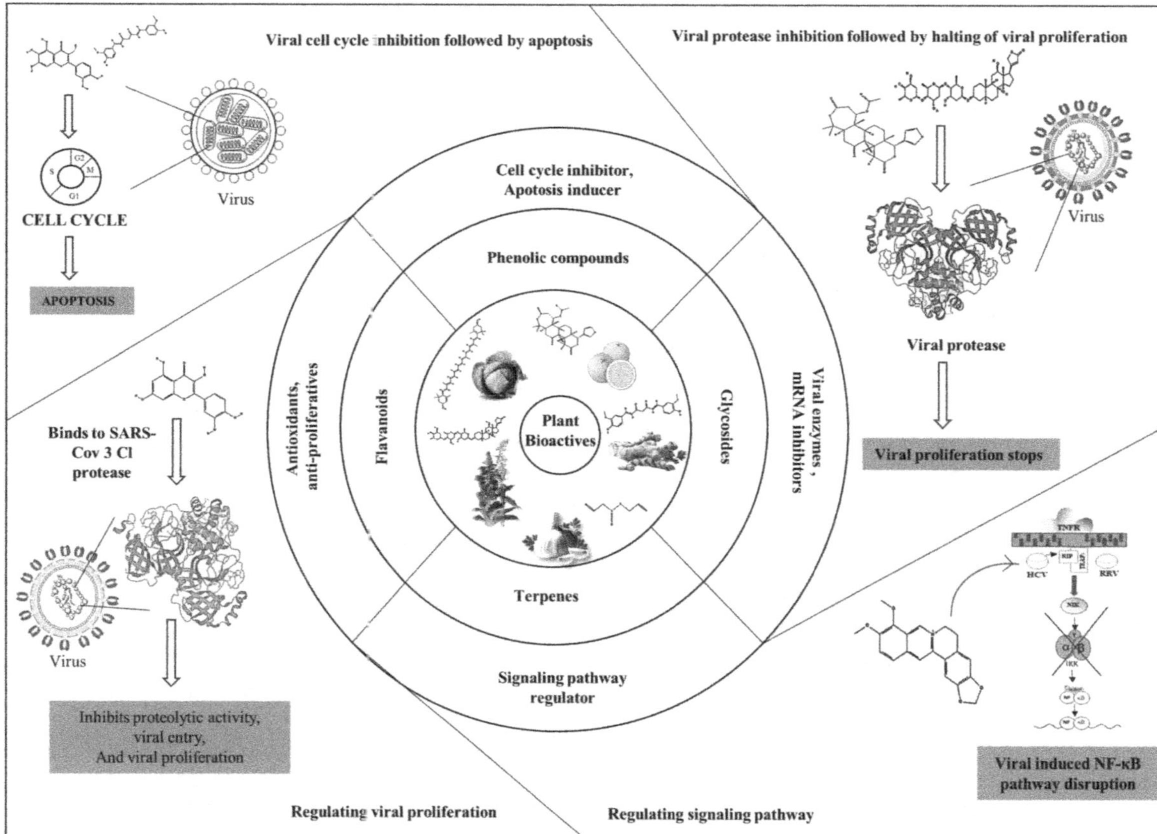

FIGURE 15.1 Graphical representation of plant bioactives and their proven antiviral mechanisms.

TABLE 15.1
List of Antiviral Plant Bioactives with Anticancer Properties with Their Mechanism of Action

Phytochemicals	Source(s)	Chemical Formula	Target Virus/Cancer	Mechanism of Action	References
Flavanoids					
Lutein	Green leafy vegetables	$C_{40}H_{56}O_2$	HBV	Potent transcriptional inhibitor of viral Fp promoter	Pang et al. (2010)
			Lung cancer	Inhibits the PI3K/AKT signalling pathway and triggers programmed cell death	Zhang et al. (2018)
Quercetin	Flavonoids	$C_{15}H_{10}O_7$	HIV	Reverse transcription inhibition	Gatto et al. (2002)
			Breast cancer	Apoptosis mediated by p53	Hashemzaei et al. (2017)
Rutin	Buckwheat, tobacco, forsythia, hydrangea, viola, etc.	$C_{27}H_{30}O_{16}$	HSV-1	Inhibition of HSV-1 replication	Ninfali et al. (2020)
			Colorectal cancer	Disturbs cell cycle and causes apoptosis	Ganeshpurkar and Saluja (2017)
Myricetin	Walnuts, onions, berries, herbs, and red grapes	$C_{15}H_{10}O_8$	HIV	Inhibits HIV-1 integrase	Cushnie and Lamb (2005)
			Hepatic, skin, pancreatic, and colon cancer	Inhibits key enzymes involved in the instigation and advancement of cancer	Semwal et al. (2016)
Genistein	Soy products	$C_{15}H_{10}O_5$	Herpes B virus	Hampers the virion production by interfering with the intracellular events happening in the late stages of viral life cycle	Le Cher et al. (2019)
			Prostate cancer	Downregulates transcription of AR (androgen receptor), followed by prostate-specific	Chae et al. (2019)

(Continued)

TABLE 15.1 (Continued)
List of Antiviral Plant Bioactives with Anticancer Properties with Their Mechanism of Action

Phytochemicals	Source(s)	Chemical Formula	Target Virus/Cancer	Mechanism of Action	References
				antigens (PSA) inhibition which are a target for AR	
Daidzein	Soybeans, kudzu, red clover.	$C_{15}H_{10}O_4$	Influenza virus	Regulates virus replication through signal transduction mechanism using the lipoxygenase products	Horio et al. (2020)
			Ovarian cancer	Induce apoptosis	Hua et al. (2018)
3,3′,4′,5,6,7-hexahydroxyflavone (Quercetagetin)	Onions, green, tea and apples	$C_{15}H_{10}O_8$	Rauscher murine leukaemia virus (RLV), HIV	Interferes with the replication of the virus; inhibits viral reverse transcriptase enzyme	Ben-Shabat et al. (2020)
Polyphenols					
Curcumin	Turmeric (*Curcuma longa* L.)	$C_{21}H_{20}O_6$	Parainfluenza virus (PIV-3), herpes simplex virus (HSV), feline infectious peritonitis virus (FIPV)	Inhibitory activity against Inosine monophosphate dehydrogenase	ZorofchianMoghadamtousi et al. (2014)
			Colorectal cancer, pancreatic cancer, prostate cancer, colon cancer	Inhibits the function of P-gp	Lopes-Rodrigues et al. (2016)
Caffeine	Coffee beans	$C_8H_{10}N_4O_2$	Herpes simplex virus	Suppresses viral protein synthesis	Shiraki and Rapp (1988)
			Melanoma	Blocks adenosine A2A receptor pathway	Tej et al. (2019)
Ellagic acid		$C_{14}H_6O_8$	Human rhinoviruses	Strongly impedes the RNA replication of human rhinovirus-4	Park et al. (2014)

(Continued)

TABLE 15.1 (Continued)
List of Antiviral Plant Bioactives with Anticancer Properties with Their Mechanism of Action

Phytochemicals	Source(s)	Chemical Formula	Target Virus/Cancer	Mechanism of Action	References
	Cranberries, strawberries, raspberries, and pomegranates		Human cervical carcinoma	cell cycle control in the G1 phase	Li et al. (2018)
Tannic acid	Tea, nettle, wood, berries	$C_{76}H_{52}O_{46}$	Human immunodeficiency virus	Inhibits HIV replication in H9 lymphocytes	Kaczmarek (2020)
			Non-small cell lung cancer	Cell cycle inhibition and induction of the intrinsic pathways of cell death	Nipin et al. (2020)
Coumarin	Tonka beans, lavender, coriander, cumin, strawberries, apricots, vanilla, cherries, Chinese cinnamon, and sweet clover	$C_9H_6O_2$	Human immunodeficiency virus type 1 (HIV-1)	HIV-1 and 2 protease inhibitors	Hassan et al. (2016)
	–	–	Breast, endometrial, and prostate cancer	Inhibits proliferation by upregulating cyclin-dependent kinase inhibitor 1A; arrests cell cycle and results in apoptosis	Thakur et al. (2015)
Carotenoids					
Lycopene	Tomato (*Solanum lycopersicum*)	$C_{40}H_{56}$	HCV	Causes oxidative DNA damage	Seren et al. (2008)
			Gastric cancer	Increases G0–G1 phase and decreased S phase	Kim and Kim (2015)

(*Continued*)

TABLE 15.1 (Continued)
List of Antiviral Plant Bioactives with Anticancer Properties with Their Mechanism of Action

Phytochemicals	Source(s)	Chemical Formula	Target Virus/Cancer	Mechanism of Action	References
Astaxanthin	Aquatic organisms with red pigment	$C_{40}H_{52}O_4$	COVID-19	Alleviates the risk of inflammatory cytokine	
			Leukaemia	Induces cellular programmed cell death in a dose-dependent method and PPAR-γ protein expression	
Terpenes					
Ursolic acid	Peels of fruits	$C_{30}H_{48}O_3$	Rotavirus	Affects lipid metabolism, ER stress	Tohmé et al. (2019)
			Lung cancer	Inhibits expression of cyclin D1 and CDK 2 and 4	Yin et al. (2018)
Glycosides					
Digoxin	Digitalis plant	$C_{41}H_{64}O_{14}$	SARS-CoV-2	Inhibits viral mRNA expression, copy number, and viral protein expression	Cho et al. (2020)
			Non-small cell lung cancer	Reduces cancer cell viability, causes DNA damage by inhibiting DNA repair mechanisms	Wang et al. (2020)
Hesperidin	Citrus fruits	$C_{28}H_{34}O_{15}$	SARS-CoV-2	Block corona virus entry	Haggag et al. (2020)
			Breast cancer	Inhibition of the expression of PD-L1 by downregulating Akt as well as NF-κB signalling	Kongtawelert et al. (2020)
Kaempherol		$C_{15}H_{10}O_6$	SARS coronavirus	Inhibits the 3a channel activity	Schwarz et al. (2014)

(Continued)

TABLE 15.1 (Continued)
List of Antiviral Plant Bioactives with Anticancer Properties with Their Mechanism of Action

Phytochemicals	Source(s)	Chemical Formula	Target Virus/Cancer	Mechanism of Action	References
	Apples, onions, leeks, citrus fruits, grapes		Breast cancer	Programmed cell death in breast cancer cells	Wang et al. (2019)
Aloe emodin	*Aloe vera*	$C_{15}H_{10}O_5$	Influenza A	Capable of inhibiting viral replication	Gansukh et al. (2018)
			Pancreatic carcinoma	Involves in apoptosis, autophagic pathways and cell cycle arrest	Du et al. (2019)
Nomilin	Citrus fruits	$C_{28}H_{34}O_9$	HIV	Inhibits HIV-protease	Battinelli et al. (2003)
			Human breast cancer	Impedes cell proliferation; triggers apoptosis and cell cycle arrest	Shanmugam et al. (2017)
Alkaloid					
Berberine	Orangeroot (*Hydrastis canadensis* L.) Berberidaceae	$C_{20}H_{18}NO_4$	SARS-CoV-2	Inhibit coronavirus RNA synthesis; inhibits viral assembly and release	Warowicka et al. (2020)
			Breast cancer	Induction of cell death	Habtemariam (2020)
Other compounds					
Agarospirol	Baizhu (*Atractylodes macrocephala*)	$C_{15}H_{26}O$	H3N2	Inhibits the proliferation of HepG2 cell	Gu et al. (2019)
Essential oils of *A. macrocephala*					
Glutathione	Avocado, asparagus, broccoli, potato, tomato, and spinach	$C_{10}H_{17}N_3$-O_6S	COVID-19	Prevents the increase of viral loads and the subsequently	Polonikov (2020)

(Continued)

TABLE 15.1 (Continued)
List of Antiviral Plant Bioactives with Anticancer Properties with Their Mechanism of Action

Phytochemicals	Source(s)	Chemical Formula	Target Virus/Cancer	Mechanism of Action	References
				releases large number of inflammatory cells into the lung	
			Breast cancer	Pro-neoplastic oxidative damages	Desideri et al. (2019)
Indole-3-carbinol	Cruciferous vegetables	C_9H_9NO	HCV	Inhibits entry and replication of the virus	Zhang et al. (2015)
			Different types of human cancers	Inhibits the activity of the over-accumulated E3 ligase	Katz et al. (2018)
Allicin	Garlic	$C_6H_{10}OS_2$	Influenza B, human rhinovirus type 2, human cytomegalovirus (HCMV), Parainfluenza virus type 3, herpes simplex type 1 and 2, vaccinia virus, and vesicular stomatitis virus	Chemical interaction with enzymes containing thiol	El-Sabre et al. (2020)
			Colorectal cancer	Interferes with cell cycle, causing apoptosis; targets tubulin and stops cell proliferation; promotes immune function, inhibits tumour vessels formation	

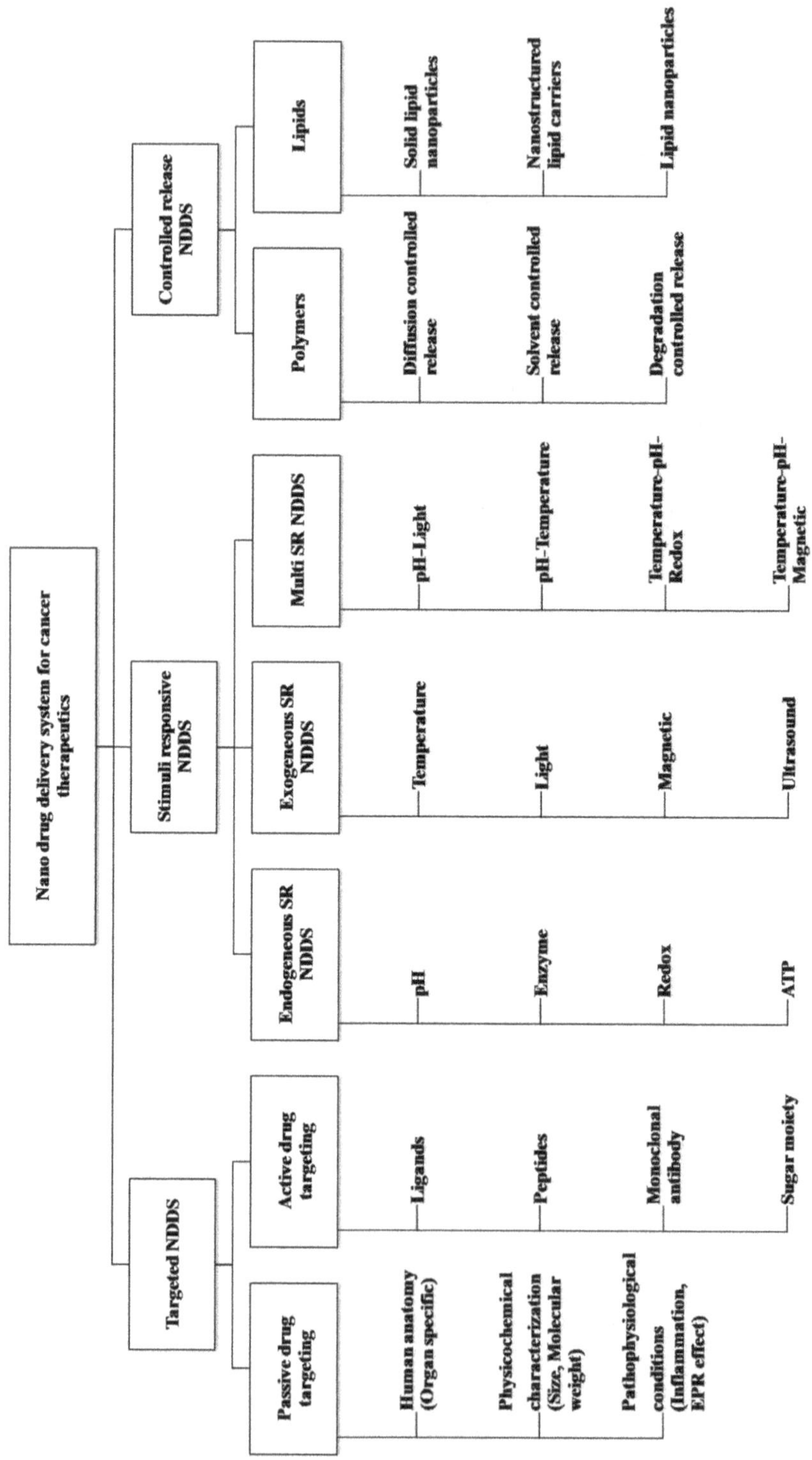

FIGURE 15.2 Graphical representation of advanced nano drug-delivery system suitable for delivery of plant bioactives for cancer therapeutics.

carriers, thereby providing tumour specificity and reduced side effects (Attia et al. 2019). For example, drug nanoparticles of 100 nm size showed a 2.5 fold greater cellular uptake in caco-2 cells compared to 1 mm drug particles (Rizvi and Saleh 2018). Nanoparticulates that satisfy size and surface characteristics might persist for a long period in the blood circulation. These nanoparticulates have a greater chance of entering the tumour tissues, thereby avoiding their elimination by body's defence mechanisms. Biodistribution, pharmacokinetics, and toxicity factors are dependent on the physicochemical nature of the nanocarrier or maybe due to the pathophysiological condition surrounding the cancer cells.

Camptothecin, a quinolone alkaloid obtained from happy tree's bark and stem (*Camptotheca acuminate*) is a topoisomerase I inhibitor with low aqueous solubility and instability that limits its therapeutic efficiency (Koo et al. 2006). Sterically stabilized micelles of PEGylated phospholipids with 14 nm size have been used as nanocarriers for delivering camptothecin. These PEGylated phospholipids can penetrate through the tumours that were leaky in nature. The passive targeting shows lesser concentration of drug toxicity to the healthy cells, even at higher concentration of drug to be delivered to the tumour cells (Koo et al. 2006). PDT facilitates drug nanopariculates delivery restricted to the interstitium of the tumours, and even does not enhance their intake by the cancerous cells. The cancerous cell intake happens by active directing of drug nanoparticulates by using overexpressed receptors on the cancerous cells (Bazak et al. 2014).

15.3.1.2 Active Drug Targeting (ADT)

Active targeting is functionalization of the surface nanocarriers using ligands such as peptides, sugar moiety, monoclonal antibody, or some ligands responsible for recognition of specific receptors on the surface of the targeted cancer cells. This kind of targeted NDDS excludes drug interaction in nontarget areas (Babazadeh et al. 2020). In active drug targeting, drugs and their delivering mechanisms are subjected to ligand attachments that help the receptor to bind at the specific target. Targeted NDDS are intended to target endocytosis-prone surface receptors (folate receptor, epidermal growth factor receptor transferrin receptor etc.,) for improved anticancer therapy. To deliver drug nanoparticles at tumour sites, folic acid is used by conjugating them on the surface of plant bioactives loaded with polymeric and lipid-based nanoparticles. Folic acid is a biological molecule, and folic acid receptors are overexpressed on the surface of cancer cells. Therefore, conjugating folic acid to nanocarriers enhances folic acid mediated targeted delivery of plant bioactives. Overexpression of folic acid receptors inside numerous carcinomas, including breast, ovary, kidney, and lung, helps in site-targeted drug delivery on tumour cells. Nanoparticles consists of drug molecules used for tumour targeting attached to the cell via endosome-dependent mechanism, thereby enhancing the intracellular concentration (Lammers et al. 2008). ADT can enhance nucleic acid cellular internalization, thereby restricting the negative side effects of drugs on nontargeted cells. Biochemical targets, physical/external stimuli, pretarget/sandwich targeting, promoter/transcriptional targeting are some active targeting approaches (Vasir et al. 2005). Nanostructures used for ADT include dendrimers, liposomes, polymeric micelles, polymeric nanoparticles, and conjugation with quantum dots.

Active NDDS is further categorized into: (i) first-order targeting that restricts delivery of plant-bioactive based NDDS to a preselected site of the target (organs, tissues), (ii) second-order targeting that delivers the drug molecules to selective types of tumour cells, avoiding normal cells, and (iii) third-order targeting that delivers the drug at the cell's intracellular sites. An example is ligand-based delivery of drug bioactives through endocytosis (Rani and Paliwal 2014). ADT for tumour cells using nanoparticles includes antibody-based targeting, antigen-based targeting, and ligand-based targeting (Bazak et al. 2015).

The advantages of targeted NDDS include simplification of drug administration protocol, administration of the plant bioactive in smaller doses to produce the desired effect, and selective targeting with fewer drug side effects (Singh and Lillard 2009; Kumari et al. 2016). However, difficulty in targeting tumour cells, maintaining the stability of dosage forms, and the lack of advanced techniques need to be seriously considered. Nanocarriers of larger size could not deliver drug molecules by EPR effect because of the leaky structure of tumour cells. Thus, targeted nanoparticles in cancer treatment involves selection of suitable nanocarrier with relevant particle size (Kumari et al. 2016). Among the two targeted NDDS, passive-targeted nanocarriers are widely available in market for the prolonged biological half-life, although premature drug release (Raza et al. 2019) is an issue that needs to be solved.

15.3.2 Stimuli-Responsive NDDS (SR-NDDS)

Stimuli-responsive NDDS may use either an external or internal stimulus specific to cancer cell microenvironment to release the drugs at a specific site. Polymers that respond to environmental stimuli like pH, temperature, enzymes, magnetic field, and solvent polarity inside the human system by undergoing physiochemical changes are used in SR-NDDS. Thermosensitive polymers [poly (N-isopropylacrylamide), poly(methyl vinyl ether), poly(N-vinyl caprolactam), elastin-like oligo- and polypeptides)] poly(N-alkylacrylamide); pH-sensitive polymers [poly(methacrylic acid)], ultrasound-sensitive polymer (ethylene vinyl acetate), and magnetic-responsive materials (co-precipitation of iron salts in polystyrene and polyacrylamide polymers; core-shell nanoparticles like magnetite (Fe_3O_4) coating on polymer, magnetoliposomes (Fe_3O_4 or maghemite (Fe_2O_3) nanocrystals encapsulated in liposomes) (Taghizadeh et al. 2015) are used for formulating stimuli responsive NDDSs. As these polymers mimic the biological systems, it is easier to make them respond to a change in mentioned stimuli, thereby hydrophobic/ hydrophilic balance, solubility, release of plant bioactive, and their conformation can be altered (Srivastava et al. 2016). There are three stimuli-responsive NDDS: endogenous, exogenous, and multi-stimuli responsive, which are discussed below,

15.3.2.1 Endogenous Stimuli-Responsive NDDS

Delivery of plant bioactives will be triggered by using the inherent biological stimuli existing around the tumour tissue environment. These stimuli might be the pH, enzyme, reactive-oxygen species, ATP-responsive, and redox responsive. Stimuli-responsive pH polymer are PLGA (Qu et al. 2017), poly (acrylamide)

(Pafiti et al. 2016), poly (diethylaminoethyl methacrylate) (Darabi et al. 2015), poly (methacrylic acid) (Shalviri et al. 2013), poly (dimethylaminoethyl methacrylate) (Lee et al. 2011), and poly (acrylic acid) (Hu et al. 2002). Example of enzyme-responsive polymers include poly dimethylsiloxane, polyethylenimine (Onaca et al. 2009), and polyethylene glycol (Kurisawa et al. 1997). For example, pH gradient between tumour and healthy proliferative cells is responsible for the rate of release of plant bioactives. The pH of normal human tissues is 7.4, whereas that of tumour tissues would be an acidic pH below 6.5. It protects the plant bioactives against degradation during transportation. It also selectively releases the plant bioactives at the tumour site (Chen et al. 2016). The major advantage of this system of NDDS is the release of plant bioactives from nanocarrier due to the triggering factor from the tumour site. The difficulty in using endogenous stimuli-responsive NDDS is the external factor control that can be inefficient in releasing the optimum quantity of drug to the tumour site. It is also difficult in evaluating the dose, duration, and frequency of drug release by an external agent.

15.3.2.2 Exogenous Stimuli-Responsive NDDS

In exogenous SR-NDDS, the release of loaded drug in polymeric NDDS will be triggered by applying an external factor like temperature, light, magnetic field, and ultrasound outside the human body. Polymers play a crucial role in exogeneous stimuli response; for example, a thermoresponsive polymer when subjected to a slight change in temperature leads to drug release due to phase transition. For photoresponsive or ultrasound responsiveness, polymers release plant bioactives only on applying specific energy irradiation (Cheng et al. 2014). Ultrasound responsive polymers are poly (methacrylic acid) (Yang et al. 2014), and poly (2-oxazoline) micelles (Salgarella et al. 2018). The electric field responsive polymers include polypyrrole (Lee et al. 2015) and poly (3,4-ethylenedioxypyrrole) (Krukiewicz et al. 2016). Temperature-responsive polymers are poly (acrylic acid-co-acrylamide) (Schmaljohann 2006) and magnetic-responsive polymers include iron-oxide magnetite (Nigam and Bahadur 2017). Graphene oxide (synthesized by pickering emulsion) is unsymmetrically attached to azidated- poly (ε-caprolactone) and (N-isopropylacrylamide-co-acrylamideco-allylamine) terpolymer to synthesize thermoresponsive nanoparticles called Janus nanostructure. Quercetin (hydrophobic) and fluorouracil (hydrophilic) bioactives are loaded inside the Janus nanoparticle. The drug release is higher at a 40 °C than at 37 °C (Khoee and Karimi 2018). A thermosensitive polymer shows an abrupt change in polymer solubility, even with a slight temperature change. For example, Poly (N-isopropyl acrylamide) has a low critical-solution temperature at 32 °C, has been shifted to the temperature of the body by adding surfactants or additives in the formulation (James et al. 2014). Likewise, a phase transition in light-sensitive polymer occurs due to the response of exposure to light. The encapsulated drugs are released under light-source irradiation, even from outside the human body. (Hsiao et al. 2012) stated that carboxymethyl-hexanoyl chitosan nanocapsules in water, when mixed along with sodium β-glycerophosphate and glycerol formed a thermo-gelling solution that exhibited a long-sustained release of ethosuximide at 37 °C. The advantage of exogenous stimuli-responsive NDDS depends on the external stimuli and not limited on the internal micro-environmental factors of the human system. However, if the nanocarrier is photosensitive, then nanocarrier penetration inside the body at the tumour site becomes difficult. The drawback of light-sensitive polymer is that it shows inconsistency in response to light caused due to the chromospheres, which are bound non- covalently on expansion or contraction of NDDS.

15.3.2.3 Multi Stimuli-Responsive (MSR) Nano Drug-Delivery System

The multi stimuli-responsive NDDS consists of combining more than one stimuli-responsive parameters inside one particular system to release the loaded plant bioactives. It improves the drug-encapsulation efficiency and bioactives' half-life; therefore, higher concentration of plant bioactives can be delivered at the targeted site. The pH-redox, pH-temperature, pH-magnetic, and temperature-redox responsiveness are MSR drug-delivery systems (Cheng et al. 2013). The pH-temperature responsive polymers are poly(N-vinylcaprolactam) and ethyl cellulose (Li et al. 2018). For example, poly(-NIPAAm-co-acrylic acid)-b-PCL (P(NIPAAm-co-AA)-b-PCL), a diblock copolymer is used for preparing temperature and pH-responding nanoparticles by encapsulating paclitaxel (30% on weight basis) and were made to release at a high temperature of 37 °C and a low pH of 6.9. Likewise, pH and redox dual-sensitive polymersomes at nano regime are prepared by increasing the solution pH to 7.4 with polyethylene glycol-SS-poly(2-(diethylamino)ethyl methacrylate) double hydrophilic diblock copolymers. And the tumour cell uptake is also enhanced through surface charge reversal at tumour pH. This enhances the release of drug in endosomal compartments with faster drug release in the nucleus and cytoplasm (Cheng et al. 2013). Despite being advantageous of using more than one stimuli response and regulating effectively the precise release into the desired target, it is complicated to design and develop such systems (Mura et al. 2013).

Among the three types of SR-NDDS, multi stimuli-responsive NDDS is advantageous over other types of SR-NDDS. Because of their multiple-response behaviour, they can simultaneously use both endogeneous and exogeneous stimuli responses like pH (endogeneous stimuli) and magnetic (exogeneous stimuli) response. The importance of SR-NDDS is it has tunable sensitivity. However, a slow response time to certain stimuli-sensitive polymers (James et al. 2014) is a limitation that needs to be addressed. Nanocarriers that react to specific stimuli are particularly important, only if the stimuli are valid for the pathology of specific diseases. Different stimuli-responsive NDDS that might be suitable for delivery of plant bioactives are listed in Table 15.2.

15.3.3 Controlled-Release NDDS

Controlled release of drug molecules to a specific site over an extended period is an important determinant in an advanced drug-delivery system. Delivering optimal concentration of plant bioactives at targeted tumour sites through controlled NDDS helps in the reduction of drug side effects and also maximizes their curative functionality (Müller et al. 2000). Controlled-release NDDS are designed for enhancing plant bioactives' pharmacological activity through selective drug delivery at a predictable rate. It helps in improving drug bioavailability by preventing

TABLE 15.2
List of Plant Bioactives Loaded Targeted NDDS with Target Site and Mechanism of Delivery for Cancer Treatment

Bioactive loaded NDDS	Plant Bioactive (Source)	Type of Cancer	Target Site	Research Findings	Reference
Hesperidin loaded Zn^{2+}sodium alginate/pectin nanocomposites	Hesperidin (Citrus fruits like orange, grapefruit, lemons, and tangerines)	Colon carcinoma	Bax proapoptotic protein induces apoptosis through mitochondrial membrane but Bcl2 anti apoptotic protein stops apoptosis through Bax expression inhibition	Demonstrated higher rate of drug release at acidic pH 2.8 relative to physiological pH 6.8, indicating Zn^{2+} sodium alginate/pectin nanocomposites are directly targeting cancer cells	Yang et al. (2020)
Chitosan coated folic acid conjugated with solid lipid nanoparticles	Resveratrol (polyphenol) (Red grapes, red wine, peanuts, cocoa powder)	Colorectal cancer	Mitochondria	Found effective only on HT-29 cancerous cells, as they had folate receptors inducing apoptotic cell death	Kumar et al. (2020)
Zein nanoparticles	Luteolin (Carrots, peppers, celery, olive oil, peppermint)	Colon cancer	Intestine	Luteolin from nanoparticles were released at a intestinal pH 6.8	Shinde et al. (2019)
Naringenin Eudraget E-100 nanoparticles	Naringenin (Cirus fruits, grapefruit, tomatoes, water mint)	Colorectal cancer	Lymphatic system	Nanoparticles of size less than 450 nm are not easily cleared by kidney, retained in the leaky cancer cells and then released the drug	Chaurasia et al. (2018)

(Continued)

TABLE 15.2 (Continued)
List of Plant Bioactives Loaded Targeted NDDS with Target Site and Mechanism of Delivery for Cancer Treatment

Bioactive loaded NDDS	Plant Bioactive (Source)	Type of Cancer	Target Site	Research Findings	Reference
Hesperetin-loaded folate-conjugated chitosan nanoparticle	Hesperetin (Citrus fruits)	Colorectal cancer	Cytoplasm	Chitosan nanoparticles are developed as core-shell model, modified to introduce folic acid and ligand for targeted drug delivery. By the process of endcocytosis chitosan folate hesperetin nano particles are engulfed and broken down by lyzosymes, thereby releasing hespertin inside the cytoplasm.	Lazer et al. (2018)
Fisetin loaded pluronic127-folic acid conjugated micelles	Fisetin (Strawberries, apples, mangoes, persimmons, kiwis, and grapes, vegetables, tomatoes, onions, and cucumbers, nuts)	Breast cancer	Targets tumour through EPR effect	Targeting happens through folate receptor, as folic acid is conjugated to the micelle structure loaded with fisetin and releases the drug due to endocytosis mechanism at lower pH 5.3	Pawar et al. (2018)
		Breast cancer		Quercetin nanoparticles impedes the cell multiplication by inducing Bax,	Aghapour et al. (2018)

(Continued)

TABLE 15.2 (Continued)
List of Plant Bioactives Loaded Targeted NDDS with Target Site and Mechanism of Delivery for Cancer Treatment

Bioactive loaded NDDS	Plant Bioactive (Source)	Type of Cancer	Target Site	Research Findings	Reference
Quercetin loaded silica nanoparticles	Quercetin (Fruits, vegetables, seeds, berries, and tea)		Oestrogen receptors, MCF7 cells	Bcl-2 or caspase mediated programmed cell death	
Zein nanoparticles	β-Carotene (Sweet potatoes, Grapefruit, apricots, dairy products, broccoli, carrots, green leafy vegetables and pink fruits)	Breast cancer	MCF-7 cells, liver	MCF-7 cells released beta carotene constantly from zein nanoparticles in lesser pH of 1.2, whereas lesser release of the drug has been observed at a pH of 7.4	Aune et al. (2012); Jain et al. (2018)
Epigallocatechin-3-gallate (EGCG) loaded folic acid-polyethylene glycol nanoparticles	Epigallocatechin Gallate (Green tea)	Breast cancer, Prostate cancer cells, Pancreatic cancer cells	MCF-7 cells	MCF-7 cells showed higher uptake of epigallocatechin-3-gallate folic acid nanoparticles to the folate receptors due to the presence of folic acid	Zeng et al. (2017)
Prostate-specific membrane-antigen (PSMA) targeted micelle	Wogonin (Roots of *Scutellaria baicalensis*)	Prostate cancer	Mitochondria	To treat prostate-specific membrane-antigen -positive prostate cancer for *in vitro* conditions are developed using 2-(3-((S)-5-amino-1-carboxypentyl) ureido) pentanedioic acid modified polyethylene glycol -cholesterol	Zhang et al. (2016)

(Continued)

TABLE 15.2 (Continued)
List of Plant Bioactives Loaded Targeted NDDS with Target Site and Mechanism of Delivery for Cancer Treatment

Bioactive loaded NDDS	Plant Bioactive (Source)	Type of Cancer	Target Site	Research Findings	Reference
				micelles containing wogonin. It increases the Bax and cytochrome c levels, leading to intrinsic programmed cell death, when compared to the wogonin loaded non-targeted micelles	
PEGylated nanoliposomes	Crocin (carotenoid pigment) (dried stigmas of *Crocus sativus*)	Colon cancer	Lymphoma sites	PEGylated nanoliposomes will passively retain in tumour region due to EPR effect and increases the therapeutic effects of liposomal anticancer agents by reducing mononuclear phagocyte system uptake	Rastgoo et al. (2013)
PEG-PLA polymer micelles	β-Lapachone Lapacho tree (*Tabebuia avellanedae*)	Colon cancer cells, lung, prostate, Breast cancer cells	Tumour tissue	The drug in PEG–PLA micelles containing β-lapachone has been released at the target site through a diffusion mechanism due to passive targeting over NQO1-overexpressing tumour cells	Blanco et al. (2007)

degradation of plant bioactives in its initial premature release stage, maintaining drug concentration in therapeutic range, thereby minimizing fluctuations in drug concentration with reduced side effects. Sustained delivery of drugs over days/weeks/months at the targeted site on a one-time administration or drug release for a longer duration in a controlled manner (Huang and Brazel 2001) is possible, based upon drugs' nature and the required dosage. Fluctuation of drug concentration in plasma can be reduced by low-dose frequency at a controlled-release rate (Lee and Yeo 2015). Drug delivery is mostly through polymer and lipid-based controlled-release NDDS targeting various types of cancer cells.

15.3.3.1 Polymers Used in Controlled-Release NDDS

The polymers used must be nonimmunogenic, nontoxic and, most importantly, water-soluble. Polymer-based, controlled-release NDDS can be formed from synthetic polymers like poly (ε-caprolactone) (PCL), PLA, PLGA, N-(2-hydroxypropyl)-methacrylamide copolymer, and poly(styrene-maleic anhydride) copolymer; and from natural polymers such as gelatin, dextran, guar gum, chitosan, and collagen. These polymers minimize drug degradation and give a prolonged time of circulation for drug molecules inside the human body. These polymers must guarantee the safe release of drugs with a good stability profile (Srivastava et al. 2016). Plant bioactives can be coated with a number of layers of polymers leading to an expected release rate. Due to erosion of polymers, loaded plant bioactives are gradually released (Sowjanya et al. 2017). Degradation rate of polymers determine the drug-release rate from the nanocarrier, which are further dependent upon molecular weight, crystallinity, composition of monomers, and end groups of the drug molecule (Fredenberg et al. 2011). The plant bioactives are gradually released from the polymer through various release mechanisms. To enable a controlled-release system, the core is loaded with the therapeutic agent or drug encapsulated by an outer shell of polymer, resulting in polymeric nanospheres. In this case, the core might be insoluble under certain conditions, like low pH, while soluble only at the desired place where the controlled release of the drug must take place. The controlled-drug release rate is dependent on the thickness of shell used for encapsulation of the therapeutic agent (Ai et al. 2003). Both controlled and prolonged release of bioactives can be altered by changing the drug/polymer ratio. Drug molecules from plant bioactives loaded polymer are released in a controlled way from their nanocarrier through diffusion-controlled release, solvent-controlled release, and degradation-controlled release mechanisms.

i. *Diffusion-controlled release*: In a core-shell NDDS, the drug molecules are encapsulated inside a core enclosed by a polymer membrane. Here, the drug release happens through dissolution of the drug inside the core, and then it is diffused through the membrane of the polymer. With the matrix type of NDDS, where plant bioactives are dispersed across a matrix of polymer network, without a polymer-membrane barrier around the drug molecules, the drug releases at a higher rate initially followed by a slower release rate.

ii. *Solvent-controlled release*: Solvent-controlled drug release happens through osmosis process of polymer swelling and controlled release of the loaded drug. Osmosis-controlled release has a polymer membrane that is semi-permeable through which water flows from outer membrane to the inside of nanocarrier (from lower to higher drug concentration). As there is continuous uptake of water inside the polymer, the polymer swells, resulting in drug release from the nanocarrier.
iii. *Degradation-controlled release*: Polymer nanocarriers that are biodegradable, like poly (amino acids), polyamides, polysaccharides, and polyesters, are used for designing such NDDS. The drug is released via enzymatic degradation or hydrolytic degradation of amide, ester, and hydrazone bonds in the polymers. Bulk degradation also occurs in matrices made of polymers like PLGA, PLA, and PCL, wherein degradation of the entire matrix occurs simultaneously (Lee and Yeo 2015).

15.3.3.2 Lipids Used in Controlled-Release NDDS

Lipid-based nanocarriers have interesting biological properties, including drug protection from the outer environment, biocompatibility, biodegradability, and capacity of entrapping hydrophobic as well as hydrophilic drugs. Lipid-based carrier properties, such as their size, charge, and surface functionality, are modified through addition of plant bioactives by altering their surface chemistry. Lipid-based nano formulations consisting of hydrophobic plant bioactives results in fine drug dispersion that enhances faster and complete dissolution of the drug molecules (Attama et al. 2012). Liposomes, polymersomes, and micelles are a class of nanocarriers used for delivering amphiphilic plant-bioactives. Likewise, solid lipid nanoparticles (SLN), nanostructured lipid carriers (NLC), and lipid-based nanoparticles have a greater area of interest for delivery of hydrophobic drugs for cancer treatment.

In SLN, burst release is controlled by temperature and surfactant concentration. When the SLN production is carried out at room temperature rather than at elevated temperature, it does not show a burst release; rather, it prevents redistribution of the drug to its aqueous phase. In NLC, an initial burst release has been observed, later with a controlled release of loaded drugs. As the outer layer shell is drug rich of liquid lipid, it causes an initial burst release due to erosion of the matrix or through diffusion. Then, slow and controlled drug release happens through the solid-lipid core (Son et al. 2017). Sesarman et al. (2019) prepared long circulating liposomes for co-delivering both curcumin and doxorubicin for treating colorectal cancer. It showed a burst drug release from the liposomes during first 24 hours, followed by a controlled drug release thereafter. Dai and his workers (2010) stated that oridonin-loaded NLCs with a mean particle size of 200 nm showed a dual-phase drug release, like initial burst release, and later a sustained drug release. Thus, it states that the drug-release rates can be enhanced by increasing the liquid-lipid content and amount of loaded drug. The controlled-release NDDS is primarily based on the lipid matrix and also on the type of surfactant, surfactant concentration, as well as production temperature (zurMühlen et al. 1998). The list of plant bioactives loaded with polymer and lipid-based controlled NDDS are provided in (Table 15.3).

TABLE 15.3
List NDDSs Suitable for Delivering Plant Bioactives with Polymers Specific to Different Types of Stimuli for Cancer Treatment

Stimuli Response	Plant Bioactive	Polymer/Bioactive Loaded NDDS	Type of Cancer	Cell Line	Findings of the Study	Reference
Redox responsive	Homoharringtonine	PLGA	Lung cancer	A549, NCI-H226, BALB/c-nude mice	Tumour cells have high glutathione, and this triggers the release of homoharringtonine at the tumour cells	Zhang et al. (2020)
pH responsive	Quercetin	Chitosan	Breast cancer	MCF-7	Quercetin release was predominant at lower pH of 5, but at pH 7, it showed lower drug release rate	de Oliveira Pedro (2018)
pH responsive	Ursolic acid	Chitosan	Cervical cancer	HeLa, CD-1 female mice	Rapid release of ursolic acid happened at pH 5, and at normal human physiological pH 7, there was no release of ursolic acid	Wang et al. (2017)
pH responsive	Quercetin	PLGA	Skin cancer	HaCaT, A431	Quercetin release rate was higher at tumour site with pH 5.5 and at pH of 7, there was not sufficient release of quercetin	Gupta et al. (2016)
Magnetic	Artemisinin	Magnetic iron oxide	Breast cancer	BALB/c mice	Artemisinin and transferrin-loaded magnetic nanoliposomes in the presence of an external magnetic field showed artemisinin release inside BALB/c mice at the tumour site.	Gharib et al. (2015)
pH responsive	Curcumin	Poly(isoprene-b-acrylic acid) deblock copolymer	acute myeloid leukaemia	HL-60	An increase in release of curcumin took place when pH decreased from 7 to pH	Jelezova (2015)

(*Continued*)

TABLE 15.3 (Continued)
List NDDSs Suitable for Delivering Plant Bioactives with Polymers Specific to Different Types of Stimuli for Cancer Treatment

Stimuli Response	Plant Bioactive	Polymer/Bioactive Loaded NDDS	Type of Cancer	Cell Line	Findings of the Study	Reference
					4.5. Higher the drug release, lower the pH.	
Thermal	Curcumin	Chitosan-g-poly(*N*-isopropylacrylamide)	Breast cancer, prostate cancer	KB, MCF-7, PC3	Above LCST (lower critical solution temperature) of 38°C, curcumin release is high due to weak hydrogen bond interaction of curcumin with the carrier system	Rejinold et al. (2011)
pH responsive	Glycyrrhetinic acid	Chitosan	Liver cancer	QGY-7703, BALB/c nude mice	Glycyrrhetinic acid release was at a higher rate at the tumour site of pH 5-5.5, whereas at physiological pH of 7, no release was observed	Tian et al. (2010)

TABLE 15.4
Plant Bioactive-Loaded Controlled-Release NDDS with Mechanism of Release for Cancer Treatment

Nanocarrier	Plant Bioactive (source)	Solubility of plant bioactive	Type of cancer	Dose and days of release	Mechanism of release from the matrix	Reference
Polymeric nano-formulation						
PLGA	Naringenin (Citrus fruits, bergamot, tomatoes and other fruits)	Hydrophobic	Pancreatic cancer	0–60 μM Release: 30 hrs.	Delivery of nanoparticles to lysosomes diffuses in the cytoplasm, resulting in sustained release of the entrapped substance, which eventually enters the nucleus, resulting in a possible cytotoxic effect of naringenin loaded PLGA nanoparticles.	Akhter et al. (2020)
	Crocetin (*Saffron stigma*)	Hydrophobic	Colorectal, Cervical, Breast, Liver, and pancreatic cancer	5 mg/mL and Release: 48 hrs	For 24 hours, the formulation is released in a diffusion mode by polymer erosion. The entire release took up to 48 hours. The first 5 hours of release is the primary drug dose and the remaining drug is released later.	Gutheil et al. (2012); Alebooye et al. (2016)
PLA	Fisetin (strawberries, apples, grapes, and cucumbers)	Hydrophobic	Colon cancer	2 mg/mL and Release: 48 hrs	PLA polymer nanoparticles have the ability to enhance tumour-specific delivery efficacy and therapeutic effect through sustained release behaviour and EPR effect. The	Feng et al. (2019)

(Continued)

TABLE 15.4 (Continued)
Plant Bioactive-Loaded Controlled-Release NDDS with Mechanism of Release for Cancer Treatment

Nanocarrier	Plant Bioactive (source)	Solubility of plant bioactive	Type of cancer	Dose and days of release	Mechanism of release from the matrix	Reference
					controlled release has been observed for four days.	
PEGylated silica nanoparticles	Genistein (soymilk, soy flour)	Hydrophobic	Colon cancer		At 37 °C, pH degradation of the polymer has an impact on the PEG-grafted nanocarrier. In HT29 cells treated with genistein PEGylated silica hybrid nanomaterial, the number of autolysosomes significantly increased compared to free genistein, resulting in increased cell damage and activation of programmed cell death by autophagy.	Pool et al. (2018)
Chitosan-glycolipid nanocarrier (polymeric hydrogels)	Fucoxanthin (Brown seaweeds, microalgae, and mostly in marine environments)	Hydrophobic	Colon cancer	10 μM and Release: 48 h	Fucoxanthin loaded in chitosan-glycolipid nanocarrier has slowly been released in a regulated manner, because of the alkaline pH and due to intestinal lysozyme. Chitosan is broken down in to N-acetyl-D-glucosamine and D-glucosamine, thereby facilitates drug release.	Yonekura et al. (2010); Ravi et al. (2018)

(Continued)

TABLE 15.4 (Continued)
Plant Bioactive-Loaded Controlled-Release NDDS with Mechanism of Release for Cancer Treatment

Nanocarrier	Plant Bioactive (source)	Solubility of plant bioactive	Type of cancer	Dose and days of release	Mechanism of release from the matrix	Reference
Silymarin encapsulated panoparticles (SNP's)	Silymarin (Seeds of *Silybum marianum*)	Hydrophobic	Human prostate cancer	0.324 mg/mL Release: 48 hrs	Diffusion of silymarin happened into the external medium from the surface layers of polymeric nnaoparticles. With hydrolysis of the ester linkage between them, PLGA is biodegraded to lactic acid and glycolic acid monomers.	Snima et al. (2014)
Lipid Based Nanoformulation						
Solid lipid nanoparticles (SLNs)	Lycopene (Guava, papaya, red fruits and vegetables like tomatoes, carrots and papaya)	Hydrophobic	Breast Cancer	40 μg/mL Release: 48h	Slow and prolonged lycopene release from SLNs is through lycopene imbibing, dissolving, and diffusion through the SLN's matrix, by slowly degenerating the lipidic matrix in the system.	Jain et al. (2017)
Nanoliposomes	β –Cryptoxanthin (Citrus fruits, orange or yellow fruits and vegetables such as oranges, corn, yellow figs and apples)	Hydrophilic	Laryngeal cancer	-	Increased water susceptibility and plant-derived compound penetration into cells with enhanced anticancer efficacy.	Leoncini et al. (2015)

Controlled-release NDDS offer numerous advantages over conventional delivery of drugs, including improved therapeutic efficiency, reduced drug toxicity and side effects, localization of therapeutic action of drugs, maintenance of drug concentration within the desired range at the target site with slow release of hydrophilic drugs, and fast release of hydrophobic drugs (Vyas and Khar 2002; Ferrari 2005). Also, reduction in drug-administration frequency and drug accumulation and the stabilized medical condition of patients due to uniform drug level (Bhowmik et al. 2012) have also been observed. Challenges in CR-NDDS may include difficulty in delivering plant bioactives with high molecular weight at higher dosage cost (Table 15.4).

15.4 CONCLUSION

As approved drugs for most viral infections are limited, mankind has always looked at plant molecules for a solution. Plant bioactives with significant antiviral activity possessed a broad-spectrum anti-tumour activity as well due to their remarkable multifunctional bioactivities and synergistic effects. Since chemotherapy has toxic drug effects on normal healthy cells, there comes a demand for developing a safer drug-delivery system. Oral delivery of most plant bioactives has bioavailability and low permeation issues, limiting their therapeutic efficacy. Hence, NDDS improve the drug's pharmacokinetic properties, apart from improving the stability of plant bioactives from gastric degradation during oral administration. Advanced NDDS like targeted, stimuli-responsive, and controlled-release NDDS have been considered suitable for cancer treatment with antiviral plant bioactives. Targeted NDDS targets on releasing plant bioactives only on the tumour cells, leaving the healthy tissues free from cytotoxicity. Stimuli-responsive NDDS mimic cellular signals and release plant bioactives to the specific tumour site, excluding the healthy cell environment. Controlled-release NDDS helps in maintaining the dosage of plant-bioactive release rate with site-targeted drug release. NDDSs are efficient in delivering plant-bioactive drugs in the desired manner with less or no negative side effects in a specific manner at the tumour site. However, certain limitations observed in the above-mentioned, advanced NDDS need to be addressed with relevant solutions to enjoy the true benefits of NDDS in delivering antiviral plant bioactives for cancer therapeutics.

REFERENCES

Aggarwal, B.B., Kumar, A. and Bharti, A.C. 2003. Anticancer potential of curcumin: preclinical and clinical studies. *Anticancer Research* 23(1/A):363–398.

Aghapour, F., Moghadamnia, A.A., Nicolini, A., et al. 2018. Quercetin conjugated with silica nanoparticles inhibits tumor growth in MCF-7 breast cancer cell lines. *Biochemical and Biophysical Research Communications* 500(4):860–865.

Ai, H., Jones, S.A., de Villiers, M.M. and Lvov, Y.M. 2003. Nano-encapsulation of furosemide microcrystals for controlled drug release. *Journal of Controlled Release* 86(1):59–68.

Akhter, M.H., Kumar, S. and Nomani, S. 2020. Sonication tailored enhance cytotoxicity of naringenin nanoparticle in pancreatic cancer: Design, optimization, and in vitro studies. *Drug Development and Industrial Pharmacy* 46(4):659–672.

Alebooye, L.F., Hafezi, G.Z., Alibolandi, M., Ebrahimian, M. and Hashemi, M. 2016. Evaluation of the effect of crocetin on antitumor activity of doxorubicin encapsulated in PLGA nanoparticles. *Nanomedicine Journal* 3(1):23–34.

Alexander, A., Patel, R.J., Saraf, S. and Saraf, S. 2016. Recent expansion of pharmaceutical nanotechnologies and targeting strategies in the field of phytopharmaceuticals for the delivery of herbal extracts and bioactives. *Journal of Controlled Release* 241:110–124.

Al-Sarraj, S., Troakes, C., Hanley, B., et al. 2021. The spectrum of neuropathology in COVID-19. *Neuropathology and Applied Neurobiology* 47(1):3–16.

Ankri, S. and Mirelman, D. 1999. Antimicrobial properties of allicin from garlic. *Microbes and Infection* 1(2):125–129.

Attama, A.A., Momoh, M.A. and Builders, P.F. 2012. Lipid nanoparticulate drug delivery systems: a revolution in dosage form design and development. *Recent Advances in Novel Drug Carrier Systems* 5:107–140.

Attia, M.F., Anton, N., Wallyn, J., Omran, Z. and Vandamme, T.F. 2019. An overview of active and passive targeting strategies to improve the nanocarriers efficiency to tumour sites. *Journal of Pharmacy and Pharmacology* 71(8):1185–1198.

Aune, D., Chan, D.S., Vieira, A.R., et al. 2012. Dietary compared with blood concentrations of carotenoids and breast cancer risk: A systematic review and meta-analysis of prospective studies. *The American Journal of Clinical Nutrition* 96(2):356–373.

Babazadeh, A., Vahed, F.M. and Jafari, S.M. 2020. Nanocarrier-mediated brain delivery of bioactives for treatment/prevention of neurodegenerative diseases. *Journal of Controlled Release* 321:211–221.

Battinelli, L., Mengoni, F., Lichtner, M., et al. 2003. Effect of limonin and nomilin on HIV-1 replication on infected human mononuclear cells. *Planta Medica* 69(10):910–913.

Bazak, R., Houri, M., El Achy, S., Hussein, W. and Refaat, T. 2014. Passive targeting of nanoparticles to cancer: A comprehensive review of the literature. *Molecular and Clinical Oncology* 2(6):904–908.

Bazak, R., Houri, M., El Achy, S., Kamel, S. and Refaat, T. 2015. Cancer active targeting by nanoparticles: A comprehensive review of literature. *Journal of Cancer Research and Clinical Oncology* 141(5):769–784.

Behl, T., Rocchetti, G., Chadha, S., et al. 2021. Phytochemicals from plant foods as potential source of antiviral agents: An overview. *Pharmaceuticals* 14(4):381.

Ben-Shabat, S., Yarmolinsky, L., Porat, D. and Dahan, A. 2020. Antiviral effect of phytochemicals from medicinal plants: Applications and drug delivery strategies. *Drug Delivery and Translational Research* 10(2):354–367.

Bhowmik, D., Gopinath, H., Kumar, B.P., Duraivel, S. and Kumar, K.S. 2012. Controlled release drug delivery systems. *The Pharma Innovation* 1(10).

Bhuvaneswari, V. and Nagini, S. 2005. Lycopene: A review of its potential as an anticancer agent. *Current Medicinal Chemistry - Anti-Cancer Agents* 5(6):627–635.

Blanco, E., Bey, E.A., Dong, Y., 2007. β-Lapachone-containing PEG–PLA polymer micelles as novel nanotherapeutics against NQO1-overexpressing tumor cells. *Journal of Controlled Release* 122(3):365–374.

Bost, P., Giladi, A., Liu, Y., et al. 2020. Host-viral infection maps reveal signatures of severe COVID-19 patients. *Cell* 181(7):1475–1488.

Chae, H.S., Xu, R., Won, J.Y., Chin, Y.W. and Yim, H. 2019. Molecular targets of genistein and its related flavonoids to exert anticancer effects. *International Journal of Molecular Sciences* 20(10):2420.

Chaurasia, S., Patel, R.R., Vure, P. and Mishra, B. 2018. Potential of cationic-polymeric nanoparticles for oral delivery of naringenin: In vitro and in vivo investigations. *Journal of Pharmaceutical Sciences* 107(2):706–716.

Chen, H., Liu, D. and Guo, Z. 2016. Endogenous stimuli-responsive nanocarriers for drug delivery. *Chemistry Letters* 45(3):242–249.

Chen, R.E., Zhang, X., Case, J.B., et al. 2021. Resistance of SARS-CoV-2 variants to neutralization by monoclonal and serum-derived polyclonal antibodies. *Nature Medicine* 27(4):717–726.

Cheng, R., Meng, F., Deng, C., Klok, H.A. and Zhong, Z. 2013. Dual and multi-stimuli responsive polymeric nanoparticles for programmed site-specific drug delivery. *Biomaterials* 34(14):3647–3657.

Cheng, W., Gu, L., Ren, W. and Liu, Y. 2014. Stimuli-responsive polymers for anti-cancer drug delivery. *Materials Science and Engineering: C* 45:600–608.

Chidambaram, M., Manavalan, R. and Kathiresan, K. 2011. Nanotherapeutics to overcome conventional cancer chemotherapy limitations. *Journal of Pharmacy and Pharmaceutical Sciences* 14(1):67–77.

Cho, J., Lee, Y.J., Kim, J.H., et al. 2020. Antiviral activity of digoxin and ouabain against SARS-CoV-2 infection and its implication for COVID-19. *Scientific Reports* 10(1):1–8.

Chou, T.H. and Liang, C.H. 2009. The molecular effects of aloe-emodin (AE)/liposome-AE on human nonmelanoma skin cancer cells and skin permeation. *Chemical Research in Toxicology* 22(12):2017–2028.

Chung, K.T., Wong, T.Y., Wei, C.I., Huang, Y.W. and Lin, Y. 1998. Tannins and human health: A review. *Critical Reviews in Food Science and Nutrition* 38(6):421–464.

Cushnie, T.T. and Lamb, A.J. 2005. Antimicrobial activity of flavonoids. *International Journal of Antimicrobial Agents* 26(5):343–356.

Dai, W., Zhang, D., Duan, C., et al. 2010. Preparation and characteristics of oridonin-loaded nanostructured lipid carriers as a controlled-release delivery system. *Journal of Microencapsulation* 27(3):234–241.

Darabi, A., Jessop, P.G. and Cunningham, M.F. 2015. One-Pot synthesis of poly ((diethylamino) ethyl methacrylate-co-styrene)-b-poly (methyl methacrylate-co-styrene) nanoparticles via nitroxide-mediated polymerization. *Macromolecules* 48(7): 1952–1958.

de Oliveira Pedro, R., Goycoolea, F.M., Pereira, S., Schmitt, C.C. and Neumann, M.G. 2018. Synergistic effect of quercetin and pH-responsive DEAE-chitosan carriers as drug delivery system for breast cancer treatment. *International Journal of Biological Macromolecules* 106:579–586.

Deng, Y., Zhang, X., Shen, H., 2020. Application of the nano-drug delivery system in treatment of cardiovascular diseases. *Frontiers in Bioengineering and Biotechnology* 7:489.

Desideri, E., Ciccarone, F. and Ciriolo, M.R. 2019. Targeting glutathione metabolism: Partner in crime in anticancer therapy. *Nutrients* 11(8):1926.

Du, Y., Zhang, J., Tao, et al. 2019. Aloe emodin exerts potent anticancer effects in MIAPaCa-2 and PANC-1 human pancreatic adenocarcinoma cell lines through activation of both apoptotic and autophagic pathways, sub-G1 cell cycle arrest and disruption of mitochondrial membrane potential (ΛΨm). *Journal of BU ON* 24(2): 746–753.

Edis, Z., Wang, J., Waqas, M.K., Ijaz, M. and Ijaz, M. 2021. Nanocarriers-mediated drug delivery systems for anticancer agents: An overview and perspectives. *International Journal of Nanomedicine* 16:1313.

El-Saber Batiha, G., Magdy Beshbishy, A., Wasef, L., et al. 2020. Chemical constituents and pharmacological activities of garlic (*Allium sativum* L.): A review. *Nutrients* 12(3):872.

Feng, C., Yuan, X., Chu, K., Zhang, H., Ji, W. and Rui, M. 2019. Preparation and optimization of poly (lactic acid) nanoparticles loaded with fisetin to improve anti-cancer therapy. *International Journal of Biological Macromolecules* 125:700–710.

Ferrari, M. 2005. Cancer nanotechnology: Opportunities and challenges. *Nature Reviews Cancer* 5(3):161–171.

Fredenberg, S., Wahlgren, M., Reslow, M. and Axelsson, A. 2011. The mechanisms of drug release in poly (lactic-co-glycolic acid)-based drug delivery systems - A review. *International Journal of Pharmaceutics* 415(1-2):34–52.

Fujimura, Y., Miura, D. and Tachibana, H. 2017. A phytochemical-sensing strategy based on mass spectrometry imaging and metabolic profiling for understanding the functionality of the medicinal herb green tea. *Molecules* 22(10):1621.

Ganeshpurkar, A. and Saluja, A.K. 2017. The pharmacological potential of rutin. *Saudi Pharmaceutical Journal* 25(2):149–164.

Gansukh, E., Gopal, J., Paul, D., et al. 2018. Ultrasound mediated accelerated Anti-influenza activity of *Aloe vera*. *Scientific Reports* 8(1):1–10.

Gansukh, E., Mya, K.K., Jung, M.,et al. 2019. Lutein derived from marigold (*Tagetes erecta*) petals triggers ROS generation and activates Bax and caspase-3 mediated apoptosis of human cervical carcinoma (HeLa) cells. *Food and Chemical Toxicology* 127:11–18.

Gatto, M.T., Falcocchio, S., Grippa, E., et al. 2002. Antimicrobial and anti-lipase activity of quercetin and its C2-C16 3-O-acyl-esters. *Bioorganic and Medicinal Chemistry* 10(2): 269–272.

Gharib, A., Faezizadeh, Z., Mesbah-Namin, S.A.R. and Saravani, R. 2015. Experimental treatment of breast cancer-bearing BALB/c mice by artemisinin and transferrin-loaded magnetic nanoliposomes. *Pharmacognosy Magazine* 11(Suppl 1):S117.

Gonzalez-Burgos, E. and Gómez-Serranillos, M.P. 2012. Terpene compounds in nature: A review of their potential antioxidant activity. *Current Medicinal Chemistry* 19(31): 5319–5341.

Gu, S., Li, L., Huang, H., Wang, B. and Zhang, T. 2019. Antitumor, antiviral, and anti-inflammatory efficacy of essential oils from *Atractylodes macrocephala* Kkoidz. produced with different processing methods. *Molecules* 24(16):2956.

Gupta, A., Kaur, C.D., Saraf, S. and Saraf, S. 2016. Formulation, characterization, and evaluation of ligand-conjugated biodegradable quercetin nanoparticles for active targeting. *Artificial Cells, Nanomedicine, and Biotechnology* 44(3):960–970.

Gutheil, W.G., Reed, G., Ray, A., Anant, S. and Dhar, A. 2012. Crocetin: An agent derived from saffron for prevention and therapy for cancer. *Current Pharmaceutical Biotechnology* 13(1):173–179.

Habtemariam, S. 2020. Recent advances in Berberine inspired anticancer approaches: From drug combination to novel formulation technology and derivatization. *Molecules* 25(6):1426.

Haggag, Y.A., El-Ashmawy, N.E. and Okasha, K.M. 2020. Is hesperidin essential for prophylaxis and treatment of COVID-19 infection?. *Medical Hypotheses* 144:109957.

Hashemzaei, M., Delarami Far, A., Yari, A., et al. 2017. Anticancer and apoptosis-inducing effects of quercetin in vitro and in vivo. *Oncology Reports* 38(2):819–828.

Hassan, M.Z., Osman, H., Ali, M.A. and Ahsan, M.J. 2016. Therapeutic potential of coumarins as antiviral agents. *European Journal of Medicinal Chemistry* 123: 236–255.

Horio, Y., Sogabe, R., Shichiri, M., et al. 2020. Induction of a 5-lipoxygenase product by daidzein is involved in the regulation of influenza virus replication. *Journal of Clinical Biochemistry and Nutrition* 66(1):36–42.

Hsiao, M.H., Larsson, M., Larsson, A., et al. 2012. Design and characterization of a novel amphiphilic chitosan nanocapsule-based thermo-gelling biogel with sustained in vivo release of the hydrophilic anti-epilepsy drug ethosuximide. *Journal of Controlled Release* 161(3):942–948.

Hu, Y., Jiang, X., Ding, Y., Ge, H., Yuan, Y. and Yang, C. 2002. Synthesis and characterization of chitosan–poly (acrylic acid) nanoparticles. *Biomaterials* 23(15): 3193–3201.

Hua, F., Li, C.H., Chen, X.G. and Liu, X.P. 2018. Daidzein exerts anticancer activity towards SKOV3 human ovarian cancer cells by inducing apoptosis and cell cycle arrest, and inhibiting the Raf/MEK/ERK cascade. *International Journal of Molecular Medicine* 41(6):3485–3492.

Huang, X. and Brazel, C.S. 2001. On the importance and mechanisms of burst release in matrix-controlled drug delivery systems. *Journal of Controlled Release* 73(2–3): 121–136.

Huda, S., Alam, M.A. and Sharma, P.K. 2020. Smart nanocarriers-based drug delivery for cancer therapy: An innovative and developing strategy. *Journal of Drug Delivery Science and Technology* 102018.

Jain, A., Sharma, G., Kushwah, V., et al. 2017. Fabrication and functional attributes of lipidic nanoconstructs of lycopene: An innovative endeavour for enhanced cytotoxicity in MCF-7 breast cancer cells. *Colloids and Surfaces B: Biointerfaces* 152:482–491.

Jain, A., Sharma, G., Kushwah, V., et al. 2018. Beta carotene-loaded zein nanoparticles to improve the biopharmaceutical attributes and to abolish the toxicity of methotrexate: A preclinical study for breast cancer. *Artificial Cells, Nanomedicine, and Biotechnology* 46(S1):402–412.

James, H.P., John, R., Alex, A. and Anoop, K. 2014. Smart polymers for the controlled delivery of drugs – A concise overview. *Acta Pharmaceutica Sinica B* 4(2): 120–127.

Jelezova, I., Drakalska, E., Momekova, D., et al. 2015. Curcumin loaded pH-sensitive hybrid lipid/block copolymer nanosized drug delivery systems. *European Journal of Pharmaceutical Sciences* 78:67–78.

Kaczmarek, B. 2020. Tannic acid with antiviral and antibacterial activity as a promising component of biomaterials - A mini review. *Materials* 13(14):3224.

Katz, E., Nisani, S. and Chamovitz, D.A. 2018. Indole-3-carbinol: A plant hormone combatting cancer. *F1000Research* 7.

Khoee, S. and Karimi, M.R. 2018. Dual-drug loaded Janus graphene oxide-based thermo-responsive nanoparticles for targeted therapy. *Polymer* 142:80–98.

Kim, M.J. and Kim, H. 2015. Anticancer effect of lycopene in gastric carcinogenesis. *Journal of Cancer Prevention* 20(2):92.

Kong, L., Li, S., Liao, Q., et al. 2013. Oleanolic acid and ursolic acid: Novel hepatitis C virus antivirals that inhibit NS5B activity. *Antiviral Research* 98(1):44–53.

Kongtawelert, P., Wudtiwai, B., Shwe, T.H., Pothacharoen, P. and Phitak, T. 2020. Inhibitory effect of Hesperidin on the expression of programmed death ligand (PD-L1) in breast cancer. *Molecules* 25(2):252.

Koo, O.M., Rubinstein, I. and Onyuksel, H. 2006. Camptothecin in sterically stabilized phospholipid nano-micelles: A novel solvent pH change solubilization method. *Journal of Nanoscience and Nanotechnology* 6(9-10):2996–3000.

Krinsky, N.I. and Johnson, E.J. 2005. Carotenoid actions and their relation to health and disease. *Molecular Aspects of Medicine* 26(6):459–516.

Krukiewicz, K., Zawisza, P., Herman, A.P., Turczyn, R., Boncel, S. and Zak, J.K. 2016. An electrically controlled drug delivery system based on conducting poly (3, 4-ethylenedioxypyrrole) matrix. *Bioelectrochemistry* 108:13–20.

Kumar, C.S., Thangam, R., Mary, S.A., Kannan, P.R., Arun, G. and Madhan, B. 2020. Targeted delivery and apoptosis induction of trans-resveratrol-ferulic acid loaded chitosan coated folic acid conjugate solid lipid nanoparticles in colon cancer cells. *Carbohydrate Polymers* 231:115682.

Kumar, S., Kashyap, P., Chowdhury, S., Kumar, S., Panwar, A. and Kumar, A. 2021. Identification of phytochemicals as potential therapeutic agents that binds to Nsp15 protein target of coronavirus (SARS-CoV-2) that are capable of inhibiting virus replication. *Phytomedicine* 85:153317.

Kumari, P., Ghosh, B. and Biswas, S. 2016. Nanocarriers for cancer-targeted drug delivery. *Journal of Drug Targeting* 24(3):179–191.

Kurisawa, M., Terano, M. and Yui, N. 1997. Double-stimuli-responsive degradation of hydrogels consisting of oligopeptide-terminated poly (ethylene glycol) and dextran with an interpenetrating polymer network. *Journal of Biomaterials Science, Polymer Edition* 8(9):691–708.

Lammers, T.G.G.M., Hennink, W.E. and Storm, G. 2008. Tumour-targeted nanomedicines: Principles and practice. *British Journal of Cancer* 99(3):392–397.

Lazer, L.M., Sadhasivam, B., Palaniyandi, K., et al. 2018. Chitosan-based nano-formulation enhances the anticancer efficacy of hesperetin. *International Journal of Biological Macromolecules* 107:1988–1998.

Le Cher, J.C., Diep, N., Krug, P.W. and Hilliard, J.K. 2019. Genistein has antiviral activity against herpes B virus and acts synergistically with antiviral treatments to reduce effective dose. *Viruses* 11(6):499.

Lee, H., Hong, W., Jeon, S., Choi, Y. and Cho, Y. 2015. Electroactive polypyrrole nanowire arrays: Synergistic effect of cancer treatment by on-demand drug release and photothermal therapy. *Langmuir* 31(14):4264–4269.

Lee, H., Son, S.H., Sharma, R. and Won, Y.Y. 2011. A Discussion of the pH-Dependent Protonation Behaviors of Poly (2-(dimethylamino) ethyl methacrylate) (PDMAEMA) and Poly (ethylenimine-ran-2-ethyl-2-oxazoline) (P (EI-r-EOz). *The Journal of Physical Chemistry B* 115(5):844–860.

Lee, J.H. and Yeo, Y. 2015. Controlled drug release from pharmaceutical nanocarriers. *Chemical Engineering Science* 125:75–84.

Leoncini, E., Nedovic, D., Panic, N., Pastorino, R., Edefonti, V. and Boccia, S. 2015. Carotenoid intake from natural sources and head and neck cancer: A systematic review and meta-analysis of epidemiological studies. *Cancer Epidemiology and Prevention Biomarkers* 24(7):1003–1011.

Li, H., Liu, K., Williams, G.R., et al. 2018. Dual temperature and pH responsive nanofiber formulations prepared by electrospinning. *Colloids and Surfaces B: Biointerfaces* 171:142–149.

Li, L.W., Na, C., Tian, S.Y., et al.2018. Ellagic acid induces HeLa cell apoptosis via regulating signal transducer and activator of transcription 3 signaling. *Experimental and Therapeutic Medicine* 16(1):29–36.

Liu, C., Huang, P., Zhao, D., et al. 2021. Effects of rotavirus NSP4 protein on the immune response and protection of the SR69A-VP8* nanoparticle rotavirus vaccine. *Vaccine* 39(2):263–271.

Lopes-Rodrigues, V., Sousa, E. and Vasconcelos, M.H. 2016. Curcumin as a modulator of P-glycoprotein in cancer: Challenges and perspectives. *Pharmaceuticals* 9(4): 71.

Mondal, A., Gandhi, A., Fimognari, C., Atanasov, A.G. and Bishayee, A. 2019. Alkaloids for cancer prevention and therapy: Current progress and future perspectives. *European Journal of pharmacology* 858:172472.

Müller, R.H., Mäder, K. and Gohla, S. 2000. Solid lipid nanoparticles (SLN) for controlled drug delivery – A review of the state of the art. *European Journal of Pharmaceutics and Biopharmaceutics* 50(1):161–177.

Mura, S., Nicolas, J. and Couvreur, P. 2013. Stimuli-responsive nanocarriers for drug delivery. *Nature Materials* 12(11):991–1003.

Nigam, S. and Bahadur, D. 2017. Dendrimer-conjugated iron oxide nanoparticles as stimuli-responsive drug carriers for thermally-activated chemotherapy of cancer. *Colloids and Surfaces B: Biointerfaces* 155:182–192.

Ninfali, P., Antonelli, A., Magnani, M. and Scarpa, E.S. 2020. Antiviral properties of flavonoids and delivery strategies. *Nutrients* 12(9):2534.

Nisar, N., Li, L., Lu, S., Khin, N.C. and Pogson, B.J. 2015. Carotenoid metabolism in plants. *Molecular Plant* 8(1):68–82.

Onaca, O., Enea, R., Hughes, D.W. and Meier, W. 2009. Stimuli-responsive polymersomes as nanocarriers for drug and gene delivery. *Macromolecular Bioscience* 9(2): 129–139.

Pafiti, K., Cui, Z., Adlam, D., Hoyland, J., Freemont, A.J. and Saunders, B.R. 2016. Hydrogel composites containing sacrificial collapsed hollow particles as dual action pH-responsive biomaterials. *Biomacromolecules* 17(7):2448–2458.

Pang, R., Tao, J.Y., Zhang, S.L., et al. 2010. In vitro antiviral activity of lutein against hepatitis B virus. *Phytotherapy Research* 24(11):1627–1630.

Park, S.W., Kwon, M.J., Yoo, J.Y., Choi, H.J. and Ahn, Y.J. 2014. Antiviral activity and possible mode of action of ellagic acid identified in *Lagerstroemia speciosa* leaves toward human rhinoviruses. *BMC Complementary and Alternative Medicine* 14(1):1–8.

Pawar, A., Singh, S., Rajalakshmi, S., Shaikh, K. and Bothiraja, C. 2018. Development of fisetin-loaded folate functionalized pluronic micelles for breast cancer targeting. *Artificial Cells, Nanomedicine, and Biotechnology* 46(S1):347–361.

Pietta, P.G. 2000. Flavonoids as antioxidants. *Journal of Natural Products* 63(7):1035–1042.

Polonikov, A. 2020. Endogenous deficiency of glutathione as the most likely cause of serious manifestations and death in COVID-19 patients. *ACS Infectious Diseases* 6(7):1558–1562.

Pool, H., Campos-Vega, R., Herrera-Hernández, M.G., et al. 2018. Development of genistein-PEGylated silica hybrid nanomaterials with enhanced antioxidant and anti-proliferative properties on HT29 human colon cancer cells. *American Journal of Translational Research* 10(8):2306.

Qu, J., Zhao, X., Ma, P.X. and Guo, B. 2017. pH-responsive self-healing injectable hydrogel based on N-carboxyethyl chitosan for hepatocellular carcinoma therapy. *Acta Biomaterialia* 58:168–180.

Rani, K. and Paliwal, S. 2014. A review on targeted drug delivery: Its entire focus on advanced therapeutics and diagnostics. *Scholars Journal of Applied Medical Sciences* 2(1C):328–331.

Rastgoo, M., Hosseinzadeh, H., Alavizadeh, H., Abbasi, A., Ayati, Z. and Jaafari, M.R. 2013. Antitumor activity of PEGylated nanoliposomes containing crocin in mice bearing C26 colon carcinoma. *Planta Medica* 79(6):447–451.

Rauf, A., Imran, M., Khan, I.A., 2018. Anticancer potential of quercetin: A comprehensive review. *Phytotherapy Research* 32(11):2109–2130.

Ravi, H., Kurrey, N., Manabe, Y., Sugawara, T. and Baskaran, V. 2018. Polymeric chitosan-glycolipid nanocarriers for an effective delivery of marine carotenoid fucoxanthin for induction of apoptosis in human colon cancer cells (Caco-2 cells). *Materials Science and Engineering: C* 91:785–795.

Raza, A., Rasheed, T., Nabeel, F., Hayat, U., Bilal, M. and Iqbal, H. 2019. Endogenous and exogenous stimuli-responsive drug delivery systems for programmed site-specific release. *Molecules* 24(6):1117.

Reddy, B.U., Mullick, R., Kumar, A., Sudha, G., Srinivasan, N. and Das, S. 2014. Small molecule inhibitors of HCV replication from pomegranate. *Scientific Reports* 4(1):1–10.

Rejinold, N.S., Sreerekha, P.R., Chennazhi, K.P., Nair, S.V. and Jayakumar, R. 2011. Biocompatible, biodegradable and thermo-sensitive chitosan-g-poly (N-isopropylacrylamide) nanocarrier for curcumin drug delivery. *International Journal of Biological Macromolecules* 49(2):161–172.

Rizvi, S.A. and Saleh, A.M. 2018. Applications of nanoparticle systems in drug delivery technology. *Saudi Pharmaceutical Journal* 26(1):64–70.

Salgarella, A.R., Zahoranová, A., Šrámková, P., et al. 2018. Investigation of drug release modulation from poly (2-oxazoline) micelles through ultrasound. *Scientific Reports* 8(1):1–13.

Saraf, S. 2010. Applications of novel drug delivery system for herbal formulations. *Fitoterapia* 81(7):680–689.

Schmaljohann, D. 2006. Thermo-and pH-responsive polymers in drug delivery. *Advanced Drug Delivery Reviews* 58(15):1655–1670.

Schwarz, S., Sauter, D., Wang, K., et al. 2014. Kaempferol derivatives as antiviral drugs against the 3a channel protein of coronavirus. *Planta Medica* 80(2–3):177.

Semwal, D.K., Semwal, R.B., Combrinck, S. and Viljoen, A. 2016. Myricetin: A dietary molecule with diverse biological activities. *Nutrients* 8(2):90.

Seren, S., Mutchnick, M., Hutchinson, D., et al. 2008. Potential role of lycopene in the treatment of hepatitis C and prevention of hepatocellular carcinoma. *Nutrition and Cancer* 60(6):729–735.

Sesarman, A., Tefas, L., Sylvester, B., et al. 2019. Co-delivery of curcumin and doxorubicin in PEGylated liposomes favored the antineoplastic C26 murine colon carcinoma microenvironment. *Drug Delivery and Translational Research* 9(1):260–272.

Shalviri, A., Chan, H.K., Raval, G., et al. 2013. Design of pH-responsive nanoparticles of terpolymer of poly (methacrylic acid), polysorbate 80 and starch for delivery of doxorubicin. *Colloids and Surfaces B: Biointerfaces* 101:405–413.

Shanmugam, H., Acharya, P., Jayaprakasha, G.K. and Patil, B.S. 2017. Nanoformulation and characterization of nomilin with different poly (lactic-co-glycolic acid) resomers and surfactants for the enhanced inhibition of α-amylase and angiotensin-converting-enzyme. *Journal of Functional Foods* 37:564–573.

Shinde, P., Agraval, H., Singh, A., Yadav, U.C. and Kumar, U. 2019. Synthesis of luteolin loaded zein nanoparticles for targeted cancer therapy improving bioavailability and efficacy. *Journal of Drug Delivery Science and Technology* 52:369–378.

Shiraki, K. and Rapp, F. 1988. Effects of caffeine on herpes simplex virus. *Intervirology* 29(4):235–240.

Singh, R. and Lillard Jr, J.W. 2009. Nanoparticle-based targeted drug delivery. *Experimental and Molecular Pathology* 86(3):215–223.

Snima, K.S., Arunkumar, P., Jayakumar, R. and Lakshmanan, V.K. 2014. Silymarin encapsulated poly (D, L-lactic-co-glycolic acid) nanoparticles: A prospective candidate for prostate cancer therapy. *Journal of Biomedical Nanotechnology* 10(4):559–570.

Son, G.H., Lee, B.J. and Cho, C.W. 2017. Mechanisms of drug release from advanced drug formulations such as polymeric-based drug-delivery systems and lipid nanoparticles. *Journal of Pharmaceutical Investigation* 47(4):287–296.

Song, J.M., Lee, K.H. and Seong, B.L. 2005. Antiviral effect of catechins in green tea on influenza virus. *Antiviral Research* 68(2):66–74.

Sowjanya, M., Debnath, S., Lavanya, P., Thejovathi, R. and Babu, M.N. 2017. Polymers used in the designing of controlled drug delivery system. *Research Journal of Pharmacy and Technology* 10(3):903.

Srivastava, A., Yadav, T., Sharma, S., Nayak, A., Kumari, A. and Mishra, N. 2016. Polymers in drug delivery. *Journal of Biosciences and Medicines* 4:69–84.

Taghizadeh, B., Taranejoo, S., Monemian, S.A., et al. 2015. Classification of stimuli–responsive polymers as anticancer drug delivery systems. *Drug Delivery* 22(2): 145–155.

Tej, G.N.V.C., Neogi, K. and Nayak, P.K. 2019. Caffeine-enhanced anti-tumor activity of anti-PD1 monoclonal antibody. *International Immunopharmacology* 77:106002.

Thakur, A., Singla, R. and Jaitak, V. 2015. Coumarins as anticancer agents: A review on synthetic strategies, mechanism of action and SAR studies. *European Journal of Medicinal Chemistry* 101:476–495.

Tian, Q., Zhang, C.N., Wang, X.H., et al. 2010. Glycyrrhetinic acid-modified chitosan/poly (ethylene glycol) nanoparticles for liver-targeted delivery. *Biomaterials* 31(17): 4748–4756.

Tohmé, M.J., Giménez, M.C., Peralta, A., Colombo, M.I. and Delgui, L.R. 2019. Ursolic acid: A novel antiviral compound inhibiting rotavirus infection in vitro. *International Journal of Antimicrobial Agents* 54(5):601–609.

Vasir, J.K., Reddy, M.K. and Labhasetwar, V.D. 2005. Nanosystems in drug targeting: opportunities and challenges. *Current Nanoscience* 1(1):47–64.

Vyas, S.P. and Khar, R.K. 2002. Controlled drug delivery concepts and advances. *Vallabhprakashan* 1:411–447.

Wang, M., Zhao, T., Liu, Y., et al. 2017. Ursolic acid liposomes with chitosan modification: Promising antitumor drug delivery and efficacy. *Materials Science and Engineering: C* 71:1231–1240.

Wang, X., Yang, Y., An, Y. and Fang, G. 2019. The mechanism of anticancer action and potential clinical use of kaempferol in the treatment of breast cancer. *Biomedicine and Pharmacotherapy* 117:109086.

Wang, Y., Ma, Q., Zhang, S., et al. 2020. Digoxin enhances the anticancer effect on non-small cell lung cancer while reducing the cardiotoxicity of adriamycin. *Frontiers in Pharmacology* 11:186.

Warowicka, A., Nawrot, R. and Goździcka-Józefiak, A. 2020. Antiviral activity of berberine. *Archives of Virology* 165(9):1935–1945.

Yang, P., Li, D., Jin, S., , et al. 2014. Stimuli-responsive biodegradable poly (methacrylic acid) based nanocapsules for ultrasound traced and triggered drug delivery system. *Biomaterials* 35(6):2079–2088.

Yang, Z., Yang, H., Dong, X., Pu, M. and Ji, F. 2020. Hesperidin loaded Zn^{2+}@SA/PCT nanocomposites inhibit the proliferation and induces the apoptosis in colon cancer cells (HCT116) through the enhancement of pro-apoptotic protein expressions. *Journal of Photochemistry and Photobiology B: Biology* 204:111767.

Yao, Y., Zhou, Y., Liu, L., et al. 2020. Nanoparticle-based drug delivery in cancer therapy and its role in overcoming drug resistance. *Frontiers in Molecular Biosciences* 7.

Yin, R., Li, T., Tian, J.X., Xi, P. and Liu, R.H. 2018. Ursolic acid, a potential anticancer compound for breast cancer therapy. *Critical Reviews in Food Science and Nutrition* 58(4):568–574.

Yonekura, L., Kobayashi, M., Terasaki, M. and Nagao, A. 2010. Keto-carotenoids are the major metabolites of dietary lutein and fucoxanthin in mouse tissues. The Journal of Nutrition 140(10):1824–1831.

Zeng, L., Yan, J., Luo, L., Ma, M. and Zhu, H. 2017. Preparation and characterization of (−)-Epigallocatechin-3-gallate (EGCG)-loaded nanoparticles and their inhibitory effects on Human breast cancer MCF-7 cells. *Scientific Reports* 7(1):1–15.

Zhang, H., Liu, X., Wu, F., et al. 2016. A novel prostate-specific membrane-antigen (PSMA) targeted micelle-encapsulating wogonin inhibits prostate cancer cell proliferation via inducing intrinsic apoptotic pathway. *International Journal of Molecular Sciences* 17(5):676.

Zhang, L. and Wang, H. 2015. Multiple mechanisms of anti-cancer effects exerted by astaxanthin. *Marine Drugs* 13(7):4310–4330.

Zhang, M.Z., Chen, Q. and Yang, G.F. 2015. A review on recent developments of indole-containing antiviral agents. *European Journal of Medicinal Chemistry* 89: 421–441.

Zhang, W.L., Zhao, Y.N., Shi, Z.Z., et al. 2018. Lutein inhibits cell growth and activates apoptosis via the PI3K/AKT/mTOR signaling pathway in A549 human non-small-cell lung cancer cells. *Journal of Environmental Pathology, Toxicology and Oncology* 37(4).

Zhang, Z., Cheng, W., Pan, Y. and Jia, L. 2020. An anticancer agent-loaded PLGA nanomedicine with glutathione-response and targeted delivery for the treatment of lung cancer. *Journal of Materials Chemistry B* 8(4):655–665.

ZorofchianMoghadamtousi, S., Abdul Kadir, H., Hassandarvish, P., Tajik, H., Abubakar, S. and Zandi, K. 2014. A review on antibacterial, antiviral, and antifungal activity of curcumin. *BioMed Research International* 2014.

zurMühlen, A., Schwarz, C. and Mehnert, W. 1998. Solid lipid nanoparticles (SLN) for controlled drug delivery–drug release and release mechanism. *European Journal of Pharmaceutics and Biopharmaceutics* 45(2):149–155.

16 Conjugated Graphene Gold Nanocomposites for Cancer Therapy

Zaira Zaman Chowdhury, Abu Nasser Faisal, Shahjalal Mohammad Shibly, Devarajan Thangadurai, Saher Islam, and Jeyabalan Sangeetha

CONTENTS

DOI: 10.1201/9781003136644-19

16.1 INTRODUCTION

Due to their distinctive nanometric sizes, nanoparticles (NPs) have emerged as prominent, broadly acceptable composites in various sectors. The upshot is that nanoparticles are chemically more reactive and exhibit more special attributes than their larger-scale counterparts. Nanodevices used for biomedical application are well-designed to play a key role in the field of medicine in the long run. A large spectrum of positive features, including increased rate of reaction, improved sensitivities, selectivity, reduced expenditure, and invasiveness (Boisseau and Loubaton 2011) are observed for NPs. The research and evidence that have been compiled reveal that nanoparticles (NPs) have the incredible opportunity to be used for disease detection, medication, prevention, and therapeutic interventions (Boisseau and Loubaton 2011). The present status of nanomaterials (NPs) applications in biomedical disciplines can be evidenced from following Table 16.1 (Al-Ani et al. 2017).

Application of NPs as anticancer agents is being substantially investigated as a strategy to resolve the significant fatality rates associated with conventional chemotherapy and radiotherapy cancer treatments (Al-Ani et al. 2017) (Table 16.2). In the near future, the emerging paradigm of cancer nanotechnologies may offer a more efficient, less aggressive method of medication that aims to offer solutions for better treatment. In that case, biochemical variations associated with pathogenicity of cancer can be identified at an early stage, and the therapeutic entities can be transported explicitly to the tumour cells, even though potentially saving the normal tissues. These breakthroughs will result in bidirectional solutions; they may be utilized for diagnosis as well as therapy, concurrently.

As a consequence of AuNPs interesting optoelectronic characteristics, along with their versatility to be tailored at multiple absorption wavelengths, they are used for a variety of applications (Moussawi and Patra 2015). The physico-chemical

TABLE 16.1
Application of Nanotechnology in Biomedical Field

Biomedical Field	Nanotechnology Approach
In vitro identification	• Biosensors • Tissue analysis, Immuno-histo-chemical stain
In vivo identification/imaging	• Ultrasound • Magnetic-resonance image development (MRI) • CT-Scan
Regenerative medication involving research on stem cells	• Tracing, separation, combination and compulsory differentiation of the stem cells
Tissue engineering	• Appropriate scaffolds made of NPs
Phototherapy	• Nano-enhanced thermal degradation of tumour cells (Photo-thermal-PTT Therapy) • Photo-dynamic PDT Therapy
Ophthalmological uses	• Intra-ocular pressure analysis • Gene therapy for retina diseases • Nano-scale ointment for eyes which can be dispersed in dry eyes
Dental uses	• Nano-fillers of spherical SiO_2 particles • Orthodontic nano robots deal with straightening of teeth
Orthopaedic applications	• Ceramic-based NPs for implantation of bones
Immune responses	• Nano-enhanced bucky balls interact with the free radicals, averting mast cells from discharging histamine
Medical instrumentation fabrication	• Nano-enhanced antibacterial fabrics
Meditation/Drug distribution	• Quantum dots as trailing agent
Surgical analysis	• Au coated nano shells for transplant surgery
Drug transport	• Nano-enhanced carriers for cancer, bacterial diseases etc.
Proteins and peptide transport	• Distribution of protein-based antigens using NPs
Gene delivery	• Aptamers

properties of AuNPs stand out in particular for the following attributes: smaller sizes, surface-charge effects, quantum effects, and electronic and optical properties. Thus, AuNPs, are reasonably stable, benign, robust, and have simple protocol for synthesis (Pattani and Tunnell 2012). Due to the distinctive localized-surface plasmon-resonance effect (LSPR) of AuNPs, different types of nano stars, nano shells, nanospheres, nano rods and nano cages of AuNPs have been synthesized during the last two millennia (Burrows et al. 2017; De Melo-Diogo et al. 2017). LSPR is an optical phenomenon that happens as a consequence of the interactions between the electrons of the conduction band and the incident light, leading to scattering and absorption of the light in AuNPs. This behaviour is influenced by the physiological dimensions of the NPs, as well as the dispersion medium (Elahi et al. 2018). When compared to

TABLE 16.2
Comparative Appraisal of Traditional Cancer Treatment with Nano-Enhanced Approach

Disadvantages of Traditional Cancer Treatment	Nano Enhanced Cancer Treatment
Hydrophobic properties of some medications restrict their efficiency	Improved stability and hydrophilic behaviour of the insoluble drugs
Recurrence of tumour	Discrimination-based therapy which eliminate cancer-stem cells
Resistance towards the drugs	Coating of nanocomposites with multiple drug resistance (MDR) inhibitors
Toxicity aspects • Side effects • Gonadal dysfunction • Continuing chemotherapy associated dysfunction	Reduced cytotoxicity and side effects • Active or passive target methods • Organised release of drugs at tumour spot specifically • Improved efficiency with lower administration regularity

nanospherical AuNPs, anisotropic AuNPs are preferable, owing to their morphological, optical, electrical, catalytic, and magnetic features (Elahi et al. 2018). AuNPs with anisotropic properties, such as Au nano stars and Au nano triangles, can absorb light in a wider range of the spectrum (Vines et al. 2019).

AuNPs are also extremely responsive to the changes in the surrounding environment, making them amazing resources for a broad array of applications (Hlapisi et al. 2019). In the field of biomedicine, the usage of AuNPs is mostly focused on its ability to distribute the controlled release of medications, as well as the therapeutic effect against tumour, cancer, viruses and bacteria. AuNPs have a greater penetration capability than standard medications, and they have a relatively lower risk of side effects (Hlapisi et al. 2019). As a result, AuNPs have the potential to be extensively used in the biomedical field, particularly in cancer therapy.

The 2D materials of graphene oxide (GO) have a single atomic layer of thickness that is a derivative of Gr and has numerous oxygenated functional groups, such as hydroxyl, epoxides, and carboxylic groups anchored with it (Pinto et al. 2013). GO, in contrast to hydrophobic graphene, is hydrophilic due to the presence of oxygen functional groups, and, as a result, it is water-soluble (Ruiz et al. 2011). The honeycomb carbon texture of GO has various imperfections, certainly located inside and also throughout the surface and edges, which are caused by oxidation (Dideikin and Vul 2019). The composition of GO makes it the perfect, viable option in the research area of biology and medicine, particularly theranostic studies, owing to its water solubility, excellent colloidal stability, reduced unit cost, scalability, thermal properties, and enlarged surface area (2630 m^2/g) (Singh et al. 2018).

GO has the extraordinary capacity to immobilize a large number of things, notably metallic particles, medicines, proteins, fluorescents, and organisms (Liu et al. 2013). Furthermore, it has been shown to exhibit intrinsic and near-infrared absorbance (NIR) qualities, which makes it effective as a photo-thermal agent. Thus, it can be applied for the treatment of cancer. The coupling of GO and AuNPs (GO–AuNPs) enhances the applicability of such composites in theranostic, bio-carrier (Khan et al. 2017), imaging (Moon et al. 2015), and sensor applications (Arvand and Gholizadeh 2013).

In this chapter, the toxicological effect of GO and AuNPs, with their bio-compatibility aspects, are discussed, with an insistence on GO owing to its superior characteristics. The subsequent section illustrates the advantages of using Gr–AuNPs composites for cancer treatments. The concluding section illustrates the eco-toxicological hazards and environmental impact of using graphene-based nanocomposites and the challenges associated with its application in the arena of biomedical research.

16.2 SYNTHESIS OF AuNPs

Among the different shape of Au nanoparticles, Au nano rods are the most widely used version for biomedical applications. Since the emergence of Au nano rods, numerous protocols have been proposed to address the inherent challenges of production, which include increased product output, resistance to minor residues, accurate leverage over the AuNPs surface chemistry, consistency, and, most specifically, a practicable method; tailored for specific functionalities (Pérez-Juste et al. 2005). Figure 16.1 illustrates all types of fabrication technique used for Au nano rods.

16.2.1 Hard-Template Approach

The hard-template approach involves electrochemical reducing of gold (Au) salt within the channels of nano-porous alumina/polycarbonate membrane acting as template (Foss et al. 1994). To summarize, the procedure begins with bombarding a trace quantity of Ag or Cu ions onto the surface of alumina membrane to create a conductive layer suitable for electro-deposition. Following that, Au ions are electrodeposited inside the nano-pores of alumina, creating Au nano rods. Although the yield and homogeneity of Au nano rods obtained using the template approach are excellent, there are significant drawbacks. It is a difficult technique to release the rods from the templates and disperse them into the suitable solvents. Additionally, Au nano rods often have relatively large diameters (>100 nm), which has a detrimental influence on their plasmonic response required for biomedical applications (Foss et al. 1993).

16.2.2 Electrochemical Approach

In this method, Au metal plates can serve as a sacrificial anode, while the cathode is a Pt plate concealed under the Ag plate. This is critical for adjusting the aspect ratio of the synthesized Au nano rods. All electrodes are submerged inside the electrolytic solution having the mixture of cationic surfactant consisting of hexa-decyl-trimethyl-ammonium bromide (CTAB) and tetra-dodecyl-ammonium

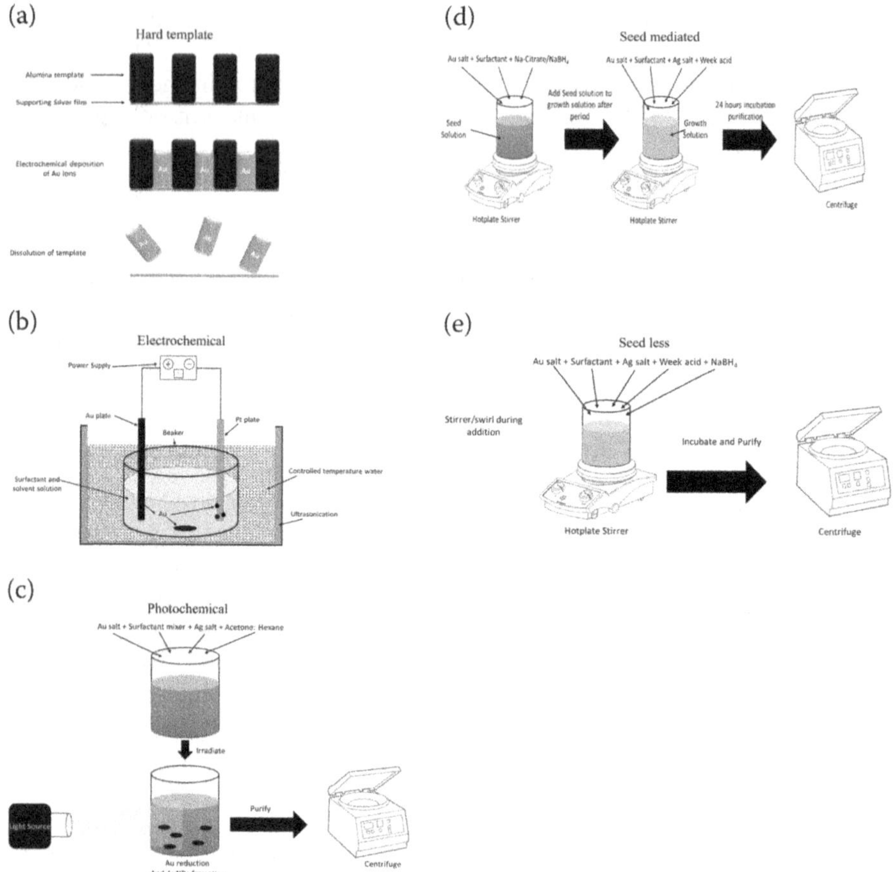

FIGURE 16.1 Fabrication technique of gold (Au) nano rods: (a) hard template, (b) electrochemical, (c) photothermal, (d) seed mediated, (e) seed less.

bromide (TDAB) (TC12AB). CTAB worked as a stabilizer to avoid agglomeration, whereas TC12AB served as a rod-generating co-surfactant in this electrochemical process. Additionally, acetone is added to destabilize the micellar structure, allowing the insertion of a co-surfactant that promotes the production of tubular CTAB micelles. Cyclohexane is introduced to promote the development of extended rod-like CTAB micelles. The electrolytic cell with this mixed solution is then immersed inside the ultrasonic bath. Sonication is necessary to disperse the Au nano rods and keep them distant from the cathode. Throughout the electrolytic process, a supply of 3 mA current is maintained for 30 minutes. The exact mechanism through which Au nano rods are formed using this approach is unknown (Pérez-Juste et al. 2005). Nevertheless, other investigators have documented that the Au anode was depleted during the process to generate $AuBr^{-4}$ ions, which were then conjugated with the surfactants and migrated toward the cathode where they were reduced (Cao et al. 2014).

16.2.3 Photochemical Synthesis

Photochemical synthesizing is identical with electrochemical synthesis, even though UV radiation is used to reduce the rods. A typical experiment begins with the preparation of a solution containing Au salts, CTAB, and TC12AB surfactants, acetone, silver nitrate and hexane, followed by bombardment of the solution using ultraviolet (UV) light at a wavelength of 254 or 300 nm, based on the aspect ratio desired. Compared to the seed-mediated approach, this method requires a longer contact period and more energy to generate Au nano rods (Huang et al. 2009; Cao et al. 2014).

16.2.4 Seedless-Growth Approach

Au nano rods are generated using a seedless one-pot synthesis approach by directly adding $NaBH_4$ with the growth solution (Zijlstra et al. 2006). Nucleation and growth take place in a single phase in the seedless approach, with asymmetrical growth commencing at the initiation stage (Xu et al. 2014). The researchers have investigated the effect of temperature on seedless Au nano rods production (Zijlstra et al. 2006). At different temperatures, the concentration of silver nitrate concentration is critical for the formation of nano rods. The length of the nano rods reduces as the temperature increases, although the breadth remains constant during the ambient temperature ranges from 25–97 °C (Zijlstra et al. 2006).

16.2.5 Seed-Mediated Approach

It can be carried out in presence or absence of $AgNO_3$ in growth medium.

16.2.5.1 Presence of $AgNO_3$ in Seed-Mediated Approach

The addition of silver nitrate ($AgNO_3$) allows the greater control over the morphology of Au nano rods, comparable to the electrochemical approach using reduced energy. $AgNO_3$ is added to the growth medium in this process to control the structural alteration of Au seeds to form rod shape. This boosts the yield of nano rods and passivates the crystal surfaces of Au nano rods (Xu et al. 2017). The lower pH of the growth fluid inhibits Ag ions reduction. Certain research has demonstrated that the aspect ratio of Au nano rods may be perfect by altering the proportion of Ag ions in the growth solution while maintaining other synthesis protocol constant (such as pH, seed solution injection, and so on). Furthermore, the ascorbic acid concentration has an effect on the yield and aspect ratio of Au nano rods.

16.2.5.2 Absence of $AgNO_3$ in Seed-Mediated Approach

Some researchers have suggested the seed-mediated technique without utilizing $AgNO_3$ (Jana et al. 2001a,b; Gao et al. 2003). In this method, Au seeds were developed using two–three steps by reducing sodium boro-hydride reduction in the presence of citric acid radicle and an optimal dosage of CTAB (0.1 M) and ascorbic acid (0.1 M). Only a limited yield of Au nano rods was achieved using this approach. Additionally, the approach was enhanced by altering the pH of the growing fluid (Pérez-Juste et al. 2005). The addition of a variety of seeds with the growing solution can result in a large production of Au nano rods.

16.3 SYNTHESIS PROTOCOLS FOR Gr–AuNPs BASED NANOCOMPOSITES

To obtain the requisite beneficial attributes, various protocols were used to produce Gr–AuNPs based composites. Moreover, Gr has been shown as an ideal support for doping AuNPs over its surface. The planar geometry facilitates the doping of AuNPs on the Gr, GO, and rGO substrates. The high specific surface area increases interlayer contact and avoids NPs agglomeration (Bai and Shen 2012; Qin et al. 2012). There are numerous configurations for Gr–based AuNPs based on shape of gold nanoparticles. Gr/GO/rGO acts as a platform for the deposition of gold onto its surface. However, Gr/GO/rGO can act as a shielding barrier that inhibits AuNPs from aggregating over graphene-enveloped/wrapped AuNPs composites (Bai and Shen 2012; Qiu et al. 2017). Figure 16.2 illustrates the classification of Gr–AuNPs synthesis techniques schematically.

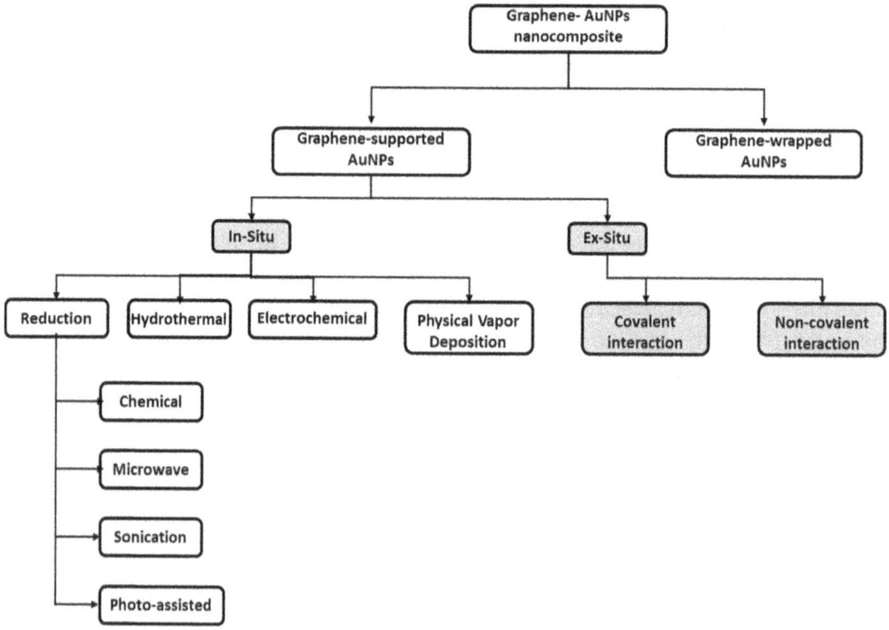

FIGURE 16.2 Classification of Gr–AuNPs based composite fabrication method.

16.3.1 Synthesis of Gr/ GO/rGO Supported Au Nanocomposites

16.3.1.1 In Situ Synthesis of Gr/ GO/rGO Supported Au Nanocomposites

In situ synthesis process can be carried out using reduction (chemical/sonication/ microwave heating/photo-induced deposition), hydrothermal, electrochemical, and physical vapour deposition methods (Pham et al. 2011; Sharma et al. 2013). Various chemical reagents can be used to carry out the reduction process (Lu et al. 2011; Zhuo et al. 2013). Additionally, it is feasible to combine reduction with the microwave-assisted (MWT) technique, which has the benefit of homogeneous and rapid warming of the slurry, enabling the retrieval of the extremely smaller products with narrower size variations (Yin et al. 2013). This moderate, straightforward, effective, and time-saving technique enables the use of an environmentally benign reductants, such as ascorbic acid to control the shape and size of the final products (Sharma et al. 2013).

Sonication and photo-aided deposition, like MWT, promote fast heating of the mixture, but the mechanism is markedly dissimilar for both circumstances. Ultrasonic reduction, as well as the photo-irradiated reduction, are both environmentally friendly, one-pot fabrication approach for the synthesis of rGO–AuNPs composites (Bai and Shen 2012). However, the autoclave atmosphere dictates whether the procedure is hydrothermal or solvothermal. Au nano-crystals, having size ranges from 10 to 20 nm, were decorated over GO without the application of any reductants or capping agents using hydrothermal approach (Qin et al. 2012). Electrochemical deposition is facilitated by the application of a voltage differential. This straightforward, environmentally friendly, cost-effective, and quick technique aims to produce dense, homogenous nanocomposites using only the precursors (Pham et al. 2011; Bai and Shen 2012).

Finally, physical vapour deposition (PVD) provides an extensive paradigm for using the pure physical process of condensation for deposition. The additional benefit of these approaches is that gold can be deposited immediately on pure graphene substrates without any chemical treatments and functionalization (Pandey et al. 2011; Bai and Shen 2012).

16.3.1.2 Ex Situ Synthesis of Gr/GO/rGO Supported Au Nanocomposites

This technique necessitates the pre-treatment of AuNPs and Gr/GO/rGO prior to implantation. The requirement for preceding modification/functionalization, as well as the use of linking chemicals, is limitations of this process. Nonetheless, this method results in more evenly dispersed and precisely regulated AuNPs (Bai and Shen 2012; Yao et al. 2013). AuNPs are attached to GO surfaces via covalent or noncovalent bonding. GO is favoured than Gr for covalent bonding because of its oxygen-functional groups (Bai and Shen 2012). This type of bonding provides a stronger interface than composites produced by in situ method, preventing AuNPs from leaking out of the graphene substrate (Pham et al. 2011). AuNPs, on the other hand, can be produced on Gr/GO/rGO using noncovalent bonding (hydrogen bonding, van der Waals couplings, electrostatic interaction and л-л stacking) (Bai and Shen 2012).

16.3.2 Synthesis of Gr/ GO/rGO Coated/Wrapped Au Nanocomposites

Earlier GO-coated/wrapped-AuNPs were synthesized using a unique process without the need of a substrate. The sample was well-dispersed and was shown to be advantageous for intracellular imaging and drug administration applications due to its nano dimensions, which surpasses in situ/ex situ synthesized GO/AuNPs composites (larger size). This one-step facile synthesis approach is fascinating for biological applications since it overrides the well-known constraints, such as cellular absorption for intracellular activities (Ma et al. 2013). The coating of Au nano rods can be carried out using in situ and ex situ coating approach. In situ wrapping involves growing of Au nano rods directly onto the GO surface, while ex situ coating involves preparation of both Au nano rods and GO individually and then blending them together. A spontaneous electrostatic interaction between GO and Au nano rods can be used to accomplish the ex situ wrapping/coating technique.

16.4 ADVANTAGES OF GR/AUNPS BASED COMPOSITES FOR CANCER TREATMENT

16.4.1 Biocompatibility

NPs need to biocompatible in order to be employed in the biological cells, and this is certainly important if the elements are to be used in medical applications. Numerous research on the cytotoxicity of the nanoparticles have revealed that the particle sizes, surface charges, and capping ligands are the most vital factors for the toxicity observed in NPs (Jones and Grainger 2009; Frohlich 2013; Fard et al. 2015; Tsolekile et al. 2017). The majority of AuNPs are synthesized using CTAB as a solvent and surfactant, which is responsible for the toxic effects (Ali et al. 2017; Allen et al. 2017). As an alternative, GO has been proven to be biocompatible, and it has been demonstrated that the coupling of AuNPs and GO yields a much more biocompatible composite when compared to untreated, pure AuNPs. The cytotoxicity of GO–AuNPs were observed using human alveolar epithelial cells and found to be hazardous. However, GO–AuNPs were much more biocompatible than pure AuNPs, even at a concentration of 200 M (Qiu et al. 2017). Au nano rod-decorated (AuNRs–PEG–GO) nanocomposites were found to be more biocompatible compared to Au nano rods when tested against carcinoma cells (Dembereldorj et al. 2014).

Plasmonic Au nano shells (AuNSs) and plasmonic Au nano roofs (AuNRs) were coated with rGO and their toxicity toward human umbilical endothelial vein cells was investigated (HUVECs). Human umbilical vein endothelial cells (HUVECs) treated with GO-coated nanocomposites for 24 hours showed no significant toxicity (Lim et al. 2013). Toxicity impact of GO–PSS–AuNPs was investigated for human pancreatic cells (Sun et al. 2017). After 24 hours, treating with several doses of GO–PSS AuNPs had shown no substantial toxicity. After 30 days of exposure, the long-term toxicity screening of the GO–AuNPs exhibited no evidence of inflammatory response, necrosis of cells, or fatality of the normal organs (Sun et al. 2017).

16.4.2 Photo-Thermal Activities

Several studies have been conducted on the photothermal characteristics of AuNPs and GO as independent entities, and both AuNPs and GO have demonstrated impressive photothermal efficacy in vivo and in vitro (Xu et al. 2013; Turcheniuk et al. 2015; Tomasella et al. 2020). However, in order to have these qualities, GO must be reduced. This is because GO has a strongly oxidized framework with a disordered conjugation, which decreases its conductance (Robinson et al. 2011; Khan et al. 2017). The photothermal feature of AuNRs is a result of the plasmonic interaction that exists on their surface. Several studies have been conducted on the coupling of AuNPs with GO to achieve effective photothermal characteristics (Lim et al. 2013; Dembereldorj et al. 2014; Song et al. 2015; Zhang et al. 2018). Earlier studies showed the photothermal effect of AuNPs–PEG–GO nanocomposites on A431 cells by exposing them to ultraviolet light (Dembereldorj et al. 2014). When subjected to a laser for 5 minutes, it was discovered that the AuNPs–PEG–GO generated heat. When exposed to light, this nanocomposite was capable of reducing dramatically the number of cells in the test tube by 40% (Dembereldorj et al. 2014). rGO–AuNPs shown superior photothermal properties when compared to untreated AuNPs, noncoated Au nano shells, and rGO–AuNOs nano shells (Lim et al. 2013). The feasibility of PEG-functionalized rGO–PEG coated with Au nano rods were coupled with Tat protein for the photothermal killing of the cancerous cells (Turcheniuk et al. 2016). The Tat protein was utilized to improve the target selectivity of the synthesized composite. In vivo experiments revealed that the combination was effective in inhibiting the proliferation of the U87MG tumour cells in mice (Turcheniuk et al. 2016).

16.5 TOXICITY ASPECTS OF GOLD NANOPARTICLES (AUNPS) AND GRAPHENE (Gr) AND ITS DERIVATIVES (GO, rGO)

16.5.1 Cytotoxicity of AuNPs

16.5.1.1 Correlation between Particle Sizes of AuNPs and Cytotoxicity

The particle size of NPs is the most important characteristic to consider for determining the degree of cytotoxicity AuNPs. Because of their potential to penetrate into the cell membrane, nanoparticles having size less than 2 nm cause considerable cytotoxicity. When the particle sizes of AuNPs are greater than 10 nm, they are able to penetrate the nucleus (Kumar and Kwang-Hyun 2016). In 2007, it was revealed that cytotoxicity is dependent on the particles sizes of AuNPs.

The cytotoxicity of Au nano-clusters is much greater than that of the AuNPs when the diameter is larger than 10 nm. AuNPs with diameter 20 nm was further more toxic for the cells of ovary of the hamster compared to the AuNPs having sizes of 14 nm (Xia et al. 2016). It was discovered that Au nano rods exhibited the greatest cytotoxicity, whereas even at high concentrations, nano hexa-pods demonstrated no appreciable cytotoxic behaviour (Montasser et al. 2017). The aggregation of AuNPs inside the cells can generate reactive oxygen species leading to cell damage (Jia et al. 2015).

16.5.1.2 AuNPs Surface Charges

Phosphocytosis of AuNPs and their distribution inside the cells are both determined by the surface charges of AuNPs. By coating AuNPs with electrolyte, the researchers have discovered that altering the surface charges of AuNPs can influence the phagocytosis in mammal cells (Facchi et al. 2017). Numerous findings have demonstrated that the cytotoxicity of positive cations of AuNPs is greater than that of the anionic and neutrally charged AuNPs. However, the link between the biocompatibility and the charge density has remained a mystery until now (Tang et al. 2018). As discovered earlier, AuNPs with positive charges are readily phagocytosed by cells and exhibit substantial cytotoxicity. It reflects that the proportion of phagocytosed AuNPs by cells is proportional to the net charge (Q_{net}) of AuNPs (Luan et al. 2016).

16.5.1.3 AuNPs Surface Modification and Cytotoxicity

Surface of AuNPs can be modified using ligands or stabilizers. This will aid in targeting the cancer cells. The variety of ligands used also can influence the cytotoxic behaviour of AuNPs (Xia et al. 2013). The CTAB-functionalized AuNPs showed a strong inhibitory effect on PC-3 tumour growth, but the AuNPs modified using polypeptide YSA- have a lower inhibitory impact on cell development. Biocompatible type ligands are usually effective for lowering the cytotoxicity of AuNPs. However, grafting with additional dendrimers over the modified interface of glutathione/AuNPs may augment the toxicity level of the cell (Chuang et al. 2017).

16.5.2 Toxicity Aspects of Graphene (Gr) and Its Derivatives (GO, rGO)

It has been demonstrated that Gr and its derivatives (GO and rGO) have a permissible biocompatibility, with some limitations (Dong and Qi 2015). Pristine graphene has been shown to have a favourable safety profile at lower concentrations, but it also has the tendency to impair cell growth owing to its mechanical properties and hydrophobicity, which aids in agglomeration and entrapment within the tissues (Liao et al. 2011; Dong and Qi 2015; Olteanu et al. 2015). As a result, hydrophilic Gr has been preferred over pure Gr in the large proportion of biomedical applications (Chang et al. 2011; Pinto et al. 2013).

In vivo cytotoxicity of GO was also investigated using a variety of animal species. The researchers observed long-term toxicity with an acute inflammation in the lungs of Kunming mice when higher intravenous dose of GO (0.4 mg) was injected. Clinical and pathological evidence indicates that GO sheets are collected in the lungs and liver, and they can be eliminated with bile (Guo and Mei 2014). On the other hand, no substantial degenerative visual alterations were observed due to intravenous administration of GO in Japanese rabbits (Guo and Mei 2014). These findings interdependently suggest that in vivo GO behaviour is dose-dependent, route-dependent, time-dependent, and dependent on the animal species used during the assays (Wang et al. 2011; Zhang et al. 2011).

Despite the fact that there are ongoing debates on GO cytotoxicity, it is stated to have dose-dependent toxicity with presumably beneficial bioactivity (Dong and Qi 2015). Furthermore, GO has overcome these barriers and increased its safety through purification and functionalization techniques. The inclusion of GO by covalent or noncovalent coupling can alter the morphology and improve the biocompatibility (Pinto et al. 2013; Guo and Mei 2014). PEG is a unique polymer in such sector, creating better soluble, safer, biocompatible, and less-poisonous composites based on GO. In conclusion, published research demonstrates that GO possesses antimicrobial characteristics (Liu et al. 2011; Ruiz et al. 2011). It can be biocompatible when properly concentrated and functionalized (Pinto et al. 2013; Guo and Mei 2014).

16.6 APPLICATION OF Gr–AuNPs IN CANCER TREATMENT

16.6.1 Detection of Cancer

Cancer is a leading reason for death globally, with an anticipated 17 million casualties in 2030, as stated by WHO (Yan et al. 2013). Early-stage, precise, and specific diagnostic methods should be devised to combat cancer. At the moment, tumour markers and cancer cells can be identified using a variety of approaches, notably chemi-luminescence, radio-immunoassays, and flow-cytometric approach (Yan et al. 2013; Zhang et al. 2014). Nevertheless, there is a growing desire for detection technologies that are easier, more precise, robust, and less costly (Zhang et al. 2014).

NPs have made significant advances in the science of biosensing, but between many of them, Gr/Au based sensors are being researched extensively owing to their synergistic impact of both Gr and AuNPs based constituents. Biosensors are often categorized according to their bio-receptor and transducer type (Kumar et al. 2015). Gr/Au based electro-chemical biosensors are the most prominent one. The

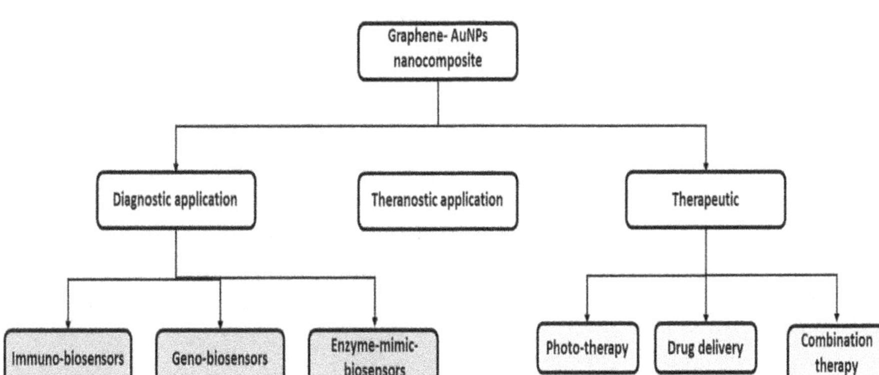

FIGURE 16.3 Application of Gr–AuNPs based composites for cancer treatment.

following section discusses several types of Gr–Au based biosensors currently being designed and employed in cancer research. They are classified here according to the category of bio-receptor employed in assembly for convenience (Figure 16.3).

16.6.1.1 Immuno-Sensors

The phrase immuno-sensor refers to the employment of immunological agents to recognize target tumour chemicals. NPs will connect the bio-elements with the surface of the electrode, hence enhancing the response and signal-amplification values (Kumar et al. 2015). In the literature, Gr–AuNPs have been utilized to manufacture a wide variety of immuno-sensors, one of which is the sandwiched type test. Gr/Au can be used as a substrate for immobilizing a huge number of antibodies for improving the conductivity, or as tracing probes containing additional bioreceptor molecules (Feng et al. 2011; Filip et al. 2015). In both cases, the manufactured biosensors exhibited enhanced sensitivity, signal amplification, and lower limit of detection (Feng et al. 2016; Zhou et al. 2016). Due to the flexibility and effectiveness of electrochemical immuno-sensing, blood samples from real-world patients were examined using the Gr/Au biosensor (Li et al. 2013; Yan et al. 2013). Although additional sample verification is required, these data strongly suggest that Gr/Au based composites are close to clinical testing and legitimate applications (Han et al. 2013).

16.6.1.2 Geno Sensors

When contrasted with traditional microarrays and other approaches, molecular markers are advisable recognition spots because they provide strong sensitivity and selectivity toward tumour cells (Yan et al. 2013; Liu et al. 2014). Using these markers can eliminate misleading results and toxicities and also can save time.

Conversely, researchers have effectively mounted graphene-based nanomaterials on the surface of the electrode and then doped them with gold moieties conjugated with thiolated DNA sequences and/or aptamers to enhance their electrochemical properties (Benvidi et al. 2015; Kumar et al. 2015; Azimzadeh et al. 2016). Enhanced conductivity and an increased number of acquisition probes existing over the Gr/Au biosensor surface tends to result in excellent specificity for supplementary sequences identification, acceleration of recombination time, and ultra-sensitivity (Benvidi et al. 2015; Filip et al. 2015).

Earlier, a nano-biosensor relying on rGO enclosed electrode with conjugated dsDNA-Au was developed (Ilkhani et al. 2016). It was used to investigate the efficacy of doxorubicin against the breast cancer gene (BRCA1) (Ilkhani et al. 2016). This type of biosensor not only enables the identification of cancer, but it also facilitates the effectiveness of anticancer drugs. These sensors can be used in the formulation of anticancer medications, and they can serve as the initial phase of drug screening by examining the interactions of the medication with the targeted DNA.

16.6.1.3 Enzyme Based Sensors

Folic acid–GO–AuNCs were conceived as a potential peroxidase alternative for cancer cell identification (Tao et al. 2013). Owing to their remarkable catalytic properties, as well as substrate selectivity, naturally occurring enzymes have long been a subject of preference in the bio-sensing process (Tao et al. 2013). Nevertheless, some disadvantages, like weather sensitivity, inactivation/denaturation, and expensive processing costs, limited their utilization. As an option, enzyme mimics have the potential to improve the performance. Gr–AuNPs provide an outstanding synthetic peroxidase-like enzymatic mimic with variable activities, low cost of production, and functionality over a wide pH range with high stability.

16.6.2 Photothermal Therapy (PTT)

This method of therapy virtually guarantees specific cancer cell annihilation by exposing the carcinoma cell/nano-composite combination under near infrared (NIR) laser, which can induce localized hyperthermia. NIR light significantly decreases incidental damage of normal tissues along the pathway, whereas heat causes cleavage of intracellular proteins and cellular malfunction (Nergiz et al. 2014). GO–AuNPs are the most extensively investigated nanoparticles in this study because they are an effective photo-thermal agent (PT). Furthermore, when Gr–AuNPs were used for PTT of cancer cells, a greater rate of heating was reported than when AuNPs were used alone (Bao et al. 2016). This effect is due to the interaction of Gr and Au plasmons, significant NIR photonic absorption, effective NIR light-to-heat transformation, and biocompatibility and rapid heat transmittance with minimum resistance across the graphene layer (Moon et al. 2015; Yang et al. 2015).

16.6.3 Photo-Dynamic Therapy (PDT)

NPs conjugated with photo-sensitizer drugs (PS) have been introduced extensively to alleviate some of the disadvantages of PS therapies, including extended tissue aggregation, cytotoxicity, insufficient solubility, and specificity (Shibu et al. 2013). Previously, the GO–AuNPs were used in conjunction with PTT and PDT. In that case, gel-like composition of amaranth extracts and GO–AuNPs were utilized to target cervical Hela carcinoma cells multimodally (Chang et al. 2015). With their well-established photothermal activities, the GO–AuNPs combination enhanced ROS production, resulting in a remarkably effective anticancer treatment (Chang et al. 2015). A sequence of photoreactions induced by light encapsulates the central, basic assumption of PDT, which generates free radicals that aid in the eradication of malignant cells (Shibu et al. 2013). A photo-sensitizer (PS) medication is bombarded in this type of cell to initiate reactions that disrupt the biomolecules at specific diffusing boundaries.

16.6.4 Drug Delivery

The Gr–AuNPs system packed with daunorubicin (DNR) drugs was discovered to induce pathological alterations in carcinoma cell membranes, allowing the release of drug inside the cells. In vivo and in vitro assays have revealed that this combination significantly reduced P-gp activity and triggered apoptosis in cancerous cells (Zhang et al. 2013). Within hepato-carcinoma (HepG2) cells, gold nanocluster-loaded rGO (rGO-GNC) was investigated as a carrier for doxorubicin (DOX) drugs. Raman spectroscopy was used to constantly monitor the functional delivery of drugs. Additionally, a synergistic trend was demonstrated when this composite was used in conjunction with DOX instead of its individual components, where the freshly formulated rGO-GNC demonstrated the capacity to stimulate karyopyknosis and inhibit HepG2 cells when ingested on its own (without the drug) at higher concentrations (Wang et al. 2015).

16.6.5 Combined Therapy

Numerous combined PTT and PDT based methods were reported using the Gr–AuNPs for cancer treatment (Chang et al. 2015). Earlier, drug-loaded Gr–AuNPs vesicles were developed (Song et al. 2015). This technique was superior in numerous respects; for example, the drug was placed on both the exterior and cavities, significantly boosting the drug-loading profile. Secondly, successive release of drugs was detected following irradiation under NIR. Acidic environment based on tumour was developed, establishing a framework for a distantly controlled drug-delivery system, when needed. Finally, the combination of chemotherapy and PTT provided an efficient cancer treatment approach. This method has been demonstrated to be effective in both in vivo and in vitro scenarios (Song et al. 2015). Likewise, a different study demonstrated the coupling of gene therapy and PTT using a Gr–AuNPs based approach (Cheng et al. 2013).

16.6.6 Theranostic Application

Multidisciplinary image-processing/therapy tools are a revolutionary sector that has the potential to significantly increase therapeutic advantages (Jin et al. 2013). The very first positive aspect that needs to be mentioned is the avoidance of numerous agent infusions inside the body, hence minimizing stress, level of toxicity, and complications related to health (Jin et al. 2013). Furthermore, Gr–AuNPs composites, due to their known enzyme-mimicking aptitude, provide a safer and effective option for imaging and treating cancer cells concurrently. A reliable quick colorimetric-detecting agent, in conjunction with a mediated treatment method based on per-oxidase-like activities, was developed earlier (Jin et al. 2013). The synthesized Gr–Si–AuNPs with cellular-targeting folic acid (GSF@AuNPs) in presence of H_2O_2 catalyzed the oxidative degradation of TMB (a per-oxidase substance). The resultant products produce a blue product or endogenous type reactive oxygen species (ROS), which results in an increased cytotoxicity level of cervical carcinoma cells (Jin et al. 2013).

16.7 ENVIRONMENTAL/ECO-TOXICOLOGICAL HAZARDS OF Gr/GO/rGO BASED NANOCOMPOSITES

Gr/GO/rGO is a novel nano-structured carbon that is undergoing rapid development. While Gr/GO/rGO and their composites have a vast range of applications in several sectors of science, they also have harmful effect on variety of biological models. Increased generation of Gr/GO/rGO based composites and their anticipated application in biomedical activities have sparked concern regarding their potential adverse effects on human health and the environment. To enhance biological applicability of Gr-based composites, it is essential to comprehend their interactions with living organisms. Although the physiological impacts of Gr-based nanocomposites have been addressed at the cellular and animal levels, the extent to which these NPs are absorbed into the human body is unspecified. Humans can be exposed to GBNs in a variety of ways, ranging from the point of manufacture through the environment (Syama and Mohanan 2016). As a result, the ecosystem's biotic and abiotic factors will be disrupted.

It is critical to examine how Gr and its derivatives-based composites will interact with membranes in the ecosystems to determine the risk posed by releasing them into the environment. There have been very few reports on the environmental impact of Gr/GO/rGO based nanocomposites. GO is suspected to be more toxic than Gr and rGO. The fate and transportation phenomenon of GO have been examined earlier (Chowdhury et al. 2013; Chowdhury et al. 2014; Zhao et al. 2014; Chowdhury et al. 2015). The effect of solar radiation on the physico-chemical characteristics, agglomeration, and precipitation of GO in aquatic ecosystems was studied earlier (Chowdhury et al. 2015). They concluded that sunlight radiation has a substantial effect on the physio-chemical characteristics of GO and their eventual transportation by lowering the nanomaterials' persistence in the environment. It is necessary to perform additional study to better understand the complex interactions between pH, natural organic matter, and other natural colloids that affect the status of photo-transformed GO. It was discussed how GO can be converted to rGO in the presence of aquatic microorganisms. Researchers have investigated how GO aggregates and resuspends in replicated natural aqueous environments (Zhao et al. 2014). The results demonstrated that the transportation and distribution of GO nanoparticles have a considerable effect on natural organic matter (NOM), divalent ions, and hydraulic systems in natural aquatic system (Hua et al. 2015).

The acute impact of GO on the microbial population in polluted water was examined (Ahmed and Rodrigues 2013). They revealed that in doses ranging from 50 to 300 mg/L of GO was harmful for microbial populations. The effluent's quality was degraded by raising the turbidity of the water and decreasing the dewaterability of the sludge. Additionally, they established that the formation of reactive oxygen species is accountable for the cytotoxicity of GO on microorganisms. Researchers have investigated the characteristic parameters of GO in the environment, including its hazardous effect, destiny, and exposures, in order to conduct a life-cycle analysis of GO-based NPs (Deng et al. 2016). Additional research is required as there is relatively limited research on the environmental risks associated with Gr-based composites and the stringent restrictions on their discharge to counteract their hazardous consequences (Liu et al. 2011; Li et al. 2013; Liu et al. 2013).

16.8 CHALLENGES AND FUTURE PERSPECTIVE OF Gr–AuNPs BASED COMPOSITES FOR CANCER TREATMENT

Despite the immense potential of Gr–AuNPs based composites as cancer therapeutic and diagnostic agents, some challenges require further experimentation (Ahmed and Rodrigues 2013). The primary hurdle for Gr–AuNPs composites is the preparation of nanomaterials (Shervedani and Amini 2014; Shi et al. 2017). Effective biosensors and drug-transportation systems typically require consistent, well-controlled synthesis protocols, which can yield AuNPs with standardized shape and dissemination over GO sheets. However, newly generated Gr–AuNPs based composites for cancer treatment are being extensively examined in vivo and in vitro research at the preclinical stage. Yet, additional attention is required to acquire correlated data for multi-species, as well as reliable estimation for long-term toxicities, tumour-targeting efficiency (Shi et al. 2017), absorption, and dispersion inside the tumour tissue in order to promote clinical practical applications (Rahmanian et al. 2017).

Although the observations from the clinical sector remain unexplored, the biosensing therapeutic arena is advancing (Li and Liu 2017). Numerous recent investigations have described the use of Gr–AuNPs based biosensors using human-derived materials (Feng et al. 2011). Additionally, a Gr–AuNPs sensor coupled with antibodies was used to precisely detect tumour cells in blood samples from patients having pancreatic, breast, and lung cancer (Yoon et al. 2013). These findings indicate that this type of biosensor has enormous potential for clinical uses, with particular emphasis on the sample type and empirical protocol assessment and validation (Shervedani and Amini 2014).

16.9 CONCLUSION

In fact, the development of Gr–AuNPs based composites is expanding at a rapid pace. This chapter provided an in-depth assessment of the synthesis protocol of Gr–AuNPs based composites and their versatile applications. The methods for synthesizing AuNPs having rod shapes were described in detail, including the hard template approach, the electrochemical technique, the photochemical, seed-mediated, and the seedless approach. Additionally, the capping of AuNPs with GO was addressed, with an emphasis on the interactions between the two components, including the mechanism to minimize AuNPs aggregation over the GO sheet. Ex situ and in situ techniques, as well as the utilization of linkers, stabilizers, and functionalization agents, were discussed. Additionally, the cytotoxicity of Gr and AuNPs was studied, revealing the composite's biocompatibility at both cellular and animal level. According to various findings, the photothermal characteristics of Gr–AuNPs indicate that Gr or its derivatives of GO and rGO not only enable AuNPs biocompatibility but also enhance their photothermal qualities. Additionally, many applications of Gr–AuNPs were reviewed, including photothermal therapy, cancer treatment, theranostics, sensors, and bio-imaging techniques. While AuNPs–Gr-based composites have demonstrated a variety of activities for cancer treatment, some challenges remain unresolved. They arise from the weak interaction of AuNPs with Gr, GO, and rGO matrix, which results in clustering and an uneven distribution of AuNPs over the graphene derivatives. The cytotoxicity of this composite requires more investigation based on the mechanism of interactions between the living cell and the composite matrix with their cellular-absorption properties.

The future of cancer therapy involves the optimal specific termination of tumour cells, with no regrowth and evolution of resistance to the treatment approach. There are currently no treatments that provide either of these alternatives. Furthermore, the preventive strategy is the way forward for nano-oncology, a word that encompasses a perception for the most effective cancer eradication. Numerous researches have demonstrated the efficacy of nano-systems to carry genetic material, administer medication, and silence genes in order to combat cancer mutations. With an understanding of these critical considerations, Gr–AuNPs based composites are unquestionably one of the most promising applications for such advancements in science and technology today. Gr–AuNPs composites have already contributed significantly to cancer treatment, but there are certainly new frontiers to be discovered.

REFERENCES

Ahmed, F. and Rodrigues, D.F. 2013. Investigation of acute effects of graphene oxide on wastewater microbial community: A case study. *Journal of Hazardous Material* 256–257:33–39.

Al-Ani, L.A., AlSaadi, M.A., Kadir, F.A., et al. 2017. Graphene/gold based nanocomposites applications in cancer diseases; efficient detection and therapeutic tools. *European Journal of Medicinal Chemistry* 139:349–366.

Ali, M.R., Rahman, M.A., Wu, Y., et al. 2017. Efficacy, long-term toxicity, and mechanistic studies of gold nano rods photothermal therapy of cancer in xenograft mice. *Proceedings of the National Academy of Sciences of the United States of America* 114: E3110–E3118.

Allen, J.M., Xu, J., Blahove, M., et al. 2017. Synthesis of less toxic gold nanorods by using dodecylethyldimethylammonium bromide as an alternative growth-directing surfactant. *Journal of Colloid and Interface Science* 505:1172–1176.

Arvand, M. and Gholizadeh, T.M. 2013. Gold nanorods–graphene oxide nanocomposite incorporated carbon nanotube paste modified glassy carbon electrode for voltammetric determination of indomethacin. *Sensors and Actuators B* 186:622–632.

Azimzadeh, M., Rahaie, M., Nasirizadeh, N., Ashtari, K. and Naderi-Manesh, H. 2016. An electrochemical nanobiosensor for plasma miRNA-155, based on graphene oxide and gold nanorod, for early detection of breast cancer. *Biosensors and Bioelectronics* 77: 99–106.

Bai, S. and Shen, X. 2012. Graphene-inorganic nanocomposites. *RSC Advances* 2:64–98.

Bao, Z., Liu, X., Liu, Y., Liu, H. and Zhao, K. 2016. Near-infrared light-responsive inorganic nanomaterials for photothermal therapy. *Asian Journal of Pharmaceutical Science* 11: 349–364.

Benvidi, A., Firouzabadi, A.D., Moshtaghiun, S.M., Mazloum-Ardakani, M. and Tezerjani, M.D. 2015. Ultrasensitive DNA sensor based on gold nanoparticles/ reduced graphene oxide/glassy carbon electrode. *Analytical Biochemistry* 484:24–30.

Boisseau, P. and Loubaton, B. 2011. Nanomedicine, nanotechnology in medicine. *Comptes Rendus Phys.* 12:620–636.

Burrows, N.D., Harvey, S., Idesis, F.A. and Murphy, C.J. 2017. Understanding the seed-mediated growth of gold nanorods through a fractional factorial design of experiments. *Langmuir* 33:1891–1907.

Cao, J., Sun, T. and Grattan, K.T. 2014. Gold nanorod-based localized surface plasmon resonance biosensors: A review. *Sensors and Actuators B* 195:332–351.

Chang, G., Wang, Y., Gong, B., et al. 2015. Reduced graphene oxide/amaranth extract/AuNPs composite hydrogel on tumor cells as integrated platform for localized and multiple synergistic therapy. *ACS Applied Materials Interfaces* 7:11246–11256.

Chang, Y., Yang, S.-T., Liu, J.-H., et al. 2011. In vitro toxicity evaluation of graphene oxide on A549 cells. *Toxicology Letters* 200:201–210.

Cheng, F.F., Chen, W., Hu, L.H., et al. 2013. Highly dispersible PEGylated graphene/Au composites as gene delivery vector and potential cancer therapeutic agent. *Journal of Materials Chemistry B* 1:4956–4962.

Chowdhury, I., Duch, M.C., Mansukhani, N.D., Hersam, M.C. and Bouchard, D. 2013. Colloidal properties and stability of graphene oxide nanomaterials in the aquatic environment. *Environmental Science Technology* 47(12):6288–6296.

Chowdhury, I., Duch, M.C., Mansukhani, N.D., Hersam, M.C. and Bouchard, D. 2014. Interactions of graphene oxide nanomaterials with natural organic matter and metal oxide surfaces. *Environmental Science Technology* 48(16):9382–9390.

Chowdhury, I., Hou, W.C., Goodwin, D., Henderson, M., Zepp, R.G. and Bouchard, D. 2015. Sunlight affects aggregation and deposition of graphene oxide in the aquatic environment. *Water Research* 78:37–46.

Chuang, S.T., Shon, Y.S. and Narayanaswami, V. 2017. Apolipoprotein E3-mediated cellular uptake of reconstituted high-density lipoprotein bearing core 3, 10, or 17 nm hydrophobic gold nanoparticles. *International Journal of Nanomedicine* 12:8495–8510.

De Melo-Diogo, D., Pais-Silva, C., Dias, D.R., Moreira, A.F. and Correia, I.J. 2017. Strategies to improve cancer photothermal therapy mediated by nanomaterials. *Advanced Healthcare Materials* 6:1700073.

Dembereldorj, U., Choi, S.Y., Ganbold, E.O., et al. 2014. Gold nanorod-assembled PEGylated graphene-oxide nanocomposites for photothermal cancer therapy. *Photochemistry and Photobiology* 90:659–666.

Deng, Y., Li, J., Qiu, M., Yang, F., Zhang, J. and Yuan, C. 2016. Deriving characterization factors on freshwater ecotoxicity of graphene oxide nanomaterial for life cycle impact assessment. *International Journal of Life Cycle Assessment* 22(2):222–236.

Dideikin, A.T. and Vul, A.Y. 2019. Graphene oxide and derivatives: the place in graphene family. *Frontiers Physics* 6:149.

Dong, H. and Qi, S. 2015. Realising the potential of graphene-based materials for bio surfaces: A future perspective. *Biosurface and Biotribology* 1:229–248.

Elahi, N., Kamali, M. and Baghersad, M.H. 2018. Recent biomedical applications of gold nanoparticles: A review. *Talanta* 184:537–556.

Facchi, D.P., Da, C.J., Bonafé, E.G., et al. 2017. Polysaccharide-based materials associated with or coordinated to gold nanoparticles: Synthesis and medical application. *Current Medicinal Chemistry* 24(25):2701–2735.

Fard, J.K., Jafari, S. and Eghbal, M.A. 2015. A review of molecular mechanisms involved in toxicity of nanoparticles. *Advanced Pharmaceutical Bulletin* 5:447.

Feng, L., Chen, Y., Ren, J. and Qu, X. 2011. A graphene functionalized electrochemical aptasensor for selective label-free detection of cancer cells. *Biomaterials* 32:2930–2937.

Feng, T., Qiao, X., Wang, H., Sun, Z., Qi, Y. and Hong, C. 2016. An electrochemical immunosensor for simultaneous point-of-care cancer markers based on the hosteguest inclusion of β-cyclodextrinegraphene oxide. *Journal of Materials Chemistry B* 4: 990–996.

Filip, J., Kasak, P. and Tkac, J. 2015. Graphene as signal amplifier for preparation of ultrasensitive electrochemical biosensors. *Chemical Papers* 69:112–133.

Foss, C.A., Hornyak, G.L., Stockert, J.A. and Martin, C.R. 1993. Optically transparent nanometal composite membranes. *Advanced Materials* 5:135–136.

Foss, C.A. Jr., Hornyak, G.L., Stockert, J.A. and Martin, C.R. 1994. Template-synthesized nanoscopic gold particles: Optical spectra and the effects of particle size and shape. *Journal of Physical Chemistry* 98:2963–2971.

Frohlich, E. 2013. Cellular targets and mechanisms in the cytotoxic action of non-biodegradable engineered nanoparticles. *Current Drug Metabolism* 14:976–988.

Gao, J., Bender, C.M. and Murphy, C.J. 2003. Dependence of the gold nanorod aspect ratio on the nature of the directing surfactant in aqueous solution. *Langmuir* 19:9065–9070.

Guo, X. and Mei, N. 2014. Assessment of the toxic potential of graphene family nanomaterials. *Journal of Food Drug Analysis* 22:105–115.

Han, X., Fang, X., Shi, A., Wang, J. and Zhang, Y. 2013. An electrochemical DNA biosensor based on gold nanorods decorated graphene oxide sheets for sensing platform. *Analytical Biochemistry* 443:117–123.

Hlapisi, N., Motaung, T.E., Linganiso, L.Z., Oluwafemi, O.S. and Songca, S.P. 2019. Encapsulation of gold nanorods with porphyrins for the potential treatment of cancer and bacterial diseases: A critical review. *Bioinorganic Chemistry and Application* 2019: 7147128.

Hua, Z., Tang, Z., Bai, X., Zhang, J., Yu, L. and Cheng, H. 2015. Aggregation and re-suspension of graphene oxide in simulated natural surface aquatic environments. *Environmental Pollution* 205:161–169.

Huang, X., Neretina, S. and El-Sayed, M.A. 2009. Gold nanorods: From synthesis and properties to biological and biomedical applications. *Advanced Materials* 21:4880–4910.

Ilkhani, H., Hughes, T., Li, J., Zhong, C.J. and Hepel, M. 2016. Nanostructured SERS-electrochemical biosensors for testing of anticancer drug interactions with DNA. *Biosensors and Bioelectronics* 80:257–264.

Jana, N.R., Gearheart, L. and Murphy, C.J. 2001a. Wet chemical synthesis of high aspect ratio cylindrical gold nano rods. *Journal of Physical Chemistry B* 105: 4065–4067.

Jana, N.R., Gearheart, L. and Murphy, C.J. 2001b. Seed-mediated growth approach for shape-controlled synthesis of spheroidal and rod-like gold nanoparticles using a surfactant template. *Advanced Materials* 13:1389–1393

Jia, H., Chen, W., Wang. X., et al. 2015. Virus-surface-mimicking surface clustering of AuNPs onto DNA-entrapped polymeric nanoparticle for enhanced cellular internalization and nanocluster-induced NIR photothermal therapy. *Advanced Science* 2(12):1500108.

Jin, Y., Wang J., Ke, H., Wang, S. and Dai, Z. 2013. Graphene oxide modified PLA microcapsules containing gold nanoparticles for ultrasonic/CT bimodal imaging guided photothermal tumor therapy. *Biomaterials* 34:4794–4802.

Jones, C.F. and Grainger, D.W. 2009. In vitro assessments of nanomaterial toxicity. *Adv. Drug Delivery Reviews* 61:438–456.

Khan, M.S., Pandey, S., Bhaisare, M.L., Gedda, G., Talib, A. and Wu, H.-F. 2017. Graphene oxide@gold nanorods for chemo-photothermal treatment and controlled release of doxorubicin in mice Tumor. *Colloids Surface B Biointerfaces* 160:543–552.

Kumar, P.J. and Kwang-Hyun, B. 2016. Comparative study of proteasome inhibitory, synergistic antibacterial, synergistic anticandidal, and antioxidant activities of gold nanoparticles biosynthesized using fruit waste materials. *International Journal of Nanomedicine* 11:4691–4705.

Kumar, S., Ahlawat, W., Kumar, R. and Dilbaghi, N. 2015. Graphene, carbon nanotubes, zinc oxide and gold as elite nanomaterials for fabrication of biosensors for healthcare. *Biosensors and Bioelectronics* 70:498–503.

Li, T. and Liu, Z. 2017. *Outlook and Challenges of Nano Devices, Sensors, and MEMS.* Cham, Switzerland: Springer.

Li, W., Li, M., Ge, S., Yan, M., Huang, J. and Yu, J. 2013. Battery-triggered ultrasensitive electrochemiluminescence detection on microfluidic paper-based immunodevice based on dual-signal amplification strategy. *Analytica Chimica Acta* 767:66–74.

Liao, K.-H., Lin, Y.-S., Macosko, C.W. and Haynes, C.L. 2011. Cytotoxicity of graphene oxide and graphene in human erythrocytes and skin fibroblasts. *ACS Applied Materials Interfaces* 3:2607–2615.

Lim, D.-K., Barhoumi, A., Wylie, R.G., Reznor, G., Langer, R.S. and Kohane, D.S. 2013. Enhanced Photothermal effect of plasmonic nanoparticles coated with reduced graphene oxide. *Nano Letters* 13:4075–4079.

Liu, F., Zhang, Y., Yu, J., Wang, S., Ge, S. and Song, X. 2014. Application of ZnO/graphene and S6 aptamers for sensitive photoelectrochemical detection of SK-BR-3 breast cancer cells based on a disposable indium tin oxide device. *Biosensors and Bioelectronics* 51:413–420.

Liu, J., Cui, L. and Losic, D. 2013. Graphene and graphene oxide as new nanocarriers for drug delivery applications. *Acta Biomaterials* 9:9243–9257.

Liu, S., Zeng, T.H., Hofmann, M., et al. 2011. Antibacterial activity of graphite, graphite oxide, graphene oxide, and reduced graphene oxide: Membrane and oxidative stress. *ACS Nano* 5:6971–6980.

Lu, W., Ning, R., Qin, X., et al. 2011. Synthesis of Au nanoparticles decorated graphene oxide nanosheets: Noncovalent functionalization by TWEEN 20 in situ reduction of aqueous chloroaurate ions for hydrazine detection and catalytic reduction of 4-nitrophenol. *Journal of Hazardous Materials* 197:320–326.

Luan, B., Friedrich, T., Zhai, J., et al. 2016. A library of AuNPs modified by RAFT polymers of different charge and chain length: High throughput synthesis and synchrotron XFM imaging using a zebrafish larvae model. *RSC Advances* 6:23550–23563.

Ma, X., Qu, Q., Zhao, Y., et al. 2013. Z. Graphene oxide wrapped gold nanoparticles for intracellular Raman imaging and drug delivery. *Journal of Materials Chemistry B* 1:6495–6500.

Montasser, M.S., Hegazi, M.M., Younes, A.M., et al. 2017. Effect of gold chloride concentration and volume on size and shape of biological synthesized gold nanoparticles (AuNPs) using red algae (*Laurencia papillosa*). *Science of Advanced Materials* 9:1105–1113.

Moon, H., Kumar, D., Kim, H., et al. 2015. Amplified photoacoustic performance and enhanced photothermal stability of reduced graphene oxide coated gold nanorods for sensitive photoacoustic imaging. *ACS Nano* 9:2711–2719.

Moussawi, R.N. and Patra, D. 2015. Synthesis of Au nanorods through prereduction with curcumin: Preferential enhancement of Au nanorod formation prepared from CTAB-capped over citrate-capped Au seeds. *Journal of Physical Chemistry C* 199: 19458–19468.

Nergiz, S.Z., Gandra, N., Tadepalli, S. and Singamaneni, S. 2014. Multifunctional hybrid nanopatches of graphene oxide and gold nanostars for ultraefficient photothermal cancer therapy. *ACS Applied Material Interfaces* 6:16395–16402.

Olteanu, D., Filip, A., Socaci, C., et al. 2015. Cytotoxicity assessment of graphene-based nanomaterials on human dental follicle stem cells. *Colloids and Surfaces B: Biointerfaces* 136:791–798.

Pandey, P.A., Bell, G.R., Rourke, J.P., et al. 2011. Physical vapor deposition of metal nanoparticles on chemically modified graphene: Observations on metalegraphene interactions. *Small* 7:3202–3210.

Pattani, V.P. and Tunnell, J.W. 2012. Nanoparticle-mediated photothermal therapy: A comparative study of heating for different particle types. *Lasers in Surgery and Medicine*, 44, 675–684.

Pérez-Juste, J., Pastoriza-Santos, I., Liz-Marzán, L.M. and Mulvaney, P. 2005. Gold nanorods: Synthesis, characterization and applications. *Coordination Chemistry Reviews* 249:1870–1901.

Pham, T.A., Choi, B.C., Lim, K.T. and Jeong, Y.T. 2011. A simple approach for immobilization of gold nanoparticles on graphene oxide sheets by covalent bonding. *Applied Surface Sciences* 257:3350–3357.

Pinto, A.M., Gonçalves, I.C. and Magalhães, F.D. 2013. Graphene-based materials biocompatibility: A review. *Colloids Surface B* 111:188–202.

Qin, Y., Li, J., Kong, Y., et al. 2012. In situ growth of Au nanocrystals on graphene oxide sheets. *Nanoscale* 6:1281–1285.

Qiu, X., You, X., Chen, X., et al. 2017. Development of graphene oxide-wrapped gold nanorods as robust nanoplatform for ultrafast near-infrared SERS bioimaging. *International Journal of Nanomedicine* 12:4349–4360.

Rahmanian, N., Eskandani, M., Barar, J. and Omidi, Y. 2017. Recent trends in targeted therapy of cancer using graphene oxide-modified multifunctional nanomedicines. *Journal of Drug Targeting* 25:202–215.

Robinson, J.T., Tabakman, S.M., Liang, Y., et al. 2011. Ultrasmall reduced graphene oxide with high near-infrared absorbance for photothermal therapy. *Journal of American Chemical Society* 133:6825–6831.

Ruiz, O.N., Fernando, K.S., Wang, B., et al. 2011. Graphene oxide: A nonspecific enhancer of cellular growth. *ACS Nano* 5:8100–8107.

Sharma, P., Darabdhara, G., Reddy, T.M., et al. 2013. Synthesis, characterization and catalytic application of Au NPs-reduced graphene oxide composites material: An eco-friendly approach. *Catalyst Communication* 40:139–144.

Shervedani, R.K. and Amini, A. 2014. Novel graphene-gold hybrid nanostructures constructed via sulfur modified graphene: Preparation and characterization by surface and electrochemical techniques. *Electrochimica Acta* 121:376–385.

Shi, J., Kantoff, P.W., Wooster, R. and Farokhzad, O.C. 2017. Cancer nanomedicine: progress, challenges and opportunities. *Nature Reviews Cancer* 17:20–37.

Shibu, E.S., Hamada, M., Murase, N. and Biju, V. 2013. Nanomaterials formulations for photothermal and photodynamic therapy of cancer. *Journal of Photochemistry and Photobiology C Photochemistry Reviews* 15:53–72.

Singh, D.P., Herrera, C.E., Singh, B., Singh, S., Singh, R.K. and Kumar, R. 2018. Graphene oxide: An efficient material and recent approach for biotechnological and biomedical applications. *Materials Science Engineering C* 86:173–197.

Song, J., Yang, X., Jacobson, O., et al. 2015. Sequential drug release and enhanced photothermal and photoacoustic effect of hybrid reduced graphene oxide-loaded ultra-small gold nanorod vesicles for cancer therapy. *ACS Nano* 9:9199–9209.

Sun, B., Wu, J., Cui, S., et al. 2017. In situ synthesis of graphene oxide/gold nanorods theranostic hybrids for efficient tumor computed tomography imaging and photothermal therapy. *Nano Research* 10:37–48.

Syama, S. and Mohanan, P.V. 2016. Safety and biocompatibility of graphene: A new generation nanomaterial for biomedical application. *International Journal of Biological Macromolecule* 86:546–555.

Tang, S., Shen, H., Hao, Y., et al. 2018. A novel cytosensor based on Pt@Ag nanoflowers and AuNPs/Acetylene black for ultrasensitive and highly specific detection of circulating tumor cells. *Biosensors and Bioelectronics* 104:72–78.

Tao, Y., Lin, Y., Huang, Z., Ren, J. and Qu, X. 2013. Incorporating graphene oxide and gold nanoclusters: A synergistic catalyst with surprisingly high peroxidase-like activity over a broad pH range and its application for cancer cell detection. *Advanced Materials* 25:2594–2599.

Tomasella, P., Sanfilippo, V., Bonaccorso, C., et al. 2020. Theranostic nanoplatforms of thiolated reduced graphene oxide nanosheets and gold nanoparticles. *Applied Sciences* 10(16):5529.

Tsolekile, N., Parani, S., Matoetoe, M.C., Songca, S.P. and Oluwafemi, O.S. 2017. Evolution of ternary I–III–VI QDs: Synthesis, characterization and application. *Nano-Structures and Nano-Objects* 12:46–56.

Turcheniuk, K., Dumych, T., Bilyy, R., et al. 2016. Plasmonic photothermal cancer therapy with gold nanorods/reduced graphene oxide core/shell nanocomposites. *RSC Advances* 6:1600–1610.

Turcheniuk, K., Hage, C.-H., Spadavecchia, J., et al. 2015. J. Plasmonic photothermal destruction of uropathogenic *E. coli* with reduced graphene oxide and core/shell nanocomposites of gold nanorods/reduced graphene oxide. *Journal of Material Chemistry B* 3:375–386.

Vines, J.B., Yoon, J.-H., Ryu, N.-E., Lim, D.-J. and Park, H. 2019. Gold nanoparticles for photothermal cancer therapy. *Frontier Chemistry* 7:167.

Wang, K., Ruan, J., Song, H., et al. 2011. Biocompatibility of graphene oxide. *Nanoscale Research Letters* 6(1):8.

Wang, X., Han, Q., Yu, N., et al. 2015. Aptamer conjugated graphene oxide/gold nanocomposites for targeted chemo-photothermal therapy of cancer cells. *Journal of Materials Chemistry B* 3:4036–4042.

Xia, C., Jia, X., Han, J., et al. 2013. Electrochemical immunosensor for simultaneous detection of multiplex cancer biomarkers based on graphene nanocomposites. *Biosensors and Bioelectronics* 50:356–361.

Xia, Y., Wu, X., Zhao, J., et al. 2016. Three dimensional plasmonic assemblies of AuNPs with an overall size of sub-200 nm for chemo-photothermal synergistic therapy of breast cancer. *Nanoscale* 8:18682–18692.

Xu, C., Yang, D., Mei, L., et al. 2013. Encapsulating gold nanoparticles or nanorods in graphene oxide shells as a novel gene vector. *ACS Applied Materials Interfaces* 5:2715–2724.

Xu, X., Zhao, Y., Xue, X., et al. 2014. Seedless synthesis of high aspect ratio gold nano rods with high yield. *Journal of Material Chemistry A* 2:3528–3535.

Xu, Y., Chen, L., Ye, X., et al. 2017. Cooperative interactions among CTA+, Br– and Ag+ during seeded growth of gold nano rods. *Nano Research* 10:2146–2155.

Yan, M., Sun, G., Liu, F., Lu, J., Yu, J. and Song, X. 2013. An aptasensor for sensitive detection of human breast cancer cells by using porous GO/Au composites and porous PtFe alloy as effective sensing platform and signal amplification labels. *Analytica Chimica Acta* 798:33–39.

Yang, L., Tseng, Y.T., Suo, G., et al. 2015. Photothermal therapeutic response of cancer cells to aptameregold nanoparticle-hybridized graphene oxide under NIR illumination. *ACS Applied Material Interfaces* 7:5097–5106.

Yao, H., Jin, L., Sue, H.-J., Sumi, Y. and Nishimura, R. 2013. Facile decoration of Au nanoparticles on reduced graphene oxide surfaces via a one-step chemical functionalization approach. *Journal of Materials Chemistry A* 1:10783–10789.

Yin, P.T., Kim, T.-H., Choi, J.-W. and Lee, K.-B. 2013. Prospects for graphene nanoparticle-based hybrid sensors. *Physical Chemistry Chemical Physics* 15:12785–12799.

Yoon, H.J., Kim, T.M., Zhang, Z., et al. 2013. Sensitive capture of circulating tumour cells by functionalized graphene oxide nano-sheets. *Nature Nanotechnology* 8: 735–741.

Zhang, G., Chang, H., Amatore, C., Chen, Y., Jiang, H. and Wang, X. 2013. Apoptosis induction and inhibition of drug resistant tumor growth in vivo involving daunorubicin-loaded graphene-gold composites. *Journal of Materials Chemistry B* 1: 493–499.

Zhang, X., Li, F., Wei, Q., Du, B., Wu, W. and Li, H. 2014. Ultrasensitive nonenzymatic immunosensor based on bimetallic goldesilver nanoclusters synthesized by simple mortar grinding route. *Sensors Actuators B: Chemicals* 194:64–70.

Zhang, X., Yin, J., Peng, C., et al. 2011. Distribution and biocompatibility studies of graphene oxide in mice after intravenous administration. *Carbon* 49:986–995.

Zhang, Z., Shi, J., Song, Z., Zhu, X., Zhu, Y. and Cao, S. 2018. A synergistically enhanced photothermal transition effect from mesoporous silica nanoparticles with gold nanorods wrapped in reduced graphene oxide. *Journal of Materials Science* 53: 1810–1823.

Zhao, J., Wang, Z., White, J.C. and Xing, B. 2014. Graphene in the aquatic environment: Adsorption, dispersion, toxicity and transformation. *Environmental Science Technology* 48(17):9995–10009.

Zhou, J., Zheng, Y., Liu, J., Bing, X., Hua, J. and Zhang, H. 2016. A paper-based detection method of cancer cells using the photo-thermal effect of nanocomposite. *Journal of Pharmaceutical and Biomedical Analysis* 117:333–337.

Zhuo, Q., Ma, Y., Gao, J., et al. 2013. Facile synthesis of graphene/metal nanoparticle composites via self-catalysis reduction at room temperature. *Inorganic Chemistry* 52 3141–3147.

Zijlstra, P., Bullen, C., Chon, J.W. and Gu, M. 2006. High-temperature seedless synthesis of gold nanorods. *Journal of Physical Chemistry B* 110:19315–19318.

17 Surface Engineered Graphene Oxide and Its Derivatives

A Prospective Nano-Structured Carbon as Anticancer Agent

Zaira Zaman Chowdhury, Abu Nasser Faisal, Shahjalal Mohammad Shibly, Devarajan Thangadurai, Saher Islam, and Jeyabalan Sangeetha

CONTENTS

DOI: 10.1201/9781003136644-20

17.1 INTRODUCTION

The use of nanoscale materials in biomedical research has elicited exciting findings since they have unique size- and shape-dependent morphology and opto-chemical properties. The higher surface area-to-volume ratio, as well as their distinctive physico-chemical properties, facilitates the development of innovative, surface-engineered structures, and system platform for versatile application (Battigelli et al. 2014). Because of its streamlined, flexible, and scalable structure with tuneable functional groups and porosities, graphene (Gr) and its family of 2D nano-carbon (GO, rGO) has attracted a lot of interest for biomedical applications compared to other nanoparticles (NPs) (Shen et al. 2012; Chen et al. 2015).

These NPs exhibit unique physio-chemical characteristics, including strong mechanical strength (Kravets et al. 2010), electrical and thermal conductivities (Suk et al. 2010), and optical sorption qualities. Graphene interacts with a broad range of biomolecules because of its delocalized electron density and considerable surface area, and it can be used for drug/gene transportation, regenerative medicine, cancer treatment, imaging, and biosensing (Shen et al. 2012; Yang et al. 2013a; Shi et al. 2014). Graphene (Gr) is hydrophobic and can exhibit more aggressive reactions at the edges owing to its substantial aromatic surface. Zig-zag edges are more unstable thermodynamically than the arm-chair edges. It is realistic to expect zig-zag edges to be better responsive than the latter one (Loh et al. 2010).

Hummers method is commonly used to oxidize the graphene to yield GO. It has epoxides, carboxylic (-COOH), and hydroxyl groups (-OH), rendering it to be further hydrophilic as well as water dispensable. The functional groups can be easily modified to accept bioactive molecules, potentially increasing the drug-delivery and loading efficiency (Sanchez et al. 2012).

Graphene can be functionalized by either covalent or noncovalent techniques to alter its physio-chemical attributes (Georgakilas et al. 2012). Covalent functionalization can be carried out via the conjugation of nucleic acids or hydrophilic polymers, the bonding of amines with the -COOH groups, sulfonylation, or the formation of reactive species including nitrenes, carbenes, radicals, and arynes (Park and Yan 2012; Chua and Pumera 2013). This method converts the hybridization of the sp^2-network of C atoms to have sp^3 configuration. Noncovalent changes, on the other hand, are characterized by Van-der-Walls forces, electrostatic forces, hydrogen networking, and π-π bonding. The noncovalent strategy has no effect on characteristic structure and π-π network of Gr. Thus, hydrophobic molecules can absorb with relatively high-loading potency through the π-π stacking onto both sides of the layer structure of the Gr atoms (Kuila et al. 2012; Sanchez et al. 2012).

Numerous graphene-based materials have been synthesized using non-covalent couplings and polymer crosslinking in order to include reactive species such as amines and hydroxyls onto their surfaces. Additionally, subsequent functionalization with cancer cell-targeting ligands might increase the drug-delivery sensitivity (Sanchez et al. 2012; Mann and Dichtel 2013). Figure 17.1 illustrates the functionalization and application of graphene (Gr), GO, and rGO in biomedical field.

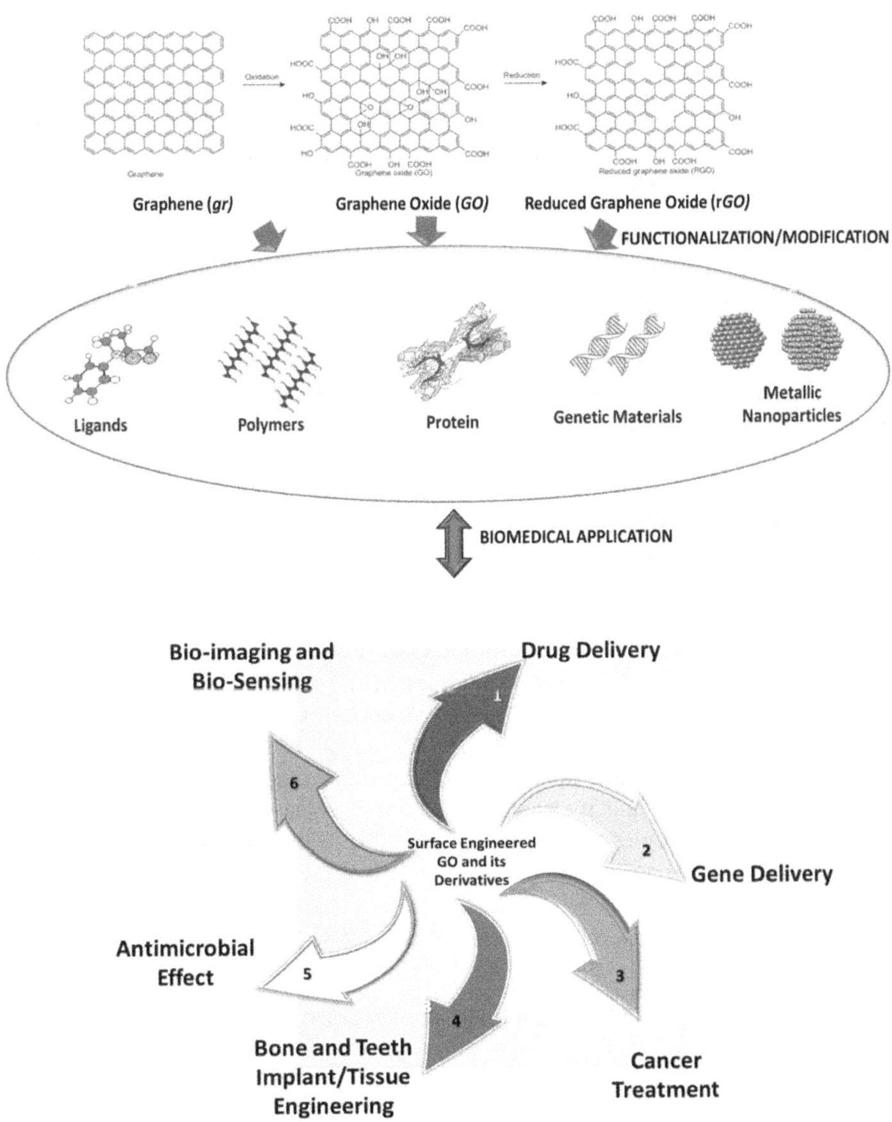

FIGURE 17.1 Surface engineered functionalized GO and its derivatives for biomedical application.

This chapter discusses the current breakthroughs of using surface-engineered, functionalized Gr, GO, and rGO substrates for medicinal applications and gene-delivery mechanisms for cancer therapy. The strengths and weaknesses of several polymer-conjugated graphene derivatives are explored in this chapter. Additionally, several techniques for coating GO and its derivatives with targeted ligands to increase its cellular sorption are also discussed. Furthermore, the drawbacks of various techniques utilizing GO-based materials as a delivery tool was highlighted. Finally, several techniques for minimizing the detrimental consequences of using GO synthesis strategies for developing biocompatible delivery platforms are illustrated as well.

17.2 SURFACE ENGINEERED GRAPHENE OXIDE (GO) AND ITS DERIVATIVES INCORPORATED NANOMEDICINES FOR CANCER PREVENTION

Cancer is a crucial human disease that affects several organs of the body. Tumour tissues can grow with uneven patterns, resulting in the formation of a network of blood vessels. This process is triggered by a variety of signals from malignant cells and epigenetic changes (Nazir et al. 2014). Conventional cancer treatments, which include chemotherapy, radiation, and surgery, face a number of difficulties, including intrinsic toxicity and lower bioavailability. Even though they may show some effectiveness, they also destroy healthy cells and induce toxicity in the patient (Kouranos et al. 2011). As a result, research and development of new nanoparticles, such as graphene (Gr), as well as their use in biomedicine have attracted considerable interest recently (Zhang et al. 2013a; Yang et al. 2015). Graphene, GO, and its derivatives, with its enhanced therapeutic and pharmacological efficacy, has evaded these previously established barriers to a certain extent (Jain and Stylianopoulos 2010). Additionally, due to the increased permeation and retention (EPR) activity, this carbonaceous nanomaterial may invariably be deposited at a greater concentration in the tumour cells than with traditional medicines. This inherent targeting of drugs results in lower systemic toxicity and efficient delivery (Nazir et al. 2014).

Cellular transportation and uptake of GO-based materials inside the cells follows a complex mechanism, and its successful entrance is dependent on lot of surface functionalization strategies (Sasidharan et al. 2012). Endocytosis is a term that refers to a generic energy-dependent entrance route for graphene. It was proposed that GO might be captured by the cancer cells via clathrin-induced endocytosis process owing to its planar 2D shape (Choi and Choy 2011; Huang et al. 2012).

17.3 GO AS VERSATILE TOOL FOR ANTICANCER MEDICINE

17.3.1 Functionalized GO and Its Derivatives for Targeted Drug Delivery

Graphene, GO, and its derivatives-based targeted medication delivery is a suitable platform for cancer treatment. The phenomenon is attributed to its toxicity and risk factors on healthy tissues. Controlled and regulated drug distribution have improved the therapeutic index GO for cancer therapy. A well-established technique is to combine graphene with specially designed ligands that are recognized by cancerous cells. Rituxan (antibody-CD20+) coupled with PEG/GO was utilized specifically to target tumour cells (Sun et al. 2008).

The growth rate of cells was inhibited from 20% to 80% for GO–PEG/DOX and GO–PEG combined Rituxan/DOX subsequently indicating a substantial increase in cytotoxicity against the cancer cells (Sun et al. 2008). Additionally, GO was coupled with a monoclonal (mAbs) antibody-integrin-αvβ3, pyropheophorbide (PPa) and PEG. The combination was utilized to persuade phototoxicity. However, the PPa-GO-mAbs assembly was successful in targeting αvβ3-positive tumour cells, with an accumulation of 80% inside the mitochondria 8 hours after the treatment (Wei et al. 2016). These antibodies seemed to be functional since their directing efficacy was exceptionally good.

Another class of commonly used targeting ligands are aptamers (Apt). The structure of aptamer contains single-lined DNA or RNA molecules (ssDNA or ssRNA). It is target-specific ligands (Dassie and Giangrande 2013). In comparison to antibodies, the ligand-aptamers excel in recognizing small substances with higher affinity and specificity. It has advantages of cost-effectiveness and adaptability with living organisms (Ilgu and Nilsen-Hamilton 2016). The drug-DOX inserts itself inside the double-lined segments of DNA via noncovalent complexation and sorption over the GO surface involving π-π bonding and hydrophobic interfaces when used in conjunction with the MUC1 aptamer. As a result, the medication-loading capacity of the system will enhance as it is targeted against the cancerous cells (Wang et al. 2015).

Dextran (DEX), a homo-polysaccharide of carbohydrate-glucose, possesses unique beneficial characteristics and excellent colloidal stability (Zhang et al. 2013b). DEX-functionalized NPs are more stable than PEG-coated NPs. The reactive site present on DEX is greater than PEG. As a result, for improving the drug-delivery efficiency to boost the cellular absorption by tumour cells, the AS1411 aptamer (targeting nucleolin) was covalently conjugated with GO-DEX-based curcumin distribution scheme (Alibolandi et al. 2017).

The GO-based drug-delivery system frequently uses folic acid (FA). The surface of cancer cells are targeted by the folate receptors. It is a tiny chemical that enables graphene to be easily targeted toward the cancerous cells. PEG-GO functionalized with FA can help in the delivery of gadolinium. Gadolinium (Gd^{+++}) is used in MRI as a tracer agent. It has limited cell selectivity and can cause fibrosis in renal systems. An MRI test revealed that GO-PEG-FA/Gd/DOX had better tumour imaging and targeting efficacy than the charged Gd^{+3} ions. FA was also conjugated with rGO/PEI-FA for targeted medication systems. rGO-PEI/FA demonstrated improved solubility in aqueous medium via π-π bonding (Wei et al. 2012).

The loading capacity of a cholesteryl hyaluronic acid (CHA) modified rGO was four times greater than unmodified rGO (Miao et al. 2013). Cellular absorption of GO/HA complex-coated with DOX drugs dramatically increased the retention of DOX drugs inside the HepG2 cancerous cells targeted by HA, relative to the nontargeted counterparts. The cell survival rate was also increased using DOX-loaded sample (Miao et al. 2013).

Furthermore, magnetic Fe_3O_4 has been suggested to have the possibility for directing GO against the cancerous cells due to its ease of collection using an external magnetic field. Fe_3O_4/graphene composites offer greater surface area as well as better stability for binding the biomolecules (Chomoucka et al. 2010; Veiseh et al. 2010; Yang et al. 2011). To increase selective drug-delivery performance, GO was functionalized with folic acid (FA) and Fe_2O_3. Here, superparamagnetic Fe_2O_3 was used for magnetic targeting, and FA was used as ligand to develop the dual drug-delivery system. When compared to a neutral carrier, this one demonstrated significantly greater cytotoxicity toward the human breast cancer cells (Yang et al. 2011; Thapa et al. 2016).

17.3.2 Functionalized GO and Its Derivatives for Nontargeted Drug Delivery

Application of Gr, GO, and rGO based nanomedicine in biomedical arena, involving the drug-delivery system, has expanded significantly in recent years. Graphene (Gr) has enlarged surface area of 2600 m^2/g, which is four times higher than any other NPs investigated so far for the drug delivery. The single-layer structure of graphene (Gr) is capable of carrying more drugs than other NPs (Liu et al. 2013). In biomedical applications, graphene is most frequently used in the format of graphene oxide (GO). Although it is water soluble, it forms clusters in physical solutions containing proteins or salts alone with cell-culture media or serum (Wojtoniszak et al. 2013). A significant complication is the development of a protein/corona complex over the GO surface (Hajipour et al. 2015; Wei et al. 2015). However, unfavourable responses affecting its bio-distribution or engagement with the immune or complementary systems may occur, resulting in acute toxicity (Mao et al. 2013). As a result, without additional functionalization or modification, it cannot be employed directly for biochemical activities of living organisms. As a result, without additional functionalization or modification, it cannot be employed directly for biochemical activities of living organisms.

Polymer-based functionalization and conjugation techniques significantly improve the in vivo circulation period and biocompatibility of GO (Nurunnabi et al. 2015). Polyethylene glycol (PEG) has received considerable interest in biomedicine. By grafting GO sheets with highly branched and biocompatible PEG, an effective drug-delivery platform using functionalized Gr was first developed. PEGylated GO was found to be nontoxic and extremely stable in biochemical fluid, including protein serum (Mao et al. 2013; Nurunnabi et al. 2015). However, PEG is a bulky and complex molecule that, due to its limited bioavailability, may substantially enhance the biomagnification of GO (Li et al. 2014a). Hydrophilic polyacrylic acid–Gr nano-sheets (PAA-GrNs) have shown excellent solubility in water and persistence in physiological system. PAA-GrNs have the capacity to adsorb aromatic compounds, water-insoluble medicines, and exhibit a significant pH dependence and effective regulated discharge (Chen et al. 2013).

Gelatine is a natural, eco-friendly, benign polymer. It has low cost and can be derived from collagen by hydrolyzing its solution. Gelatine is extremely biodegradable, biocompatible, and has lower immunogenic tendencies. Gelatine conjugation can lower the cytotoxicity of Gr (Nezhadi et al. 2009; Santoro et al. 2014). Gelatine also serves as a key capping agent for graphene stabilization. In distilled water and several physiological fluids, biocompatible gelatin-GrNSs demonstrated good dispensability and stability. Furthermore, gelatin–graphene nano-sheets were used as carriers for anticancer medicines to improve cellular absorption and drug delivery (An et al. 2013). Toxicity studies on MCF-7 cells revealed that the gelatin–based Gr conjugate was safe, even at higher concentrations (200 mg/m). The doxorubicin/gelatin–GrNs composite was highly toxic to MCF-7 cells and demonstrated gelatin-mediated release profile in vitro, making it an excellent therapeutic agent (Liu et al. 2011).

The pluronic F127-modified Gr was also discovered to be efficient for capturing doxorubicin (DOX). The resultant sample showed higher drug loading effectiveness. Inclusion of PF127 provided the Gr nano-sheet with excellent water solubility and physiological resilience. The hybridization also exhibited pH-responsive release of drugs (Hu et al. 2012).

Chitosan (CS) is a natural polymer that is responsive toward stimulation. It has reversible solubility, depending on the pH of the solution. Because of its muco-adhesive characteristics, chitosan can enhance the bioavailability of orally delivered oligonucleotides, proteins, plasmids, and peptides (Sonia and Sharma 2011; Tabujew and Peneva 2014). CS has a positive charge over its surface. This phenomenon can enhance the cellular absorption of GO–CS–camptothecin with the cell membranes having negative charges via electrostatic attraction.

Resistance toward a drug is a significant reason for unsuccessful chemotherapy. Thus, chemo-photothermal therapy has been proposed as a viable way to alleviate the problem (Tran et al. 2015). Functionalized, biocompatible GO can be utilized as a multipurpose medication-delivery vector. Hypocrellin A/SN-38 were co-loaded onto GO. The system showed higher cytotoxic effect rather than the individual compound. Doxorubicin (DOX) and irinotecan co-loaded with GO was stabilized with poloxamer188 for prolonged blood circulation under the laser irradiation (Tran et al. 2015). A brief overview on covalent and noncovalent functionalization of GO is illustrated in Table 17.1.

17.4 GO AS MULTIPURPOSE TOOL FOR GENE DELIVERY

17.4.1 Functionalized GO and Its Derivatives for Targeted Gene Delivery

As previously stated, chitosan (CS) was used as a vector for gene transfer and as a transporter for drugs. GO treated with CS has the potential to reduce its cytotoxicity while increasing its biocompatibility and capacity to carry the genes. Hepatic binder protein (HBP) is a type of glycoprotein receptor (ASGP/R) found over the hepatocytic membrane that recognizes galactose groups inside the de-sialylated glycoproteins. Thus, GO functionalized with lacto-sylated

TABLE 17.1
A Brief Overview on Covalent and Noncovalent Functionalization of GO

Functionalization	Functionalized Sample	Mechanism	Major Findings	References
Covalent functionalization of GO	GO-PEG	PEG was terminated using amine and was attached with -COOH containing GO sheet using amidation reaction	• Enhanced the solubility and biocompatibility of GO • Enhanced stability and drug loading aptitude	Du et al. (2014)
	GO-PEG	PEG was terminated using amine and was attached with -COOH containing GO sheet using amidation reaction	• Decreased the synthesis of immunoglobulin E (IgE) which is antigen-specific • Boost up the activity of T cell-antigen-specific • Biocompatible • Did not disturb the metabolic activities of the splenocytes	Wu et al. (2014)
	PEG-SS-GO	The polymer PEG was terminated using amine and was attached with -COOH containing GO sheet disulphide interaction (SS)	• Biocompatible • Biodegradable after the disruption of di-sulphide linkages	Li et al. (2014b)
	mPEG-SS-nGO	Succinate-activated mPEG-COOH and cystamine, was used to produce mPEG-SS-NH_2 and then it was used for functionalization of GO to have mPEG-SS-nGO	• Boost up the stability of functionalized GO inside the organism • Higher drug stacking • Faster release of drug for cancerous cells	Wen et al. (2012)
	n-GO-FA	FA coupled together sulfonated nGO (nano graphene oxide)	• Higher drug loading rate • Competence of multiple anticancer drugs delivery	Zhang et al. (2010)
	Au doped FA-GO	Folic acid (FA) was covalently attached with -SO_3H groups of sulfonated-GO. Au was deposited over it.	• Tumour targeting • Improved photothermal effects for cancer treatment	Chauhan et al. (2016)
	HA-GO	-COOH containing GO was functionalized with ADH-HA to yield the HA-GO	• Targeted prevention of breast cancer	Hwang et al. (2017)
	Hydrogel of HA-GO	PEG-GO-NH_2 was modified with catechol to-HA to yield the hydrogel	• Relatively higher drug stacking efficacy	Byun and Lee (2014)

(Continued)

TABLE 17.1 (Continued)
A Brief Overview on Covalent and Noncovalent Functionalization of GO

Functionalization	Functionalized Sample	Mechanism	Major Findings	References
	GO-PEI/UCNP	PEI and GO interacted using amide bond and UCNP was loaded over it	• Excellent imaging • DOX loading ability • Less toxicity toward human endothelial cells • Higher solubility and stability inside the physiological fluid	Yan et al. (2013)
	GO-Dex	Dex was terminated by Amine and conjugated with –COOH group containing GO	• Higher drug loading aptitude	Alibolandi et al. (2017)
	L-Cysteine/GO	–COOH containing GO was functionalized using L-cysteine and was modified with Thiol	• Biocompatible	Mu et al. (2015)
	PVP-GO	GO was coated by PVP	• Delayed apoptosis • Improved the antiphagocytic aptitude of GO	Zhi et al. (2013)
	GO-CMC-FIHA	The –OH groups of the GO was bonded with carboxymethyl cellulose (CMC). HA and FI was conjugated with the aforementioned sample.	• Targeted delivery of DOX to HeLa cells • pH-responsive drug delivery	Yang et al. (2016)
Non-covalent functionalization of GO	HA-GO	DOX loaded GO was functionalized with folic acid (FA)	• pH dependent discharge of drugs	Ma et al. (2017a)
	GO-BSA-FA	-COOH groups of folic acid (FA) was bonded with $-NH_2$ group of BSA, GO was attached with it.	• Improved biocompatibility and stability • Toxicity reduction	Song et al. (2014)
	PVA-pGO	Polydopamine-functionalized GO was modified using PVP	• Better thermal and electrical properties • Stability • Enhanced water blockade	Ma et al (2017b)
	nGO-PLA-PEG	PLA-PEG copolymer synthesized using ring opening reaction and was attached with nano GO	• Enhanced stability • Controlled release of paclitaxel drugs	Angelopoulou et al. (2015)

CS oligo-saccharide containing quaternary ammonium units may be used to improve the delivery of DNA inside the hepato-carcinoma (QGY-7703) cell matrix of humans (Cao et al. 2015). Additionally, using 1-pyrene-methylamine hydrochloride, GO may be grafted with folic acid (FA) to form a GO-PEG-FA/$PyNH_2$ combination for cancer therapy and intracellular release of sRNA. PEG was utilized to enhance the solubility of this gene vector in serum and physiological media. The paradigm of multi-functionalization strategy, provided GO/PEI–PEG–FA/si-Stat-3 based system in order to achieve increased si-Stat-3 translation efficacy and subsequently decreased the toxicity. The results indicated that increasing the PEI weight ratios ramped up the transfection aptitude. Additionally, folate targeting facilitated the introduction of si-Stat-3 into SMMC-7721 cells with intent to lower the protein levels (Wang et al. 2016).

17.4.2 Functionalized GO and Its Derivatives for Nontargeted Gene Delivery

Legitimate genetic recombination for treating human diseases, as well genetic abnormalities such as cancer, Parkinson's disease, and cystic fibrosis, necessitates the use of a vector that safeguards the DNA from degradation and enables a higher transfection efficacy (Zhang et al. 2011a). Immune system destruction, cytotoxicity, chromosomal inclusion, mutagenesis, and high input costs are the potential drawbacks associated with viral vectors. Cationic lipids, cationic polymers, and cationic peptides are regarded as nonviral gene carriers, which have relatively fewer restrictions (Guo and Huang 2011; Kim and Kim 2014). Polyethylenimine (PEI) had already been widely studied as a nonviral gene carrier due to its strong binding tendency with the nucleic acid molecules and rapid cellular absorption. It protects against disintegration and enhances endosomal liberation of the RNA or DNA (Jaeger et al. 2012; Ren et al. 2012). PEI has limited biocompatibility as well as significant toxicity. Because GO has negative charges over its surface, electrostatic repulsion between the negative charges of the phosphate structure of the DNA molecule frequently takes place. This must be mitigated by surface functionalization with the cationic polymers, such as polyethylenimine (PEI). As a result, PEI-functionalized GO may be able to solve these limitations, resulting in an efficient gene-delivery mechanism. Branched polyethylenimine (bPEI) has reduced molecular weight and lower toxicity but inadequate transfection efficiency.

The efficacy of PEI-GO transmission was compared to lipofectamin, a popular nonviral type gene therapy system. The results revealed the unique characteristic of GO in comparison to lipofectamin, which caused an inflammatory immune response and exhibited poor transfection efficacy (Choi et al. 2016). rGO was earlier stabilized using an amphiphilic polymer of PEG (PL-PEG) having phospholipid and octa-arginine (R8) and its performance was analyzed in presence of peptide (CPPs), which can penetrate the cells. The siRNA-infused nanocarrier was used to infiltrate the cells of MCF-7. The gene-suppression capacity was assessed using the specified gene model (Imani et al. 2016).

17.5 COMBATING CYTOTOXICITY OF GO AND ITS DERIVATIVES: PLATFORM TO ENSURE BIOCOMPATIBILITY

As previously stated, GO and its derivatives have stimulated considerable interest in biomedical disciplines. Unfortunately, serious concerns have been raised regarding its cytotoxicity. Numerous variables govern the toxicity of GO (Yue et al. 2012; Ali-Boucetta et al. 2013; Duan et al. 2015), involving charge density, concentration, lateral dimension, surface structure, impurities, corona effect and modification/functionalization. Correspondingly, various mechanisms for GO cytotoxicity have been proposed, including inflammatory responses (Orecchioni et al. 2016), mitochondria and DNA damage (Liu et al. 2013), necrosis (Qu et al. 2013), autophagy (Chen et al. 2014) and apoptosis (Li et al. 2012), as well as ROS production, which results in oxidative damage inside the body. This is a precursor for ageing, mutagenesis, and carcinogenesis (Chen et al. 2016). To maximize the distinctive properties of GO, numerous functionalization strategies have been developed to reduce its cytotoxicity. Possible toxicity mechanism of GO and its derivatives are illustrated in Figure 17.2.

Functionalization of GO with synthetic and natural polymeric compounds, including pluronic, PEG, gelatin, chitosan (CS), polyacrylic acid (PAA), and dextran (DEX), has been indicated as the primary method for decreasing cytotoxic effects (Makharza et al. 2013; Yang et al. 2015). Additionally, an in vitro cytotoxicity investigation using the breast cancer cell lines (EMT6) were carried out where pEGylated GO was used. However, the rate of cell survival was above 95%, even after using 100 g/mL concentration (Makharza et al. 2013). Additionally, there was no evidence of substantial pro-inflammatory cytokine production following the treatment with GO–PEG (Feito et al. 2014). Chitosan (CS) is a suitable polymer that can inhibit the haemolytic activity of red blood cells in the presence of GO. GO can induce haemolytic activity due to interaction between the negative charge of GO with the positive charge of lipid (phosphatidylcholine) present inside the outer membrane of the red blood cells (RBC). Additionally, gelatin-modified GO exhibited minimal cytotoxicity against A549 cells when the concentration was around 300 mg/mL (An et al. 2013). Additionally, the cytotoxicity of GO-DEX was compared with the pure GO using various concentrations of GO-DEX and GO. Only extremely high concentrations of GO-DEX (300 μg/mL) decreased the cell viability. Uncoated GO, on the other hand, demonstrated a significant cytotoxicity, even at low doses (Alibolandi et al. 2017).

GO has been shown to cause blood clots upon intravenous administration if it is not modified. Additionally, it was observed that unaltered GO may accumulate in the lungs for an extended period of time, resulting in pulmonary oedema and granuloma development in this organ (Kiew et al. 2016). Further research evaluated the cytotoxicity of various GO-based materials, including poly (acrylic acid)-modified GO (GO-PAA), aminated GO (GO-NH_2), poly(acrylamide)-modified GO (GO-PAM), and PEGylated GO. The toxicity profile of polymer-functionalized graphene oxide (GO) was compared with unmodified graphene oxide (GO). Relatively poor protein adsorption aptitude limited their interactions with the macrophage. Functionalization of GO with PEG and polyacrylic acid (PAA) produced GO having reduced toxicity, compared to other methods. Additionally, the

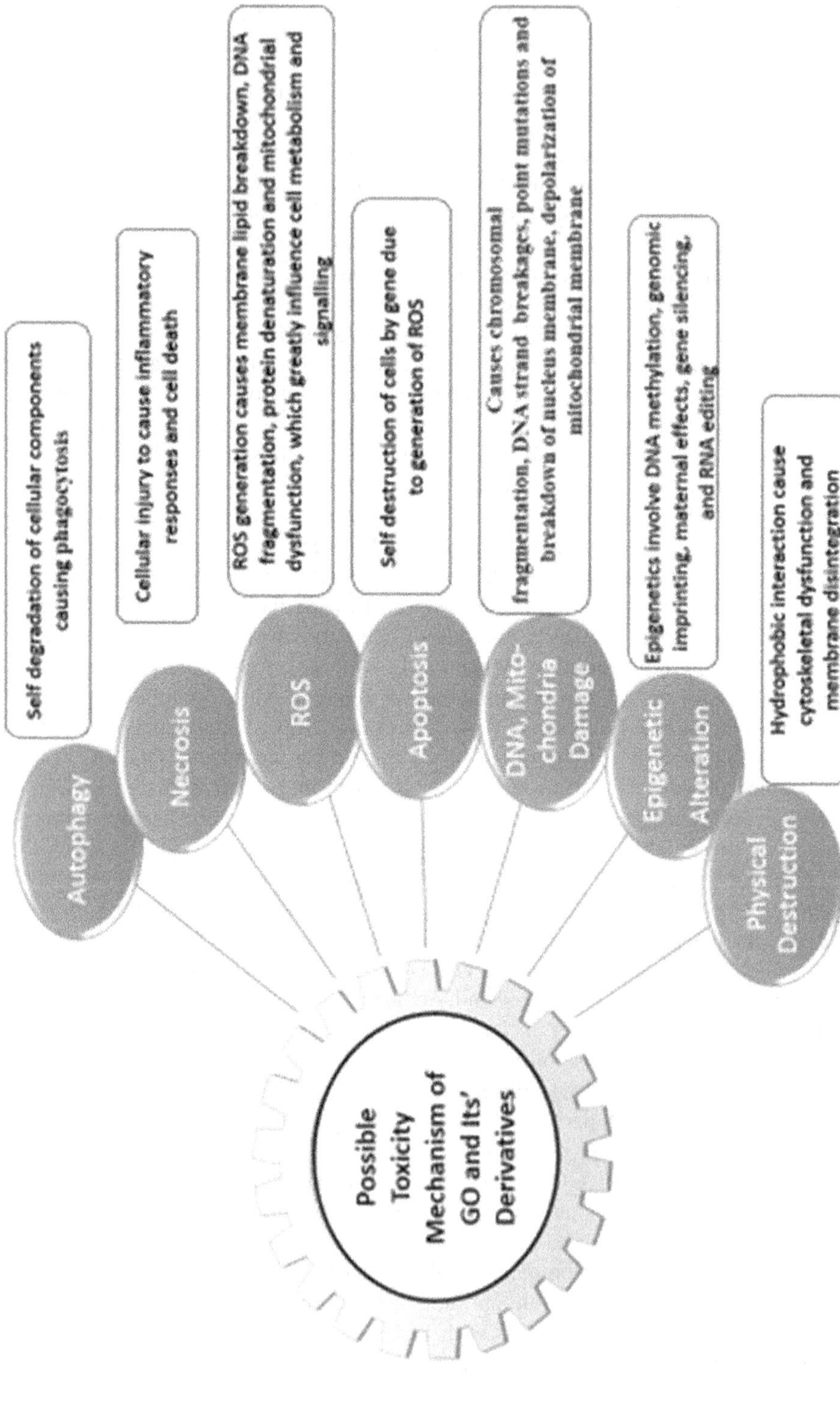

FIGURE 17.2 Possible toxicity mechanism of GO and its derivatives.

toxicity of the substance to the key organs was evaluated. The results revealed that the mice treated with GO-PAA had the minimum dysfunction of lung and liver when related to animals treated with unmodified or other polymer-treated samples (Xu et al. 2016). Thus, optimum surface-functionalization strategy of GO can reduce its toxicity significantly in vivo and in vitro. Thus, this approach should be improved. This will enable GO and its derivatives to act as a suitable delivery platform for treatment of cancer.

17.6 FACTS BEHIND THE TOXICITY PROFILE OF GO AND ITS DERIVATIVES

17.6.1 Dimensional Aspect

It has been demonstrated that nanoparticles (NPs) with diameters less than 100 nm may pass through the cell, particles less than 40 nm might penetrate the nucleus, and particles smaller than 35 nm might pass the barriers of brain/blood (Liu et al. 2013; Mytych and Wnuk 2013). One research study found that GO (556, 588 and 148 nm) did not penetrate A549 cells and did not appear to have any cytotoxic effects. It is possible that the lowest size of graphene (Gr) will produce the greatest amount of toxic effect when its diameter ranges between 100 and 500 nm. However, when it is smaller than 40 nm, the tiniest sizes might be the safest. A small molecule such as rGO having dimension of 114 nm may penetrate inside the nucleus of human hMSCs cells and cause abnormalities in chromosome structure leading to DNA disintegration at lower concentrations (0.1 to 1.0 mg/mL) within 1 hour. Conversely, rGO sheets having width of 3.8–0.4 nm demonstrated no significant genotoxicity after 24 hours when present with hMSCs, even after usage of higher dosage (100 mg/mL) (Akhavan et al. 2012).

17.6.2 Concentration

The application of graphene-based materials (Gr, GO, rGO) has been found to produce dosage-dependent cytotoxicity in living organisms. It can cause kidney and liver damage, granuloma on lungs, reduced cell sustainability, and eventually, apoptosis of the cells (Singh 2016). When administered at lower and intermediate doses (0.1 mg and 0.25 mg), GO did not cause apparent toxicity in mice, but it did cause chronic toxicity when administered at a high dose of 0.4 mg (Wang et al. 2010). As GO concentration was increased, it was accumulated in the lysosomes, mitochondria, endoplasm, and the nucleus of the cells and the body organism failed to flush it out. In several studies, rGO was shown to produce apoptosis-mediated cell death at lower dose within a shorter duration of time, but the necrosis became more frequent when the duration and dose were increased (Wang et al. 2010; Chatterjee et al. 2014).

17.6.3 Charge

GO internalization was minimal for nonphagocytes. This was owing to the electrostatic repulsion between the negative charges GO and the cellular surface (Yue et al. 2012). According to some researchers, NPs can internalize into the nonphagocytic cells due to the presence of negative charge. It can attach itself with the accessible cationic sites over the surface of the cells and can be replaced by scavenger receptors (Chatterjee et al. 2014).

17.6.4 Surface Structure

GO and its derivatives have a wide range of surface-functional groups. For instance, pristine Gr is known to exhibit hydrophobic behaviour. Due to presence of –COOH groups, GO is partly hydrophobic (Hasan et al. 2010), and rGO is partially hydrophilic (Bagri et al. 2010). These can alter the structure and function of the protein and cell membrane due to having strong molecular interactions with the cells (Sanchez et al. 2012). rGO can bind with the membrane of the cells, resulting in activation of the mitochondria which stimulate the receptors. Overall, the phenomenon leads to the apoptosis of the cells (Jaworski et al. 2013). Due to existence of oxygen containing functional groups with smooth, hydrophilic surface of GO, it has less hazardous aspect than rGO (Sasidharan et al. 2012). The double-stranded DNA has base pairs where GO can penetrate to disrupt the transfer of genomic component at the cellular scale, which could be the primary reason for mutagenic impact of GO (Liu et al. 2013; Seabra et al. 2014).

17.6.5 Functionalization

Coating GO with chitosan (CS) significantly reduced its haemolytic activities in blood (Wu et al. 2015). Additionally, the polyethylene glycol (PEG) coating successfully mitigated GO-induced chronic tissue damage and lowered GO agglomeration and accumulation inside the liver, lungs, and spleen. It also accelerated the elimination of GO, GO-DEX (Dextran), and fluorinated GO (FGO) (Zhang et al. 2011b; Romero-Aburto et al. 2013; Li et al. 2014b). PEG-GO showed noncytotoxic behaviour toward the variety of cell cultures, including breast cancer cells (MCF-7), colon cancer cells in colon (HCT-116), glioblastoma cells (U87MG), carcinoma cells (OVCAR-3) inside human ovary, and lympho-blastoid cells (RAJI) receiving doses up to 100 g/mL (Chang et al. 2011; Feng and Liu 2011).

17.6.6 Impurities

Purity of the NPs is critical since residual interfering metals may be accountable for reported toxicity rather than the NPs itself, resulting in contradictory results for cytotoxicity of GO (Nezakati et al. 2014). Typically, produced GO often contain significant concentrations of Mn^{2+} and Fe^{2+}, two potentially carcinogenic metal ions. The unintended discharge of these ions from conventionally synthesized GO might result in excessive level of cytotoxicity and DNA fragmentation (Wu et al. 2015).

17.6.7 Protein/Corona

The higher amount of free-surface charges of NPs makes them particularly adapted for the synthesis of "coronas" with protein molecules in living organisms

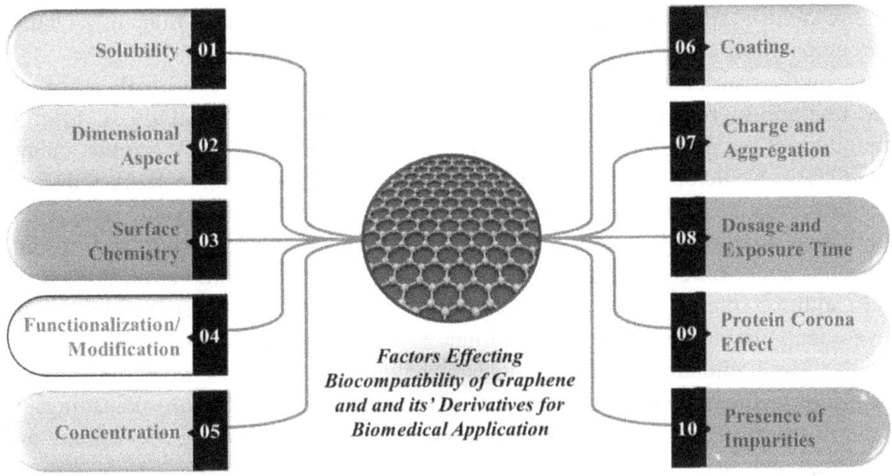

FIGURE 17.3 Factors influencing biocompatibility of graphene and its derivatives.

(Dell'Orco et al. 2010; Eudald et al. 2010). It has been proposed that the protein corona has an effect on the distribution, circulation, toxicity, and clearance of NPs. GO can form corona protein/GO complexes by adsorbing serum plasma protein. The complex can perform an essential role for determining the bio-kinetic behaviour in vivo studies of GO. The complex can further control the adherence of GO with the immune and endothelium cells by interacting with them in an intricate way. Figure 17.3 illustrates the factors that are responsible for possible toxicities of GO and its derivatives.

17.7 RESEARCH GAPS AND FUTURE STUDIES

In order to make conclusions concerning the possible hazards of GO, there is currently inadequate literature available. Some investigators have asserted that GO materials are biocompatible, focusing on their biomedical applications (Chang et al. 2011; Sahu et al. 2012), while others have documented unfavourable biological reactions, including cytotoxicity (Akhavan et al. 2012; Wen et al. 2015). Several factors, especially the diverse-analysis groups, different cellular or animal types, and various physico-chemical properties of GO might have contributed to the inconsistent data. In the case of GO being investigated for in vivo studies inside the organs of humans for related biochemical reactions, the aspect of biocompatibility should be considered. Supplementary comprehensive and precise investigations regarding the toxicity of GO are needed.

Firstly, it is essential to conduct comprehensive physico-chemical analyzes in all future investigations regarding the toxicity of GO and its derivatives. Aspects of GO-based materials, including their particle size, shape, charge, surface area, functionalization/modifications, pureness, and aggregation, should be described in detail during the trials (Hsieh and Chen 2011; Mu et al. 2012). A description of the synthesis process should also be included because the oxidative materials that are produced during the modification process have the potential to significantly change the composition of graphene oxide (GO) to a greater extent.

Secondly, variable qualitative parameters, factors, and empirical technique selection might result in significant discrepancies in data based on a different laboratory (Ema et al. 2012; Zhang et al. 2012). GO toxicity, for example, cannot be reliably predicted using the MTT test because of the spontaneous reduction reactions, which results in a misleading positive signal. As a result, acceptable alternative analyzes, including the ROS assay, trypan blue rejection testing, and tetrazolium salt-based reagent (WST-8) should be used (Bitounis et al. 2013).

In addition, the choice of tumour cell lines is critical since the selected cell lines have a tendency to be sensitive or immune to chemotherapy based on their genetic history. The identical GO nanoparticles can induce a variety of diverse responses according to the origin of the cells that they come from. To avoid misleading positive or negative data, it is necessary to choose cell lines that are appropriate and have high stability.

Fourth, the route of administration of GFNs is critical in toxicity studies. The varying delivery routes will result in a diversity of toxicological responses (Fu et al. 2015). As a result, the route and duration of exposure must be carefully selected in accordance with the research objectives. Despite the fact that intranasal drug delivery is frequently employed to examine the neurotoxicity of nanoparticles (Kwon et al. 2013), this administration technique has only been used in limited studies to investigate the toxicity of GO. Toxicological impact of GO-based materials inside the neurological system is limited, and the mechanistic approach by which they act is unknown also. Thus, it needs additional investigation in the future.

A fifth significant problem that requires consideration is the long-term fate of GO and its derivatives when they enter into the body or are absorbed by the cells. The majority of recent studies have focused on the short-term toxicity level, while long-term toxic impact has received little attention (Xu et al. 2016). The biocompatibility of a functionalized/modified GO surface can also be improved. However, the persistence of the coating materials for longer periods of time must be considered (Schriver et al. 2013). If the coating of the surface ultimately degrades, the toxicity of the GO-based materials might be considerably different from the findings of the short-term exposure.

Sixth, it is necessary to identify and characterize more precise signalling pathways involved in the pathogenesis of GO-based materials' toxicity. Recent research has revealed and extensively recognized several typical toxicity pathways of GO-based materials, including apoptosis, autophagy, and ROS generation. Conversely, these processes have merely been characterized in broader contexts. Further research regarding the particular signalling passageways inside the practical system is required. The signalling pathways associated with the toxicity of other nanoparticles may also provide an idea regarding the GO-based derivatives. For example, nano-epigenetics has been addressed in several studies on nanoparticles, and it has been proven to be effective in analyzing the minimal toxic side effects of GO-based materials. Subsequent research has also revealed that GFNs can produce epigenetic and genomic alterations, which may increase the risk of physiological toxicity and carcinogenicity (Soldano et al. 2010).

17.8 CONCLUSION

Graphene GO and its derivatives pioneered the development of a series of surface-engineered carbon having a wide range of dynamic characteristics that can produce variety of biochemical responses. GO has a large surface area and the highest absorption capacity, making it an excellent candidate for tumour-combination therapy using multiple medicines and a laser treatment process. Some speculative conclusions can be drawn from the literature covered in this chapter. For instance, it is established that chitosan-functionalized GO can enhance medication potency without raising the dose of the chemotherapeutic agent.

Another noteworthy aspect of chitosan-functionalized GO is that its encapsulation efficiency varies according to the chemical composition and reactivity of GO-CS with that individual drug. Additionally, it had been demonstrated that the targeted techniques outperformed nontargeted approaches. Applying aptamer-conjugated GO to deliver the chemotherapeutic drugs can intercalate inside the double-strand oligo-nucleotides, leading to spectacular outcomes.

The drug could enter the double-stranded areas by noncovalent bonding or sorption inside the GO matrix using π-π and hydrophobic interactions networking. Consequently, when the organism is targeted against cancer, its medication-processing capacity should enhance. As a result, in these instances, aptamer as a targeting tool for GO-based therapy is recommended. Additionally, by elucidating the process through which GO and its derivatives enters the cell, highly efficient administration systems that rely on clathrin receptor selectivity can be devised.

Recombination of GO and its derivatives with biocompatible poly-cations is postulated for a gene-delivery system. This combination may alleviate their constraints and offer delivery methods that are more advantageous for gene packing than simple GO or poly-cations. It can improve the process of gene transfection. Utilizing these special features of GO, however, requires certain precautions regarding its cytotoxicity. To create a much more biocompatible framework, GO needs to be functionalized using effective polymeric agents. For synthesizing a very biocompatible and biodegradable GO-based distribution system, gelatin-functionalization is recommended. Although GO has shown potential in biomedicine, particularly in the delivery of therapeutic materials for treating cancer, there is currently no GO-based nanomedicine for large-scale medical trials. It appears that successful adaptation of this technique for experimental practise will necessitate more exploration regarding its longstanding toxicity aspect with subsequent elimination from the living organism. This chapter will aid in addressing the safety concerns associated with GO prior to their clinical and medicinal applications, which will be critical for the additional commercialization of GO in biomedical activities.

REFERENCES

Akhavan, O., Ghaderi, E. and Akhavan, A. 2012. Size-dependent genotoxicity of graphene nanoplatelets in human stem cells. *Biomaterials* 33(32):8017–8025.

Alibolandi, M., Mohammadi, M., Taghdisi, S.M., Ramezani, M. and Abnous K. 2017. Fabrication of aptamer decorated dextran coated nano-graphene oxide for targeted drug delivery. *Carbohydrate Polymer* 155:218–229.

Ali-Boucetta, H., Bitounis, D., Raveendran-Nair, R., Servant, A., Van den Bossche, J. and Kostarelos, K. 2013. Purified graphene oxide dispersions lack in vitro cytotoxicity and in vivo pathogenicity. *Advanced Healthcare Materials* 2(3):433–441.

An, J., Gou, Y., Yang, C., Hu, F. and Wang, C. 2013. Synthesis of a biocompatible gelatin functionalized graphene nanosheets and its application for drug delivery. *Materials Science and Engineering* C 33(5):2827–2837.

Angelopoulou, A., Voulgari, E. and Diamanti, E.K., et al. 2015. Graphene oxide stabilized by PLA-PEG copolymers for the controlled delivery of paclitaxel. *European Journal of Pharmaceutics and Biopharmaceutics* 93:18–26.

Bagri, A., Mattevi, C., Acik, M., Chabal, Y.J., Chhowalla, M. and Shenoy, V.B. 2010. Structural evolution during the reduction of chemically derived graphene oxide. *Nature Chemistry* 2(7):581–587.

Battigelli, A., Menard-Moyon, C. and Bianco, A. 2014. Carbon nanomaterials as new tools for immunotherapeutic applications. *Journal of Materials Chemistry A* 2(37): 6144–6156.

Bitounis, D., Ali-Boucetta, H., Hong, B.H., Min, D.H. and Kostarelos, K. 2013. Prospects and challenges of graphene in biomedical applications. *Advanced Materials* 25(16): 2258–2268.

Byun, E. and Lee, H. 2014. Enhanced loading efficiency and sustained release of doxorubicin from hyaluronic acid/graphene oxide composite hydrogels by a mussel-inspired catecholamine. *Journal of Nanoscience Nanotechnology* 14(10):7395–7401.

Cao, X., Zheng, S., Zhang, S., et al. 2015. Functionalized graphene oxide with hepatocyte targeting as anti-tumor drug and gene intracellular transporters. *Journal of Nanoscience and Nanotechnology* 15(3):2052–2059.

Chang, Y., Yang, S.T., Liu, J.H., et al. 2011. In vitro toxicity evaluation of graphene oxide on A549 cells. *Toxicology Letters* 200(3):201–210.

Chatterjee, N., Eom, H.J. and Choi, J. 2014. A systems toxicology approach to the surface functionality control of graphene-cell interactions. *Biomaterials* 35:1109–1127.

Chauhan, G., Chopra, V., Tyagi, A., Rath, G., Sharma, R.K., and Goyal, A.K.2016. Gold nanoparticles composite-folic acid conjugated graphene oxide nanohybrids for targeted chemo-thermal cancer ablation: In vitro screening and in vivo studies. *European Journal of Pharmaceutical Science* 96:351–361.

Chen, G.Y., Chen, C.L., Tuan, H.Y., et al. 2014. Graphene oxide triggers toll-like receptors/ autophagy responses in vitro and inhibits tumor growth in vivo. *Advanced Healthcare Materials* 3(9):1486–1495.

Chen, M., Yin, J. and Liang, Y. 2016. Oxidative stress and immunotoxicity induced by graphene oxide in Zebrafish. *Aquatic Toxicology* 174:54–60.

Chen, Y., Qi, Y. and Liu, B. 2013. Polyacrylic acid functionalized nanographene as a nanocarrier for loading and controlled release of doxorubicin hydrochloride. *Journal of Nanomaterials* 2013:16.

Chen, Y., Tan, C., Zhang, H. and Wang, L. 2015. Two-dimensional graphene analogues for biomedical applications. *Chemical Society Reviews* 44(9):2681–2701.

Choi, H.Y., Lee, T.-J. and Yang, G.-M. 2016. Efficient mRNA delivery with graphene oxide-polyethylenimine for generation of footprint-free human induced pluripotent stem cells. *Journal of Controlled Release* 235:222–235.

Choi, S.-J. and Choy J.-H. 2011. Layered double hydroxide nanoparticles as target-specific delivery carriers: Uptake mechanism and toxicity. *Nanomedicine* 6(5):803–814.

Chomoucka, J., Drbohlavova, J., Huska, D., Adam, V., Kizek, R. and Hubalek, J. 2010. Magnetic nanoparticles and targeted drug delivering. *Pharmacology Research* 62(2): 144–149.

Chua, C.K. and Pumera, M. 2013. Covalent chemistry on graphene. *Chemical Society Reviews* 42(8):3222–3233.

Dassie, J.P. and Giangrande, P.H. 2013. Current progress on aptamer-targeted oligonucleotide therapeutics. *Therapeutic Delivery* 4(12):1527–1546.

Dell'Orco, D., Lundqvist, M., oslakovic, C. , Cedervall, T. and Linse, S. 2010. Modeling the time evolution of the nanoparticle-protein corona in a body fluid. *PLoS One* 5(6): e10949-e.

Du, L., Wu, S., Li, Y., Zhao, X., Ju, X., and Wang, Y.2014. Cytotoxicity of PEGylated graphene oxide on lymphoma cells. *Biomedical Materials and Engineering* 24(6): 2135–2141.

Duan, G., Kang, S.-G., and Tian, X. 2015. Protein corona mitigates the cytotoxicity of graphene oxide by reducing its physical interaction with cell membrane. *Nanoscale* 7(37):15214–15224.

Ema, M., Aoyama, H., Arima, A., et al. 2012. Historical control data on prenatal developmental toxicity studies in rabbits. *Congenital Anomalies* 2(3):155–161.

Eudald, C., Tobias, P., Albert, D., Gertie, J.O. and Victor P. 2010. Time evolution of the nanoparticle protein corona. *ACS Nano* 4(7):3623–3632.

Feito, M., Vila, M., Matesanz, M., et al. 2014. In vitro evaluation of graphene oxide nanosheets on immune function. *Journal of Colloid and Interface Science* 432: 221–228.

Feng, L. and Liu, Z. 2011. Graphene in biomedicine: Opportunities and challenges. *Nanomedicine* 6(2):317–324.

Fu, C., Liu, T., Li, L., Liu, H., Liang, Q. and Meng, X. 2015. Effects of graphene oxide on the development of offspring mice in lactation period. *Biomaterials* 40:23–31.

Georgakilas, V., Otyepka, M., Bourlinos, A.B., et al. 2012. Functionalization of graphene: Covalent and non-covalent approaches, derivatives and applications. *Chemical Society Reviews* 112(11):6156–6214.

Guo, X. and Huang, L. 2011. Recent advances in nonviral vectors for gene delivery. *Accounts of Chemical Research* 45(7):971–979.

Hajipour, M.J., Raheb, J., Akhavan, O., et al. 2015. Personalized disease specific protein corona influences the therapeutic impact of graphene oxide. *Nanoscale* 7(19): 8978–8994.

Hasan, S.A., Rigueur, J.L., Harl, R.R., et al. 2010. Transferable graphene oxide films with tunable microstructures. *ACS Nano* 4(12):7367–7372.

Hsieh, C.T. and Chen, W.Y. 2011. Water/oil repellency and work of adhesion of liquid droplets on graphene oxide and graphene surfaces. *Surface and Coatings Technology* 205(19):4554–4561.

Hu, H., Yu, J., Li, Y., Zhao, J. and Dong, H. 2012. Engineering of a novel pluronic F127/graphene nanohybrid for pH responsive drug delivery. *Journal of Biomedical Material Research Part A* 100(1):141–148.

Huang, J., Zong, C., Shen, H., et al. 2012. Mechanism of cellular uptake of graphene oxide studied by Surface-Enhanced Raman Spectroscopy. *Small* 8(16):2577–2584.

Hwang, D.W., Kim, H.Y., Li F., et al. 2017. In vivo visualization of Endogenous miR-21 using hyaluronic acid-coated graphene oxide for targeted cancer therapy. *Biomaterials* 121:144–154.

Ilgu, M. and Nilsen-Hamilton, M. 2016. Aptamers in analytics. *Analyst* 141(5):1551–1568.

Imani, R., Shao, W., Taherkhani, S., Emami, S.H., Prakash, S. and Faghihi, S. 2016. Dualfunctionalized graphene oxide for enhanced siRNA delivery to breast cancer cells. *Colloids and Surfaces* B 147:315–325.

Jaeger, M., Schubert, S., Ochrimenko, S., Fischer, D. and Schubert, U.S. 2012. Branched and linear poly (ethylene imine)-based conjugates: synthetic modification, characterization, and application. *Chemical Society Reviews* 41(13):4755–4767.

Jain, R.K. and Stylianopoulos, T. 2010. Delivering nanomedicine to solid tumors. *Nature Reviews Clinical Oncology* 7(11):653–664.

Jaworski, S., Sawosz, E., Grodzik, M., et al. 2013. In vitro evaluation of the effects of graphene platelets on Glioblastoma multiforme cells. *International Journal of Nanomedicine* 8:413–420.

Kiew, S.F., Kiew, L.V., Lee, H.B., Imae, T. and Chung, L.Y. 2016. Assessing biocompatibility of graphene oxide-based nanocarriers: A review. *Journal of Controlled Release* 226:217–228.

Kim, H. and Kim, W.J. 2014. Photothermally controlled gene delivery by reduced graphene oxide–Polyethylenimine nanocomposite. *Small* 10(1):117–126.

Kouranos, V., Dimopoulos, G., Vassias, A. and Syrigos, K.N. 2011. Chemotherapy-induced neutropenia in lung cancer patients: The role of antibiotic prophylaxis. *Cancer Letters* 313(1):9–14.

Kravets, V., Grigorenko, A., Nair, R., et al. 2010. Spectroscopic ellipsometry of graphene and an exciton-shifted Van Hove peak in absorption. *Physical Review B* 81(15): 155413:1–6.

Kuila, T., Bose, S., Mishra, A.K., Khanra, P., Kim, N.H. and Lee, J.H. 2012. Chemical functionalization of graphene and its applications. *Progress in Materials Science* 57(7):1061–1105.

Kwon, J.T., Seo, G.B., Jo, L.M., et al. 2013. Aluminium nanoparticles induce ERK and p38MAPK activation in Rat brain. *Toxicology Research* 29:181–185.

Li, B., Zhang, X.Y., Yang, J.Z., et al. 2014a. Influence of polyethylene glycol coating on biodistribution and toxicity of nanoscale graphene oxide in mice after intravenous injection. *International Journal of Nanomedicine* 9:4697–4707.

Li, Y., Feng, L., Shi, X., et al. 2014b. Surface coating-dependent cytotoxicity and degradation of graphene derivatives: Towards the design of non-toxic, degradable nano-graphene. *Small* 10(8):1544–1554.

Li, Y., Liu, Y., Fu, Y., et al. 2012. The triggering of apoptosis in macrophages by pristine graphene through the MAPK and TGF-beta signalling pathways. *Biomaterials* 33(2): 402–411.

Liu, J. Cui, L. and Losic, D. 2013. Graphene and graphene oxide as new nanocarriers for drug delivery applications. *Acta Biomaterials* 9(12):9243–9257.

Liu, K., Zhang, J.-J., Cheng, F.-F., Zheng, T.-T., Wang, C. and Zhu J.-J. 2011. Green and facile synthesis of highly biocompatible graphene nanosheets and its application for cellular imaging and drug delivery. *Journal of Materials Chemistry* 21(32): 12034–12040.

Liu, Y., Luo, Y., Wu, J., et al. 2013. Graphene oxide can induce in vitro and in vivo mutagenesis. *Scientific Reports* 3(3469):1–8.

Loh, K.P., Bao, Q., Ang, P.K. and Yang, J. 2010. The chemistry of graphene. *Journal of Materials Chemistry* 20(12):2277–2289.

Ma, J., Pan. J., Yue, J., Xu, Y., and Bao, J. 2017b. High performance of poly(dopamine) functionalized graphene oxide/poly(vinyl alcohol) nanocomposites. *Applied Surface Science* 427(4):428–436.

Ma, N., Liu, J., He, W., et al. 2017a. Folic acid-grafted bovine serum albumin decorated graphene oxide: An efficient drug carrier for targeted cancer therapy. *Journal of Colloid Interface Science* 490:598–607.

Makharza, S., Cirillo, G. and Bachmatiuk, A. 2013. Graphene oxide-based drug delivery vehicles: Functionalization, characterization, and cytotoxicity evaluation. *Journal of Nanoparticle Research* 15(12):2099. 10.1007/s11051-013-2099-y.

Mann, J.A. and Dichtel, W.R. 2013. Noncovalent functionalization of graphene by molecular and polymeric adsorbates. *Journal of Physical Chemistry Letters* 4(16): 2649–2657.

Mao, H., Chen, W. and Laurent, S. 2013. Hard corona composition and cellular toxicities of the graphene sheets. *Colloids Surfaces B Biointerfaces* 109:212–218.

Miao, W., Shim, G., Kang, C.M., et al. 2013. Cholesteryl hyaluronic acid-coated, reduced graphene oxide nanosheets for anti-cancer drug delivery. *Biomaterials* 34(37): 9638–9647.

Mu, L., Gao, Y. and Hu, X. 2015. L-cysteine: A biocompatible, breathable and beneficial coating for graphene oxide. *Biomaterials* 52:301–311.

Mu, Q., Su, G., Li, L., et al. 2012. Size-dependent cell uptake of protein-coated graphene oxide nanosheets. *ACS Applied Materials and Interfaces* 4(4):2259–2266.

Mytych, J. and Wnuk, M. 2013. Nanoparticle technology as a double-edged sword: Cytotoxic, genotoxic and epigenetic effects on living cells. *Journal of Biomaterials and Nanobiotechnology* 4:53–63.

Nazir, S., Hussain, T., Ayub, A., Rashid, U. and MacRobert, A.J. 2014. Nanomaterials in combating cancer: Therapeutic applications and developments. *Nanomedicine: Nanotechnology, Biology and Medicine* 10(1):19–34.

Nezakati, T., Cousins, B.G. and Seifalian, A.M. 2014. Toxicology of chemically modified graphene based materials for medical application. *Archives of Toxicology* 88(11): 1987–2012.

Nezhadi, S.H., Choong, P.F., Lotfipour, F. and Dass, C.R. 2009. Gelatin-based delivery systems for cancer gene therapy. *Journal of Drug Targeting* 17(10):731–738.

Nurunnabi, M., Parvez, K., Nafiujjaman, M., et al. 2015. Bio-application of graphene oxide derivatives: Drug/gene delivery, imaging, polymeric modification, toxicology, therapeutics and challenges. *RSC Advances* 5(52):42141–42161.

Orecchioni, M., Menard-Moyon, C., Delogu, L.G. and Bianco, A. 2016. Graphene and the immune system: Challenges and potentiality. *Advanced Drug Delivery Reviews* 105:163–175.

Park, J. and Yan, M. 2012. Covalent functionalization of graphene with reactive intermediates. *Accounts of Chemical Research* 46(1):181–189.

Qu, G., Liu, S., Zhang, S., et al. 2013. Graphene oxide induces toll-like receptor 4 (TLR4)-dependent necrosis in macrophages. *ACS Nano* 7(7):5732–5745.

Ren, T., Li, L., Cai, X., Dong, H., Liu, S. and Li, Y. 2012. Engineered polyethylenimine/graphene oxide nanocomposite for nuclear localized gene delivery. *Polymer Chemistry* 3(9):2561–2569.

Romero-Aburto, R., Narayanan, T.N., Nagaoka, Y., et al. 2013. Fluorinated graphene oxide: A new multimodal material for biological applications. *Advanced Materials* 25(39): 5632–5637.

Sahu, A., Choi, W.I. and Tae, G. 2012. A stimuli-sensitive injectable graphene oxide composite hydrogel. *Chemical Communication* 48(47):5820–5822.

Sanchez, V.C., Jachak, A., Hurt, R.H. and Kane, A.B. 2012. Biological interactions of graphene-family nanomaterials: An interdisciplinary review. *Chemical Research in Toxicology* 25(1):15–34.

Santoro, M., Tatara, A.M. and Mikos, A.G. 2014. Gelatin carriers for drug and cell delivery in tissue engineering. *Journal of Control Release* 190:210–218.

Sasidharan, A., Panchakarla, L.S., Sadanandan, A.R., et al. 2012. Hemocompatibility and macrophage response of pristine and functionalized graphene. *Small* 8(8):1251–1263.

Schriver, M., Regan, W., Gannett, W.J., Zaniewski, A.M., Crommie, M.F. and Zettl, A. 2013. Graphene as a long-term metal oxidation barrier: Worse than nothing. *ACS Nano* 7:5763–5768.

Seabra, A.B., Paula, A.J., de Lima, R., Alves, O.L. and Duran N. 2014. Nanotoxicity of graphene and graphene oxide. *Chemical Research in Toxicology* 27(2):159–168.

Shen, H., Zhang, L., Liu, M. and Zhang, Z. 2012. Biomedical applications of graphene. *Theranostics* 2(3):0020283–0020294.

Shi, S., Chen, F., Ehlerding, E.B. and Cai, W. 2014. Surface engineering of graphene-based nanomaterials for biomedical applications. *Bioconjugate Chemistry* 25(9):1609–1619.

Singh, Z. 2016. Applications and toxicity of graphene family nanomaterials and their composites. *Nanotechnology, Science and Applications* 9:15–28.

Soldano, C., Mahmood, A. and Dujardin, E. 2010. Production, properties and potential of graphene. *Carbon* 48(8):2127–2150.

Song, E., Han, W. and Li, C., et al. 2014. Hyaluronic acid-decorated graphene oxide nano-hybrids as nano-carriers for targeted and pH-responsive anticancer drug delivery. *ACS Applied Materials & Interfaces* 6(15):11882–11890.

Sonia, T. and Sharma, C.P. 2011. Chitosan and its derivatives for drug delivery perspective. Chitosan for biomaterials. *Advanced Polymer Science* 2011:23–53.

Suk, J.W., Piner, R.D., An, J. and Ruoff, R.S. 2010. Mechanical properties of monolayer graphene oxide. *ACS Nano* 4(11):6557–6564.

Sun, X., Liu, Z., Welsher, K., et al. 2008. Nanographene oxide for cellular imaging and drug delivery. *Nano Research* 1(3):203–212.

Tabujew, I. and Peneva, K. 2014. Functionalization of cationic polymers for drug delivery applications. In *Cationic Polymers in Regenerative Medicine*, eds. S. Samal and P. Dubruel, 1–29. Cambridge: Royal Society of Chemistry.

Thapa, R.K., Choi, Y., Jeong, J.H., et al. 2016. Folate-mediated targeted delivery of combination chemotherapeutics loaded reduced graphene oxide for synergistic chemo-photothermal therapy of cancers. *Pharmaceutical Research* 33(11): 2815–2827.

Tran, T.H., Nguyen, H.T., Pham, T.T., et al. 2015. Development of a graphene oxide nanocarrier for dual-drug chemo-phototherapy to overcome drug resistance in cancer. *ACS Applied Materials & Interfaces* 7(51):28647–28655.

Veiseh, O., Gunn, J.W. and Zhang, M. 2010. Design and fabrication of magnetic nanoparticles for targeted drug delivery and imaging. *Advance Drug Delivery Reviews* 62(3):284–304.

Wang, C., Wang, X., Lu, T., et al. 2016. Multi-functionalized graphene oxide complex as a plasmid delivery system for targeting hepatocellular carcinoma therapy. *RSC Advances* 6(27):22461–22468.

Wang, K., Jing, R., Song, H., et al. 2010. Biocompatibility of graphene oxide. *Nanoscale Research Letters* 6(1):1–8.

Wang, X., Han, Q., Yu, N., et al. 2015. Aptamer–conjugated graphene oxide–gold nanocomposites for targeted chemo-photothermal therapy of cancer cells. *Journal of Materials Chemistry B* 3(19):4036–4042.

Wei, G., Yan, M., Dong, R., et al. 2012. Covalent modification of reduced graphene oxide by means of diazonium chemistry and use as a drug delivery system. *Chemistry European Journal* 18(46):14708–14716.

Wei, X.Q., Hao, L.Y. and Shao, X.R. 2015. Insight into the interaction of graphene oxide with serum proteins and the impact of the degree of reduction and concentration. *ACS Applied Materials & Interfaces* 7(24):13367–13374.

Wei, Y., Zhou, F., Zhang, D., Chen, Q. and Xing, D. 2016. A graphene oxide based smart drug delivery system for tumor mitochondria-targeting photodynamic therapy. *Nanoscale* 8(6):3530–3538.

Wen, H., Dong, C., Dong, H., et al. 2012. Engineered redox-responsive PEG detachment mechanism in PEGylated nano-graphene oxide for intracellular drug delivery. *Small* 8(5):760–769.

Wen, K.P., Chen, Y.C., Chuang, C.H., Chang, H.Y., Lee, C.Y. and Tai, N.H. 2015. Accumulation and toxicity of intravenously-injected functionalized graphene oxide in mice. *Journal of Applied Toxicology* 35(10):1211–1218.

Wojtoniszak, M., Urbas, K., Perużyńska, M., Kurzawski, M., Droździk, M., Mijowska, E. 2013. Covalent conjugation of graphene oxide with methotrexate and its antitumor activity. *Chemical Physics Letter* 568:151–156.

Wu, H.Y., Lin, K.J., Wang, P.Y., et al. 2014. Polyethylene glycol-coated graphene oxide attenuates antigen-specific IgE production and enhanced antigen-induced T-cell reactivity in ovalbumin-sensitized BALB/c mice. *International Journal of Biomedicine* 9:4257–4266.

Wu, S.Y., An, S.S. and Hulme, J. 2015. Current applications of graphene oxide in nanomedicine. *International Journal of Nanomedicine* 10(S):9–24.

Xu, M., Zhu, J., Wang, F., et al. 2016. Improved in vitro and in vivo biocompatibility of graphene oxide through surface modification: poly (acrylic acid)-functionalization is superior to PEGylation. *ACS Nano* 10(3):3267–3281.

Yan, L., Chang, Y.N., Zhao, L., et al. 2013. The use of polyethylenimine-modified graphene oxide as a nanocarrier for transferring hydrophobic nanocrystals into water to produce water-dispersible hybrids for use in drug delivery. *Carbon* 57(3):120–129.

Yang, H., Bremner D.H., Lei, T., et al. 2016. Carboxymethyl chitosan-mediated synthesis of hyaluronic acid-targeted graphene oxide for cancer drug delivery. *Carbohydrate Polymer* 135:72–78.

Yang, K., Feng, L. and Liu, Z. 2015. The advancing uses of nano-graphene in drug delivery. *Expert Opinion on Drug Delivery* 12(4):601–612.

Yang, K., Feng, L., Shi, X. and Liu, Z. 2013a. Nano-graphene in biomedicine: Theranostic applications. *Chemical Society Reviews* 42(2):530–547.

Yang, X., Wang, Y., Huang, X., et al. 2011. Multifunctionalized graphene oxide based anticancer drug-carrier with dual-targeting function and pH-sensitivity. *Journal of Materials Chemistry* 21(10):3448–3454.

Yue, H., Wei, W., Yue, Z., et al. 2012. The role of the lateral dimension of graphene oxide in the regulation of cellular responses. *Biomaterials* 33(16):4013–4021.

Zhang, G., Zeng, X. and Li, P. 2013a. Nanomaterials in cancer-therapy drug delivery system. *Journal of Biomedical Nanotechnology* 9(5):741–750.

Zhang, H., Gruner, G. and Zhao, Y. 2013b. Recent advancements of graphene in biomedicine. *Journal of Materials Chemistry B* 1(20):2542–2567.

Zhang, L., Lu, Z., Zhao, Q., Huang, J., Shen, H. and Zhang, Z. 2011a. Enhanced chemotherapy efficacy by sequential delivery of siRNA and anticancer drugs using PEI-grafted graphene oxide. *Small* 7(4):460–464.

Zhang, L., Xia, J., Zhao, Q., et al. 2010. Functional graphene oxide as a nanocarrier for controlled loading and targeted delivery of mixed anticancer drugs. *Small* 6(4): 537–544.

Zhang, S., Yang, K., Feng, L. and Liu, Z. 2011b. In vitro and in vivo behaviors of dextran functionalized graphene. *Carbon* 49(12):4040–4049.

Zhang, W., Wang, C., Li, Z., et al. 2012. Unraveling stress-induced toxicity properties of graphene oxide and the underlying mechanism. *Advanced Materials* 24(39): 5391–5397.

Zhi, X., Fang, H., Bao, C., et al. 2013. The immunotoxicity of graphene oxides and the effect of PVP-coating. *Biomaterials* 34(21):5254–5261.

18 Antiviral Nanomaterials as Potential Targets for Malaria Prevention and Treatment

Kantrol Kumar Sahu, Sunita Minz, Madhulika Pradhan, Monika Kaurav, and Krishna Yadav

CONTENTS

DOI: 10.1201/9781003136644-21

18.1 INTRODUCTION

Malaria is expected to be the most widespread and irresistible infection in the world, causing a predominantly serious health problem in Asia, Africa, and the Americas (Biamonte et al. 2013). According to the World Health Organization (WHO), in 2018 alone, there were an expected 228 million instances of malaria around the world, which caused about 405,000 demises (World Malaria Report 2019). About 93% of the cases and demises happened in an African Province of a region the WHO describes as the most elevated concern of malaria morbus. Children 5 years of age and under are the extremely susceptible group, addressing 67% of all malarial demises around the world (World Malaria Report 2019). Five distinctive species of *plasmodium* family can manifest malaria in people, including *P. falciparum*, *P. ovale*, *P. vivax*, *P. malariae*, and *P. knowlesi* (Greenwood et al. 2008). The malarial cycle is regulated by apicomplexial parasites of the *plasmodium* category that are spread to the mammalian host by the bite of a female Anopheles mosquito. Malaria is a deadly virus and uses human cellular machinery for its growth. The life cycle of these parasites is multifarious (Figure 18.1), switching back and forth between a mammalian and an invertebrate host, and contains three formative phases: the pre-erythrocytic or liver phase, the erythrocytic or blood phase, and the sporogonic or mosquito phase (Prudêncio et al. 2006; Cowman et al. 2016).

Numerous drugs have been used in the management of malaria, including chloroquine, hydroxychloroquine, doxycycline, quinine, atovaquone, and so forth

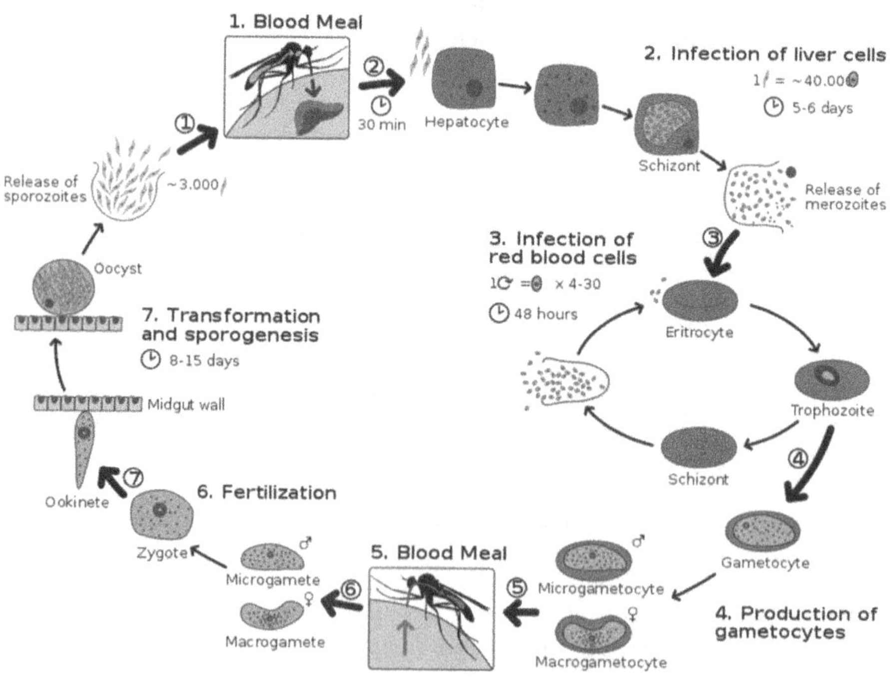

FIGURE 18.1 Illustration showing the life cycle of the malaria parasite.

(Talapko et al. 2019). On the other hand, various endeavours have embrace the advancement of an effective vaccine for malaria, an aspiring objective that to date has not succeeded. Without appropriate inoculation to malaria, the WHO advises vector control to lessen the odds of mosquito bites and chemoprevention to stifle infection (World Malaria Report 2015, 2019). Because the existing pipeline of plasmodial antagonist is probably not going to be adequate to compensate the malaria-eradication schema, it is critical to find new treatment modalities with broad application to stop the growth and transmission of parasites (Prudêncio et al. 2006; Cowman et al. 2016).

Since drug development is expensive and tedious, alternate methodologies are important to restructure the discovery procedure of novel malaria therapeutics. Drug repurposing is a technique focused on the recognizable proof of different utilizations for already approved or trial medications that are out of the field of reference of the first clinical sign (Ashburn and Thor 2004; Pushpakom et al. 2019). As a matter of fact, their pharmacokinetic and pharmacodynamics parameters have already been known and can be strategies to target other disorders in search of new pharmacological activity.

Antiviral drugs, i.e. antiretroviral (ARVL), are the most potential candidates for to be repurposed as antimalarial therapeutics. In a few places around the globe, *plasmodium* parasites and HIV have a topographical similarity (Fontinha et al. 2020). In this sense, ARVL therapeutics used to treat HIV have been tested through in vitro and in vivo investigations in rats as well as human models for their efficacy against the *plasmodium* liver stage. In addition to these treatments in ARVL therapy, protease inhibitors (PIs) have also been the utmost choice of therapeutic for *plasmodium* parasites. PIs including saquinavir, ritonavir, lopinavir, and indinavir seem to have been effective in the hepatic phases of the *P. berghei* rodent parasites (Machado et al. 2017). In addition, lopinavir tended to affect the liver step of the *P. falciparum* parasite capable of infecting humans at a clinically relevant dose (Hobbs et al. 2009). Both PIs that demonstrate effectiveness against hepatic *plasmodium* parasite infections were additionally evaluated in vitro and in vivo against the subsequent blood phase of *P. falciparum*, *P. vivax*, and *P. berghei* (Fontinha et al. 2020) infections, with lopinavir seeming to have the most important activity in this category of ARVL. Other than affecting abiogenetic blood stages, saquinavir, lopinavir, and ritonavir also repressed *P. falciparum* parasites with gametocytogenesis (Peatey et al. 2010). This was subsequently supported by the second study, which found that saquinavir and lopinavir had decreased gametocyte development and had *P. falciparum* prevented transmission of activity (Hobbs et al. 2013).

Despite the inaccessibility of explicit antiviral therapeutics for repurposing as antimalarial, utilization of most antiviral medications is also connected to their restricted solubility, short half-life time, and deficient penetration to determined anatomic compartments. Appropriately, a perpetual effort is required to improve the physico-chemical qualities of existing antiviral medications (Cojocaru et al. 2020). Since nanomaterials show significant physical and chemical assets, high surface zone-to-volume proportion, and expanded reactivity, new methodologies for antiviral delivery and repurposing may be achieved by using blends of nanomaterials with available therapeutics. Nanotechnology has arisen as quite possibly the most encouraging advancement because of its capacity to deliver therapeutics effectively, tending to the limits of conventional antiviral medications for repurposing. It has not only assisted us with defeating issues identified with solubility and toxicity of medications, it has also granted extraordinary properties to drugs, expanding their potency and selectivity toward target cells (Lembo et al. 2018). Nanomaterials offer novel physico-chemical attributes that have associated advantages for drug delivery as idyllic devices for treatment. As of now, various sorts of nanomaterials, specifically liposomes, nanoparticles, nanospheres, nanosuspensions, nanoemulsions, and nanogel, were investigated either in vitro or in vivo for drug conveyance of antiviral agents to repurpose them for the treatment of malaria with possibilities to be interpreted in clinical practice. The applications of antiviral nanomaterials as potential targets for malaria prevention and treatment are emphasized in this section. It covers a wide range of nanomaterials and fabrication processes, as well as their administration routes, uptake mechanisms, and therapeutic ability.

18.2 CURRENT METHODS OF PREPARATION OF NANOPARTICLES

Depending on the type of materials used in the preparation, such as polymers, lipids, and metals, several methods for preparing nanoparticles have been published in the literature. The following section discusses some of the most used polymeric and lipid nanoparticle preparation techniques.

18.2.1 Preparation of Polymeric Nanoparticles

18.2.1.1 Nanoprecipitation

Nanoprecipitation technique was developed and patented by Fessi et al. (1989). This method is based on the precipitation mechanism, in which the polymer precipitates in a nonsolvent. The polymer is dispersed into the aqueous phase (nonsolvent phase) under mild stirring after being dissolved in a water-soluble organic solvent (solvent phase). The organic solvent evaporates at room temperature or by the action of rotavapor, resulting in the formation of nanoparticles (NPs) in aqueous suspension. After that, the aqueous phase is removed through ultracentrifugation and freeze-drying. The drug molecules, a film-former/polymeric material, one/more lipophilic surfactant, and one/more lipophilic surfactant, are usually found in the solvent phase organic solvents (Miladi et al. 2016).

The most commonly employed solvents are ethanol, acetone, hexane or methylene chloride. Water is used widely as nonsolvent (or aqueous phase). During preparation of aqueous phase, hydrophilic excipients must be added to the nonsolvent phases. Polymeric materials used could be of synthetic, semisynthetic, or natural origin. Further, to improve the stability of the NPs and to avoid its aggregation, surfactants could be added into the formulation (Minost et al. 2012).

18.2.1.2 Milling Technique

This method involves different techniques, such as wet milling, milling media, and high-pressure homogenization for the preparation of nanoparticles (Loh et al. 2015). In wet milling, drug and excipients are suspended in a liquid medium and are milled using beads to get suspension of NPs. The nano suspension produced is subsequently dried using freeze-drying or spray drying. This technique is well suited for potent therapeutic moieties, as well as for the drugs' high moisture content. Media milling and high-pressure homogenization are two common top-down approaches for production of drug NPs without using beads. Micro fluidizer and piston-gap homogenizer are employed for high-pressure homogenization. The size of NPs is controlled by homogenization speed, characteristic of drug, number of cycles, and temperature. This technique offers robust processing, economic, and easily scalable.

18.2.1.3 Extrusion Method

It's a simple, scalable, and cost-effective method for transforming aqua-phobic drugs into nanoparticles. This is a continuous method in which starting material is pumped at high pressure and temperature, resulting in uniformly sized NPs (Maniruzzaman et al. 2012). Hydrophobic nanoparticles are made using the nanoporous membrane-extrusion (NME) process. The four main techniques used in NME are vesicle extrusion, membrane emulsification, precipitation extrusion, and biological-membrane extrusion. These methods are used to make a range of NPs, including liposomes, emulsions, nanoparticles, nanotubes, and nanofibers (Guo et al. 2018). A downside of this method is the clogging of raw-material filtration residue at the feeding site of the channel, which can block the channel easily.

18.2.1.4 Supercritical Fluid Technology

This is one of the most common methods for making polymeric nanoparticles because it doesn't require high temperatures and doesn't use organic solvents. This technique is a green and novel method for increasing the solubility of poorly soluble medicines, improving polymer plasticization, and modifying surfaces, nanosizing, and nanocrystals. Fluid is converted to gas using this method when the density of the liquid is greater than its critical temperature and pressure (Chakravarty et al. 2019). The use of supercritical CO_2 and H_2O in the preparation of nanomaterials is common.

Surfactants and capping agents can be used to monitor the size and surface modification of NPs under supercritical fluid (SCF) conditions. The SCF technology employs a variety of approaches (Yeo and Kiran 2005), including:

- Static supercritical fluid process
- Rapid expansion of supercritical solutions
- Particles from gas-saturated solutions
- Gas anti-solvent process

18.2.1.5 Salting-Out Method

The salting-out approach was created to address the drawbacks of both the emulsification-solvent evaporation and nanoprecipitation techniques. The polymer and drug are dissolved in a water-miscible organic solvent, such as acetone, and then emulsified into an aqueous gel with a salting-out agent (magnesium chloride, calcium chloride, magnesium acetate, and sucrose) and a colloidal stabilizator (Mendoza-Muñoz et al. 2012).

After that, a large quantity of aqueous medium is applied to the emulsion (o/w) to enable the solvent to diffuse into the aqueous phase and form nanoparticles. Crossflow filtration is used to eliminate the salting-out agent and solvent. This method offers various advantages, including high efficiency, ease of scale-up, and utility for heat-sensitive substances. The disadvantages include the need to extensively wash NPs and the limited applicability for lipophilic drugs.

18.2.1.6 Ionic Gelation Technique

The NPs are formed by crosslinking an ionic polymer with a counterionic material using the ionic gelation technique. To build cross-links, if the polymer material is negatively charged, the counterionic material must be positively charged. Electrostatic interactions between the counterionic material and the polymer cause crosslinking. Ionic gelation, which involves converting the substance from a liquid to a gel at room temperature, follows cross-linking. The cluster is formed in the pre-gel process because of a reaction between the polymer and the gelling agent, and then stabilized by forming a complex with polyelectrolyte (Krishnamoorthy and Mahalingam 2015). Improper surface morphology, fragibility of NPs, and high poly dispersibility index are the major drawbacks of this method.

18.2.1.7 Sonication Method

This approach involves simply dissolving the drug and polymer in an organic solvent to make NPs. The resulting mixture is sonicated under controlled conditions, the solvent is filtered out, and the residue is washed several times to obtain drug-loaded NPs (Shi et al. 2012). For the dispersion of noninert and non-functionalized metal nanoparticles such as Cu, Mn, and Al, the sonication approach is favoured. The release of metal from metal nanoparticles is accelerated by sonication, which can be further accelerated by adding solubilizing agents like bovine serum albumin (Pradhan et al. 2016a).

18.2.2 Preparation of Lipid Nanoparticles

18.2.2.1 High-Pressure Homogenization (HPH)

HPH is the most popular method for creating lipid NPs in the pharmaceutical industry. This technique has many benefits, including improved drug stability, increased drug loading, no use of organic solvents, and large-scale manufacturing (Corrias and Lai 2011). HPH could be carried out in two ways, i.e. cold homogenization and hot homogenization.

Hot homogenization takes place at a temperature above the lipid's melting point. In this technique, the drug is mixed with melted lipid and the resultant mixture is then dispersed into an aqueous phase under homogenization. Lipid NPs are produced upon solidification of the nano-emulsion at room temperature (Naseri et al. 2015). This method is well suited for lipophilic and thermostable drugs, but it has several drawbacks, including drug loss in the aqueous phase and a complicated crystallization process.

High-pressure milling of suspension is used in the cold homogenization process, where lipid microparticles are first manufactured by melting the lipid with the drug and then allowed to solidify rapidly using liquid nitrogen or dry ice. The nanoparticles are made by homogenizing prepared microparticles in a cold aqueous surfactant using high-speed stirring below room temperature. This method generates large particles with a broad size distribution. (Singh et al. 2020).

18.2.2.2 Solvent-Emulsification Evaporation (SEE)

This technique uses a lipid phase containing lipid excipients, and the therapeutic moiety is solubilized in a volatile organic solvent. Other excipients, such as a stabilizing agent, are dissolved to make an aqueous phase. To obtain a small droplet-formed emulsion (O/W), the lipid phase is applied to the aqueous phase, followed by high-energy homogenization or sonication. Furthermore, the organic solvent is removed by evaporation or decreased pressure. This process subsequently results in the formation of lipid NPs. Ethyl acetate is an extensively used organic solvent because it offers better toxicological profile as compared to earlier used solvents, such as dichloromethane and chloroform (Pradhan et al. 2017, 2020). The NPs are washed repeatedly with distilled water by ultracentrifugation, and then the pellets of lipid NPs are collected. The physico-chemical properties of prepared nanoparticles are influenced by experimental parameters, such as processing temperature, solvent-evaporation method, aqueous-phase length, surfactant quantity, and polymer-molecular weight (Pradhan et al. 2016b). This is well suited for lipophilic and thermolabile drugs because this technique does not involve thermal stress. However, this method has limitations, such as time consumption and coalescence during the solvent evaporation (Singh et al. 2016).

18.2.2.3 Emulsification and Solvent Displacement Method

This method is a variation of SEE. The drug and lipid excipients are dissolved in a partly water-miscible organic solvent, and the resulting mixture is then saturated with water to reach thermodynamic equilibrium between the two liquids. Since the organic solvent is partly miscible with water, it diffuses into it, causing lipid NPs to precipitate from the dispersed process. This technique can generate NPs with a size of less than 100 nm and a low polydispersity index. Further, the solvent from the system could be eliminated by evaporation or filtration. Nonionic surfactants lead to the formation of larger particles than ionic surfactants (Pineda-Reyes et al. 2021). This technique offers many advantages, such as high encapsulation efficiency, good reproducibility, avoidance of homogenization, easy scale-up, and narrow particle-size distribution. Process parameters such as polymer and surfactant nature and concentration, type of solvent, agitation speed and time, type of stirrer, temperature of dilution water, and viscosity of external phase affect the properties of prepared nanoparticles. The main disadvantages of this method are the removal of large amounts of water from suspension and the risk of hydrophilic drug leakage into the aqueous external process during emulsification.

18.3 ADMINISTRATION OF ANTIVIRAL DRUGS

In the field of drug repurposing to target antiviral drugs for antimalarial activity, new or superior formulations for manufacturing techniques are considered necessary to expand their efficacy. Novel drug-delivery system-based approaches are utilized to reduce dose and dosing frequency, reduce side effects, as well as improve therapeutic efficiency. Modern research work in the field of nanotechnology-based drug-delivery systems has gained extreme attention and seems to provide next-generation drug-delivery methodologies to improvise antiviral drug therapy (Cojocaru et al. 2020).

Nanoparticulate drug-delivery systems have been anticipated to improve drugs' pharmacokinetic drawbacks, such as solubility, absorption, and bioavailability profile. Nanoparticulate drug-delivery systems could be used for local as well as systemic drug delivery of antiviral drugs. In the case of systemic administration, i.e. intravenous administration of nano-sized drug formulations is required to maintain it in the bloodstream circulation and to cross over the pulmonary capillaries. To overcome the uptake mechanism by the reticulo-endothelial system (RES) and to increase the circulation half-life of nanocarriers, strategies such as sheathings or surface-decorated (PEG coated) nano-carriers have been employed (Lembo and Cavalli 2010).

Also, various types of nanoparticulate systems specifically provide characteristics that are more appropriately suitable for some administration routes, such as topical, nasal, oral, ocular, and pulmonary. Due to this concern, various types of nanoparticulate carrier systems have been researched, formulated by different methodologies to encapsulate antiviral drugs, characterized, and administered via different delivery routes. These delivery systems are mostly made up of natural and biodegradable polymers to provide physical and biological compatibility to ensure consumers' safety and to reduce toxicity. The administration of antiviral drug-based nanocarriers via possible delivery routes are discussed in detail here (Lembo and Cavalli 2010).

18.3.1 ORAL ROUTE

The oral route is the one of the most preferred, accessible, and prevalent routes of drug administration for the attainment of significant therapeutic level of drugs. However, some major limitations are associated with oral drug administration, such as the requirement of higher drug dose for effective pharmacological effect due to respective obstacles such as stringent gastrointestinal milieu, biological membrane barrier action, chemical instability of drugs, and hepatic first-pass metabolism, thus patient compliance is also compromised. In this regard, to overcome all above discussed problems along with achievement of efficient therapeutic response with minimal patient related side effects, a number of novel drug-delivery system-based techniques and methods has been employed for the delivery of antiviral drugs. For example, a lipid-based nano-emulsion system (emulsifier coated nanosized lipid droplets) has been formulated with acyclovir to improve its solubility and oral bioavailability (Sapra et al. 2013). Similar types of delivery system-based formulations for oral antiviral drug administration are discussed in Table 18.1 (Patel and Sawant 2007; Sankar et al. 2012; Sapra et al. 2013; Kumar et al. 2015; Venkatesh et al. 2015; Kumar et al. 2016; Öztürk and Yurtdaş Kirimlioğlu 2019).

TABLE 18.1
List of Antiviral Drug-Based Delivery Systems Administered via Different Routes

Drug	Delivery System	Route of Administration	Significant Result	Reference
Zidovudine (AZT)	AZT loaded lactoferrin nanoparticles (NPs)	Oral	Improvement of pharmacokinetic profile (bioavailability and tissue distribution of zidovudine), reduced toxicity	Kumar et al. (2015)
Lamivudine (LAM)	LAM loaded polylactic-co-glycolic acid (PLGA) NPs	Oral	Improved stability, release rate, drug bioavailability	Öztürk and Yurtdaş Kirimlioğlu (2019)
Zidovudine-Lamivudine	PLGA, poly(lactic acid) PLA, Poly (methyl methacrylate) PMMA, methyl methacrylate-sulfopropylmethacrylate (MMA-SPM) NPs	Oral	Improved drug release rate kinetics with reduced side effect	Sankar et al. (2012)
Nelfinavir mesylate (NFV)	NFV loaded PLGA NPs	Oral	Enhanced drug bioavailability and reduced dosing frequency	Venkatesh et al. (2015)
Efavirenz (EFV)	Efavirenz-loaded lactoferrin NPs	Oral	Improved oral bioavailability and improved pharmacokinetic profile, no organ related toxicity	Kumar et al. (2016)
Acyclovir (ACV)	Lipid nanoemulsion system	Oral	Improved drug solubility and bioavailability	Sapra et al. (2013)
ACV	ACV loaded self-micro-emulsifying drug-delivery system (SMDDS)	Oral	Sustained release drug profile	Patel and Sawant (2007)
ACV	ACV loaded poly-N-vinyl-2-pyrrolidone (PVP) liposomes gel	Nasal	Due to muco-adhesion contact time at mucosal surface increase and facilitate drug absorption	Alsarra et al. (2008)
ACV	ACV loaded L-α-dipalmitoyl phosphocholine and cholesterol liposomes in PVP gel	Nasal	Via nasal route 3 times increased bioavailability as compared to oral route	Amel (2001)
Azidothymidine	Azidothymidine loaded transferrin (Tf) anchored PEGylated NPs	Brain	Enhanced uptake in brain as compared to unmodified nanoparticles	Mishra et al. (2006)

(Continued)

TABLE 18.1 (Continued)
List of Antiviral Drug-Based Delivery Systems Administered via Different Routes

Drug	Delivery System	Route of Administration	Significant Result	Reference
Saquinavir (SQV)	SQV loaded Tf-conjugated quantum rods (QR)	Brain	Significant uptake QR-Tf-Saquinavir in brain microvascular endothelial cells (BMVECs), significant enhancement in the crossing ability and observed reduction in HIV-1 viral replication in the PBMCs.	Mahajan et al. (2010)
AZT and lamivudine (3TC)	AZT and 3TC loaded polybutylcyanoacrylate (PBCA) and methylmethacrylate-sulfopropylmethacrylate (MMA-SPM) polymeric NPs	Brain	100% increase in the BBB permeability for both drugs due to the carrier mediated system	Kuo and Chen (2006)
Stavudine (D4T), delavirdine (DLV), and SQV	D4T, DLV and SQV loaded polybutylcyanoacrylate (PBCA) NPs,methylmethacrylate-sulfopropylmethacrylate (MMA-SPM) NPs, and solid lipid nanoparticles (SLNs)	Brain	Enhancement in the BBB permeability (3-16 folds)	Kuo and Su (2007)
Azidothymidine 5'-triphosphate (AZTTP)	AZTTP loaded magnetic liposomal formulation	Brain	Permeability and transport of AZTTP magnetic liposomes was observed 3-fold higher than AZTTP alone	Saiyed et al. (2010)
Saquinavir mesylate	Nanoemulsion	Intranasal (nose to brain)	Detectable amount of saquinavir mesylate was observed after intranasal administration as compared to intravenous route	Mahajan et al. (2010)
Efavirenz	Efavirenz loaded PEO–PPO polymeric micelles	Intranasal (nose to brain)	Detectable increased accumulation of Efavirenz micelles was observed and 5 times more relative exposure index in CNS as compared to intravenous administration	Chiappetta et al. (2013)

(Continued)

TABLE 18.1 (Continued)
List of Antiviral Drug-Based Delivery Systems Administered via Different Routes

Drug	Delivery System	Route of Administration	Significant Result	Reference
ACV	Bovine serum albumin (BSA) NPs	Ocular	No cytotoxicity effect and in vitro transcorneal permeation studies results revealed that via BSA-NPs ACV could permeate into the corneal cells significantly as compared to aqueous ACV solutions	Suwannoi et al. (2017)
ACV	ACV loaded Biotin-12 hydroxystearic acid (B-12HS) micellar formulation	Ocular	Good ocular biocompatibility with no cytotoxicity effect and promising delivery of biotinylated lipid prodrug	Vadlapudi et al. (2014)
Acyclovir L-aspartate (L-Asp-ACV) and acyclovir D-aspartate (D-Asp-ACV)	Amino acid prodrug approach	Ocular	Observed enhanced transport of L-Asp-ACV in the corneal space due to amino acid transporter mediated transport of amino acid derived form	Majumdar et al. (2009)
ACV	SLNs and nanostructured lipid carriers (NLCs)	Ocular	Results revealed that NLCs faster permeation in the excised cornea due to its enhanced corneal penetration properties	Seyfoddin and Al-Kassas (2013)
Ganciclovir (GCV)	GCV loaded Ophthalmic liposomes	Ocular	Higher in vitro transcorneal permeability of GCV liposomes and Ocular tissue distribution of GCV-liposomes was 2 to 10 folds higher (sclera, cornea, iris, lens, and vitreous humour) as compared to solution form	Shen and Tu (2007)
ACV	ACV loaded chitosan nanospheres (NS)	Topical	Improved permeability with improved antiviral activity	Donalisio et al. (2018)
Penciclovir (PCV) and ACV	1% PCV and 3% ACV cream	Topical	Improved results observed in patients under treatment	Lin et al. (2002)
GCV	Borneol-modified GCV-loaded SLNs	Parenteral	Borneol-GCV-SLN enhances the transport of GCV to the brain	Ren et al. (2012)
Antiretroviral drugs	Mannosylated PLGA NPs	Parenteral	Results observed that mannosylated PLGA NPs are promising candidate for brain delivery	Patel and Sawant (2007)

18.3.2 Nasal Route

The nasal route of drug administration has been primarily exploited for delivering the drug into the vicinity of the nasal mucosa surface. However, this route offers several advantages due to excessive permeability of nasal mucosal epithelium for higher molecular weight drug candidates (≤1000 Da) with rapid drug absorption rate, but the therapeutic effectiveness of administered drugs depends on mucus secretions, enzymes presence, mucus-secretion rate mucociliary-clearance rate, and deposition and uptake of drugs in nasal mucosa. Various novel drug formulations (mucoadhesive systems) are efficiently effective in the enhancement of drug facilitation in nasal mucosa along with stability and protection from nasal enzymes (Durai 2015). Mucoadhesive system based antiviral drug delivery via nasal route discussed in Table 18.1 (Amel 2001; Alsarra et al. 2008).

18.3.3 Brain Route

In the delivery of drugs to brain, blood-brain barrier (BBB) has been a major obstacle. The BBB acts as a diffusive barrier and impedes transition of most of the compounds rather than essentials into the brain blood. To cross over BBB and to deliver drugs in brain-space, nanocarriers have been employed. They enhance BBB permeability and drug deposition in the brain due to its nano-size and lipophilic characteristics through various mechanisms, such as via passive targeting to increase the local drug concentration gradient, receptor-mediated endocytosis of drugs, and blockage of its efflux transporters (Dong 2018). Mainly, few types of nanocarriers have been explored for the antiretroviral drug delivery to the brain; they include polymeric nanoparticles, dendrimers, lipid-based carriers, and micelle-based systems discussed in Table 18.1 (Kuo and Chen 2006; Mishra et al. 2006; Kuo and Su 2007; Mahajan et al. 2010; Saiyed et al. 2010).

18.3.4 Nose-Brain Route

In recent years, the use of the nose-to-brain route (intranasal delivery) of drug administration to effectively deliver therapeutics to the brain has achieved considerable significance. This noninvasive drug-delivery method directly delivers the therapeutics into the brain via bypassing the BBB. The various preclinical and clinical studies have shown that via intranasal delivery, the effective transport of drugs directly from the nasal cavity to the brain can be achieved through the olfactory epithelium (Md et al. 2015). In a study, nanoemulsion of saquinavir mesylate was prepared for brain delivery. Results show that a high in vivo saquinavir concentration level was achieved by nanoemulsion after intranasal administration, as compared to intravenous administration (Mahajan et al. 2014). Similar results were also observed in another study (Chiappetta et al. 2013) (Table 18.1).

18.3.5 Ocular Route

To develop drug-delivery systems for ocular delivery is a very challenging task due to natural static (physical barriers) and dynamic barriers (blinking and tear secretions), as well as the anatomical characteristics of the eye, which block the entry of harmful chemicals, as well as drug substances, into the vicinity of eye. In this regard, consideration of all above factors is necessary for designing of ocular drug-delivery systems so that actual drug dosing and high drug absorption can be achieved. During the past decade, many studies were conducted to explore the liposomes-based antiviral drug delivery. Due to the biodegradable and nontoxic characteristics of nanosized colloidal drug carriers to the cells, phosphatidylcholine-cholesterol-dimethyl dioctadecyl liposomes had been utilized for acyclovir delivery to the eyes (Gote et al. 2019). Delivery systems used for antiviral drug delivery via the ocular route is discussed in Table 18.1.

18.3.6 Topical and Transdermal Route

These routes of drug administration provide the local effect in a noninvasive manner using a smaller drug amount for the targeted therapeutic effect. These routes also offer several advantages, such as improved patient compliance, reduced toxicity, and potential for multiple targets and systemic drug distribution (Hassan et al. 2016). The main obstacle in topical delivery is stratum corneum. This barrier is overcome using various physical practices, such as electroporation, sonophoresis, iontophoresis, and chemical practices, such as employing alcohols, terpenes, alkanes, phospholipids, esters, surfactants, and amines used to induce drug diffusion across skin surface along with novel delivery systems approaches (liposomes, niosomes, transferosomes, nanoparticles, etc.) (Hassan et al. 2016). Via topical and transdermal antiviral drug-delivery systems, protection against the entry and spread of pathogens can be achieved. To improve the topical delivery of antiviral drugs, such as acyclovir, micro- and nanoemulsion-based topical formulations have been developed and show significant improved effects (Shishu et al. 2009; Schwarz et al. 2012). Similar studies' results are discussed in Table 18.1 (Lin et al. 2002; Donalisio et al. 2018).

18.3.7 Parenteral Route

Drug delivery of antiviral drugs via the parenteral route would cause many severe problems, such as thrombophlebitis at the application site and high-dose crystal formation in the kidney, which may lead to renal failure and renal damage (Fleischer and Johnson 2010; Seedat and Winnett 2012; Lu et al. 2014; Zhang et al. 2014). Recently, novel intravenous (IV) polylactic coglycolic acid (PLGA) stealth nanoparticles have been developed to increase antiviral drugs' plasma half-life, circulation time, and poor bioavailability issues (Kamel et al. 2009; Gupta et al. 2013). Similar types of novel delivery systems administrated via the parenteral route are listed in Table 18.1 (Ren et al. 2012; Patel and Sawant 2007).

18.3.8 Vaginal Route

Another noninvasive route of drug delivery is offered by the vaginal cavity, which avoids pain, tissue damage, and probable infections, as well as toxicity-associated side effects associated with the parenteral route. This route of drug delivery also bypasses hepatic first-pass metabolism and is fully equipped with a well-developed blood supply for drugs systemic delivery too. Additionally, the vagina cavity provides great permeability and drug absorption for certain drug candidates based on their physico-chemical characteristics, including molecular weight, dissolution, and ionization properties via active and passive transport mechanisms (Ferguson and Rohan 2011).

18.4 UPTAKE AND PROCESSING OF ANTIVIRAL NANOPARTICLES

Antimalarials are used extensively and at random, resulting in the growth of drug-resistant parasites. The use of nanomaterials containing antimalarials, which are capable of controlling and treating malaria through targeted delivery, will help to prevent or delay the production of resistance (Aditya et al. 2013; Borgheti-Cardoso et al. 2020). The key emphasis in nanomedicine science, putting this idea into practice, poses a significant challenge in the development of malaria drugs. The reason for this challenge is due to the parasite's intracellular and strewn location (Aditya et al. 2013).

Antimalarial drugs must overcome a variety of barriers, including host red cell membrane, the parasitephorous vacuolar membrane, and the parasite plasma membrane. Solutes and macromolecules move through these multiple membrane barriers on their way to the parasite. Therefore, it is necessary to understand how solutes pass from the intracellular parasite to the external solution without entering the erythrocyte cytosol. Solutes must first enter the erythrocyte via the red blood cell membrane before being picked up by parasites located intracellular. They will then enter the parasite either by endocytosis or by being transported across the parasitophorous vacuolar membrane and the parasite plasma membrane, in that order. However, the possibility of parallel routes for solutes to meet the parasite without going into the cytosol of the erythrocyte cannot be ruled out. Endocytosis does not occur in mature erythrocytes, and no other pathways for macromolecule uptake by erythrocytes are known (Haldar and Uyetake 1992; Pouvelle et al. 1994).

But red blood cells infected with *plasmodium* species may endocytose. In mature red cells, a scaffold-like protein that lies underneath the membrane prevents endocytosis. Malaria parasites, on the other hand, rearrange the lipids in the membrane of red blood cells and proteins are intercalated through the red cell cytosol underneath the membrane and is capable of destroying spectrin, thus, mature RBCs may endocytose (Tokuyasu et al. 1979; Kilejian 1984; Burns and Pollack 1988).

Absorption of drugs requires effective transport through membrane barriers. Adsorption of nanoparticles on tissue and cells is easy. Due to their ultra-small size, they have large surface area; as a result, nanoparticles have high concentration and extended nanoparticles residence and payload at the site (Liu et al. 2013).

Transcellular pathways are currently thought to be one of the most important for nanoparticles absorption pathways. Endocytosis is one of the most popular methods for transporting nanoparticles through cell membranes. After opsonin, the cells first accept nanoparticles through a set of cell-surface receptors. The plasma membrane is then stimulated to generate tiny vesicles, and the invaginated small vesicles are then isolated from the plasma membrane and divided into cells; fused together with the lysosome, drugs are released after being hydrolyzed by enzymes. This endocytosis is further classified into two distinct forms, i.e. phagocytosis and pinocytosis, which are primarily determined by nanoparticle properties and target cells. The majority of nanocarriers are liquid dispersions that can pass through biological membrane barriers, primarily through pinocytosis.

Pinocytosis is classified as caveolae-mediated endocytosis, clathrin-mediated endocytosis clathrin- and caveolae-independent endocytosis, and micropinocytosis (Sun et al. 2019) (Table 18.2). Caveolar-mediated endocytosis is a clathrin-independent endocytotic mechanism involving bulb-shaped caveolae. Caveolae are 50-60 nm plasma membrane invaginations. Caveolae are formed by caveolins, which are integral membrane proteins, and cavins are peripheral membrane proteins. Clathrin-mediated endocytosis possesses complex protein machinery that transiently assembles on the plasma membrane and creates clathrin-coated endocytic vesicles. This machinery selects and concentrates cargo molecules and shapes the membrane into a vesicle.

Macropinocytosis is a mechanism in which a significant proportion of the fluid phase is absorbed in comparison to the cell volume. Most nanoparticles enter red blood cells through clathrin-mediated endocytosis, which is one of the most important pathways (Conner and Schmid 2003).

18.5 THERAPEUTIC APPROACH FOR MALARIA TREATMENT

Recent viral outbreaks in the world, such as Coronavirus, Ebola, influenza, and dengue, have highlighted the urgent need for new antiviral therapeutics to save public health. One such epidemic is malaria, which exists from the past; its complete cure still needs attention. Considerable efforts have focused on developing antiviral nanoparticles for the treatment of this pathogenic disease. Some of these nanoparticles have been summarized, as follows:

1. Phyto-synthesized silver-NPs of a*rtemisia nilagirica* aqueous leaf filtrate would be promising antiviral therapeutics for malaria control. These NPs exhibit strong larvicidal and pupicidal action against developmental stages of

TABLE 18.2
Types of Cellular Uptake Mechanism

Endocytotic Pathway	Cellular Uptake Mechanism	Engulf Particles
Non receptor-mediated endocytosis	Phagocytosis	0.5–10 μm
	Macropinocytosis	100–5000 nm
Receptor-mediated endocytosis	Clathrin-mediated endocytosis	approx. 120 nm
	Caveolae-mediated endocytosis	approx. 80 nm

An. stephensi and *Ae. aegypti*. Using these cost-effective and biodegradable natural resources techniques is an affordable method for prevention of not only malaria, but also dengue, chikungunya, and filariasis. Moreover, it is also a rapidly synthesised, viable, recyclable, nontoxic, and stable approach for treating this disease (Nalini et al. 2017).

2. Green-synthesized silver nanoparticles of *annona muricata* plant leaf extract is a powerful larvicidal as compared to other plant parts. It is active against third instar larvae of three medically important mosquitoes, i.e. *Aedes aegypti*, *Anopheles stephensi*, and *Culex quinquefasciatus*. Prepared NPs were tested with its different concentration and compared with aqueous crude leaf extract for larvicidal action. Significant larval mortality with lowest LC50 and LC90 values was observed after the treatment with NPs for 24 hours. Comparatively better larvicidal efficacy was shown by green-synthesized AgNPs rather than other plant parts, and as it is more toxic than aqueous crude extract. Silver NPs so prepared were also heat stable from 20 to 70°C and nontoxic. Thus, it can be a suitable and effective approach to control malaria in any aquatic system (Santhosh et al. 2015).
3. Synthetic molecules are one such approach for new directions in drug discovery. Bioactive hybrid molecules may be the improved version of their parent compounds. Designing and de novo synthesis of a series of betulinic acid/betulin based dimer and hybrid compounds carrying ferrocene and/or artesunic acid moieties opens new hope of its treatment. These newer synthesised molecules have been tested in vitro and show the increased cytotoxic properties against malaria parasites (growth inhibition of 3D7-strain *P. falciparum*-infected erythrocytes). The betulinic acid/betulin and artesunic acid hybrids 11 and 12 (in nanomolar and micromolar concentration) have found more potent and effective against *P. falciparum* among the series of hybrids/dimmers. Moreover, these molecules showed activity against human cytomegalovirus (HCMV) also. These activities were compared to the reference drug ganciclovir, where lowest EC50 values are noted in the range of 0.24–0.88 μM against HCMV. Significant potency has been showed by hybridization of betulin/betulinic acid and artesunic and thus can be suggested for new directions in drug discovery (Karagoz et al. 2018).
4. Artesunate is one of the well-known antimalarial drugs, but its efficacy and potency can be increased by formulating it as iron oxide-fortified artesunate nanoparticles. These nanoparticles are more selective and provide iron-radical-mediated irreversible damage of parasite biomolecules. The nanoparticle so designed and synthesized that it releases Fe^{2+} in pH-dependent sustained release manner in food vacuole of parasites. This will increase reactive oxygen species (ROS) attack and result in enhanced toxicity to *P. falciparum*. An experimental model has been designed in *Plasmodium berghei* ANKA-infected swiss albino mice and wild type and artemisinin-resistant *Plasmodium falciparum* (R539T) grown in O^{+ve} human blood. Whereas, the ring-survival assay method was performed for evaluating the efficacy of nanoparticles against resistant parasite. Various physiochemical and biochemical parameters were also employed to determine the NPs

internalization, Fe^{2+} release, ROS generation, and attack of parasitic biomolecules. Results showed the significant growth retardation and damage to *P. falciparum* at very low concentration of NPs i.e. $1/8^{th}$ concentration of artesunate IC_{50}. All these clinical assessments proved that these NPs would be a promising and potent therapeutic agent in future toward multidrug-resistant and artemisinin-resistant malaria in humans (Kannan et al. 2019).

5. Cyclodextrines (CD) containing nanoparticles demonstrated good prospective as nanocarriers for drug delivery. These CD complexes are highly biocompatible and can interact with a wide range of molecules for easy functionalization to tune specific interactions. It increases solubility and oral bioavailability of low water-soluble molecules and thus would be a novel treatment option for already known lipophilic antimalarial drugs (Santos-Magalhães and Mosqueira 2010). Sulfated form of cyclodextrins showed significant action against *Plasmodium falciparum* in mM range. 7–17 sulphates containing compounds interact with anion transport protein, AE1, and arrest the growth of parasite. In the mouse model, it also inhibits the entry of *P. berghei* merozoite and thus reduces parasitemia infection caused by *P. berghei*. The result suggests that two plasmodium species are affected by these compounds by arresting the invasion pathway of merozoites. However, future study and development is still required to confirm, target, and understand the merozoite invasion process, which can lead to formulate better antimalarial therapeutics (Crandall et al. 2007).
6. Nanoparticles containing mesoporous silica has proved to be a potential nanocarrier system for variety of drug delivery with its high intracellular uptake ability. It has high loading capacity along with adjustable pore size and volume (Mehmood et al. 2017). Mesoporous MCM-41 is modified form of these nanoparticles for increased carrier ability of antimalarial drugs (Amolegbe et al. 2018). Beck et al. (1992) synthesised MCM-41 as silicate/aluminosilicate mesoporous nanoparticles, which is sieved molecularly and has higher bioavailability. This is useful for encapsulating the drug, which has short half-life in blood such as artesunate (Figure 18.2).

In last decades, nanoparticulate formulation and treatment of malaria has received increased attention. Moreover, it provides promising pre-clinical results also, but to understand the nature and mechanism of nanoparticles this area still needs research of interest. It will not only overcome one another's limitation but also leads to development of better therapeutics to target different stages of infection and cure malaria.

18.6 CONCLUSION AND FUTURE PROSPECTS

After the application of chloroquine and hydroxychloroquine for the management of COVID-19, the repurposing of drugs is now emerging and pre-established drugs are been investigated for their other pharmacological utility. Compared to other infections, malaria is the most suffering cause of death in a large amount of populace. Around 228 million people on the globe suffer from malaria. It is the

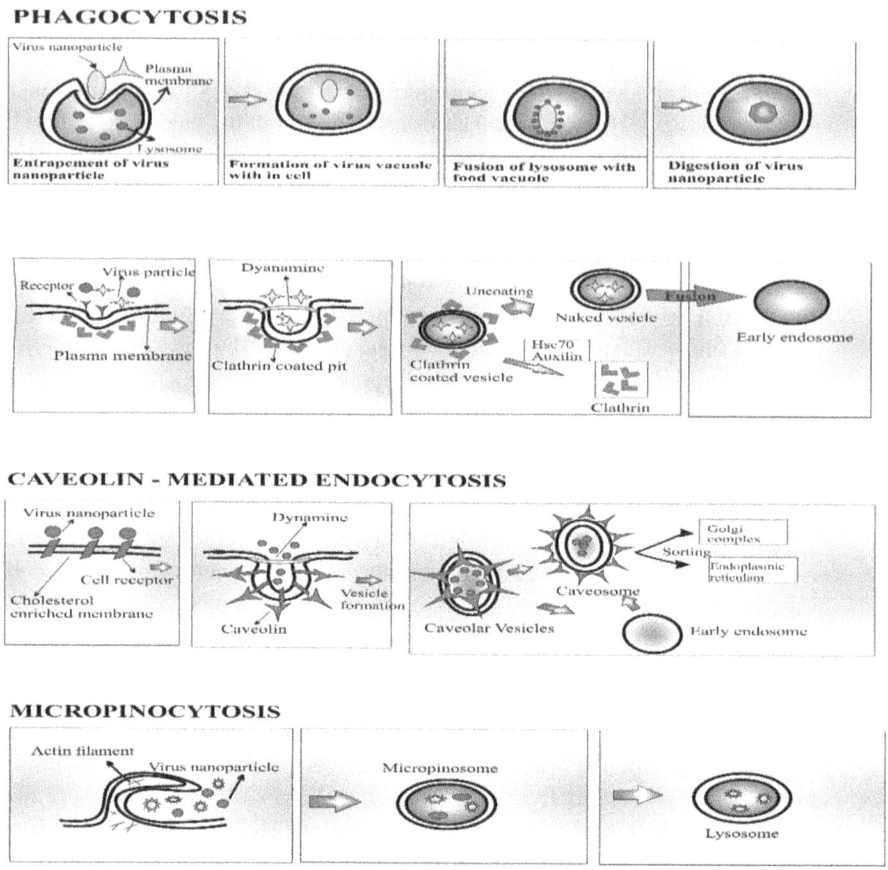

FIGURE 18.2 Cellular uptake mechanism of nanoparticles.

deadliest of viruses and uses human-cellular machinery for its growth, leaving an unhealthy human body behind. Numerous drugs have been used in the management of malaria, including chloroquine, hydroxychloroquine, doxycycline, quinine, atovaquone, and so forth. The inefficacy of vaccines and the limited option of therapeutics for the anticipation and management of malaria has forced the researchers to focus on other alternatives. Among existing classes of therapeutics, antiretroviral has proven to be a significant and potential candidate for their repurposing as an antimalarial. Despite the effectiveness of these drugs, limited drug choice, unpredictable and undesired side effects, and unconventional physico-chemical attributes limit their use, which could be compensated by the application of nanomaterial. The investigations included in this chapter have indicated the potential of antiviral drugs as antimalarial and their various aspects nanomaterials.

Drug repurposing is an undeniably well-known methodology in the development of therapeutics devices to battle illnesses or disorders for which novel therapeutic agents are needed. This repurposing is especially alluring when managing malaria and other disregarded illnesses that apply their financial burden in the absolute poorest regions of various countries. It has been contended that one of the critical boundaries to malaria therapy and chemoprevention lies in the hesitance of the pharmaceutical industry to put resources into the development of medications that would not merely restrict marketing possibilities. The expense decreases related to repurposing existing therapeutics, rather than creating drugs again, may in this manner help overcome a portion of the conservative obstacles in the method of the development of the existing antagonist of plasmodial pharmacological resources where the nanomaterials add on the novelty to effectively treat such condition. Notwithstanding ongoing advancement in the quest for a compelling immunization against malaria, with specific accentuation on antibody applicants that focus on the different phases of the *Plasmodium* life cycle, such an immunization stays elusive. In that capacity, the requirement for different procedures to forestall and treat malaria stays as squeezing as could be expected. In this specific circumstance, novel drugs that manage the cost of the effective chemoprophylaxis, treatment, and revolutionary therapeutics of *Plasmodium* disease keep on being earnestly required. While this may involve the improvement of novel chemotypes for antiviral-antimalarial intrusion, nanomaterial offers a promising option in contrast to conventional medication-advancement pipelines and can assist quick advance resolution with adequately combatting this distressing infection.

REFERENCES

Aditya, N.P., Vathsala, P.G., Vieira, V., Murthy, R.S.R. and Souto, E.B. 2013. Advances in nanomedicines for malaria treatment. *Advances in Colloid and Interface Science* 201-202:1–17.

Alsarra, I.A., Hamed, A.Y. and Alanazi, F.K. 2008. Acyclovir liposomes for intranasal systemic delivery: Development and pharmacokinetics evaluation. *Drug Delivery* 15: 313–321.

Amel, Y.H. 2001. *Development and Evaluation of Liposomal Mucoadhesive Gel for Nasal Delivery of Acyclovir*. Dissertation, Riyadh, Saudi Arabia: King Saud University.

Amolegbe, S.A., Hirano, Y., Adebayo, J.O., et al. 2018. Mesoporous silica nanocarriers encapsulated antimalarials with high therapeutic performance. *Scientific Reports* 8:378.

Ashburn, T.T. and Thor, K.B. 2004. Drug repositioning: Identifying and developing new uses for existing drugs. *Nature Reviews. Drug Discovery* 3(8):673–683. 10.1038/nrd1468

Beck, J.S., Vartuli, J.C., Roth, W.J., et al. 1992. A new family of mesoporous molecular sieves prepared with liquid crystal templates. *Journal of the American Chemical Society* 114:10834–10843.

Biamonte, M.A., Wanner, J. and Le Roch, K.G. 2013. Recent advances in malaria drug discovery. *Bioorganic and Medicinal Chemistry Letters* 23(10):2829–2843. 10.1016/j.bmcl.2013.03.067

Borgheti-Cardoso, L.N., San Anselmo, M., Lantero, E., et al. 2020. Promising nanomaterials in the fight against malaria. *Journal of Materials Chemistry B* 8:9428–9448.

Burns, E.R. and Pollack, S.P. 1988. Falciparum infected erythrocytes are capable of endocytosis. *In Vitro Cellular & Developmental Biology* 24:481–486.

Chakravarty, P., Famili, A., Nagapudi, K. and Al-Sayah, M.A. 2019. Using supercritical fluid technology as a green alternative during the preparation of drug delivery systems. *Pharmaceutics* 11(12). 10.3390/pharmaceutics11120629

Chiappetta, D.A., Hocht, C., Opezzo, J.A.W. and Sosnik, A. 2013. Intranasal administration of antiretroviral-loaded micelles for anatomical targeting to the brain in HIV. *Nanomedicine* 8(2):223–237.

Cojocaru, F.-D., Botezat, D., Gardikiotis, I., et al. 2020. Nanomaterials designed for antiviral drug delivery transport across biological barriers. *Pharmaceutics* 12(2). 10.3390/pharmaceutics12020171

Conner, S.D. and Schmid, S.L. 2003. Regulated portals of entry into the cell. *Nature* 422:37–44.

Corrias, F. and Lai, F. 2011. New methods for lipid nanoparticles preparation. *Recent Patents on Drug Delivery and Formulation* 5(3):201–213. 10.2174/187221111797200597

Cowman, A.F., Healer, J., Marapana, D. and Marsh, K. 2016. Malaria: Biology and disease. *Cell* 167(3), 610–624. 10.1016/j.cell.2016.07.055

Crandall, I.E., Szarek, W.A., Vlahakis, J.Z., et al. 2007. Sulfated cyclodextrins inhibit the entry of *Plasmodium* into red blood cells: implications for malarial therapy. *Biochemical Pharmacology* 73:632–642.

Donalisio, M., Leone, F., Civra, A., et al. 2018. Acyclovir-loaded chitosan nanospheres from nano-emulsion templating for the topical treatment of herpesviruses infections. *Pharmaceutics* 10(2):46. doi: 10.3390/pharmaceutics10020046.

Dong, X. 2018. Current strategies for brain drug delivery. *Theranostics* 8(6):1481–1493. doi:10.7150/thno.21254.

Durai, R.D. 2015. Drug delivery approaches of an antiviral drug: A comprehensive review. *Asian Journal of Pharmaceutics* 9:1–12.

Ferguson, L.M. and Rohan, L.C. 2011. The importance of the vaginal delivery route for antiretrovirals in HIV prevention. *Therapeutic Delivery* 2(12):1535–1550. doi: 10.4155/tde.11.126.

Fessi, H., Puisieux, F., Devissaguet, J.P., Ammoury, N. and Benita, S. 1989. Nanocapsule formation by interfacial polymer deposition following solvent displacement. *International Journal of Pharmaceutics* 55(1):R1–R4. 10.1016/0378-5173(89)90281-0

Fleischer, R. and Johnson, M. 2010. Acyclovir nephrotoxicity: A case report highlighting the importance of prevention, detection, and treatment of acyclovir-induced nephropathy. *Case Reports in Medicine* 602783:1–3.

Fontinha, D., Moules, I. and Prudêncio, M. 2020. Repurposing drugs to fight hepatic malaria parasites. *Molecules* 25(15). 10.3390/molecules25153409

Gote, V., Sikder, S., Sicotte, J. and Pal, D. 2019. Ocular drug delivery: Present innovations and future challenges. *Journal of Pharmacology and Experimental Therapeutics* 370(3): 602–624. doi: 10.1124/jpet.119.256933.

Greenwood, B.M., Fidock, D.A., Kyle, D.E., et al. 2008. Malaria: Progress, perils, and prospects for eradication. *The Journal of Clinical Investigation* 118(4):1266–1276. 10.1172/JCI33996

Guo, P., Huang, J., Zhao, Y., Martin, C.R., Zare, R.N. and Moses, M.A. 2018. Nanomaterial Preparation by extrusion through nanoporous membranes. *Small 14*(18):1703493. 10.1002/smll.201703493

Gupta, S., Agarwal, A., Gupta, N.K., Saraogi, G., Agrawal, H. and Agrawal, G.P. 2013. Galactose decorated PLGA nanoparticles for hepatic delivery of acyclovir. *Drug Development and Industrial Pharmacy* 39(12):1866–1873.

Haldar, K. and Uyetake, L. 1992. The movement of fluorescent endocytic tracers in *Plasmodium falciparum* infected erythrocytes. *Molecular and Biochemical Parasitology* 50:161–177.

Hassan, H., Adam, S.K., Othman, F., Shamsuddin, A.F. and Basir, R. 2016. Antiviral nano-delivery systems: Current trends in acyclovir administration. *Journal of Nanomaterials* 4591634:1–8. 10.1155/2016/4591634

Hobbs, C.V., Tanaka, T.Q., Muratova, O., et al. 2013. HIV treatments have malaria gametocyte killing and transmission blocking activity. *The Journal of Infectious Diseases* 208(1):139–148. 10.1093/infdis/jit132

Hobbs, C.V., Voza, T., Coppi, A., et al. 2009. HIV protease inhibitors inhibit the development of preerythrocytic-stage plasmodium parasites. *The Journal of Infectious Diseases* 199(1):134–141. 10.1086/594369

Kamel, A.O., Awad, G.A.S., Geneidi, A.S. and Mortada, N.D. 2009. Preparation of intravenous stealthy acyclovir nanoparticles with increased mean residence time. *AAPS PharmSciTech* 10(4):1427–1436.

Kannan, D., Yadav, N., Ahmad, S., et al. 2019. Pre-clinical study of iron oxide nanoparticles fortified artesunate for efficient targeting of malarial parasite. *EBioMedicine* 45: 261–277.

Karagoz, A.C., Leidenberger, M., Hahn, F., et al. 2018. Synthesis of new betulinic acid/ betulin-derived dimers and hybrids with potent antimalarial and antiviral activities. *Bioorganic and Medicinal Chemistry* doi: 10.1016/j.bmc.2018.11.018.

Kilejian, A. 1984. The biosynthesis of the knob protein and a 65000 dalton histidine-rich polypeptide of *Plasmodium falciparum. Molecular and Biochemical Parasitology* 12:185–194.

Krishnamoorthy, K. and Mahalingam, M. 2015. Selection of a suitable method for the preparation of polymeric nanoparticles: Multi-criteria decision making approach. *Advanced Pharmaceutical Bulletin* 5(1):57–67. 10.5681/apb.2015.008

Kumar, P., Lakshmi, Y.S., C. B. Golla, K. and Kondapi, A.K. 2015. Improved safety, bioavailability and pharmacokinetics of zidovudine through lactoferrin nanoparticles during oral administration in rats. *PLoS One* 10(10):e0140399. doi: 10.1371/journal.pone.0140399.

Kumar, P., Lakshmi, Y.S. and Kondapi, A.K. 2016. An oral formulation of efavirenz-loaded lactoferrin nanoparticles with improved biodistribution and pharmacokinetic profile. *HIV Medicines* 8(7):452–462. doi: 10.1111/hiv.12475.

Kuo, Y.C. and Chen, H.H. 2006. Effect of nanoparticulate polybutylcyanoacrylate and methylmethacrylate-sulfopropylmethacrylate on the permeability of zidovudine and lamivudine across the in vitro blood-brain barrier. *International Journal of Pharmaceutics* 327(1-2):160–169. doi: 10.1016/j.ijpharm.2006.07.044.

Kuo, Y.C. and Su, F.L. 2007. Transport of stavudine, delavirdine, and saquinavir across the blood-brain barrier by polybutylcyanoacrylate, methylmethacrylate-sulfopropylmethacrylate, and solid lipid nanoparticles. *International Journal of Pharmaceutics* 340(1-2):143–152. doi: 10.1016/j.ijpharm.2007.03.012.

Lembo, D. and Cavalli, R. 2010. Nanoparticulate delivery systems for antiviral drugs. *Antiviral Chemistry & Chemotherapy* 21(2):53–70. doi: 10.3851/IMP1684.

Lembo, D., Donalisio, M., Civra, A., Argenziano, M. and Cavalli, R. 2018. Nanomedicine formulations for the delivery of antiviral drugs: A promising solution for the treatment of viral infections. *Expert Opinion on Drug Delivery* 15(1):93–114. 10.1080/17425247.2017.1360863

Lin, L., Chen, X.S., Cui, P.G., et al. 2002. Topical Penciclovir Clinical Study Group. Topical application of penciclovir cream for the treatment of herpes simplex facialis/ labialis: A randomized, double-blind, multicentre, aciclovir-controlled trial. *Journal of Dermatological Treatment* 13(2):67–72. doi: 10.1080/095466302317584412.

Liu, Y., Wang, X.Q., Ren, W.X., et al. 2013. Novel albendazole-chitosan nanoparticles for intestinal absorption enhancement and hepatic targeting improvement in rats. *Journal of Biomedical Materials Research Part B* 101B:998–1005.

Loh, Z.H., Samanta, A.K. and Sia Heng, P.W. 2015. Overview of milling techniques for improving the solubility of poorly water-soluble drugs. *Asian Journal of Pharmaceutical Sciences* 10(4):255–274. 10.1016/j.ajps.2014.12.006

Lu, H., Han, Y.J., Xu, J.D., Xing, M.W. and Chen, J. 2014. Proteomic characterization of acyclovir-induced nephrotoxicity in a mouse model. *PLoS One* 9:e103185.

Machado, M., Sanches-Vaz, M., Cruz, J.P., Mendes, A.M. and Prudêncio, M. 2017. Inhibition of *Plasmodium* hepatic infection by antiretroviral compounds. *Frontiers in Cellular and Infection Microbiology* 7:329. 10.3389/fcimb.2017.00329

Mahajan, H.S., Mahajan, M.S., Nerkar, P.P. and Agrawal, A.A. 2014. Nanoemulsion-based intranasal drug delivery system of saquinavir mesylate for brain targeting. *Drug Delivery*. 21(2):148–154.

Mahajan, S.D., Roy, I., Xu, G., et al. 2010. Enhancing the delivery of anti retroviral drug "Saquinavir" across the blood brain barrier using nanoparticles. *Current HIV Research* 8(5):396–404.

Maniruzzaman, M., Boateng, J.S., Snowden, M.J. and Douroumis, D. 2012. A review of hot-melt extrusion: process technology to pharmaceutical products. *ISRN Pharmaceutics* 2012:436763. 10.5402/2012/436763

Md, S., Mustafa, G., Baboota, S. and Ali, J. 2015. Nanoneurotherapeutics approach intended for direct nose to brain delivery. *Drug Development and Industrial Pharmacy* 41(12):1922–1934.

Mehmood, A., Ghafar, H., Yaqoob, S., Gohar, U.F. and Ahmad, B. 2017. Mesoporous silica nanoparticles: a review. *Journal of Developing Drugs* 6:174. doi:10.4172/2329-6631.1000174.

Mendoza-Muñoz, N., Quintanar-Guerrero, D. and Allémann, E. 2012. The impact of the salting-out technique on the preparation of colloidal particulate systems for pharmaceutical applications. *Recent Patents on Drug Delivery and Formulation* 6(3): 236–249. 10.2174/187221112802652688

Miladi, K., Sfar, S., Fessi, H. and Elaissari, A. 2016. Nanoprecipitation process: from particle preparation to in vivo applications. In *Polymer Nanoparticles for Nanomedicines*, 17–53. Springer International Publishing. 10.1007/978-3-319-41421-8_2

Minost, A., Delaveau, J., Bolzinger, M.-A., Fessi, H. and Elaissari, A. 2012. Nanoparticles via nanoprecipitation process. *Recent Patents on Drug Delivery and Formulation*, 6(3):250–258. 10.2174/187221112802652615

Mishra, V., Mahor, S., Rawat, A., et al. 2006. Targeted brain delivery of AZT via transferrin anchored pegylated albumin nanoparticles. *Journal of Drug Targeting* 14(1):45–53. doi: 10.1080/10611860600612953.

Nalini, M., Lena, M., Sumathi, P. and Sundaravadivelan, C. 2017. Effect of phyto-synthesized silver nanoparticles on developmental stages of malaria vector, *Anopheles stephensi* and dengue vector, *Aedes aegypti. Egyptian Journal of Basic and Applied Sciences*. 10.1016/j.ejbas.2017.04.005.

Naseri, N., Valizadeh, H. and Zakeri-Milani, P. 2015. Solid lipid nanoparticles and nanostructured lipid carriers: structure, preparation and application. *Advanced Pharmaceutical Bulletin* 5(3):305–313. 10.15171/apb.2015.043

Öztürk, A.A. and Yurtdaş Kirimlioğlu, G. 2019. Preparation and in vitro characterization of lamivudine loaded nanoparticles prepared by acid and/or ester terminated PLGA for effective oral anti-retroviral therapy. *Journal of Research in Pharmacy* 23(5):897–913.

Patel, B.K., Parikh, R.H. and Patel, N. 2014. Targeted delivery of mannosylated-PLGA nanoparticles of antiretroviral drug to brain. *International Journal of Nanomedicine* 15:97–100. doi: 10.2147/IJN.S124692.

Patel, D. and Sawant, K.K. 2007. Oral bioavailability enhancement of acyclovir by self-microemulsifying drug delivery systems (SMEDDS). *Drug Development and Industrial Pharmacy* 33(12):1318–1326.

Peatey, C.L., Andrews, K.T., Eickel, N., et al. 2010. Antimalarial asexual stage-specific and gametocytocidal activities of HIV protease inhibitors. *Antimicrobial Agents and Chemotherapy* 54(3):1334–1337. 10.1128/AAC.01512-09

Pineda-Reyes, A.M., Hernández Delgado, M., Zambrano-Zaragoza, M., et al. 2021. Implementation of the emulsification-diffusion method by solvent displacement for polystyrene nanoparticles prepared from recycled material. *RSC Advances* 11(4): 2226–2234. 10.1039/D0RA07749F

Pouvelle, B., Gormley, J.A. and Taraschi, T.F. 1994. Characterization of trafficking pathways and membrane genesis in malaria-infected erythrocytes. *Molecular and Biochemical Parasitology* 66:83–96.

Pradhan, M., Alexander, A., Singh, M.R., et al. 2020. Statistically optimized calcipotriol fused nanostructured lipid carriers for effectual topical treatment of psoriasis. *Journal of Drug Delivery Science and Technology* 102168. 10.1016/j.jddst.2020.102168

Pradhan, S., Hedberg, J., Blomberg, E., Wold, S. and Odnevall Wallinder, I. 2016b. Effect of sonication on particle dispersion, administered dose and metal release of non-functionalized, non-inert metal nanoparticles. *Journal of Nanoparticle Research* 18(9):285. 10.1007/s11051-016-3597-5

Pradhan, M., Singh, D. and Singh, M.R. 2016a. Influence of selected variables on fabrication of Triamcinolone acetonide loaded solid lipid nanoparticles for topical treatment of dermal disorders. *Artificial Cells, Nanomedicine and Biotechnology* 44(1):392–400. 10.3109/21691401.2014.955105

Pradhan, M., Singh, D. and Singh, M.R. 2017. Fabrication, optimization and characterization of Triamcinolone acetonide loaded nanostructured lipid carriers for topical treatment of psoriasis: Application of Box Behnken design, in vitro and ex vivo studies. *Journal of Drug Delivery Science and Technology* 41:325–333. 10.1016/j.jddst.2017.07.024

Prudêncio, M., Rodriguez, A. and Mota, M.M. 2006. The silent path to thousands of merozoites: The Plasmodium liver stage. *Nature Reviews Microbiology* 4(11):849–856. 10.1038/nrmicro1529

Pushpakom, S., Iorio, F., Eyers, P.A., et al. 2019. Drug repurposing: Progress, challenges and recommendations. *Nature Reviews Drug Discovery* 18(1):41–58. 10.1038/nrd.2018.168

Ren, J., Zou, M., Gao, P., Wang, Y. and Cheng, G. 2012. Tissue distribution of borneol-modified ganciclovir-loaded solid lipid nanoparticles in mice after intravenous administration. *European Journal of Pharmaceutics and Biopharmaceutics* 83(2):141–148. doi: 10.1016/j.ejpb.2012.10.018.

Saiyed, Z.M., Gandhi, N.H. and Nair, M.P. 2010. Magnetic nanoformulation of azidothymidine 5'-triphosphate for targeted delivery across the blood-brain barrier. *International Journal of Nanomedicine* 5:157–166. doi:10.2147/ijn.s8905.

Sankar, V., Madhura Keerthi, L. and Nilaykumar, P. 2012. Formulation and in-vitro evaluation of zidovudine- lamivudine nanoparticles. *Indian Journal of Pharmaceutical Education and Research* 46:192–196.

Santhosh, S.B., Yuvarajan, R. and Natarajan, D. 2015. *Annona muricata* leaf extract-mediated silver nanoparticles synthesis and its larvicidal potential against dengue, malaria and filariasis vector. *Parasitology Research* doi:10.1007/s00436-015-4511-2.

Santos-Magalhães, N.S. and Mosqueira, V.C.F. 2010. Nanotechnology applied to the treatment of malaria. *Advance Drug Delivery Review* 62:560–575.

Sapra, B., Thatai, P., Bhandari, S., Sood, J., Jindal, M. and Tiwary, A. 2013. A critical appraisal of microemulsions for drug delivery - Part I. *Therapeutic Delivery* 4(12): 1547–1564.

Schwarz, J.C., Klang, V., Karall, S., Mahrhauser, D., Resch, G.P. and Valenta, C. 2012. Optimisation of multiple W/O/W nanoemulsions for dermal delivery of acyclovir. *International Journal of Pharmaceutics* 435(1):69–75.

Seedat, A. and Winnett, G. 2012. Acyclovir-induced acute renal failure and the importance of an expanding waist line. *BMJ Case Reports*. doi:10.1136/bcr-2012-006264.

Shi, L.-E., Fang, X.-J., Zhang, Z.-L., et al. 2012. Preparation of nano-ZnO using sonication method and its antibacterial characteristics. *International Journal of Food Science and Technology* 47(9):1866–1871. https://doi.org/10.1111/j.1365-2621.2012.03043.x

Shishu, G., Rajan, S. and Kamalpreet. 2009. Development of novel microemulsion-based topical formulations of acyclovir for the treatment of cutaneous herpetic infections. *AAPS Pharm-SciTech* 10(2):559–565.

Singh, B., Yang, S., Krishna, A. and Sridhar, S. 2020. Nanoparticle formulations of poly (ADP-ribose) polymerase inhibitors for cancer therapy. *Frontiers in Chemistry* 8:594619. 10.3389/fchem.2020.594619

Singh, D., Pradhan, M., Shrivastava, S., Murthy, S.N. and Singh, M.R. 2016. Skin autoimmune disorders: Lipid biopolymers and colloidal delivery systems for topical delivery. In *Nanobiomaterials in Galenic Formulations and Cosmetics*, ed. A.M. Grumezescu, 257–296. William Andrew Publishing. 10.1016/B978-0-323-42868-2.00011-5

Sun, Y., Chen, D., Pan, Y., et al. 2019. Nanoparticles for antiparasitic drug delivery. *Drug Delivery* 26(1):1206–1221.

Talapko, J., Škrlec, I., Alebić, T., Jukić, M. and Včev, A. 2019. Malaria: The past and the present. *Microorganisms* 7(6). 10.3390/microorganisms7060179

Tokuyasu, K.T., Schekman, R. and Singer, S.J. 1979. Domains of receptor mobility and ndocytosis in the membranes of neonatal human erythrocytes and reticulocytes are deficient in spectrin. *Journal of Cell Biology* 80:481–486.

Venkatesh, D.N., Baskaran, M., Karri, V.V., Mannemala, S.S., Radhakrishna, K. and Goti, S. 2015. Fabrication and in vivo evaluation of Nelfinavir loaded PLGA nanoparticles for enhancing oral bioavailability and therapeutic effect. *Saudi Pharmaceutical Journal* 23(6):667–674. doi:10.1016/j.jsps.2015.02.021.

World Malaria Report. 2019. *World Malaria Report, 2019*. WHO Regional Office for Africa. https://www.who.int/publications/i/item/9789241565721. Accessed 27 March 2021

World Malaria Report 2015. 2019. Geneva, Switzerland: World Health Organization.

Yeo, S.-D. and Kiran, E. 2005. Formation of polymer particles with supercritical fluids: A review. *The Journal of Supercritical Fluids* 34:287–308. 10.1016/j.supflu.2004.10.006

Zhang, X., Li, Y., Zhou, H., et al. 2014. Plasma metabolic profiling analysis of nephrotoxicity induced by acyclovir using metabonomics coupled with multivariate data analysis. *Journal of Pharmaceutical and Biomedical Analysis* 97:151–156.

Index

Note: *Italicized* page numbers refer to figures, **bold** page numbers refer to tables

D

E

F

G

H

N

O

P

Q

R

S

W

X

Z